Mycobacteria Protocols

METHODS IN MOLECULAR BIOLOGY™

John M. Walker, SERIES EDITOR

METHODS IN MOLECULAR BIOLOGY™

Mycobacteria Protocols

Second Edition

Edited by

Tanya Parish and Amanda Claire Brown

Barts and The London
Queen Mary's School of Medicine
and Dentistry
London, UK

 Humana Press

Editors
Tanya Parish
Barts and The London
Queen Mary's School of Medicine
  and Dentistry,
London, UK
t.parish@qmul.ac.uk

Amanda Claire Brown
Barts and The London
Queen Mary's School of Medicine
  and Dentistry,
London, UK
a.c.brown@qmul.ac.uk

Series Editor
John M. Walker
University of Hertfordshire
Hatfield, Hertfordshire,
UK

ISBN: 978-1-58829-889-8
ISSN: 1064-3745
DOI: 10.1007/978-1-59745-207-6

e-ISBN: 978-1-59745-207-6
e-ISSN: 1940-6029

Library of Congress Control Number: 2008933364

Preface

The mycobacteria include a number of important human and animal pathogens and pose major problems worldwide in terms of global health and economies. Tuberculosis poses a significant threat to global health by infecting and killing millions annually. Leprosy has not yet been eradicated, and other infections, such as Buruli ulcer and opportunistic infections associated with immunodeficiency, are on the rise. For these reasons, the need for methods to study the biology of the mycobacteria and to improve diagnostic, therapeutic, and preventative reagents is still a priority.

It has been nearly 10 years since the first edition of *Mycobacteria Protocols*. The response to the first edition was both surprising and pleasing. Many readers commented on how the book had helped them with a tricky problem or allowed them to branch into new areas, and newcomers to the field were able to avoid many common pitfalls and progress quickly. Mycobacteria are difficult organisms to work with, but the availability of the tips and tricks developed by a multitude of scientists over many years has been received very positively. Within my own laboratory, this book has proved invaluable for neophytes. During the time since it was published, research into the mycobacteria has continued to expand, and the number of scientists studying these problematic bacteria has increased. New methods have been developed, and older methods have been refined, making a second edition timely.

In this second edition, we have tried to include a range of methods, from the basics of subcellular fractionation, strain typing, and determining minimum inhibitory concentrations (MICs) to more advanced methods using specialized growth conditions and whole genome, transcriptome, and proteome analyses. Although we cannot include all methods used in the modern research laboratory, we hope that this edition will provide a useful primer for new mycobacterial researchers and stimulate new avenues for established scientists.

In developing this edition, we have revised and updated some of the most used methods from the original version and incorporated a number of wholly new methods, some of which make use of our expanded horizons in the post–genomic age. Several methods are described for *Mycobacterium tuberculosis*, as this is now the most widely studied species, but can be easily adapted for other members of the genus.

Updated and revised methods include techniques for subcellular fractionation, notably the isolation of genomic DNA and RNA—techniques that are now more useful than ever given the availability of whole genome microarrays. Methods for identifying deletions in genomic DNA and for assaying genome-wide gene expression patterns are presented to make use of these fractions. Analysis of the proteome by two-dimensional gel electrophoresis is also included. These methods will allow the exploitation of the information afforded by the availability of the whole genome sequence for several mycobacterial species. Such a concerted sequencing effort has provided a plethora of information, although this can sometimes be overwhelming. *MycoDB* is an online resource designed to facilitate genomic analyses of *Mycobacterium* spp. and related genera, and a chapter explaining this powerful engine is included.

Mycobacteria have a characteristic lipid-rich cell wall. This outer layer is responsible for a number of their features, including the acid-fast property used to identify the genus. Cell wall composition has implications in many other areas, from immune recognition to insensitivity to antibiotics. A number of techniques are presented to allow the reader to analyze the composition of the cell wall including the complex lipoglycans, which include the immunomodulatory molecule lipoarabinomannan. Lipids plays an important role in mycobacteria, and methods for analyzing lipid biosynthesis and location as well as metabolism are detailed. The cell wall forms a strong barrier to many antimicrobials, and the use of transport assays to measure permeability is discussed.

Genetic techniques for manipulating the genome of mycobacteria have greatly improved since the past edition. Introducing recombinant DNA into mycobacteria is not as straightforward as for other bacterial species, so an updated version of the electroporation chapter is included. Much has happened in the plasmid field, and an updated review of the most commonly used vectors is given together with methods for using the only available temperature-sensitive plasmid, and two chapters deal with expression of genes in mycobacteria using inducible plasmid systems. Homologous recombination has been used to construct mutants for many years, and two methods are described here. Other methods for creating mutants have also been developed, and phage transposon mutagenesis is described. The identification of mycobacterial promoters and analysis of gene expression is a growing area of research, and a chapter that describes the use of two reporter systems, powerful tools for investigating transcriptional regulation in response to environmental signals, is included. A great deal of interest in identifying essential genes has arisen, and two techniques are provided for checking the essentiality of a given gene. The first method, CESTET, can rapidly identify essential genes. The second method, gene switching, can also be used to study the role of essential genes further.

The mycobacteria include several pathogens, so there is a need for strain identification in molecular epidemiology and genome variation studies. Methods for typing strains using insertion sequence, IS6110, and the newer

PCR-based method of analyzing repeat units (MIRU/VNTR) are given. The determination of MICs can be technically difficult, so protocols that can be used with clinical isolates and an Alamar blue–based method that could be applied to the development of novel inhibitors are included. A method for looking at molecular drug resistance using a scanning-frame oligonucleotide microarray is also included.

We have included a chapter describing the use of a chemostat to grow mycobacterial cultures under highly defined conditions, which is useful for studying environmental responses. One particular environment of interest to mycobacteriologists is the intracellular environment, as several mycobacterial species can survive and multiply inside eukaryotic cells. A model for this is provided in a chapter dealing with survival and gene expression in amoebae.

We hope that this book helps to promote and stimulate further research into these intriguing, important, and most fractious of bacteria. We would like to thank all the authors who have made this second edition possible and hope that these updated and new protocols will continue to serve the mycobacterial research community as a useful resource.

Tanya Parish
Amanda Claire Brown

Contents

Contributors

Anita G. Amin, MS
Department of Microbiology, Immunology and Pathology, Colorado State
University, Fort Collins, Colorado, USA

Shiva K. Angala, MS
Department of Microbiology, Immunology and Pathology, Colorado State
University, Fort Collins, Colorado, USA

Joanna Bacon, PhD
TB Research Programme, CEPR, Health Protection Agency, Porton Down,
Salisbury, UK

John T. Belisle, PhD
Department of Microbiology, Immunology and Pathology, Colorado State
University, Fort Collins, Colorado, USA

Luiz E. Bermudez, MD
Biomedical Sciences, College of Veterinary Medicine, Oregon State University,
Corvallis, Oregon, USA

Apoorva Bhatt, PhD
School of Biosciences, University of Birmingham, Edgbaston, Birmingham, UK

Wilbert Bitter, PhD
Medical Microbiology, VU University Medical Centre, Amsterdam,
The Netherlands

Amanda Claire Brown, PhD
Centre for Infectious Disease, Institute of Cell and Molecular Science, Barts
and The London, Queen Mary's School of Medicine and Dentistry,
London, UK

T.J. Brown, PhD
Health Protection Agency Mycobacterium Reference Unit, Barts
and The London, Queen Mary's School of Medicine and Dentistry,
London, UK

P.D. Butcher, PhD
Division of Cellular and Molecular Medicine, St. George's University
of London, London, UK

Paul Carroll, PhD
Centre for Infectious Disease, Institute of Cell and Molecular Science,
Barts and The London, Queen Mary's School of Medicine and Dentistry,
London, UK

Delphi Chatterjee, PhD
Department of Microbiology, Immunology and Pathology, Colorado State
University, Fort Collins, Colorado, USA

Roy R. Chaudhuri, PhD
Department of Veterinary Medicine, Cambridge Veterinary School,
University of Cambridge, Cambridge, UK

Vladimir E. Chizhikov, PhD
Center for Biologics Evaluation and Research, U.S. Food and Drug
Administration, Rockville, Maryland, USA

Dean C. Crick, PhD
Department of Microbiology, Immunology and Pathology, Colorado State
University, Fort Collins, Colorado, USA

Mamadou Daffé, PhD
Department of Molecular Mechanisms of Mycobacterial Infections, Institute
of Pharmacology and Structural Biology, UMR 5089 of the National Centre
for Scientific Research (CNRS) and University Paul Sabatier (Toulouse III),
Toulouse France

Steven Denkin, PhD
Department of Molecular Microbiology and Immunology,
Bloomberg School of Public Health, Johns Hopkins University, Baltimore,
Maryland, USA

F.A. Drobniewski, PhD, MBBS
Health Protection Agency Mycobacterium Reference Unit, Barts
and The London, Queen Mary's School of Medicine and Dentistry,
London, UK

Rosangela Frita, BSc
Department of Pathology and Infectious Diseases, The Royal Veterinary
College, London, UK

Nicolaas C. Gey van Pittius, PhD
Department of Biomedical Sciences, Faculty of Health Sciences, Stellenbosch
University, Cape Town, South Africa

Renan Goude, PhD
Centre for Infectious Disease, Institute of Cell and Molecular Science,
Barts and The London, Queen Mary's School of Medicine and Dentistry,
London, UK

Christophe Guilhot, PhD
Centre National de la Recherche Scientifique and Université Paul Sabatier,
Institut de Pharmacologie et de Biologie Structurale, Département Mécanismes
Moléculaires des Infections Mycobactériennes, Toulouse, France

Kristine Hagens, TA (Technician)
Department of Immunology, Max Planck Institute for Infection Biology,
Berlin, Germany

Melanie Harriff, PhD
Vollum Institute for Biomedical Research, Oregon Health and Sciences
University, Portland, Oregon, USA

Kim A. Hatch, MSc
TB Research Programme, CEPR, Health Protection Agency, Porton Down,
Salisbury, UK

Preston J. Hill, BSc
Department of Microbiology, Immunology and Pathology, Colorado State
University, Fort Collins, Colorado, USA

J. Hinds, PhD
Division of Cellular and Molecular Medicine, St. George's University
of London, London, UK

William R. Jacobs Jr., PhD
Department of Microbiology and Immunology, Howard Hughes Medical
Institute, Albert Einstein College of Medicine, Bronx, New York, USA

Jade James, BSc
Centre for Infectious Disease, Institute of Cell and Molecular Science, Barts
and The London, Queen Mary's School of Medicine and Dentistry,
London, UK

Stefan H.E. Kaufmann, Prof., PhD
Department of Immunology, Max Planck Institute for Infection Biology,
Berlin, Germany

Sharon L. Kendall, PhD
Department of Pathology and Infectious Diseases, The Royal Veterinary
College, London, UK

Julia Kreuzeder, PhD
Department of Immunology, Max Planck Institute for Infection Biology,
Berlin, Germany

Marie-Antoinette Lanéelle, PhD
Department of Molecular Mechanisms of Mycobacterial Infections, Institut
of Pharmacology and Structural Biology, UMR 5089 of the National Centre
for Scientific Research (CNRS) and University Paul Sabatier (Toulouse III),
Toulouse France

Reiling P. Liao, MS
Seattle Biomedical Research Institute, Seattle, Washington, USA

Spencer B. Mahaffey, BSc
Department of Microbiology, Immunology and Pathology, Colorado State
University, Fort Collins, Colorado, USA

Wladimir Malaga, PhD
Centre National de la Recherche Scientifique and Université Paul Sabatier,
Institut de Pharmacologie et de Biologie Structurale, Département Mécanismes
Moléculaires des Infections Mycobactériennes, Toulouse, France

Riccardo Manganelli, PhD
Department of Histology, Microbiology and Medical Biotechnologies,
University of Padua, Padua, Italy

Jens Mattow, PhD
Department of Immunology, Max Planck Institute for Infection Biology,
Berlin, Germany

Farahnaz Movahedzadeh, PhD
Institute for Tuberculosis Research, University of Illinois at Chicago (UIC),
Chicago, Illinois, USA

Michael Niederweis, PhD
Department of Microbiology, University of Alabama at Birmingham,
Birmingham, Alabama, USA

V.N. Nikolayevskyy, PhD
Health Protection Agency Mycobacterium Reference Unit, Barts
and The London, Queen Mary's School of Medicine and Dentistry,
London, UK

Giorgio Palù, MD
Department of Histology, Microbiology and Medical Biotechnologies,
University of Padua, Padua, Italy

Tanya Parish, PhD
Centre for Infectious Disease, Institute of Cell and Molecular Science,
Barts and The London, Queen Mary's School of Medicine and Dentistry,
London, UK

Damien Portevin, PhD
Mycobacterial Division Research, National Institute for Medical Research,
London, UK

Roberta Provvedi, PhD
Department of Biology, University of Padua, Padua, Italy

David M. Roberts, PhD
Seattle Biomedical Research Institute, Seattle, Washington, USA

Eric J. Rubin, MD, PhD
Department of Immunology and Infectious Diseases, Harvard School of Public
Health, Boston, Massachusetts, USA

Tige R. Rustad, PhD
Seattle Biomedical Research Institute, Seattle, Washington, USA

Anthony A. Ryan, PhD
Centenary Institute of Cancer Medicine and Cell Biology, University
of Sydney, Sydney, Australia

Christopher M. Sassetti, PhD
Department of Molecular Genetics and Microbiology, University
of Massachusetts Medical School, Worcester, Massachusetts, USA

Ulrich E. Schaible, PhD
London School of Hygiene and Tropical Medicine, London, UK

David R. Sherman, PhD
Seattle Biomedical Research Institute, Seattle, Washington, USA

Libin Shi, PhD
Colorado State University, Department of Microbiology, Immunology
and Pathology, Fort Collins, Colorado, USA

M. Sloan Siegrist, PhD
Department of Immunology and Infectious Diseases, Harvard School of Public
Health, Boston, Massachusetts, USA

Frank Siejak
Department of Immunology, Max Planck Institute for Infection Biology,
Berlin, Germany

Frederick A. Sirgel, PhD
Division of Molecular Biology and Human Genetics, MRC Centre
for Molecular and Cellular Biology, DST/NRF Centre of Excellence
for Biomedical TB Research, Stellenbosch University, Tygerberg, South Africa

Houhui Song, PhD
Department of Microbiology, University of Alabama at Birmingham,
Birmingham, Alabama, USA

Jordi B. Torrelles, PhD
Department of Internal Medicine and Division of Infectious Diseases,
The Ohio State University, Columbus, Ohio, USA

James A. Triccas, PhD
Discipline of Infectious Diseases and Immunology, University of Sydney,
Sydney, Australia

David Paul van Helden, PhD
MRC Centre for Molecular and Cellular Biology, DST/NRF Centre
of Excellence for Biomedical TB Research, Stellenbosch University, Tygerberg,
South Africa

Dmitriy V. Volokhov, PhD
Center for Biologics Evaluation and Research, U.S. Food and Drug
Administration, Rockville, Maryland, USA

S.J. Waddell, PhD
Division of Cellular and Molecular Medicine, St. George's University
of London, London, UK

Robin M. Warren, PhD
Division of Molecular Biology and Human Genetics, MRC Centre
for Molecular and Cellular Biology, DST/NRF Centre of Excellence
for Biomedical TB Research, Stellenbosch University, Tygerberg, South Africa

Paul Robert Wheeler, PhD
Tuberculosis Research Group, Veterinary Laboratories Agency (Weybridge),
Addlestone, Surrey, UK

Ian J.F. Wiid, PhD
Division of Molecular Biology and Human Genetics, MRC Centre
for Molecular and Cellular Biology, DST/NRF Centre of Excellence
for Biomedical TB Research, Stellenbosch University, Tygerberg, South Africa

Frank Wolschendorf, PhD
Department of Microbiology, University of Alabama at Birmingham,
Birmingham, Alabama, USA

Ying Zhang, MD, PhD
Department of Molecular Microbiology and Immunology, Bloomberg School
of Public Health, Johns Hopkins University, Baltimore, Maryland, USA

Chapter 1
Isolation of *Mycobacterium* Species Genomic DNA

John T. Belisle, Spencer B. Mahaffey and Preston J. Hill

Abstract A myriad of methods has been reported for the isolation of genomic DNA from *Mycobacterium* spp.; some methods use mechanical disruption of the bacterial cells, whereas others use some form of chemical or enzymatic lysis. Regardless of the approach, the end points remain efficient breaking of the complex mycobacterial cell wall and release of high-quality DNA that is suitable for manipulation and analyses by molecular genetic techniques. This chapter providers detailed methods for the large and small isolation of mycobacterial genomic DNA.

Keywords Cell lysis · DNA isolation · genomic DNA · *Mycobacterium*

1.1 Introduction

In the nearly 10 years since we contributed a chapter on DNA isolation to the original volume of *Mycobacteria Protocols* in this series [1], there has been a significant shift in the reasons why individual investigators need to isolate genomic DNA and in the use of the DNA. Many of the early studies that used genomic DNA of *Mycobacterium* spp. required large-scale genomic preparations for the generation of recombinant genomic libraries [2, 3, 4, 5], the search for native plasmids [6, 7, 8], or non–sequence-based analyses of genomes [9, 10]. With the sequencing of multiple *Mycobacterium* spp. genomes [11, 12, 13, 14], the multiple uses and approaches of PCR amplification [15], and the use of high-throughput methodologies to identify important mycobacterial gene products [16, 17, 18], the requirement for large preparations of mycobacterial DNA has been replaced by the need to rapidly isolate DNA from small samples such as a single colony. Moreover, with the use of rolling circle

J.T. Belisle
Mycobacteria Research Laboratories, Department of Microbiology, Immunology, and Pathology, Colorado State University, Fort Collins, CO 80523, USA
e-mail: jbelisle@colostate.edu

T. Parish, A.C. Brown (eds.), *Mycobacteria Protocols*,
doi: 10.1007/978-1-59745-207-6_1, © Humana Press, Totowa, NJ 2008

1

amplification, it is possible to amplify whole mycobacterial genomes such that extensive DNA analyses can be performed with limiting amounts of genomic material [19].

Regardless of whether large-scale or small-scale genomic DNA preparations are needed, the mycobacterial cell wall remains a hindrance to the efficient isolation of mycobacterial DNA. *Mycobacterium* spp. are endowed with a unique cell wall composed of a covalently attached complex of peptidoglycan, arabinogalactan, and mycolic acids [20]. In addition, an array of glycolipids, lipoglycans, and unique apolar lipids form an outer leaflet that closely associates with the cell wall mycolic acids [20]. This structure of tightly packed lipophilic molecules and highly branched polysaccharides is largely responsible for the low permeability of the mycobacterial cell envelope and results in a formidable protective barrier [20]. Because of this unique cellular envelope, the standard methods for isolating DNA from Gram-negative [21] and Gram-positive [22, 23] bacteria are not optimal for mycobacteria.

Several methods, with varying approaches to achieve efficient cell lysis, are reported for the isolation of genomic DNA from *Mycobacterium* spp. Protocols employing mechanical or physical disruption of mycobacterial cells include homogenation with glass [5, 24] or zirconium beads [25, 26], rapid nitrogen decompression [27], French pressing [28], and grinding of cells on dry ice [29]. Although effective for cell lysis, drawbacks to these techniques are shearing of the genomic DNA, the need for specialized equipment, and the increased potential for the generation of aerosols when working with pathogenic mycobacteria. Other methods have used lysozyme for enzymatic degradation of the cell wall. However, to obtain reasonable yields of DNA, this approach requires the organic extraction [30] or enzymatic degradation [31, 32] of the lipophilic cell wall components so that the mycobacterial peptidoglycan is accessible to lysozyme. In addition to the rigid cell wall, mycobacteria possess an abundance of free lipoglycans and polysaccharides [20] that partition with the aqueous soluble DNA. For the most part, researchers have not been concerned with the removal of these contaminants; however; others have used cetyltrimethylammonium bromide (CTAB) to selectively precipitate these complex carbohydrates [33] that can interfere with enzymatic digestion of genomic DNA [34].

In our laboratory, two separate methods are used for the isolation of mycobacterial genomic DNA. The first of these is a modification of a procedure developed by S. Bardarov at the Albert Einstein College of Medicine of Yeshiva University and is similar to the method reported by Patel et al. [30]. This method is used for the large-scale production of high-quality genomic DNA and employs organic extractions and lysozyme treatment for cell lysis. The second method is used for rapid isolation of genomic DNA on a small scale such as from a single colony [35]. This method has great utility in isolating DNA for PCR analyses and rapid screening for events such as transposon mutagenesis [17].

1.2 Materials

1.2.1 Growth of Mycobacterium *spp.* (see *Note 1*)

1. Glycerol alanine salts (GAS) broth [36]. This medium is prepared by adding 0.3 g Bacto casitone, 50 mg ferric ammonium citrate, 4.0 g K_2HPO_4, 2.0 g citric acid, 1.0 g L-alanine, 1.2 g $MgCl_2$-6 H_2O, 0.6 g K_2HPO_4, 2.0 g NH_4Cl, 1.8 mL 10 M NaOH, 10 mL glycerol, and 982 mL deionized water. The pH is adjusted to 6.6 with 10 M NaOH, and the medium is sterilized by autoclaving. For the growth of *Mycobacterium bovis*, the addition of 1% sodium pyruvate to the medium is required (*see* **Note 2**).
2. Middlebrook 7 H9 broth. Prepare by adding 4.7 g powdered 7 H9 medium (Becton Dickinson, Sparks, MD) and 2 mL glycerol to 900 mL deionized water and autoclaving for 15 min. Cool the sterile medium to 55°C and add 100 mL OADC supplement (*see* **Note 3** for the preparation of OADC supplement).
3. Middlebrook 7 H11 agar. Prepare by adding 21 g powdered 7 H11 medium (Difco) and 5 mL glycerol to 900 mL deionized water. Heat at 100°C until the agar dissolves and sterilize by autoclaving. Cool to 55°C, and add 100 mL OADC supplement immediately before pouring plates (*see* **Note 3**).

1.2.2 Large-Scale Enzymatic Lysis of Mycobacterial Cells

1. Tris-EDTA (TE) Buffer: 10 mM Tris-Cl (pH 8.0), 1 mM EDTA; sterilize by autoclaving.
2. Chloroform/methanol 2:1 (*see* **Note 4**).
3. 1 M Tris-base, pH 9.0; sterilize by autoclaving.
4. Lysozyme: 10 mg/mL stock solution in water, filter-sterilized.
5. 10% w/v SDS, sterile.
6. Proteinase K: 10 mg/mL stock solution in water, filter-sterilized.
7. Phenol/chloroform/isoamyl alcohol 25:24:1 (*see* **Notes 4** and **5**).
8. Chloroform/isoamyl alcohol 24:1 (*see* **Note 4**).
9. DNase-free RNase (Roche, Mannheim, Germany): 500 µg/mL (*see* **Note 6**).
10. Sterile 250 mL centrifuge bottles.
11. Sterile 50 mL Teflon Oak Ridge centrifuge tubes (*see* **Note 7**).
12. Tabletop centrifuge.
13. High-speed centrifuge.
14. Platform shaker or rocker.

1.2.3 Small-Scale Lysis of Mycobacterial Cells

1. Milli-Q water, sterilize by autoclaving.
2. 10% w/v SDS solution, sterile.

3. Proteinase K: 10 mg/mL stock solution in water, filter sterilized.
4. 5 M NaCl, sterilize by autoclaving.
5. 10% cetyltrimethylammonium bromide (CTAB) in 0.7 M NaCl.
6. Chloroform/isoamyl alcohol (24:1) (*see* **Note 4**).
7. Sterile 1.7 mL microcentrifuge tubes with screw caps.
8. Water bath, set at 80°C.
9. Thermomixer.
10. Microcentrifuge.

1.2.4 Precipitation of Mycobacterial Genomic DNA

1. 3 M sodium acetate (pH 5.2); sterilize by autoclaving.
2. Isopropanol.
3. 70% ethanol; store at –20°C.
4. TE buffer.

1.2.5 Genomic DNA Quantification

1. TE Buffer; sterilize by autoclaving.
2. Matched quartz cuvettes or UV-transparent plastic cuvettes (1.5 mL, 5-mm path width).
3. UV spectrophotometer.
4. NanoDrop ND-1000 Spectrophotometer (NanoDrop Technologies, Wellington, DE).

1.3 Methods

1.3.1 Growth of **Mycobacterium** *spp. (*see *Note 8)*

The methods described are for the cultivation of *Mycobacterium* spp. from a 1-mL frozen glycerol stock.

1.3.1.1 Fast-Growing *Mycobacterium* spp. (*see* **Notes 9 and 10**)

1. For small-scale DNA preparation: Inoculate a 7H11 agar plate with a loopful of bacteria from a 1-mL frozen stock (in 20% glycerol). Incubate at 37°C for 2 to 3 days.
2. For large-scale DNA preparation: Inoculate 9 mL GAS or 7H9 broth with 1 mL frozen stock and incubate for 2 to 3 days at 37°C with agitation. Transfer 10 mL culture into 90 mL broth and incubate for 2 to 3 days at 37°C with shaking. Use the 100 mL culture as the inoculum for a 1-L culture in a similar fashion.

1.3.1.2 Slow-Growing *Mycobacterium* spp. (*see* Notes 9 and 10)

1. For small-scale DNA preparation: Inoculate a 7H11 agar plate with an inoculation loop full of bacteria from the 1-mL frozen stock. Incubate at 37°C for 14 to 21 days.
2. For large-scale purification: Scale up cultures of slow-growing *Mycobacterium* spp in the same manner as the rapid growing *Mycobacterium* spp. (*see* Section 1.3.1.1, step 2), except the incubation periods should be for 14 to 21 days (*see* Note 11).

1.3.2 Large-Scale Enzymatic Lysis of Mycobacterial Cells (see Note 12)

1. Harvest the cells from 1 L culture by centrifugation in 250-mL centrifuge bottles at $3000 \times g$ for 15 min.
2. Decant the culture supernatant and suspend each cell pellet in 25 mL TE buffer and pool into a single centrifuge bottle. Centrifuge the cells as above and remove the supernatant. One liter of *Mycobacterium tuberculosis* cultured for 14 days in GAS medium generally yields 8 g (wet weight) of cells.
3. **Optional:** Freeze the cell pellet at –80°C for a minimum of 4 h, preferably overnight. A freeze-thaw step is not required; however, it results in the weakening of the cell wall and more efficient lysis.
4. Thaw and suspend the cells in 5 mL TE buffer.
5. Transfer the suspended cell pellet into a sterile 50-mL Teflon Oak Ridge centrifuge tube.
6. Add an equal volume of chloroform/methanol (2:1) to the suspension and rock at room temperature for at least 1 hour. This removes a substantial quantity of the cell wall lipids, thus allowing for more efficient cellular lysis (*see* Note 13).
7. Centrifuge the suspension at $2500 \times g$ for 20 min to generate separate phases. This causes the bacteria to form a tight band at the organic-aqueous interface. Decant both the organic and aqueous layers, being careful to leave the tightly packed bacterial band in the tube.
8. Place the uncapped tube containing the delipidated cells at 55°C for 10 to 15 min to remove traces of the organic solvents that can interfere with lysozymal activity.
9. Add 5 mL TE buffer and suspend the cells by vortexing vigorously. The previous organic extraction causes excessive clumping of the mycobacteria, making the pellet difficult to suspend.
10. Increase the pH of the cell suspension by the addition of 0.1 vol 1 M Tris-base (pH 9.0).
11. Add 0.01 vol lysozyme stock solution and incubate at 37°C for 12 to 16 h. It is important not to vortex the mixture after addition of lysozyme as shearing of genomic DNA may occur.

12. Add 10 µL DNase-free RNase solution and incubate at 37°C for 30 min.
13. Add 0.1 vol 10% SDS solution and 0.01 vol of Proteinase K stock solution to the cell lysate. Mix by inverting several times and incubate at 55°C for 3 h. At this point, the resulting suspension should be homogeneous and extremely viscous (*see* **Note 14**).
14. Extract contaminating proteins by the addition of an equal volume of phenol/chloroform/isoamyl alcohol (25:24:1). Mix gently for 30 min and centrifuge at 12,000 × *g* for 30 min at room temperature.
15. Carefully transfer the aqueous layer to a sterile 50-mL Teflon Oak Ridge tube (*see* **Note 15**).
16. Remove residual phenol by extracting the aqueous layer with an equal volume of chloroform/isoamyl alcohol (24:1) for 10 min with gentle mixing. Centrifuge as before and transfer to a new tube.

1.3.3 Small-Scale Lysis of Mycobacterial Cells

1. Add 0.5 mL sterile water to a 1.7-mL microcentrifuge tube.
2. Transfer one to two colonies of *Mycobacterium* spp. from a 7 H11 agar plate into the 1.7-mL microcentrifuge tube.
3. **Optional:** Freeze the cell suspension at –80°C for 4 to 16 h. A freeze-thaw step is not required; however, it results in the weakening of the cell envelope and more efficient lysis of the cell.
4. To inactivate the cells, place the cell suspension in an 80°C water bath for 30 min.
5. Cool to room temperature and add 70 µL 10% SDS solution and 50 µL Proteinase K stock to solution.
6. Place the cell suspension on a Thermomixer that is prewarmed to 60°C. Incubate with slow shaking at 60°C for 1 h.
7. While the cell suspension is incubating, preheat the solutions of 5 M NaCl and 10% CTAB to 60°C (*see* **Note 16**).
8. Add 100 µL 5 M NaCl and 100 µL 10% CTAB to the Proteinase K–treated cell suspension (*see* **Note 17**).
9. Incubate at 60°C for an additional 15 min with slow shaking, and freeze at –80°C for 15 min.
10. Warm the sample to room temperature. Incubate at 60°C with slow shaking for 15 min.
11. Freeze sample a final time at –20°C for 30 min to 16 h (*see* **Note 18**).
12. Warm the cell lysate to room temperature, add 700 µL chloroform/isoamyl alcohol (24:1), and invert 20 to 25 times. Make sure the organic and aqueous components mix to form a homogenous white-opaque solution.
13. Centrifuge at 13,000 × *g* for 10 min at room temperature.
14. Carefully transfer the aqueous layer to a sterile 1.7-mL microcentrifuge tube (*see* **Note 15**).

1.3.4 Precipitation of Mycobacterial Genomic DNA

1. Precipitate the DNA (from Sections 1.3.2 or 1.3.3) by adding 0.1 vol 3 M sodium acetate (pH 5.2) and 1 vol isopropanol to the aqueous extracts from the lysis procedures. Invert the tube slowly to mix and place at 4°C for at least 1 h (*see* **Note 19**).
2. Centrifuge the solution at 12,000 × g for 30 min at 4°C to pellet the DNA. Remove the supernatant, and wash the DNA pellet with cold (–20°C) 70% ethanol.
3. Centrifuge the DNA, remove the 70% ethanol, and allow the pellet to air dry.
4. Dissolve the pellet of genomic DNA in TE buffer. For the large-scale preparation, add 10 mL TE, and for the small-scale preparation, add 20 to 50 μL of TE (*see* **Note 20**).

1.3.5 Genomic DNA Quantification

1.3.5.1 Determination of DNA Concentration and Purity by UV Spectroscopy

1. Prepare a dilution of genomic DNA with TE buffer. Generally, a 1:50 or 1:100 dilution is adequate.
2. To one quartz cuvette of a matched set, add the diluted DNA. The second cuvette is filled with TE buffer and used as the blank.
3. Measure and record the absorbance of the diluted DNA solution at 260 nm and 280 nm.
4. Calculate the concentration of DNA using the formula $(A_{260})(50\ \mu g/mL)(\text{dilution factor}) = \text{concentration}\ (\mu g/mL)$ of undiluted genomic DNA. For double-stranded DNA, 1 absorbance unit at 260 nm is equal to 50 μg/mL DNA.
5. The purity of the DNA is determined by calculating the A_{260}/A_{280} ratio. For pure double-stranded DNA, an A_{260}/A_{280} ratio is 1.8 [21]. An absorbance ratio of 1.7 to 2.0 is considered acceptable (*see* **Note 21**).

1.3.5.2 Determination of DNA Concentration Using a NanoDrop ND-1000 Spectrophotometer

1. On a computer linked to the NanoDrop spectrophotometer, start the software package ND-1000 and select the "Nucleic Acid" option.
2. Begin instrument initialization by cleaning the pedestals using a lint-free wipe and 70% ethanol solution (*see* **Note 22**).
3. Complete initialization by placing a 1-μL drop of UltraPure water on the bottom pedestal; lower the upper pedestal so that it connects with the upper pedestal.
4. In the NanoDrop program, select and click the "OK" icon.
5. Open and clean the pedestals as previously described.

6. Blank the instrument by placing 1 µL TE buffer on the bottom pedestal, lower the upper pedestal, and in the NanoDrop program select and click the "Blank" icon. The blank process should take 5 to 10 s.

7. Open and clean the pedestals.

8. Place 1 µL DNA solution on the bottom pedestal, lower the upper pedestal, enter the sample name in the NanoDrop program, and select and click the "Measure" button to determine the DNA concentration (*see* **Note 23**). The measurement of the DNA concentration takes 5 to 10 s.

9. Open and clean the pedestals as previously described.

1.4 Notes

1. The media described in this chapter are those commonly used in our laboratory. Other media for the growth of *Mycobacterium* spp., such as Sautons [37] or Proskauer-Beck [38], can be used.

2. Cultivation of *M. bovis* requires a medium containing 1% sodium pyruvate. *M. bovis* efficiently uses pyruvate as a carbon source, and poor or no growth is observed on medium containing more than 1% glycerol [37, 39].

3. Oleic acid-albumin-dextrose-catalase (OADC) supplement is commercially available or can be made in the following manner. To 460 mL double-distilled water, add 4.05 g NaCl, 25.0 g albumin fraction V, 10.0 g glucose, 20 mg catalase, and 15 mL oleic acid solution [120 mL double-distilled water, 2.4 mL 6 M NaOH, 2.4 mL oleic acid (Sigma, St. Louis, MO)]. Adjust the pH to 7.0, stir for 1 h to completely solubilize the albumin, and filter sterilize in 100-mL aliquots. Check the sterility of the OADC by incubating overnight at 37°C. Add 100 mL of this supplement per liter of Middlebrook medium. Because of the high protein content of this supplement, it cannot be autoclaved and should be added to autoclaved media after it has cooled to at least 55°C.

4. Phenol and chloroform are considered extremely toxic, and proper precautions need to be taken in their handling. These solvents must be used in an appropriate fume hood, and eye protection, laboratory coats, and gloves should be worn.

5. Phenol used in phenol/chloroform/isoamyl alcohol (25:24:1) is buffered with Tris to a pH of 7.8. This pH allows for effective solubilization of double-stranded DNA in the aqueous layer during biphasic partitioning with organic solvents. To buffer phenol, first melt molecular biology grade (99% pure) phenol at 55°C. Always contain the phenol bottle in a secondary container to avoid spills. After the phenol has liquefied, add an equal volume of 0.5 M Tris-HCl, pH 8.0 and mix. Allow the solution to sit at room temperature until a biphase forms, and remove the aqueous layer (top). Repeat this procedure with 0.1 M Tris-Cl, pH 8.0, until the pH of the phenol phase is 7.8. Equilibrate with an equal volume of 0.1 M Tris-HCl, pH 8.0, containing 0.2% 8-hydroxyquinolone. Store at 4°C in a light-inhibiting bottle for up to 1 month. Mixtures of phenol/chloroform/isoamyl alcohol (25:24:1) are commercially available and are shipped with buffering solution that is added prior to use.

6. RNase that is free of DNase is commercially available or can be prepared easily as follows: dissolve RNase A at a concentration of 10 mg/mL in 0.01 M sodium acetate, pH 5.2, and heat to 100°C for 15 min. Cool to room temperature and add 0.1 vol 1 M Tris-HCl, pH 7.4. Store at −20°C.

7. The 50-mL high-speed centrifuge tubes must be made of Teflon. This material is resistant to the organic solvents used in the DNA isolation procedures.

8. Cultivation of avirulent *Mycobacterium* spp. such as *Mycobacterium smegmatis* or *Mycobacterium phlei* is carried out under Biosafety Level 2 (BSL-2) conditions. Virulent species such as *M. tuberculosis* and *M. bovis* are Biosafety Level 3 (BSL-3) organisms and must be handled as such. These organisms should not be removed from a BSL-3 facility until they are completely inactivated. Consult the Centers for Disease Control and Prevention (CDC) guidelines [40] or other national regulatory authorities and institutional biosafety committees for BSL designations for other mycobacteria.

9. The conditions described for growth of fast- and slow-growing *Mycobacterium* spp. are those used for *M. smegmatis* and *M. tuberculosis*, respectively. The growth rate of other species within these two groups may vary. Thus, growth curves should be performed on other *Mycobacterium* spp. to determine when mid- to late-log phase growth is obtained.

10. Some procedures report the addition of 1.0% glycine to the medium at mid-log phase growth. A high concentration of glycine in the medium results in weakening of the cell wall by inhibiting crosslinking of the peptidoglycan [41].

11. Frozen stocks of slow-growing *Mycobacterium* spp. should be cultured on a solid medium to obtain a sufficient amount of cells for inoculation of broth cultures. Insufficient inoculum will result in low yields of pelleted bacilli.

12. Enzymatic lysis of mycobacterial cells is described for a large-scale purification. However, this method can be scaled down for isolation of smaller quantities of DNA.

13. After performing the chloroform/methanol (2:1) lipid extraction, the peptidoglycan of the bacilli is further exposed and therefore more sensitive to cleavage by lysozyme treatment. This extraction also inactivates the bacilli so that subsequent steps with BSL-3 species can be performed under BSL-2 conditions.

14. If at this point the suspension is not homogeneous or viscous, increase the SDS concentration to 2%, add another 0.01 vol Proteinase K stock solution, mix by inverting several times, and incubate for an additional hour at 55°C.

15. When transferring solubilized DNA solutions, it is important not to cause mechanical shearing of the DNA. To avoid this, solutions should be transferred with a large bore pipette or micropipette tip.

16. CTAB at the concentration used will precipitate at room temperature. Thus it is critical that it is maintained at 60°C including while it is being added to the cell lysate.

17. CTAB is added to assist in the removal of residual proteins and contaminating polysaccharides that are abundant in *Mycobacterium* spp. The CTAB-macromolecular complexes are separated from the genomic DNA with the chloroform/isoamyl alcohol extraction.

18. After the 10% CTAB treatment and subsequent freeze/thaw cycles, the bacilli are inactivated so that subsequent steps with BSL-3 species can be performed under BSL-2 conditions.

19. We have observed that precipitation with isopropanol results in much cleaner DNA than does precipitation with ethanol. In many instances, the use of ethanol has resulted in the coprecipitation of a yellow waxy substance, presumably polar lipids not removed by organic extraction. Such material, which can interfere with enzymes used to manipulate the DNA, is not present when isopropanol is used.

20. Precipitated genomic DNA is difficult to dissolve. Thus, after the addition of TE, place the tube at 4°C for 12 to 16 h. Brief warming to 37°C may also help dissolve the DNA. Ideally, the concentration of the suspended DNA should be approximately 1 mg/mL.

21. An A_{260}/A_{280} ratio of greater than 2.0 indicates contamination with protein. If such a value is obtained, dilute the DNA to a concentration of 100 to 250 µg/mL with TE and repeat the phenol/chloroform/isoamyl alcohol (25:24:1) and chloroform/isoamyl alcohol (24:1) extractions as previously described. Phenol contamination may also result in a poor A_{260}/A_{280} ratio. To determine if phenol is present, measure the absorbance of the diluted DNA from 200 nm to 300 nm in 10-nm increments. For pure DNA, the graph of these values will indicate an initial peak between 210 nm and 220 nm and a second peak at 260 nm. If the

absorbance does not decline between 220 nm and 260 nm, then phenol is the likely contaminant. The presence of contaminating polysaccharides will have little effect on the A_{260}/A_{280} ratio; however, these products will inhibit digestion with restriction endonucleases. Some contaminating carbohydrates are removed by reprecipitation of the DNA with isopropanol. Selective precipitation with CTAB will remove a large majority of contaminating polysaccharides. The NaCl concentration of a DNA solution must be 0.5 M or greater to avoid precipitation of DNA with CTAB [34].

22. The NanoDrop ND-1000 spectrophotometer measures samples using fiberoptics. Thus, it is important to thoroughly clean the pedestals prior to the measurement of subsequent samples to remove any potential contamination that would interfere with the fiberoptic readings.

23. The range of the NanoDrop is 2 to 3700 ng/μL dsDNA, 2 to 2400 ng/μL ssDNA, and 2 to 3000 ng/μL RNA. However, if an accurate reading cannot be obtained using 1 μL, up to 2 μL of sample can be used; conversely, the genomic DNA may also be diluted as needed. The software will provide the A_{260}/A_{280} ratio, the A_{260}/A_{230} ratio, absolute absorbance of the sample, and the sample concentration.

References

1. Belisle, J. T. & Sonnenberg, M. G. (1998). Isolation of genomic DNA from mycobacteria. *Methods Mol Biol* **101**, 31–44.
2. Andersen, A. B., Worsaae, A. & Chaparas, S. D. (1988). Isolation and characterization of recombinant lambda gt11 bacteriophages expressing eight different mycobacterial antigens of potential immunological relevance. *Infect Immun* **56**, 1344–51.
3. Young, D. B., Kent, L. & Young, R. A. (1987). Screening of a recombinant mycobacterial DNA library with polyclonal antiserum and molecular weight analysis of expressed antigens. *Infect Immun* **55**, 1421–5.
4. Clark-Curtiss, J. E., Jacobs, W. R., Docherty, M. A., Ritchie, L. R. & Curtiss, R., III. (1985). Molecular analysis of DNA and construction of genomic libraries of *Mycobacterium leprae*. *J Bacteriol* **161**, 1093–102.
5. Jacobs, W. R., Jr., Kalpana, G. V., Cirillo, J. D., Pascopella, L., Snapper, S. B., Udani, R.A., Jones, W., Barletta, R. G. & Bloom, B. R. (1991). Genetic systems for mycobacteria. *Methods Enzymol* **204**, 537–55.
6. Hull, S. I., Wallace, R. J., Jr., Bobey, D. G., Price, K. E., Goodhines, R. A., Swenson, J. M. & Silcox, V. A. (1984). Presence of aminoglycoside acetyltransferase and plasmids in *Mycobacterium fortuitum*. Lack of correlation with intrinsic aminoglycoside resistance. *Am Rev Respir Dis* **129**, 614–8.
7. Crawford, J. T. & Bates, J. H. (1984). Restriction endonuclease mapping and cloning of *Mycobacterium intracellulare* plasmid pLR7. *Gene* **27**, 331–3.
8. Crawford, J. T., Cave, M. D. & Bates, J. H. (1981). Characterization of plasmids from strains of *Mycobacterium avium-intracellulare*. *Rev Infect Dis* **3**, 949–52.
9. Baess, I. & Bentzon, M. W. (1978). Deoxyribonucleic acid hybridization between different species of mycobacteria. *Acta Pathol Microbiol Scand [B]* **86**, 71–6.
10. Roberts, M. C., McMillan, C. & Coyle, M. B. (1987). Whole chromosomal DNA probes for rapid identification of *Mycobacterium tuberculosis* and *Mycobacterium avium* complex. *J Clin Microbiol* **25**, 1239–43.
11. Stinear, T. P., Seemann, T., Pidot, S., Frigui, W., Reysset, G., Garnier, T., Meurice, G., Simon, D., Bouchier, C., Ma, L., Tichit, M., Porter, J. L., Ryan, J., Johnson, P. D., Davies, J. K., Jenkin, G. A., Small, P. L., Jones, L. M., Tekaia, F., Laval, F., Daffe, M., Parkhill, J. & Cole, S. T. (2007). Reductive evolution and niche adaptation inferred from the genome of *Mycobacterium ulcerans*, the causative agent of Buruli ulcer. *Genome Res* **8**, 8.

12. Cole, S. T., Eiglmeier, K., Parkhill, J., James, K. D., Thomson, N. R., Wheeler, P. R., Honore, N., Garnier, T., Churcher, C., Harris, D., Mungall, K., Basham, D., Brown, D., Chillingworth, T., Connor, R., Davies, R. M., Devlin, K., Duthoy, S., Feltwell, T., Fraser, A., Hamlin, N., Holroyd, S., Hornsby, T., Jagels, K., Lacroix, C., Maclean, J., Moule, S., Murphy, L., Oliver, K., Quail, M. A., Rajandream, M. A., Rutherford, K. M., Rutter, S., Seeger, K., Simon, S., Simmonds, M., Skelton, J., Squares, R., Squares, S., Stevens, K., Taylor, K., Whitehead, S., Woodward, J. R. & Barrell, B. G. (2001). Massive gene decay in the leprosy bacillus. *Nature* **409**, 1007–11.
13. Cole, S. T., Brosch, R., Parkhill, J., Garnier, T., Churcher, C., Harris, D., Gordon, S. V., Eiglmeier, K., Gas, S., Barry, C. E., 3rd, Tekaia, F., Badcock, K., Basham, D., Brown, D., Chillingworth, T., Connor, R., Davies, R., Devlin, K., Feltwell, T., Gentles, S., Hamlin, N., Holroyd, S., Hornsby, T., Jagels, K., Krogh, A., McLean, J., Moule, S., Murphy, L., Oliver, K., Osborne, J., Quail, M. A., Rajandream, M. A., Rogers, J., Rutter, S., Seeger, K., Skelton, J., Squares, R., Squares, S., Sulston, J. E., Taylor, K., Whitehead, S. & Barrell, B. G. (1998). Deciphering the biology of *Mycobacterium tuberculosis* from the complete genome sequence. *Nature* **393**, 537–44.
14. Li, L., Bannantine, J. P., Zhang, Q., Amonsin, A., May, B. J., Alt, D., Banerji, N., Kanjilal, S. & Kapur, V. (2005). The complete genome sequence of *Mycobacterium avium* subspecies *paratuberculosis*. *Proc Natl Acad Sci U S A* **102**, 12344–9.
15. Bartlett, J. M. S. & Striling, D. (2003). *PCR Protocols*. 2 nd ed. Humana Press, Totowa, NJ.
16. Lane, J. M. & Rubin, E. J. (2006). Scaling down: a PCR-based method to efficiently screen for desired knockouts in a high density *Mycobacterium tuberculosis* picked mutant library. *Tuberculosis (Edinb)* **86**, 310–3.
17. Lamichhane, G., Zignol, M., Blades, N. J., Geiman, D. E., Dougherty, A., Grosset, J., Broman, K. W. & Bishai, W. R. (2003). A postgenomic method for predicting essential genes at subsaturation levels of mutagenesis: application to *Mycobacterium tuberculosis*. *Proc Natl Acad Sci U S A* **100**, 7213–8.
18. Cox, J. S., Chen, B., McNeil, M. & Jacobs, W. R., Jr. (1999). Complex lipid determines tissue-specific replication of *Mycobacterium tuberculosis* in mice. *Nature* **402**, 79–83.
19. Groathouse, N. A., Brown, S. E., Knudson, D. L., Brennan, P. J. & Slayden, R. A. (2006). Isothermal amplification and molecular typing of the obligate intracellular pathogen *Mycobacterium leprae* isolated from tissues of unknown origins. *J Clin Microbiol* **44**, 1502–8.
20. Brennan, P. J. & Nikaido, H. (1995). The envelope of mycobacteria. *Annu Rev Biochem* **64**, 29–63.
21. Sambrook, J., Fritsch, E. F. & Maniatis, T. (1989).*Molecular Cloning: A Laboratory Manual*. 2 nd ed. Cold Spring Harbor Laboratory Press, New York.
22. Caparon, M. G. & Scott, J. R. (1991). Genetic manipulation of pathogenic streptococci. *Methods Enzymol* **204**, 556–86.
23. Hoch, J. A. (1991). Genetic analysis in *Bacillus subtilis*. *Methods Enzymol* **204**, 305–20.
24. Murray, A., Winter, N., Lagranderie, M., Hill, D. F., Rauzier, J., Timm, J., Leclerc, C., Moriarty, K. M., Gheorghiu, M. & Gicquel, B. (1992). Expression of Escherichia coli beta-galactosidase in *Mycobacterium bovis* BCG using an expression system isolated from *Mycobacterium paratuberculosis* which induced humoral and cellular immune responses. *Mol Microbiol* **6**, 3331–42.
25. Barrera, L. F., Skamene, E. & Radzioch, D. (1993). Assessment of mycobacterial infection and multiplication in macrophages by polymerase chain reaction. *J Immunol Methods* **157**, 91–9.
26. Hurley, S. S., Splitter, G. A. & Welch, R. A. (1987). Rapid lysis technique for mycobacterial species. *J Clin Microbiol* **25**, 2227–9.
27. Yandell, P. M. & McCarthy, C. (1980). Isolation of deoxyribonucleic acid from *Mycobacterium avium* by rapid nitrogen decompression. *Infect Immun* **27**, 368–75.

28. Yoshimura, H. H., Graham, D. Y., Estes, M. K. & Merkal, R. S. (1987). Investigation of association of mycobacteria with inflammatory bowel disease by nucleic acid hybridization. *J Clin Microbiol* **25**, 45–51.

29. Imai, T., Ohta, K., Kigawa, H., Kanoh, H., Taniguchi, T. & Tobari, J. (1994). Preparation of high-molecular-weight DNA: application to mycobacterial cells. *Anal Biochem* **222**, 479–82.

30. Patel, R., Kvach, J. T. & Mounts, P. (1986). Isolation and restriction endonuclease analysis of mycobacterial DNA. *J Gen Microbiol* **132**, 541–51.

31. Belisle, J. T., Pascopella, L., Inamine, J. M., Brennan, P. J. & Jacobs, W. R., Jr. (1991). Isolation and expression of a gene cluster responsible for biosynthesis of the glycopeptidolipid antigens of *Mycobacterium avium*. *J Bacteriol* **173**, 6991–7.

32. Whipple, D. L., Le Febvre, R. B., Andrews, R. E., Jr. & Thiermann, A. B. (1987). Isolation and analysis of restriction endonuclease digestive patterns of chromosomal DNA from *Mycobacterium paratuberculosis* and other Mycobacterium species. *J Clin Microbiol* **25**, 1511–5.

33. Baess, I. (1974). Isolation and purification of deoxyribonucleic acid from mycobacteria. *Acta Pathol Microbiol Scand [B] Microbiol Immunol* **82**, 780–4.

34. Ansubel, F. M., Brent, R., Kingston, R. E., Moore, D. D., Seidman, J. G., Smith, J. A. & Struhl, K., eds. (2007). *Current Protocols in Molecular Biology*. John Wiley & Sons, New York.

35. van Soolingen, D., Hermans, P. W., de Haas, P. E., Soll, D. R. & van Embden, J. D. (1991). Occurrence and stability of insertion sequences in *Mycobacterium tuberculosis* complex strains: evaluation of an insertion sequence-dependent DNA polymorphism as a tool in the epidemiology of tuberculosis. *J Clin Microbiol* **29**, 2578–86.

36. Takayama, K., Schnoes, H. K., Armstrong, E. L. & Boyle, R. W. (1975). Site of inhibitory action of isoniazid in the synthesis of mycolic acids in *Mycobacterium tuberculosis*. *J Lipid Res* **16**, 308–17.

37. Collins, F. M., Wayne, L. G. & v, M. (1974). The effect of cultural conditions on the distribution of *Mycobacterium tuberculosis* in the spleens and lungs of specific pathogen-free mice. *Am Rev Respir Dis* **110**, 147–56.

38. Youmans, G. P. & Karlson, A. G. (1947). Streptomycin sensitivity of tubercle bacilli—studies on recently isolated tubercle bacilli and the development of resistance to streptomycin in vivo. *Am Rev Tuberculosis* **55**, 529–535.

39. Sommers, H. M. & Good, R. C. (1985). *Mycobacterium*. In *Manual of Clinical Microbiology* (Lennette, E. H., Balows, A., Hausler, J. & Shadomy, H. J., eds.), pp. 217. American Society for Microbiology, Washington, D.C.

40. Centers for Disease Control and Prevention and National Institutes of Health. (1999). *Biosafety in Microbiological and Biomedical Laboratories*. 4th ed. HHS Publication No. (CDC) 93-8395, pp. 102–106. U.S. Department of Health and Human Services, Public Health Service. Centers for Disease Control and Prevention and National Institutes of Health, Bethesda, MD.

41. Sedlaczek, L., Gorminski, B. M. & Lisowska, K. (1994). Effect of inhibitors of cell envelope synthesis on beta-sitosterol side chain degradation by *Mycobacterium* sp. NRRL MB 3683. *J Basic Microbiol* **34**, 387–99.

Chapter 2
Isolation of Mycobacterial RNA

Tige R. Rustad, David M. Roberts, Reiling P. Liao and David R. Sherman

Abstract This chapter describes two protocols for isolating total RNA from mycobacteria: one for extraction from *in vitro* cultures and one for extraction from *in vivo*. In these protocols, RNA is liberated from mycobacteria by disruption with small glass beads in the presence of Trizol to stabilize the RNA. The RNA is further purified with DNAse treatment and RNeasy columns. This protocol leads to microgram quantities of RNA from log-phase cultures.

Keywords cell disruption · extraction · *in vivo* · *in vitro* · RNA

2.1 Introduction

In nearly all living cells, ribonucleic acid (RNA) exists simply to convert the information stored in DNA into a functional form, generally as proteins. Even though RNA is an indirect measure of proteins, RNA analysis is central to modern molecular biology. RNA-based methods such as Northern blotting, transcriptome profiling by microarray [1], and quantitative real-time PCR [2] provide an unparalleled opportunity to interrogate cellular processes such as development and responses to environmental perturbations. The speed and ease of these methods make RNA an indispensable material for high-throughput analysis.

RNA extraction is a relatively simple process that takes only a few hours of labor over 2 days. And yet, RNA extraction is often a dreaded chore, particularly in a lab that works with mycobacteria. Why? Because RNA is fragile, mycobacteria are not, and RNA released from bacilli faces a hostile world of thermal and enzymatic breakdown. RNases are extremely common in the environment, highly stable, and they can reduce months of careful experimental preparation to a tube of useless nucleotides in seconds.

D.R. Sherman
Seattle Biomedical Research Institute 307 Westlake Seattle, WA 98119
e-mail: david.sherman@ski.org

T. Parish, A.C. Brown (eds.), *Mycobacteria Protocols*,   13
doi: 10.1007/978-1-59745-207-6_2, © Humana Press, Totowa, NJ 2008

Success, readily achieved, depends on preparation. The simple protocol outlined below can be used to isolate large quantities of RNA from mycobacterial *in vitro* cultures using only a limited amount of specialized equipment and easily available materials. With a few modifications, this protocol can be used to isolate RNA from macrophages or infected animal tissue. Our laboratory has used this protocol or subtle variants hundreds of times over many experiments to isolate grams of mycobacterial RNA [3, 4, 5]. The primary key to success with this protocol is keeping it simple, quick, and consistent. As is traditional in this book, we will include many footnotes with observations, tips, and tricks.

2.2 Materials

2.2.1 Isolation of Mycobacterium tuberculosis RNA from In Vitro Culture

2.2.1.1 Log-Phase Culture of *Mycobacterium tuberculosis*

1. ADC (albumin-dextrose-catalase) supplement (Becton Dickinson, Franklin Lakes, NJ).
2. Middlebrook 7 H9 medium (Becton Dickinson) supplemented with 10% v/v ADC or other mycobacterial growth medium.
3. Roller apparatus (Stovall Life Science, Inc., Greensboro, NC).
4. 50-mL conical tubes (e.g., Falcon).
5. 250-mL, 500-mL, and 2-L roller bottles (*see* **Note 1**).

2.2.1.2 RNA Extraction

1. Trizol solution (Invitrogen Corporation, Carlsbad, CA).
2. Lysing matrix B supplied in 2-mL screw-cap tubes (QBiogene, Inc., Irvine, CA) (*see* **Note 2**).
3. Fastprep 120 homogenizer (Qbiogene).
4. Phase lock gel (Eppendorf North America, Inc., Westbury, NY).
5. Chloroform/isoamyl alcohol 24:1 (CIA).
6. Isopropanol.
7. High Salt Solution; 0.8 M Na citrate, 1.2 M NaCl.
8. RNAse-free water.
9. 75% ethanol solution; make up with RNAse-free water.

2.2.1.3 RNA Purification and On-Column DNase Digestion

1. RNeasy RNA purification kit (Qiagen Inc., Valencia, CA).
2. 95% ethanol solution (dilute with RNAse-free water).

3. On-column DNase kit (Qiagen). Prepare DNase I stock by dissolving 1500 U DNase I in 550 μL RNAse-free water and mix gently by inverting the tube. Make into single-use aliquots and store at –20°C. A thawed aliquot is stable at 2°C to 8°C for 6 weeks.

2.2.1.4 RNA Quantification

1. TE buffer (10 mM Tris pH 7.5, 1 mM EDTA)
2. Spectrophotometer equipped to read at wavelengths 260 nm and 280 nm.
3. Gel electrophoresis equipment.

2.2.2 Isolation of Mycobacterial RNA from Tissue

2.2.2.1 Tissue Homogenization

1. 13-mL round-bottom polypropylene tubes (e.g., Falcon).
2. Trizol solution (Invitrogen Corporation).
3. OMNI TH tissue homogenizer with serrated disposable tips (Omni International, Marietta, GA).

2.2.2.2 Extraction of *Mycobacterium tuberculosis* RNA

1. Trizol solution (Invitrogen Corporation).
2. Lysing matrix B supplied in 2-mL screw-cap tubes (QBiogene) (*see* **Note 2**).
3. Fastprep 120 homogenizer (Qbiogene).
4. RNAse-free tubes.
5. Chloroform/isoamyl alcohol 24:1 (CIA).
6. Phase lock gel (Eppendorf North America, Inc.).
7. Isopropanol.
8. RNAse-free water.
9. 75% ethanol solution; make up with RNAse-free water.

2.2.2.3 Qiagen RNeasy Column Purification and On-Column Digestion

1. Qiagen RNeasy kit (Qiagen Inc.).
2. RNAse-free tubes.

2.3 Methods

Isolation of high-quality RNA from mycobacteria depends on limiting the amount of time the RNA is susceptible to degradation. These protocols are therefore best done rapidly. Wherever possible, prepare tubes, tips, and equipment in advance. For pathogenic species of mycobacteria, the bulk of the first part of this protocol is done inside a BL-3 (biosafety level 3) facility due to the risk of aerosols from the vortexing, centrifuging, and bead beating. However, because Trizol rapidly sterilizes the pellets, only the first step of the RNA extraction needs to be carried out in the safety cabinet.

2.3.1 Isolation of Mycobacterium tuberculosis RNA from In Vitro Culture

2.3.1.1 Log-Phase Culture of Mycobacterium tuberculosis

1. To obtain log-phase bacilli, inoculate 25 mL Middlebrook 7 H9-ADC broth with *M. tuberculosis* in a 250-mL roller bottle (*see* **Notes 1 and 3**).
2. Grow to early- to mid-log phase (optical density, or O.D., of 0.1 to 0.2) at 37°C with constant rolling.
3. Centrifuge cultures for 5 min at 2000 × g, at room temperature.
4. Pipette off supernatant and immediately process for RNA or freeze on dry ice (*see* **Note 4**).

2.3.1.2 RNA Extraction

1. Add 1 mL Trizol to the cell pellets and transfer to ice (*see* **Note 2**).
2. Resuspend pellets by vortexing until there are no visible clumps.
3. Add the suspension to a screw-top tube containing Lysing Matrix B and place on ice (*see* **Note 2**).
4. To disrupt the cells and liberate the RNA, shake for 30 s at maximum speed (6.5) in the Fastprep machine.
5. Place on ice for 30 s to cool sample. Repeat steps 3 to 4 two more times (*see* **Note 5**).
6. Centrifuge samples for 1 min at 16,000 × g in a Microfuge.
7. Remove Trizol solution to a 2-mL Heavy Phase Lock Gel I snap cap tube containing 300 μL chloroform/isoamyl alcohol.
8. Mix by inverting rapidly for 15 s, and place on ice. Once all the samples are transferred, continue inverting periodically for 2 min.
9. Centrifuge for 5 min at 16,000 × g in a Microfuge, remove aqueous layer (above the phase lock, volume ∼540 μL), and add to a 1.5-mL tube containing 270 μL isopropanol and 270 μL high salt solution (*see* **Note 6**).
10. Invert several times and spray outside of tube with a mycobactericidal agent and remove from the BL-3. Precipitate at 4°C overnight (*see* **Note 7**).
11. Centrifuge for 10 min at 4°C and remove isopropanol.
12. Add 1 mL 75% EtOH, invert several times, centrifuge for 5 min at 16000 × g, and decant ethanol (*see* **Notes 8 and 9**).
13. Add 100 μL RNAse-free water.

2.3.1.3 RNA Purification and On-Column DNase Digestion (*see* Note 10)

This is a variant of the Qiagen protocol for RNA purification with the RNeasy columns, simplified slightly for our purposes.

1. Add 350 µL RLT buffer. *N.B.*: Remember to add 10 µL β-mercaptoethanol per milliliter of RLT prior to use (*see* Note 11).
2. Vortex. Add 265 µL 95% EtOH to each sample. Mix by vortexing briefly.
3. Transfer mixture to an RNeasy spin column, centrifuge for 15 s, and transfer column to a new 2-mL collection tube.
4. Add 350 µL buffer RW1 (*see* Note 11), centrifuge for 15 s, and discard flow through.
5. Add 70 µL buffer RDD to 10-µL aliquot of DNase I stock solution and pipette directly onto the column membrane (*see* Note 11). Allow digestion to continue at room temperature for 15 min to 1 h (*see* Note 12).
6. Add 350 µL buffer RW1, centrifuge for 15 s.
7. Add 500 µL RPE buffer, centrifuge for 15 s, discard flow-through. Add 500 µL additional RPE and centrifuge for 2 min. Discard flow-through and spin for 1 min to dry completely (*see* Note 13).
8. Transfer to a 1.5-mL collection tube, elute with 40 µL RNase-free water, centrifuge 1 min (*see* Note 14) and recover eluate.

2.3.1.4 RNA Quantification

1. Dilute 1 µL RNA sample in 74 µL TE and measure absorbance at 260 and 280 nm to determine RNA concentration and purity (*see* Note 15).
2. Using the RNA concentrations estimated by the spectrophotometer reading, run 1 µg RNA on a 2% w/v agarose gel.
3. Run gel for 45 min at 100 V. Ribosomal bands should be clear, concise, and relatively equal (*see* Note 16 and Fig. 2.1).

2.3.2 Isolation of Mycobacterial RNA from Tissue (see Note 17)

The initial steps of this protocol should be performed in a strict biosafety environment when processing pathogenic mycobacterial species due to the high probability of aerosolization during the tissue homogenization process.

2.3.2.1 Tissue Homogenization

1. Place infected tissue in a 13-mL round-bottom tube and weigh.
2. Add 1 mL Trizol reagent per 50 to 100 mg tissue. For less than 10 mg tissue, add 0.8 mL Trizol reagent.

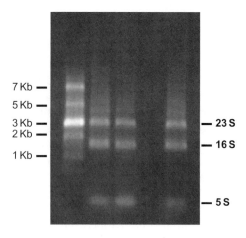

Fig. 2.1 Nondenaturing
"check" gel of *Mycobacterium
bovis* RNA isolated from
infected rabbit tissue samples

3. Thoroughly homogenize up to 5 mL of sample using the OMNI TH with serrated disposable tips (*see* **Note 18**) in 13-mL round-bottom tubes. For amounts >5 mL, it is recommended that the sample be split into two. Make sure the tissue is completely homogenized before proceeding.
4. Centrifuge homogenate at $3200 \times g$ for 10 min at room temperature (RT) to separate out the intact bacteria from eukaryotic cell lysate.
5. Decant the supernatant to a fresh tube and save if eukaryotic RNA extraction is desirable. Take care not to disturb the pellet containing the bacteria.

2.3.2.2 Extraction of *Mycobacterium tuberculosis* RNA

The steps below are very similar to the protocol for extraction of RNA from *in vitro* culture above.

1. Add 0.2 mL Trizol per milliliter of Trizol used in the tissue homogenization step to the bacterial cell pellet. Transfer up to 1 mL Trizol suspension to a 2-mL screw-cap tube of Lysing Matrix B (*see* **Note 2**).
2. Homogenize the cells 3 times using intervals of 30 s at speed 6.5 in a Fastprep 120. Incubate samples 30 s on ice in between disruptions to allow tubes to cool (*see* **Note 5**).
3. Centrifuge for 1 min at $>10,000 \times g$ to pellet the beads.
4. Transfer supernatant to a 1.5-mL RNAse-free tube containing 300 μL chloroform per mL of Trizol used in the bead-beating step.
5. Shake tube vigorously for 15 s and transfer contents to a (precentrifuged) 2 mL Heavy Phase Lock Gel.
6. Continue to vigorously invert tube for 2 min. At this point, the sample tube can be decontaminated and removed from the biosafety area.

 7. Centrifuge for 10 min at 16,000 × g for 10 min at 2°C to 8°C (*see* **Note 19**).
 8. Transfer the aqueous phase (~540 μL from 1 mL Trizol) to a 1.5-mL RNAse-free tube.
 9. Precipitate RNA by adding 0.5 mL isopropanol per milliliter Trizol used to resuspend the bacterial pellet.
10. Mix the sample by repeated inversion and centrifuge *briefly* to bring down any contents from the lid.
11. Incubate sample at 15°C to 30°C for 10 min.
12. Centrifuge the sample at 16,000 × g for 10 min at 2°C to 8°C.
13. The RNA pellet should be visible on the side and bottom of tube. Carefully aspirate the supernatant
14. Add 0.2 mL Trizol per milliliter Trizol used in the initial tissue homogenization and gently pipette up and down to dissolve the RNA pellet.
15. Add 300 μL chloroform per milliliter Trizol and shake tube vigorously for 15 s.
16. Transfer contents to a (precentrifuged) 2 mL Heavy Phase Lock Gel.
17. Continue to vigorously invert tube for 2 min.
18. Centrifuge at 16,000 × g for 10 min at 2°C to 8°C.
19. Transfer the aqueous phase to a 1.5-mL RNAse-free tube.
20. Add 1 volume 70% EtOH or 0.7 volume 100% EtOH. Invert tube immediately to mix.

2.3.2.3 Qiagen RNeasy Column Purification and On-Column Digestion

 1. Apply the entire sample (up to 700 μL) to an RNeasy minicolumn placed in a 2-mL collection tube.
 2. Centrifuge for 15 s at 16,000 × g and discard the flow-through.
 3. Add 350 μL RW1 buffer and centrifuge at 16,000 × g for 15 s (*see* **Note 11**). Discard flow-through.
 4. Add 500 μL RPE and wash the column by centrifuging at 16,000 × g for 15 s.
 5. Discard flow-through and add another 500 μL RPE to the column. Centrifuge at 16,000 × g for 15 s and discard flow-through.
 6. Thoroughly dry the column by centrifuging at full speed for 1 min.
 7. Add 10 μL DNase I stock to 70 μL RDD (Qiagen) and mix by gently inverting the tube. Pipette all 80 μL of the mix directly onto the column and incubate at RT for 15 min.
 8. Add 350 μL RW1 and centrifuge at 16,000 × g for 15 s. Discard flow-through.
 9. Add 500 μL RPE and wash the column by centrifuging at 16,000 × g for 15 s.
10. Discard flow-through and add another 500 μL RPE to the column.
11. Centrifuge at 16,000 × g for 15 s and discard flow-through, or place column into a new 2-mL collection tube.

12. Thoroughly dry the column by centrifuging at full speed for 1 min.
13. Transfer column to a 1.5-mL RNase-free tube and keep cap open to allow membrane to air dry for 2 min.
14. To elute the RNA, add 50 µL RNase-free water to the center of the membrane (do not close column cap). Allow the membrane to soak for 2 min. Close cap and centrifuge at max speed for 1 min. Store RNA at –80°C.
15. Quantify RNA as in Section 2.3.1.4.

2.4 Notes

1. Culture volumes of up to 20% of the total container volume can be aerobically grown in roller bottles.
2. It is possible to use screw-cap tubes with sterile 0.2-µm beads in the place of Lysing Matrix beads. However, we prefer to use the commercially available Lysing Matrix B tubes for convenience. Similarly, it is possible to formulate a noncommercial guanidium thiocyanate solution to replace Trizol as a stabilization reagent.
3. The protocol is written for isolation of RNA from log-phase culture to provide the highest yield. This protocol has been used to isolate RNA from late log, hypoxic, and many other culture conditions.
4. We routinely leave pellets frozen at –80°C, sometimes for as long as 2 years with no discernible degradation or loss of RNA, as long as they are frozen rapidly after centrifugation.
5. Samples will heat during processing, which may damage the RNA. Cooling the samples on ice between each round of bead beating minimizes RNA damage.
6. Note that this protocol is slightly different from the method used to extract RNA from tissue culture: the phase separation is performed in a room-temperature centrifuge, instead of at 2°C to 8°C, which we have found to be sufficient for *in vitro* isolation. However, if the phase separation is not clean, centrifugation in a refrigerated centrifuge may improve resolution of the phases.
7. We have found that precipitating in a freezer can decrease the yield of RNA. Precipitating for as long as 3 days at 4°C seems to result in little or no loss or degradation of the RNA.
8. Make certain that the pellet is not decanted with the supernatant. After the isopropanol extraction, a small, often translucent pellet may be visible. After the ethanol wash, a pellet is almost always visible. We typically decant by simply pouring off the supernatant. However, the risk of losing the pellet can be reduced by suctioning off the supernatant with a pipeteman or RNase-free Pasteur pipette. When washing, it is not necessary to resuspend the pellet entirely.
9. RNA can be air dried for 5 to 15 min to remove the last traces of ethanol, resuspended in RNAse-free water, and frozen at –80°C for later use. The RNA is not clean at this step, but for some experiments this crude extract is sufficient. Do not dry RNA extensively, as it can be very difficult to resuspend.
10. The RNeasy column purification is optional. If column purification is skipped, an additional chloroform extraction step can be added immediately after step 5 of Section 2.3.1.2, and any commercially available DNAse can be used. We use the RNeasy columns for simplicity and consistently higher-purity samples, but the added expense of the columns may make them a dispensable step.
11. Supplied with RNeasy kit by manufacturer (Qiagen).
12. The protocol from Qiagen recommends a 15-min incubation. However, RT-qPCR results suggest that extending the incubation to an hour reduced DNA contamination.

The length of DNase treatment will depend on the downstream application and the RNA purity required.

13. Residual ethanol from RPE can interfere with downstream reactions.

14. For RT-qPCR, we have found that it is helpful to have an additional DNAse step.

15. The 260/280 ratios are notoriously poor at detecting protein contamination [6]. If yields are very low, the 260/280 readings may be below the level of detection of spectrophotometers and will be thereby skewed.

16. Standard gel electrophoresis is usually sufficient. However, some degradation often occurs in the gel. If degradation is a serious concern, RNA can be analyzed on a denaturing gel containing 20% formaldehyde. Formaldehyde gels require buffer recirculation and ventilation. It is almost certain that there will be some amount of RNA degradation and DNA contamination in a typical sample. However, most applications allow for this fact, particularly microarray analysis where cDNA synthesis and lack of amplification reduce the effects of degradation and DNA contamination.

17. Isolation of good-quality mycobacterial RNA from infected tissue relies primarily on (1) the time it takes to get the tissue into an RNA stabilization reagent, (2) the type of stabilization reagent used, and (3) the method used to preserve the tissue. Different stabilization reagents include Trizol reagent (Invitrogen), RNAlater (Ambion, Austin, TX), or any of the various available guanidine-based solutions. Methods of tissue preservation include immediate freezing of the tissue with or without prior submersion of the tissue in a stabilization reagent. Ideally, infected tissue would be placed quickly into a stabilization reagent, homogenized, and immediately processed. Immediate processing of tissue is often not convenient or not possible due to technical restrictions. In these cases, the tissue should be submerged in stabilization reagent and frozen in liquid nitrogen or dry ice/ethanol for subsequent processing.

18. Adding too much Trizol will not adversely affect RNA stability, but adding too little can lead to stabilization failure. Typically, 5 mL Trizol can be used for tissues weighing 10 to 500 mg.

19. Phase separation in the presence of Trizol should be performed at 2°C to 8°C as indicated. Separations performed at higher temperatures may lead to DNA and protein contamination of the aqueous phase. This does not appear to be a problem with RNA extraction from *in vitro* cultures.

References

1. Butcher PD. Microarrays for *Mycobacterium tuberculosis*. Tuberculosis (Edinb) 2004;84(3-4):131–7.

2. Delogu G, Sanguinetti M, Pusceddu C, et al. PE_PGRS proteins are differentially expressed by *Mycobacterium tuberculosis* in host tissues. Microbes Infect 2006;8(8):2061–7.

3. Roberts DM, Liao RP, Wisedchaisri G, Hol WG, Sherman DR. Two sensor kinases contribute to the hypoxic response of *Mycobacterium tuberculosis*. J Biol Chem 2004;279(22):23082–7.

4. Guinn KM, Hickey MJ, Mathur SK, et al. Individual RD1-region genes are required for export of ESAT-6/CFP-10 and for virulence of *Mycobacterium tuberculosis*. Mol Microbiol 2004;51(2):359–70.

5. Park HD, Guinn KM, Harrell MI, et al. Rv3133c/dosR is a transcription factor that mediates the hypoxic response of *Mycobacterium tuberculosis*. Mol Microbiol 2003;48(3):833–43.

6. Sambrook J, Russell D. Molecular cloning: A laboratory manual. Cold Spring Harbor, Cold Spring Harbor Laboratory Press, 2001;3(III):A8.20–21.

Chapter 3
Lipoglycans of *Mycobacterium tuberculosis*: Isolation, Purification, and Characterization

Libin Shi, Jordi B. Torrelles and Delphi Chatterjee

Abstract In this chapter, we describe in detail the steps involved in isolation and characterization of lipoglycans from *Mycobacterium tuberculosis* and *Mycobacterium smegmatis*. In addition, procedures involved in structural analysis such as immunoblotting with mAb CS-35 or CS-40, gas chromatography, gas chromatography/mass spectrometry, nuclear magnetic resonance spectroscopy, and endoarabinanase digestion followed by high-pH anion exchange chromatography and two-dimensional gel electrophoresis are presented.

Keywords cell wall · lipoarabinomannan (LAM) · lipoglycans · *M. smegmatis* · *M. tuberculosis* · *Mycobacterium*

3.1 Introduction

The most prominent components intercalated in the cell wall of mycobacteria and associated with the outer lipid layer are the phosphatidyl-*myo*-inositol (PI)-based lipoglycans. These lipoglycans are biosynthetically related and classified into three classes: phosphatidyl-*myo*-inositol mannosides (PIMs), lipomannan (LM), and lipoarabinomannan (LAM) [1]. These are ubiquitous among all mycobacterial species and are implicated in host pathogen interactions.

LAM is an extremely heterogeneous lipoglycan with three distinct structural domains including a PI anchor, a branched mannan backbone, and a branched and extended arabinan (Fig. 3.1). The PI anchor is composed of a *myo*-inositol phosphoryl diacylglycerol substituted at the 2-position with a single mannopyranose (Man*p*) and at the 6-position with the mannan core [1]. The mannan core consists of a linear $\alpha(1\rightarrow6)$ Man*p* chain with intermittent substitution of a single $\alpha(1\rightarrow2)$ Man*p* branching residue [1]. The arabinan domain is assembled mostly into stretches of -[α-D-Ara*f*$(1\rightarrow5)$-α-D-Ara*f*]$_n$- with critically

D. Chatterjee
Department of Microbiology, Immunology, and Pathology, 1682 Campus Delivery,
Colorado State University, Fort Collins, CO 80523, USA
e-mail: delphi@lamar.colostate.edu

T. Parish, A.C. Brown (eds.), *Mycobacteria Protocols*,
doi: 10.1007/978-1-59745-207-6_3, © Humana Press, Totowa, NJ 2008

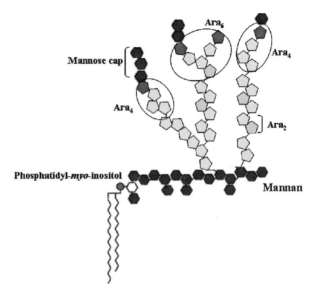

Fig. 3.1 Schematic visualization of the structure of mannose-capped lipoarabinomannan (ManLAM). The *myo*-inositol phosphoryl diacylglycerol is substituted at the 2-position with a single mannopyranose (Man*p*) and at the 6-position with the mannan core. The mannan core consists of a linear $\alpha(1\rightarrow6)$ Man*p* chain with intermittent substitution of a single $\alpha(1\rightarrow2)$ Man*p* branching residue. The arabinan domain is assembled mostly into stretches of -[α-D-Ara*f*($1\rightarrow5$)-α-D-Ara*f* (Ara$_2$)]$_n$- with critically spaced 3,5-α-Ara*f*-branched sites. There are two distinct nonreducing terminal motifs, a linear β-D-Ara*f*($1\rightarrow2$)-α-D-Ara*f*($1\rightarrow5$)-α-D-Ara*f*($1\rightarrow5$)-α-D-Ara*f* (known as Ara$_4$ motif) and a branched [β-D-Ara*f*($1\rightarrow2$)-α-D-Ara*f*]$_2$-3,5-α-D-Ara*f*($1\rightarrow5$)-α-D-Ara*f* (known as Ara$_6$ motif) that typify any kind of LAM. The exact branching pattern of the mannan core and the arabinan proximal to the ManLAM mannan is not known and is depicted as a shaded ellipse

spaced 3,5-α-Ara*f*-branched sites. There are two distinct nonreducing terminal motifs, a linear β-D-Ara*f*($1\rightarrow2$)-α-D-Ara*f*($1\rightarrow5$)-α-D-Ara*f*($1\rightarrow5$)-α-D-Ara*f* (known as Ara$_4$ motif) and a branched [β-D-Ara*f*($1\rightarrow2$)-α-D-Ara*f*]$_2$-3,5-α-D-Ara*f*($1\rightarrow5$)-α-D-Ara*f* (known as Ara$_6$ motif) that typify any kind of LAM [2, 3, 4].

There are three kinds of LAM distinguished by modification on their nonreducing termini. Slow-growing mycobacteria such as *Mycobacterium tuberculosis, Mycobacterium leprae, Mycobacterium bovis* BCG, *Mycobacterium avium, Mycobacterium kansasii, Mycobacterium xenopi*, and *Mycobacterium marinum* [3, 4, 5, 6, 7, 8] have ManLAM. ManLAM is characterized by the presence of short $\alpha(1\rightarrow2)$ Man*p* chains capping the nonreducing Ara*f* termini. Fast-growing mycobacteria such as *Mycobacterium smegmatis* (often used as an experimental laboratory model) contain LAM that is largely uncapped with a small proportion of *myo*-inositol phosphate caps and termed PILAM [9, 10]. LAM in *Mycobacterium chelonae* is devoid of any modification and is termed AraLAM [8].

LAM exhibits a wide spectrum of immunobiological functions and is generally considered to be a crucial factor in mycobacterial pathogenesis. ManLAM is thought to be anti-inflammatory, whereas PILAM is thought to be proinflammatory. ManLAM has been implicated in inhibition of phagosomal maturation, apoptosis, interferon-γ signaling, and interleukin-12 secretion in phagocytic cells [11, 12, 13]. It has also been suggested that the nature of capping on LAM is a major structural feature in determining how the immune system is modulated [11], and a recent publication suggests that dendritic cell–specific intercellular adhesion molecule-3-grabbing non-integrin (DC-SIGN) may act as a pattern recognition receptor and discriminate between *Mycobacterium* species through selective recognition of the mannose caps on LAM [14]. In addition, a major surface macrophage receptor that interacts with ManLAM is the mannose receptor (MR) [15]. It has been shown that *M. tuberculosis* interacts with MR through the mannose caps of ManLAM, directing the bacillus to its natural niche with the macrophage (i.e., a phagosome with limited fusion with lysosomes) [16].

In this chapter, we describe a method for the extraction and purification of *M. tuberculosis* ManLAM, LM, and PIMs and outline the steps involved in structural analysis of ManLAM applying a combination of immunoblotting with mAb CS-35 or CS-40, gas chromatography (GC), gas chromatography/mass spectrometry (GC/MS), nuclear magnetic resonance (NMR) spectroscopy, and endoarabinanase digestion followed by high-pH anion exchange chromatography (HPAEC) and two-dimensional gel electrophoresis for ManLAM isoform resolution. In addition, we introduce the methods of extraction of LAM, LM, and PIMs from *M. smegmatis* both from large-scale culture and colonies.

3.2 Materials

3.2.1 *Extraction of* M. tuberculosis *Lipoglycans (ManLAM, LM, and PIMs)*

1. *M. tuberculosis* $H_{37}R_v$ frozen ($-80°C$) stock in 1-mL aliquots.
2. Middlebrook 7H9 broth (Difco, USA) containing 0.2% (v/v) glycerol, 15 mM NaCl, 0.05% (v/v) Tween 80, and 10% (v/v) albumin-dextrose-catalase (ADC; Difco) supplement.
3. Chloroform:methanol:H_2O (10:10:3, v/v/v).
4. Endotoxin-free water.
5. 8% Triton X-114 in sterile endotoxin free PBS (*see* **Note 1**).
6. 3 mg/mL Pepstatin A in ethanol.
7. 1 mg/mL Leupeptin in ethanol.
8. 1 mg/mL PMSF in isopropanol.

9. Breaking buffer: add 23.3 μL 3 mg/mL Pepstatin A, 50 μL 1 mg/mL Leupeptin, 200 μL 1 mg/mL PMSF, 300 μg DNAse I, and 330 μg RNAse I to 100 mL 8% TrtionX-114.
10. 95% ethanol.
11. Proteinase K solution: 2 mg/mL in 10 mM Tris-HCl (pH 7.5), 20 mM $CaCl_2$, 50% (v/v) glycerol.
12. 15% SDS-polyacrylamide gel (80 × 70 mm, WXL; 0.75 mm thick): 6% acrylamide stacking gel and 15% acrylamide resolving gel.
13. 10X running buffer: dissolve 30.3 g Tris (base), 144 g glycine, and 10 g SDS in Milli-Q water, make up to1 L. Dilute to 1X in Milli-Q water before use.
14. Gel running apparatus: Mighty small II, SE250/SE260 (Hoefer Phamacia Biotech Inc., USA).
15. Sterile 250-mL conical flask.
16. Two sterile 4-L conical flasks.
17. 30-mL Oak Ridge Teflon FEP tubes (Nalgene, USA).
18. 13 × 100 mm glass tubes (Kimble Glass, Inc., USA).
19. Adjustable-tilt rocking platform shaker (Labnet International Inc., USA).
20. Reacti-ThermIII Heating/Stirring Module Triple-Block (Pierce, USA).
21. Slide-A-Lyzer dialysis cassette with 3.5-kDa MWCO (molecular weight cut off) (Pierce).
22. Speed Vac SC 110A vacuum concentrator (Savant, USA).
23. MSE Soniprep 150, high-frequency 23-kHz generator (Sanyo Gallenkamp PLC, UK).
24. French Press (Thermo Scientific, USA) (aperture pressure change = 12,500 psi).

3.2.2 Purification of ManLAM, LM, and PIMs

1. 15 mg of lipoglycans extracted from *M. tuberculosis* $H_{37}R_v$.
2. Column running buffer: 10 mM Tris-HCl, 0.2 M NaCl, 0.02% (w/v) sodium azide, 0.25% (w/v) deoxycholic salt, and 1 mM EDTA, pH 8.0.
3. Dialysis buffer: 10 mM Tris-HCl, 0.2 M NaCl, 0.02% (w/v) sodium azide, and 1 mM EDTA, pH 8.0.
4. 1 M NaCl.
5. Endotoxin-free water.
6. High-performance liquid chromatography system fitted with a Sephacryl S-200 HiPrep 26/60 column in tandem with a Sephacryl S-100 HiPrep 16/60 column (Amersham Bioscience, UK) equilibrated with column running buffer.
7. 0.8-μm and 0.2-μm syringe filter units.
8. Slide-A-Lyzer dialysis cassette with 3.5-kDa MWCO.

9. 15% SDS-polyacrylamide gel, 10X running buffer, and gel running apparatus (*see* Section 3.2.1, items 12, 13, and 14).
10. Freeze dry/shell freeze system (Labconco Corp., USA).

3.2.3 Polyacrylamide Gel Electrophoresis Followed by Periodic Acids-Schiff (PAS) Staining

1. 5 to 10 μg of purified ManLAM.
2. PageRuler Prestained Protein Ladder, broad range: 10 to 170 kDa (Fermentas, USA).
3. 5X sample buffer: 250 mM Tris-HCl (pH 6.8), 10% (w/v) SDS, 30% (v/v) glycerol, 5% (v/v) β-mercaptoethanol, and 0.02% (w/v) bromophenol blue.
4. 15% SDS-polyacrylamide gel, 10X running buffer, and gel running apparatus (*see* Section 3.2.1, items 12, 13, and 14).
5. First fixative: 40% (v/v) methanol, 10% (v/v) acetic acid in distilled H_2O.
6. Second fixative: 5% (v/v) methanol, 7% (v/v) acetic acid in distilled H_2O.
7. 0.7% (w/v) periodic acid in first fixative.
8. 2.5% (v/v) glutaraldehyde solution in distilled H_2O.
9. 0.0025% (w/v) a-dithiothreitol (DTT) in distilled H_2O.
10. 0.1% (w/v) silver nitrate in distilled H_2O.
11. 3% (w/v) sodium carbonate in distilled H_2O.
12. 37% (v/v) formaldehyde.
13. 50% (w/v) citric acid in distilled H_2O.
14. Adjustable-tilt rocking platform shaker.

3.2.4 Immunoblotting Using mAb CS-35 or mAb CS-40

1. 5 μg purified ManLAM.
2. 5X sample buffer (*see* Section 3.2.3, item 3).
3. 15% SDS-polyacrylamide gel, 10X running buffer, and gel running apparatus (*see* Section 3.2.1, items 12, 13, and 14).
4. 0.2-μm nitrocellulose membrane (Whatman, UK).
5. TE22 Mini Tank transphor unit (GE Healthcare, USA).
6. Transfer buffer: 25 mM Tris (base), 182 mM glycine, 20% (v/v) methanol.
7. Square Petri dish: 100 × 15 mm (BD Falcon, USA).
8. TBS buffer: 10 mM Tris-HCl, 0.87% (w/v) NaCl, pH 8.0.
9. Blocking buffer: 2% (w/v) bovine serum albumin (BSA) in TBS buffer.
10. mAbs CS-35 and CS-40 (1:20, available through http://www.cvmbs.colostate.edu/mip/leprosy/index.html): add 0.5 mL mAbs into 9.5 mL blocking buffer (*see* **Note 2**).
11. Alkaline phosphatase anti-mouse IgG conjugate (Sigma, USA): diluted to 1:5,000 (v/v) with 10 mL TBS buffer.

12. 5-bromo-4-chloro-3-indolyl phosphate/nitro blue tetrazolium (Sigma Fast BCIP/NBT tablet, B5655-25TAB; Sigma): dissolve 1 tablet in 10 mL Milli-Q water.

3.2.5 Analysis of Monosaccharide Composition Using Gas Chromatography

1. Monosaccharides stock solution: 0.5 mM of each of nine monosaccharides (rhamnose, fucose, arabinose, ribose, xylose, mannose, galactose, glucose and *myo*-inositol) in Milli-Q water.
2. Internal standard (*scyllo*-inositol): 5.5 mM in Milli-Q water.
3. 25 µg purified ManLAM.
4. 13 × 100 mm glass tubes with Teflon-lined screw cap (Kimble Glass Inc.).
5. Trifluoroacetic acid [TFA; assay ≥ 99.5% (T), Fluka, USA]: 2 M in Milli-Q water.
6. Reducing reagent: 10 mg/mL sodium borodeuteride (98 atom% D; Sigma) in 1 M aqueous ammonium hydroxide (NH_4OH; 28% NH_3 in water, 99.99+%; Aldrich, USA)/95% ethanol (1:1, v/v).
7. Acetic acid, glacial (99.99+%; Aldrich).
8. 10% (v/v) glacial acetic acid in methanol (A.C.S. reagent; Fisher Scientific, USA).
9. Acetic anhydride (98%, A.C.S. reagent; Sigma-Aldrich, USA).
10. Chloroform (A.C.S. reagent; Fisher Scientific).
11. Reacti-ThermIII Heating/Stirring Module Triple-Block.
12. Speed Vac SC 110A vacuum concentrator.
13. Hewlett Packard gas chromatography model 5890 fitted with a SP-2380 column (30-m × 0.25-mm inner diameter) (Hewlett Packard, USA).

3.2.6 Structural Characterization by Gas Chromatography/Mass Spectrometry

1. 50 µg purified ManLAM.
2. Sodium hydroxide (NaOH) pellets (99.99%, semiconductor grade; Aldrich).
3. Dimethyl sulfoxide (DMSO) (A.C.S. reagent; Fisher Scientific).
4. Iodomethane (reagent plus, 99.5%; Aldrich).
5. 0.2 oz. clear glass mortar and pestle set (Fisher Scientific).
6. Ice-cold Milli-Q water.
7. Items 4 to 12 in Section 3.2.5.
8. ThermoQuest Trace Gas Chromatograph 2000 (ThermoQuest, USA) with DB-5 column (10-m × 0.18-mm inner diameter, 0.18-µm film thickness; J&W Scientific, USA) connected to a GCQ/ Polaris MS mass detector (ThermoQuest).

3.2.7 Nuclear Magnetic Resonance Spectroscopy

1. 4 to 5 mg purified ManLAM.
2. Deuterium oxide (D_2O; 99.9% D; Cambridge Isotope Laboratories, Inc., USA).
3. Deuterium oxide (D_2O; 100.0 atom% D; Aldrich).
4. NMR sample tubes (Wilmad/Aldrich, USA).
5. Freeze dry/shell freeze system.
6. Two-dimensional ^{1}H-^{13}C heteronuclear single quantum correlation spectroscopy (HSQC) NMR spectra acquired on a Varian Inova 500 MHz NMR spectrometer using the supplied Varian pulse sequences.

3.2.8 Cellulomonas *Enzyme Digestion Followed by High-pH Anion Exchange Chromatography*

1. 10 µg purified ManLAM.
2. *Cellulomonas gelida* secreted endoarabinanase: 1 U [1 U of endoarabinanase is defined as the endoarabinanase activity necessary to obtain the complete arabinosyl oligomers release on 10 µg LAM in 20 µL 25 mM phosphate buffer (pH 7.0) at 37°C for 12 h] [17].
3. Speed vac SC 110A vacuum concentrator.
4. Analytical high-pH anion exchange chromatography performed on a Dionex liquid chromatography system fitted with a Carbopac PA-1 column. The oligoarabinosides are detected with a pulse-amperometric detector (PAD-II) (Dionex, USA).

3.2.9 Extraction of LAM, LM, and PIMs from M. smegmatis

1. *M. smegmatis* strain mc^{2}155 frozen (–80°C) stock in 1-mL aliquots.
2. Middlebrook 7 H11 agar (Difco) containing 0.5% (v/v) glycerol, 15 mM NaCl, 10% oleic acid-albumin-dextrose-catalase (OADC; Difco) supplement.
3. Middlebrook 7 H9 broth (*see* Section 3.2.1, item 2).
4. 95% ethanol.
5. 50% ethanol in Milli-Q water.
6. Chloroform: methanol (2:1, v/v).
7. Proteinase K (*see* Section 3.2.1, item 11).
8. Silicone fluid SF96/50 dimethyl silicone fluid (Thomas Scientific, USA).
9. 1-µL sterile inoculating loop (NUNC, USA).
10. Polypropylene sterile 50-mL culture tube (Greiner Bio-one, USA).
11. Two sterile 4-L conical flasks.
12. 30-mL Oak Ridge Teflon FEP tubes.

13. Pyrex round-bottom flask 100-mL 24/40.
14. Pyrex round-bottom flask 200-mL 24/40.
15. Spectra/Por molecular porous membrane tubing with 3.5-kDa MWCO (Spectrum Laboratories, Inc., USA).
16. MSE Soniprep 150, high-frequency 23-kHz generator.
17. Rotary evaporator RE 500 and water bath (Yamato Corporation, USA).
18. Adjustable-tilt rocking platform shaker.
19. Freeze dry/shell freeze system.
20. Speed Vac SC 110A vacuum concentrator.

3.2.10 Colony Extraction for Screening of LAM, LM, and PIMs from M. smegmatis

1. *M. smegmatis* strain mc^{2}155 frozen (–80°C) stock in 1-mL aliquots.
2. Middlebrook 7H11 agar (*see* Section 3.2.9, item 2).
3. 1-μL sterile inoculating loop.
4. Chloroform: methanol: H_2O (v/v/v, 10:10:3).
5. PBS-saturated phenol (*see* **Note 3**). Thaw 100 mL phenol (redistilled, 99+%) in 50°C water bath. Transfer the phenol in 200-mL dark color bottle, add 100 mL PBS (pH 8.0) and stir gently for several hours at room temperature to mix. Let the mixture stand at 4°C overnight. Discard the top aqueous layer. Repeat twice. Add 20 mL PBS to the phenol. Store at 4°C.
6. Chloroform.
7. Spectra/Por molecular porous membrane tubing with 3.5-kDa MWCO.

3.2.11 Two-dimensional Gel Electrophoresis for Isoform Resolution

1. 12 μg purified ManLAM.
2. Isoelectric focusing (IEF) sample buffer: 7 M urea, 5% (v/v) ampholyte 4.5–5.4 and ampholyte 3–10 (Amersham Biosciences), 2% (v/v) Nonidet P-40, 5% (v/v) β-mercaptoethanol.
3. First dimension focusing tube gel: 6% polyacrylamide isoelectric gel (2-mm inside diameter, 70-mm length) containing 1.6% (v/v) ampholytes.
4. Upper electrode buffer (catholyte): 20 mM NaOH.
5. Lower electrode buffer (anolyte): 20 mM H_3PO_4.
6. Transfer buffer: 2.9% (w/v) SDS, 71 mM Tris-HCl (pH 6.8), and 0.003% (w/v) bromophenol blue.
7. PageRuler Prestained Protein Ladder, broad range: 10 to 170 kDa.

8. Second dimension slab gel: 15% SDS-polyacrylamide gel (80 × 70 mm, WXL; 1.5 mm thick) with 6% acrylamide stacking gel and 15% acrylamide resolving gel.
9. 10X running buffer using for the second dimension gel running and gel running apparatus (*see* Section 3.2.1, items 13 and 14).
10. Speed Vac SC 110A vacuum concentrator.

3.3 Methods

3.3.1 Extraction of M. tuberculosis Lipoglycans (ManLAM, LM, and PIMs)

To obtain the high quality and quantity of *M. tuberculosis* ManLAM, LM, and PIMs to address biochemical and biological studies, we have modified our extraction and purification protocols [4]. The initial delipidation step uses chloroform, methanol, and H_2O (10:10:3, v/v/v) to remove the peripheral noncovalent linked lipids of the mycobacterial cell wall. A brief sonication step with a probe is used prior to disruption with a French press. We have substituted the original hot-phenol partition step with proteinase K treatment to eliminate proteins, as residual phenol in the preparations may interfere with the biological assays.

1. Inoculate 50 mL Middlebrook 7 H9 broth in a 250-mL conical flask with 1 mL frozen *M. tuberculosis* stock (*see* **Note 4**).
2. Grow for 7 days at 37°C with shaking at 180 rpm.
3. Inoculate 4 L Middlebrook 7 H9 broth in two 4-L conical flasks with 50 mL cells culture.
4. Grow for 7 days at 37°C with shaking at 180 rpm.
5. Harvest cells from 4-L cultures.
6. Transfer cells to 30-mL Oak Ridge Teflon FEP tube.
7. Delipidate cells using 10 mL chloroform: methanol:H_2O (10:10:3, v/v/v) for every 1 g of wet cells.
8. Gently rock the tubes using the rocking platform shaker at room temperature for 2 h.
9. Centrifuge at 27,000 × g for 20 min.
10. Save the organic supernatant for further lipid analyses.
11. Repeat steps 7 to 10 twice for the pellet. At the end of this step, work can be performed outside the pathogen containment laboratory, as the *M. tuberculosis* bacilli are killed by delipidation.
12. Dry the delipidated cells under a gentle stream of air at room temperature to eliminate residual organic solvent (*see* **Note 5**).
13. Grind the dried residual biomass to obtain fine powder.
14. Add a minimal amount (about 1 to 2 mL) of breaking buffer.
15. Freeze/thaw delipidated cells three times.

16. Add 30 mL breaking buffer and sonicate the delipidated cells using 9.5-mm probe for 6 cycles (60 s on and 90 s off with cooling) at 4 µm amplitude.
17. Further break cells using French press for 8 cycles (*see* **Note 6**).
18. Transfer the suspension into 30-mL Oak Ridge tubes.
19. Add an equal volume of breaking buffer.
20. Gently rock the tubes overnight at 4°C.
21. Centrifuge at 27,000 × *g* at 4°C for 1 h (*see* **Note 7**).
22. Transfer the clear supernatant to 30-mL Oak Ridge Teflon FEP tubes.
23. Keep the cell debris at 4°C.
24. Incubate the supernatant at 37°C to separate the viscous detergent phase from the aqueous phase (*see* **Note 8**).
25. Centrifuge at 27,000 × *g* at room temperature for 15 min.
26. Carefully remove the aqueous layer (upper phase) into the tubes containing the cell debris.
27. Keep the first detergent phase at 4°C.
28. Add Triton X-114 to the aqueous layer with the cell debris to achieve a final concentration of 8% (v/v).
29. Gently rock the tubes at 4°C for 2 h.
30. Repeat the detergent extraction twice as above described (steps 21 to 29).
31. After the third extraction, keep the pellet (cell debris) to further perform the mycolyl-arabinogalactan-peptidoglycan (mAGP) complex purification.
32. Combine all detergent phases (first + second + third detergent layers) and precipitate the extracted lipoglycoconjugates with cold 95% ethanol (1:10, v/v) overnight at –20°C.
33. Centrifuge at 27,000 × *g* for 20 min.
34. Collect all the precipitate (lipoglycoconjugates) and transfer it into 13 × 100 mm glass tubes.
35. Dry the precipitate using vacuum concentrator.
36. Digest with 2 mg/mL of proteinase K at 37°C overnight.
37. Dialyze against the endotoxin free water for 24 h using Slide-A-Lyzer Cassette with 3.5 kDa MWCO.
38. Transfer the sample to a 13 × 100 mm glass tubes.
39. Dry the sample containing the mixture of ManLAM, LM, and PIMs using vacuum concentrator.
40. Run a 15% SDS-polyacrylamide gel followed by PAS staining (*see* Section 3.3.3) to confirm that the sample contains protein-free soluble lipoglycans (ManLAM, LM, and PIMs). Otherwise, repeat the proteinase K digestion step.

3.3.2 Purification of ManLAM, LM, and PIMs

Size-exclusion chromatography is performed on a Rainin SD 200 series liquid chromatography HPLC system fitted with a Sephacryl S-200 HiPrep 26/60 column in tandem with a Sephacryl S-100 HiPrep 16/60 column.

1. Dissolve 15 mg lipoglycans extracted from *M. tuberculosis* $H_{37}R_v$ with 5 mL column running buffer.
2. Filter through 0.8-μm and 0.2-μm syringe filter units.
3. Centrifuge at 3200 × *g* for 10 min prior to injection to HPLC.
4. Use the size-exclusion chromatography system to separate ManLAM, LM, and PIMs. After collecting the column void volume, elute sample at a flow rate of 1 mL/min and collect about 100 fractions of 2 mL (*see* **Note 9**).
5. Monitor the elution profile by running 15% SDS-polyacrylamide gel followed by PAS staining (*see* Fig. 3.2).
6. Collect and pool fractions accordantly containing purified ManLAM, LM, and PIMs based on the results of PAS staining.
7. Dialyze against the dialysis buffer using Slide-A-Lyzer Cassette with 3.5-kDa MWCO at 37°C for 24 h.
8. Dialyze against 1 M NaCl at room temperature for 24 h.
9. Dialyze against the endotoxin free H_2O (*see* **Note 1**) at room temperature for 24 h (changing H_2O every 6 h).
10. Transfer the dialyzed fractions to a 100-mL round-bottom flask.
11. Dry the sample using freeze-dry/shell freeze system.

3.3.3 Polyacrylamide Gel Electrophoresis Followed by PAS Staining

Visualize ManLAM, LM, and PIMs using PAS staining after running 15% SDS-polyacrylamide gel.

1. Dry 5 to 10 μg of purified ManLAM using vacuum concentrator.
2. Dissolve the sample in 8 μL Milli-Q water and 2 μL of 5X sample buffer.
3. Heat at 100°C for 5 min.
4. Load 5 μL PageRuler Prestained Protein Ladder and the sample onto a 15% SDS- polyacrylamide gel.
5. Run the gel at 10 mA/gel to begin and increase to 15 mA/gel when the dye front has run through the stacking gel (the dye becomes a thin line).
6. After electrophoresis, transfer the gel to a dish containing 100 mL of the first fixative and rock for a minimum of 45 min.

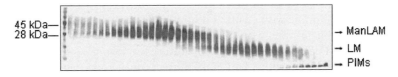

Fig. 3.2 Monitoring elution profile of the fractions containing containing ManLAM, LM, and PIMs eluted from a Sephacryl S-200 HiPrep 26/60 column in tandem with a Sephacryl S-100 HiPrep 16/60 column (Amersham Bioscience) equilibrated with column running buffer. Fractions are monitored by running 15% SDS-polyacrylamide gel followed by PAS staining

7. Discard the first fixative and add 100 mL 0.7% periodic acid solution for 7 min.
8. Discard the periodic acid solution and add 100 mL of the second fixative for 5 min.
9. Discard the second fixative and add 50 mL 2.5% glutaraldehyde for 5 min.
10. Carefully remove the glutaraldehyde and wash with 100 mL distilled water 4 times for 10 min each (*see* **Note 10**).
11. Add 100 mL 0.0025% DTT for 6 min.
12. Discard DTT solution and add 100 mL 0.1% silver nitrate for 5 min.
13. Rinse the gel very quickly with 100 mL distilled water.
14. Develop the gel in 200 mL 3% sodium carbonate containing 100 μL 37% formaldehyde (*see* **Note 11**).
15. Stop development by adding 10 mL 50% citric acid for 10 min when the desired contrast between the sample and background is reached.
16. Rinse the stained gel with distilled water before mounting the gel.

3.3.4 Immunoblotting Using mAb CS-35 or mAb CS-40

To look for the presence of the immunoreactive terminal motifs in ManLAM, Western blot is performed using mAbs CS-35 and mAbs CS-40. mAb CS-35 was generated against *M. leprae* and has been shown to cross-react with ManLAM from *M. tuberculosis* as well as other mycobacteria. The primary epitope of recognition is the terminal-branched Ara$_6$ motif present in both LAM and AG [18]. mAb CS-40, on the other hand, is raised against ManLAM from *M. tuberculosis* Erdman [19] and reacts preferentially with the mannose-capped LAMs, although the precise epitope of recognition has not been established.

1. Electrophorese 5 μg ManLAM in 15% SDS-polyacrylamide gel (*see* Section 3.3.3, items 1 to 5).
2. After electrophoresing, transfer the material from the gel to a nitrocellulose membrane using a Tank transphor unit with transfer buffer at 50 V for 1 h.
3. Put the membrane in a square Petri dish and block the membrane with 15 mL blocking buffer at room temperature for 1 h (or at 4°C overnight).
4. Discard the blocking buffer and incubate the membrane with 10 mL mAb CS-35 or mAb CS-40 in blocking buffer at room temperature for 1 h.
5. Wash the membrane with 15 mL TBS buffer 3 times for 5 min each.
6. Incubate the membrane with 10 mL alkaline phosphatase anti-mouse IgG conjugate in TBS buffer at room temperature for 45 min.
7. Wash the membrane with 15 mL TBS buffer 3 times for 5 min each.
8. Detect the reactivity by using 10 mL BCIP/NBT substrate solution until the purple broad band appears.
9. Stop the development by decanting off the substrate solution and washing the membrane with distilled water (Fig. 3.3).

Fig. 3.3 Immunoblotting analyses of ManLAM. ManLAM was transferred onto nitrocellulose membranes and probed with (A) mAb CS-35 and (B) mAb CS-40. (A) Lane 1, molecular weight; lane 2, *M. tuberculosis* $H_{37}R_v$ ManLAM. (B) Lane 1, *M. tuberculosis* $H_{37}R_v$ ManLAM; lane 2, *M. tuberculosis* Erdman ManLAM

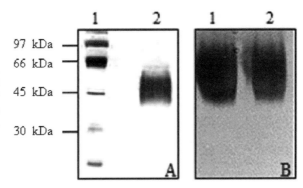

3.3.5 Analysis of Monosaccharide Composition Using Gas Chromatography

To analyze the monosaccharide composition of the isolated lipoglycans, it is necessary to convert the lipoglycans to alditol acetate after hydrolysis, reduction, and acetylation. After injecting the alditol acetate sample into the head of the chromatographic column, the sample is transported through the column by the flow of inert, gaseous mobile phase. The differential partitioning into the stationary phase allows the monosaccharide composition in the sample to be separated in time and space.

1. Add 50 µL of the monosaccharides stock solution and 10 µL of the internal standard to a 13 × 100 glass tube (*see* **Note 12**).
2. Aliquot 25 µg purified LAM and 10 µL of the internal standard in a second 13 × 100 mm glass tube.
3. Dry the samples using vacuum concentrator.
4. Add 250 µL 2 M TFA and heat at 120°C for 2 h.
5. Dry the samples under a gentle stream of air at room temperature.
6. Add 200 µL reducing reagent and stand at room temperature for 2 h.
7. Stop the reaction by dropwise adding 50 µL glacial acetic acid.
8. Dry the samples under a gentle stream of air.
9. Add 100 µL 10% glacial acetic acid in methanol and mix thoroughly.
10. Dry the samples under a gentle stream of air.
11. Repeat steps 9 and 10 three times.
12. Add 100 µL acetic anhydride and heat at 100°C for 1 h.
13. Dry the samples under a gentle stream of air.
14. Add 2 mL chloroform and 1 mL Milli-Q water and mix thoroughly.
15. Centrifuge at 3200 × g for 5 min.
16. Discard the upper aqueous layer.
17. Add 1 mL Milli-Q water to the lower organic layer and mix thoroughly.
18. Repeatsteps 15 to 17 until the lower organic layer looks clear.

19. Transfer the alditol acetate-containing organic layer to 13×100 mm glass tubes.
20. Dry the samples under a gentle stream of air.
21. Dissolve each sample in 50 µL chloroform.
22. Inject 1 µL into the head of the chromatographic column using an initial temperature of 50°C for 1 min, increasing to 170°C by 30°C/min followed by 270°C by 5°C/min.
23. Analyze the monosaccharide composition based on a single *myo*-inositol residue per LAM chain.

3.3.6 Structural Characterization by Gas Chromatography/ Mass Spectrometry

To determine the linkage between carbohydrate residue constituents of the mycobacterial lipoglycans, DMSO/NaOH permethylation followed by alditol acetate conversion is performed for GC/MS analysis. DMSO/NaOH permethylation replaces hydroxyl and acetyl esters groups with methyl ether groups. Then, the alditol acetate conversion allows us to determine linkage positions. For lipoglycans, native and especially peracetylated ones, the fatty acyl substituents tend to be partially replaced if perform directly the permethylation, therefore, it may require a prior deacylation step to ensure complete permethylation.

1. Aliquot 50 µg purified LAM in a 13×100 mm glass tube.
2. Dry the sample completely using a vacuum concentrator.
3. Place five pellets (about 500 mg) of NaOH in a dry mortar.
4. Add approximately 3 mL DMSO (not exposed to air) to the mortar using a glass pipette.
5. Use a glass pestle to grind the NaOH pellets until the slurry is formed (*see* **Note 13**).
6. Add 200 µL DMSO/NaOH slurry to the completely dried sample with a glass pipette and mix thoroughly.
7. Quickly add 100 µL iodomethane and mix vigorously for 10 min at room temperature (*see* **Note 14**).
8. Quench the reaction by slowly and dropwise adding 1 mL ice-cold Milli-Q water with constant shaking to lessen the effects of the highly exothermic reaction.
9. Add 2 mL chloroform and mix thoroughly.
10. Centrifuge at $3200 \times g$ for 5 min.
11. Discard the upper aqueous layer.
12. Add 1 mL Milli-Q water to the lower organic layer and mix thoroughly.
13. Repeat steps 10 to 12 until the lower organic layer looks clear.
14. Transfer the lower organic layer to another 13×100 mm glass tube.
15. Dry the sample under a gentle stream of air.

Table 3.1 Methylation Analysis on ManLAM from *M. tuberculosis* $H_{37}R_v$

Full name of the partially methylated alditol acetate	Abbreviation name of the glycosyl residue	ManLAM from *M. tuberculosis* $H_{37}R_v$ (mol%)
2,3,5-tri-*O*-Me-1,4-di-*O*-Ac-arabinitol	Terminal Ara*f*	3.6
3,5-di-*O*-Me-1,2,4-tri-*O*-Ac-arabinitol	2-Linked Ara*f*	6.4
2,3-di-*O*-Me-1,4,5-tri-*O*-Ac-arabinitol	5-Linked Ara*f*	36
2-*O*-Me-1,3,4,5-tetra-*O*-Ac-arabinitol	3,5-Linked Ara*f*	10
2,3,4,6-tetra-*O*-Me-1,5-di-*O*-Ac-mannitol	Terminal Man*p*	14
3,4,6-tri-*O*-Me-1,2,5-tri-*O*-Ac-mannitol	2-Linked Man*p*	8
2,3,4-tri-*O*-Me-1,5,6-tri-*O*-Ac-mannitol	6-Linked Man*p*	5
3,4-di-*O*-Me-1,2,5,6-tetra-*O*-Ac-mannitol	2,6-Linked Man*p*	9

The linkage profile of arabinose and mannose residues obtained by structural characterization of ManLAM by GC/MS. The conclusion is derived from m/z of fragments obtained from partially methylated alditol acetates (*see* Section 3.3.6, item 19).

16. Carry out the preparation of alditol acetate on sample (*see* Section 3.3.5, items 4 to 20) without adding internal standard.
17. Dissolve the methylated alditol acetate derivatives in 50 µL chloroform.
18. Perform GC/MS at an initial temperature of 60°C for 1 min, increasing to 130°C at 30°C/min and finally to 280°C at 5°C/min.
19. Establish linkage pattern of ManLAM based on m/z values of partially methylated alditol acetates of mannoses and arabinoses by GC/ MS (*see* Table 3.1)

3.3.7 Nuclear Magnetic Resonance Spectroscopy

Although this technique requires sophisticated instrumentation, it is not necessary to have an in-house NMR instrument. Most institutions will have some mode of NMR services available. This is the only nondestructive way of visualizing presence/absence of contaminants in a certain preparation of LAM. For a quick checking of sample purity, less than 500 mg can be examined with a 1D 1 H-NMR. For detailed structural analyses of LAM, 2D ^{1}H-^{13}C HSQC, heteronuclear multiple bond correlation (HMBC) NMR experiments are required. This is only possible with 4 to 5 mg of LAM as described below.

1. Lyophilize 4 to 5 mg of purified ManLAM.
2. Add 500 µL 99.9 atom% D_2O and mix thoroughly.
3. Lyophilize the sample using freeze-dry/shell freeze system.
4. Repeat steps 2 and 3 twice.
5. Dissolve the sample in 500 µL 100.0 atom% D_2O.
6. Centrifuge at 3200 × *g* for 5 min.
7. Transfer the supernatant to the NMR tube.
8. Analyze using NMR spectrometer. The HSQC data is acquired with a 7-kHz window for proton in F2 and a 15-kHz window for carbon in F1. The total

recycle time is 1.65 s between transients. Composite pulse, globally optimized alternating phase rectangular pulse (GARP), decoupling should be applied to carbon during proton acquisition. Pulsed field gradients should be used throughout for artifact suppression but are not used for coherence selection. The data set consists of 1000 complex points in t2 by 256 complex points in t1 using States-TPPI. Forward linear prediction is used for resolution enhancement to expand t1 to 512 complex points. Cos^2 weighting functions are matched to the time domain in both t1 and t2, and zero-filling is applied to both t1 and t2 before the Fourier transform. The final resolution is 3.5 Hz/ point in F2 and 15 Hz/pt in F1.

9. Analyze the data (*see* Fig. 3.4).

3.3.8 Cellulomonas Enzyme Digestion Followed by High-pH Anion Exchange Chromatography

The arabinan of LAM can be digested by endoarabinanase from *C. gelida* to yield di- arabinosides (Ara_2), tetra-arabinosides (Ara_4) and hexa-arabinosides (Ara_6). For ManLAM such as that from *M. tuberculosis* $H_{37}Rv$, the major

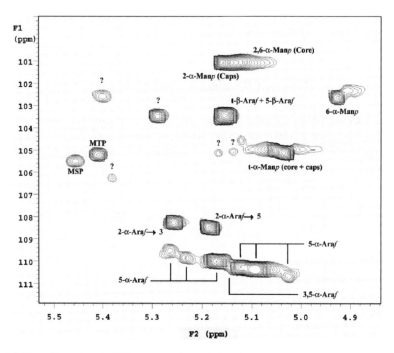

Fig. 3.4 Two-dimensional NMR spectra of ManLAM from *M. tuberculosis* $H_{37}R_v$. The two-dimensional NMR ^1H-^{13}C HSQC spectra of ManLAM was acquired in D_2O. Only the expanded anomeric regions are shown

Fig. 3.5 The nonreducing terminal oligoarabinosyl structural motifs of ManLAM. The formation of the characteristic mannose-capped linear Ara$_4$ and branched Ara$_6$ motifs (*boxed*) by endoarabinanase cleavages are indicated by vertical down-arrows. The remainder of the arabinan chains are mostly digested into Ara$_2$

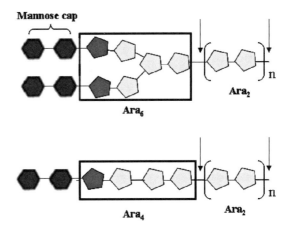

products are Ara$_2$, Man$_2$Ara$_4$, and Man$_4$Ara$_6$ because of terminal mannose substitution (*see* Fig. 3.5).

1. Dry 10 μg purified ManLAM using vacuum concentrator.
2. Add 10 μL Milli-Q water and 1U of *Cellulomonas* arabinanase and mix thoroughly.
3. Incubate at 37°C overnight to ensure complete digestion.
4. Heat at 100°C for 2 min to inactivate the enzyme.
5. Add 30 μL Milli-Q water to make 50 μL of the total volume.
6. Analyze the digestion products directly using Dionex analytical HPAEC performed on a Dionex chromatography system fitted with a Dionex Carbopac PA-1 column. Detect the oligoarabinosides with a pulse-amperometric detector (PAD-II) (*see* Fig. 3.6).

3.3.9 Extraction of LAM, LM, and PIMs from M. smegmatis

The method for extraction of LAM, LM, and PIMs from *M. smegmatis* is different from *M. tuberculosis* because of the different distribution of lipoglycans between these two strains.

1. Streak the frozen *M. smegmatis* stock cells on Middlebrook 7H11 agar plate using a sterile inoculating loop.
2. Grow for 2 days at 37°C.
3. Inoculate four 10-mL cultures of Middlebrook 7H9 broth in 50-mL polypropylene culture tubes with 4 colonies of *M. smegmatis*.
4. Grow for 2 days at 37°C with shaking at 150 rpm.
5. Inoculate 4 L Middlebrook 7H9 broth in two 4-L conical flask with 40 mL culture.
6. Grow for 2 days at 37°C with shaking at 150 rpm.
7. Harvest cells from 4 L culture.

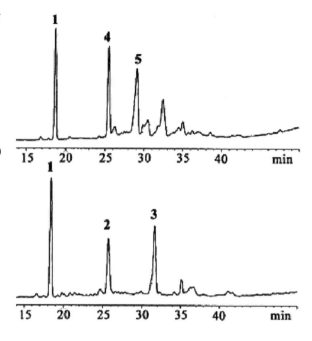

Fig. 3.6 HPAEC profiles of the endoarabinanase digestion products of *M. tuberculosis* $H_{37}R_v$ ManLAM (Top) and after α-mannosidase digestion (Bottom). Peak 1, Ara_2; peak 2, Ara_4; peak 3, Ara_6; peak 4, Man_2Ara_4; peak 5, Man_4Ara_6. The large peak at around 6 min in (Bottom) is mannose residue released by α-mannosidase

8. Transfer cells to a 30-mL Oak Ridge Teflon FEP tube.
9. Add 25 mL 95% ethanol preheated to 80°C.
10. Incubate at 80°C for 15 min and mix 3 times during the incubation.
11. Cool the sample to the room temperature.
12. Centrifuge at 3200 × g for 15 min.
13. Save the supernatant for further lipid analysis.
14. Dry the pellet under a gentle stream of air at the room temperature overnight.
15. Grind the pellet to obtain fine powder.
16. Add 25 mL chloroform:methanol (2:1, v/v) to the pellet.
17. Gently rock the tubes using the rocking platform shaker at room temperature for 2 h.
18. Centrifuge at 3200 × g for 15 min.
19. Transfer the supernatant, which includes some phospholipids and apolar PIMs, into a 200-mL Pyrex round-bottom flask.
20. Repeat steps 16 to 19 twice.
21. Evaporate the combined supernatant using rotary evaporator for further analysis.
22. Dry the pellet under a gentle stream of air.
23. Grind the pellet to obtain fine powder.
24. Add a minimum volume ((about 1 to 2 mL) of Milli-Q water.
25. Freeze/thaw delipidated cells three times.
26. Add 30 mL Milli-Q water.

27. Sonicate the delipidated cells using 9.5-mm probe for 10 cycles (60 s on and 90 s off with cooling) at 4 μm amplitude.
28. Transfer the sonicated cells to a 100-mL Pyrex round-bottom flask with a stir bar.
29. Add an equal volume of 95% ethanol.
30. Reflux at 80°C for 2 h with slow stirring (*see* **Note 15**).
31. Cool the sample to room temperature.
32. Transfer the sample to 30-mL Oak Ridge tubes.
33. Centrifuge at 3200 × g for 15 min.
34. Transfer the supernatant to a 200-mL round-bottom flask.
35. Resuspend the pellet in 30 mL 50% ethanol, transfer the sample to the 100-mL round-bottom flask, and add 20 mL 50% ethanol.
36. Repeat steps 30 to 34.
37. Save the pellet for further analysis of mAGP.
38. Centrifuge the combined supernatant at 27,000 × g for 40 min.
39. Evaporate the supernatant using rotary evaporator to remove the ethanol.
40. Perform steps 36 to 40 in Section 3.3.1 except using Spectra/Por molecular porous membrane tubing with 3.5-kDa MWCO for dialysis.
41. Purify the LAM, LM, and PIMs using the same method as described in Section 3.3.2 (*see* **Note 16**).

3.3.10 Colony Extraction for Screening of LAM, LM, and PIMs from M. smegmatis

The method for extraction of LAM, LM, and PIMs from two to three colonies is a facile way of screening lipoglycan profiles in mutants. This procedure is not suited for large-scale extractions/isolation.

1. Streak frozen *M. smegmatis* stock cells on Middlebrook 7 H11 agar plate using a sterile inoculating loop.
2. Grow for 2 days at 37°C.
3. Pick up two to three colonies from the plate using a sterile inoculating loop and transfer to a microcentrifuge tube.
4. Add 200 μL chloroform: methanol:H_2O (10:10:3, v/v/v).
5. Incubate at 55°C for 30 min (*see* **Note 17**).
6. Centrifuge at 18,000 × g for 5 min.
7. Discard the supernatant.
8. Add 200 μL PBS-saturated phenol and 200 μL Milli-Q water to the pellet.
9. Incubate at 80°C for 2 h.
10. Cool the sample to the room temperature.
11. Add 100 μL chloroform and mix thoroughly.
12. Centrifuge at 18,000 × g for 15 min.

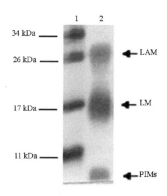

Fig. 3.7 SDS-polyacrylamide analysis of the LAM/LM/PIMs from wild-type *M. smegmatis* as visualized using PAS staining. Lane 1, molecular weight markers; lane 2, wild-type *M. smegmatis*

13. Transfer the upper aqueous layer containing the soluble LAM, LM, and PIMs to a microcentrifuge tube without cap.
14. Microdialyze using Spectra/Por molecular porous membrane tubing with 3.5-kDa MWCO against the flowing distilled water at room temperature overnight.
15. Transfer the dialyzed sample to a microcentrifuge tube.
16. Monitor the profile of LAM, LM, and PIMs by PAS staining after running 15% SDS-polyacrylamide gel (*see* Section 3.3.3 and Fig. 3.7).

3.3.11 Two-dimensional Gel Electrophoresis for LAM Isoform Resolution

Two-dimensional gel electrophoresis is achieved by a first-dimensional isoelectric focusing (IEF) separation followed by second-dimensional SDS-polyacrylamide gel. During the isoelectric focusing, LAM can be resolved into several discrete isoforms at different pH. Each isoform on the second dimension carry the same spread in size.

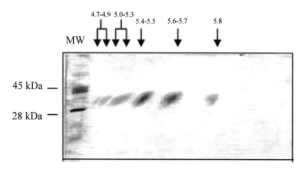

Fig. 3.8 Two-dimensional gel electrophoresis of ManLAM from *M. tuberculosis* $H_{37}R_v$. Up to 12 μg of ManLAM sample (minimum amount required to obtain separation) were applied to the IEF gel. The pI value of each band is determined by IEF pH gradient profile (pI 4.7 to 5.8). MW, molecular weight

1. Dry 12 μg purified ManLAM using the vacuum concentrator.
2. Dissolve the sample in 20 μL IEF sample buffer and denature at room temperature for 3 h or 4°C overnight.
3. Load the sample on the first dimension electrophoresis tube gel and run at 1000 V for 3 h using degassed 20 mM NaOH as the catholyte buffer and 20 mM of H_3PO_4 as the anolyte buffer.
4. Expel the tube gel into the transfer buffer.
5. Incubate the tube gel in the transfer buffer at the room temperature for 30 min (*see* **Note 18**).
6. Lay the tube gel on to the 15% SDS-polyacrylamide slab gel. Place the acidic end in on left and the basic end on the right.
7. Load 10 μL PageRuler Prestained Protein Ladder.
8. Run at 75 V for 20 min and then increase to 100 V.
9. Stop the electrophoresis when the blue dye reaches the bottom of the gel.
10. Visualize the profile by PAS staining (*see* Section 3.3.3 and Fig. 3.8).

3.4 Notes

1. PILAM and ManLAM share properties similar to endotoxin from Gram-negative bacteria. Therefore, caution must be taken during preparations, and endotoxin-free water needs to be used for preparation of the buffer for the sizing-column chromatography. However, this is not feasible and cost effective in case of large-scale preparations for structural analysis. For the latter, double-distilled water should replace endotoxin-free water.
2. Dilution may be variable according to the titer of the antibody.
3. When thawing phenol, leave the cap slightly open. Carry out each step in the chemical hood and use glass pipettes.
4. Live pathogenic mycobacteria should be handled in a suitable containment laboratory (e.g., biosafety level (BSL)-3 or γ-irradiated cells in a BSL-2 laboratory).
5. Dry the sample under a gentle stream of air using Reacti-ThermIII Heating Module.
6. Use acid-fast staining to confirm about 90% of cell breakage.
7. It is important to obtain a clear supernatant to avoid any carry over from the cell pellet (mAGP). Therefore, if the supernatant is not clear, repeat centrifugation step.
8. Triton X-114 is miscible with water at 4°C, however, a biphase is obtained at 37°C and the detergent phase is the lower layer.
9. Open column chromatography system is an alternative to the HPLC.
10. Frequently observed effects from exposure of glutaraldehyde include skin sensitivity resulting in dermatitis, and irritation of the eyes and nose with accompanying rhinitis. So be careful to transfer glutaraldehyde solution to a container for further specific disposal.
11. Add a small amount of the sodium carbonate/formaldehyde solution. Quickly remove the solution when the color changes to brown. Then, add the rest of the solution to allow development.
12. The alditol acetate derivatives will provide a profile of monosaccharides to calibrate the instrument.
13. This step should be done quickly to avoid excessive absorption of moisture from the atmosphere.
14. Add another 100 μL iodomethane to react for additional 10 min to avoid undermethylation of LAM.

15. Use silicone fluid (SF96/50 dimethyl silicone fluid) to keep the temperature at 80°C and cold running water flowing through the condenser to cool down the system. This will allow sample reflux within the system.
16. Before purification of LAM, LM, and PIMs from *M. smegmatis*, hydrophobic interaction chromatography is commonly used to eliminate neutral components derived from the outer layer of the mycobacterial cell wall (i.e., arabinomannan, glucan).
17. For colony extraction of ManLAM, LM, and PIMs from *M. tuberculosis*, incubate at 80°C for 1 h after addition of chloroform, methanol, and H_2O. Perform steps 1 to 5 in the BSL-3 laboratory.
18. The acidic end will be yellow and the basic end will be blue during the first 10 to 15 min after tube gel treatment with the transfer buffer.

Acknowledgments This work was supported by National Institutes of Health grant AI 37139 and in part from grant AI 52439.

References

1. Chatterjee, D., and Khoo, K. H. (1998) Mycobacterial lipoarabinomannan: an extraordinary lipoheteroglycan with profound physiological effects. *Glycobiology 8*, 113–20.
2. Chatterjee, D., Khoo, K.-H., McNeil, M. R., Dell, A., Morris, H. R., and Brennan, P. J. (1993) Structural definition of the non-reducing termini of mannose-capped LAM from *Mycobacterium tuberculosis* through selective enzymatic degradation and fast atom bombardment-mass spectrometry. *Glycobiology 3*, 497–506.
3. Prinzis, S., Chatterjee, D., and Brennan, P. J. (1993) Structure and antigenicity of lipoarabinomannan from *Mycobacterium bovis* BCG. *J. Gen. Microbiol. 139*, 2649–2658.
4. Chatterjee, D., Lowell, K., Rivoire, B., McNeil, M., and Brennan, P. J. (1992) Lipoarabinomannan of *Mycobacterium tuberculosis*. Capping with mannosyl residues in some strains. *J. Biol. Chem. 267*, 6234–6239.
5. Venisse, A., Berjeaud, J.-M., Chaurand, P., Gilleron, M., and Puzo, G. (1993) Structural features of lipoarabinomannan from *Mycobacterium bovis* BCG. Determination of molecular mass by laser desorption mass spectrometry. *J. Biol. Chem. 268*, 12401–12411.
6. Nigou, J., Gilleron, M., Cahuzac, B., Bounéry, J. D., Herold, M., Thurnher, M., and Puzo, G. (1997) The phosphatidyl-*myo*-inositol anchor of the lipoarabinomannans from *Mycobacterium bovis* bacillus Calmette Guerin—Heterogeneity, structure, and role in the regulation of cytokine secretion. *J. Biol. Chem. 272*, 23094–23103.
7. Torrelles, J. B., Khoo, K. H., Sieling, P. A., Modlin, R. L., Zhang, N., Marques, A. M., Treumann, A., Rithner, C. D., Brennan, P. J., and Chatterjee, D. (2004) Truncated structural variants of lipoarabinomannan in *Mycobacterium leprae* and an ethambutol-resistant strain of *Mycobacterium tuberculosis*. *J. Biol. Chem. 279*, 41227–41239.
8. Guerardel, Y., Maes, E., Elass, E., Leroy, Y., Timmerman, P., Besra, G. S., Locht, C., Strecker, G., and Kremer, L. (2002) Structural study of lipomannan and lipoarabinomannan from *Mycobacterium chelonae*. Presence of unusual components with alpha 1,3-mannopyranose side chains. *J. Biol. Chem. 277*, 30635–30648.
9. Khoo, K.-H., Dell, A., Morris, H. R., Brennan, P. J., and Chatterjee, D. (1995) Inositol phosphate capping of the nonreducing termini of lipoarabinomannan from rapidly growing strains of *Mycobacterium*. *J. Biol. Chem. 270*, 12380–12389.
10. Gilleron, M., Himoudi., Adam, O., Constant, O., Constant, P., Vercellone, A., Riviere, M., and Puzo, G. (1997) *Mycobacterium smegmatis* phosphatidylinositols-glyceroarabinomannans. *J. Biol. Chem. 272*, 117–124.
11. Nigou, J., Gilleron, M., and Puzo, G. (2003) Lipoarabinomannans: from structure to biosynthesis. *Biochimie 85*, 153–166.

12. Briken, V., Porcelli, S. A., Besra, G. S., and Kremer, L. (2004) Mycobacterial lipoarabinomannan and related lipoglycans: from biogenesis to modulation of the immune response. *Mol. Microbiol. 53*, 391–403.
13. Fratti, R. A., Chua, J., and Deretic, V. (2003) Induction of p38 mitogen-activated protein kinase reduces early endosome autoantigen 1 (EEA1) recruitment to phagosomal membranes. *J. Biol. Chem. 278*, 46961–46967.
14. Maeda, N., Nigou, J., Herrmann, J. L., Jackson, M., Amara, A., Lagrange, P. H., Puzo, G., Gicquel, B., and Neyrolles, O. (2003) The cell surface receptor DC-SIGN discriminates between Mycobacterium species through selective recognition of the mannose caps on lipoarabinomannan. *J. Biol. Chem. 278*, 5513–5516.
15. Schlesinger, L. S., Kaufman, T. M., Iyer, S., Hull, S. R., and Marchiando, L. K. (1996) Differences in mannose receptor-mediated uptake of lipoarabinomannan from virulent and attenuated strains of *Mycobacterium tuberculosis* by human macrophages. *J. Immunol. 157*, 4568–4575.
16. Kang, P. B., Azad, A. K., Torrelles, J. B., Kaufman, T. M., Beharka, A., Tibesar, E., DesJardin, L. E., and Schlesinger, L. S. (2005) The human macrophage mannose receptor directs Mycobacterium tuberculosis lipoarabinomannan-mediated phagosome biogenesis. *J. Exp. Med. 202*, 987–999.
17. McNeil, M. R., Robuck, K. G., Harter, M., and Brennan, P. J. (1994) Enzymatic evidence for the presence of a critical terminal hexa-arabinoside in the cell walls of *Mycobacterium tuberculosis*. *Glycobiology 4*, 165–173.
18. Kaur, D., Lowary, T. L., Vissa, V. D., Crick, D. C., and Brennan, P. J. (2002) Characterization of the epitope of anti-lipoarabinomannan antibodies as the terminal hexaarabinofuranosyl motif of mycobacterial arabinans. *Microbiology 148*, 3049–3057.
19. Chatterjee, D., Hunter, S. W., McNeil, M., and Brennan, P. J. (1992) Lipoarabinomannan. Multiglycosylated form of the mycobacterial mannosylphophatidylinositols. *J. Biol. Chem. 267*, 6228–6233.

Chapter 4
Analyzing Lipid Metabolism: Activation and β-Oxidation of Fatty Acids

Paul Robert Wheeler

Abstract There is massive gene replication predicted for the activation of fatty acids and their entry into the β-oxidation cycle for fatty acid oxidation. These two steps in fatty acid metabolism are catalyzed by FadD and FadE enzymes with 36 genes predicted for each of these respective activities in *Mycobacterium tuberculosis*. Here we present methods for the cell-free assay of types of enzymes in live bacteria, as well as for fatty acid oxidation overall.

Keywords β-oxidation · FadD · FadE · fatty-acid-CoA ligase (EC 6.2.1.3) · fatty acid metabolism · fatty acid oxidation · fatty acyl-CoA dehydrogenase (EC 1.3.99.3) · fatty acyl-CoA synthase (EC 6.2.1.3) · mycobacteria · *Mycobacterium tuberculosis*

4.1 Introduction

Genomic analysis of *Mycobacterium tuberculosis* predicts that more than 80 genes are involved in the β-oxidation cycle for fatty acid oxidation, even though there are only five enzyme activities needed to complete the cycle [1]. Two classes of genes seem particularly overrepresented, the *fadD* and *fadE* genes. Fortunately, both classes are amenable to enzyme assay enabling functional characterization studies to be carried out. This chapter provides protocols for the assay of these two classes of enzyme.

However, a researcher interested in the genes for fatty acid metabolism more generally may need to investigate whole cells. Therefore, methods are also given to follow fatty acid oxidation in intact mycobacteria by measuring the release of radiolabeled CO_2 from labeled fatty acids. This methodology would enable some information to be obtained, for example, when the genes of interest are *echA* or *fadA* genes, which also encode the

P.R. Wheeler
Tuberculosis Research Group, Veterinary Laboratories Agency (Weybridge), New Haw,
Addlestone, Surrey, KT15 3NB, UK
e-mail: p.wheeler@vla.defra.gsi.gov.uk

T. Parish, A.C. Brown (eds.), *Mycobacteria Protocols*,
doi: 10.1007/978-1-59745-207-6_4, © Humana Press, Totowa, NJ 2008

remaining enzymes of β-oxidation, but which are more difficult to assay. One issue emphasized in this chapter is the use of a range of substrates, as fatty acids may be used differentially. Because there is little scope to assay EchA or FadA with a range of substrates, literature references are provided as a guide to their assay.

4.1.1 Oxidation of Fatty Acids by Intact Mycobacteria

A protocol for following fatty acid oxidation is given here, as it is a potentially useful tool in the postgenomic area for analyzing the widest range of genes predicted to be involved in lipid metabolism. This would be the best choice of method to try if the gene(s) under investigation are annotated for lipid catabolism but not necessarily in β-oxidation. Given the possibility that the often replicated genes in lipid metabolism may encode a series of enzymes that are involved in the differential utilization of fatty acids, the protocol described is best done with a range of fatty acids. Fatty acids with the carboxyl carbon ($[1\text{-}^{14}C]$) labeled will have $^{14}CO_2$ released from the first turn of the β-oxidation cycle of fatty acid oxidation. This activity has been measured, with a range of fatty acids, to distinguish species of mycobacteria [2, 3] and provide a rapid correlate of viability, for example in the early BACTEC technology [4, 5] and even in suspensions of that most moribund of mycobacteria, *Mycobacterium leprae* [6, 7, 8].

4.1.2 Enzyme Assays for FadD-Encoded Enzymes

Most of the *fadD* genes are described as encoding "probable long-chain-fatty-acid-CoA ligase." But which fatty acids, and do all the *fadD* gene products use the same coenzymes? FadD enzymes activate fatty acids, thus can potentially be involved in both anabolic and catabolic processes. An interesting start to this postgenomic quest has been made, at least for the *fadD* genes, some of which encode enzymes that form acyl-adenylates, rather than coenzyme A (CoA) esters [9]. This immediately raises an important issue in the assay of the FadD enzymes: the standard spectrophotometric assay measures the production of AMP released from the reaction

$$\text{Fatty acid} + \text{ATP} + \text{CoA} = \text{Fatty acyl-CoA} + \text{CoA} + \text{AMP} + \text{PPi.}$$

FadDs that form acyl-adenylates sequester the AMP, so it cannot be assayed. Instead, a radioactive method must be used to trace the formation of the fatty acyl product directly.

The standard assays described in this chapter use palmitic acid to show whether fatty acyl-CoA synthase activity is present. To start to address the questions raised above ("which fatty acids? which coenzymes? what do the FadDs all do?"), a range of fatty acids should be tested. Fatty acids can readily be purchased from 2 to 30 carbons, fully saturated or with up to four double bonds. Because so many fatty acids can be tried, an ELISA plate assay is given as an alternative to the "traditional" assay in cuvettes. Testing a range of fatty acids could be informative in crude extracts of native enzyme material, perhaps from mutants compared with their parent strain, or from different environmental conditions. It could also be informative in testing cloned FadD proteins [9]. With individual enzymes, it is possible to investigate the optimum concentrations of each fatty acid, CoA, and ATP. The enzyme kinetics of each FadD might distinguish enzymes with the same substrates but with different affinities and catalytic efficiencies.

4.1.3 Enzyme Assays for FadE-Encoded Enzymes

FadE enzymes (fatty acyl CoA dehydrogenases) are the first committed step in β-oxidation. Like *fadD,* there are 35 or 36 genes predicted to encode these enzymes in the *M. tuberculosis* complex genomes. Therefore, like in the assays for FadD, the most amount of information will be obtained by following the protocol with a range of substrates. For FadE, the substrates are fatty acyl CoA esters, which limits the number of experiments one can do because their chemical synthesis is not trivial [10], especially with long-chain fatty acids. Fewer are commercially available (~15 fatty acyl-CoAs up to 24 carbons) than are free fatty acids, and those that are commercially available are much more expensive. The kinetic data with fatty acyl-CoA substrates are more easily interpreted than are kinetic data with fatty acids, as their CoA esters are much more compatible with the aqueous assay system.

4.1.4 Enzyme Assays for EchA and FadA-Encoded Enzymes: Key Literature References

Although there are multiple genes—*echAs*—several *fadA* genes annotated for the remaining enzymes of the β-oxidation complex, assays for them are not given in this chapter because it is difficult to follow the overall strategy of assaying with a variety of substrates and coenzymes to reveal any differences between isoenzymes encoded by different genes. The ability to synthesize CoA esters chemically [10] is essential, as these enzymes use β-oxidation cycle *intermediates*—enoyl-CoAs and ketoacyl-CoAs—as their substrates, and very few of these are available commercially. Assays for these enzymes that work in slow-growing bacteria can be found in Ref. 6.

4.2 Materials

4.2.1 Oxidation of Fatty Acids by Intact Bacteria

1. A suitable culture of the mycobacteria being studied (see **Note 1**).
2. Scintillation vials with inserts (see **Note 2**).
3. 0.025% (w/v) tyloxapol in Milli-Q water. Make a 12.5% stock solution. Store for up to 3 months at 4°C.
4. Incubation buffer: 50 mM HEPES, 5 mM Mg SO$_4$, 1 mM NaH$_2$PO$_4$, adjust to pH 7.0 with 12 M KOH.
5. [1-^{14}C] labeled palmitic acid (at ~50 Ci/mol) and a range of labeled fatty acids (see **Note 3**).
6. 1 M NaOH.
7. Scintillation fluid compatible with NaOH (e.g., Fluoran Flow; VWR Merck Lutterworth, UK).

4.2.2 Assay for Fatty Acyl-CoA Synthases and FadD Enzymes

4.2.2.1 To Detect Acyl-CoA Synthase Only, by Detecting AMP Formation

1. Cell-free preparation with FadD enzyme activity in 50 mM Tris-HCl, pH 7.4 (see **Note 4**). Make fresh on day of assay or store at –20°C (see **Note 5**).
2. Cuvettes (1-cm pathlength, ~1 mL volume) transparent at 340-nm wavelength (alternative: ELISA plate, 96 flat-bottomed wells).
3. 0.5 M MgCl$_2$. Keeps 1 year stored at room temperature.
4. 0.5 M Tris-HCl, pH 7.4, adjusted with 6 M HCl. Keeps 1 year stored at room temperature.
5. 20 mM dithiothreitol in Milli-Q water. Make fresh every day and store on ice.
6. 50 mM ATP in 25 mM MgCl$_2$. Make fresh every day and store on ice.
7. 40 mM coenzyme A, Li salt in Milli-Q water. Make fresh every day and store on ice.
8. 5 mM Triton X-100 in Milli-Q water. Keeps for 1 month stored at 4°C.
9. 1 mM palmitic acid (potassium salt), and a range of other fatty acids as potassium salts (see **Note 3**), in 5 mM Triton X-100. Keeps for 1 month stored at 4°C.
10. Coupling enzyme mix: L-lactate dehydrogenase, pyruvate kinase, and adenylate kinase each at 500 units/mL (see **Note 6**) in 20 mM Tris-HCl, pH 7.4. Make the mix fresh every day and store on ice. The individual enzymes are stable for 1 year stored at 4°C.

11. 20 mM phospho*enol*pyruvate (tricyclohexammonium salt is recommended) in Milli-Q water (*see* **Note 6**). Make fresh every day and store on ice.

12. 7.5 mM β-NADH (*see* **Note 7**) (sodium salt) in 0.5 mM KOH (*see* **Note 8**). Make fresh every day and store on ice.

4.2.2.2 To Detect Acyl-CoA Synthase and Related Activities of FadD Enzymes Using Radioactivity

1. Thin-layer chromatography (TLC) plates: glass-backed silica gel 60 F254 plates, 10 cm × 10 cm, with concentrating zone (Merck 13748).

2. Butan-1-ol.

3. Glacial acetic acid.

4. [1-^{14}C] palmitic acid (at ~50 Ci/mol), and a range of other [1-^{14}C] fatty acids (*see* **Note 3**).

5. 5 mM Triton X-100 in Milli-Q water. Keeps for 1 month stored at 4°C.

6. 20 mM dithiothreitol in 0.5 M Tris-HCl, pH 7.4. Make fresh every day and store on ice.

7. 50 mM ATP (Mg salt) in 25 mM MgCl$_2$. Make fresh every day and store on ice.

8. Cell-free preparation with FadD enzyme activity in 50 mM Tris-HCl, pH 7.4 (*see* **Note 4**). Make fresh on day of assay or store at –20°C (*see* **Note 5**).

9. 20 mM coenzyme A, Li salt in Milli-Q water. Make fresh every day and store on ice.

10. Butan-1-ol:acetic acid:Milli-Q water (15:5:8). Make fresh every day and store at room temperature.

4.2.3 Assay for FadE Enzyme Activity: Fatty Acyl CoA Dehydrogenase

1. Cell-free preparation with FadE enzyme activity in 0.2 M potassium phosphate buffer (pH 7.0) with 5 mM dithiothreitol (DTT) (*see* **Note 9**). Make fresh on day of assay or store at –20°C (*see* **Note 5**).

2. 0.2 M potassium phosphate buffer (pH 7.0) with 5 mM dithiothreitol.

3. 0.5 M potassium phosphate buffer (pH 8.0).

4. 6 mM Na$_2$ EDTA.

5. 0.3 mM palmitoyl-CoA and a range of other fatty acyl-CoAs.

6. 0.6 mM 2,6-dichlorophenol-indophenol (DCPIP) (*see* **Note 7**). Make fresh every day. Keep on ice until needed and in the dark or at least in low light (wrapping in silver foil is acceptable).

7. 14 mM phenazine methosulfate (PMS). Make fresh every day. Keep on ice and in low light.

8. 2 mM flavin adenine dinucleotide (FAD) (*see* **Note 10**). Make fresh every day. Keep on ice and in low light (same as for DCPIP).

4.3 Methods

4.3.1 Oxidation of Fatty Acids by Intact Bacteria

1. Prepare a series of Buddemeyer vials as follows: place an 0.5-mL tube vial (which will contain the incubation mixture) inside a scintillation vial with a screw-top (*see* **Note 2**).
2. To each inner vial, add 0.2 μCi of one fatty acid. Add unlabeled fatty acid to make the total amount 5 nmol. Prepare triplicates of each fatty acid as a minimum for statistical analysis (*see* **Note 11**).
3. Allow to dry. This can be accelerated using a stream of nitrogen.
4. Add 400 μL 1 M NaOH to each outer vial (i.e., the scintillation vial).
5. Harvest a culture of mycobacteria (*see* **Note 1**), keeping it at the temperature used for growth. Wash once in Milli-Q water prewarmed to the temperature used for growth.
6. Resuspend the pellet at 5 mg (dry weight bacteria)/mL in incubation buffer. This can be achieved by making a suspension of 100 mL culture grown to optical density at 600 nm wavelength (OD_{600}) = 0.5 in 3 mL (*see* **Note 12**).
7. Check the weight of bacteria in the washed suspension by taking a sample (1 mL) and determining its dry weight: the safest way to do this is in a preweighed polytetrafluoroethane (PTFE)-capped Pyrex tube (16 mm diameter × 125 mm), heating at 80°C for 30 min, and freeze-drying. If you can reliably correlate OD_{600} to dry weight of bacteria, you may only need to do this once (*see* **Note 12**).
8. Add 200 μL washed suspension in incubation buffer to the inner (incubation) vial of each Buddemeyer vessel. Swirl gently; take care not to splash the incubation mixture into the alkali.
9. Incubate at the temperature used for growth for 1 h.
10. Remove the insert containing the incubation mixture leaving the alkali in the scintillation vial (*see* **Note 13**).
11. Add a scintillation fluid that is compatible with aqueous solutions of alkalis to the scintillation vials and determine the radioactivity in each scintillation vial by scintillation counting (*see* **Note 14**). Calculate all results as dpm /mg (dry wt.) bacteria.

4.3.2 Assay for Fatty Acyl-CoA Synthases and FadD Enzymes

4.3.2.1 To Detect Acyl-CoA Synthase Only, by Detecting AMP Formation

1. Prepare the test reaction containing the following components in a 1-mL cuvette (*see* **Note 15**):

   (a) 100 μL of 1 mM fatty acid in 5 mM Triton X-114
   (b) 90 μL of 50 mM Tris-HCl, pH 7.4

(c) 100 μL of 20 mM DTT
(d) 100 μL of 50 mM ATP (in 25 mM MgCl$_2$)
(e) 20 μL of coupling enzyme mix
(f) 20 μL of 7.5 mM β-NADH
(g) 50 μL of 20 mM phospho*enol*pyruvate
(h) 100 μL of cell-free preparation
(i) Milli-Q water to a final volume of 950 μL

2. Prepare the control reaction containing the following components in a 1-mL cuvette:

(a) 100 μL of 5 mM Triton X-114
(b) 90 μL of 50 mM Tris-HCl, pH 7.4
(c) 100 μL of 20 mM DTT
(d) 100 μL of 50 mM ATP (in 25 mM MgCl$_2$)
(e) 20 μL of coupling enzyme mix
(f) 20 μL of 7.5 mM β-NADH
(g) 50 μL of 20 mM phospho*enol*pyruvate
(h) 100 μL of cell-free preparation
(i) Milli-Q water to a final volume of 950 μL

3. Warm to the assay temperature of 30°C.
4. Follow fall in A_{340} for 5 min (up to 10 min if it is not a constant rate by 5 min) in both cuvettes.
5. Add 50 μL of 40 mM coenzyme A (CoA) to each cuvette to start the reaction. Follow the fall in A_{340} until it reaches no more than 0.3.
6. Calculate activity in enzyme units, using maximum rates of ΔA_{340}/min (*see* **Notes 6 and 7**):

Before CoA was added:
$\Delta a = \Delta A_{340}$/min with complete mix
$\Delta b = \Delta A_{340}$/min with mix with fatty acid omitted
After CoA was added:
$\Delta c = \Delta A_{340}$/min with complete mix
$\Delta d = \Delta A_{340}$/min with mix with fatty acid omitted

$$\text{Enzyme units} = \frac{[\Delta c - \Delta d] - [\Delta a - \Delta b]}{12.46 \times \text{vol (mL)} \times \text{pathlength (cm)}}.$$

The correction for volume applies if the volume is not 1 mL (*see* **Note 15**). If the pathlength the light travels through is not 1 cm (it is in all standard cuvettes, but it will have to be measured as the depth of the assay mixture in ELISA plate wells), divide by the pathlength in centimeters as shown in the Equation. If different concentrations of substrates are used, it will be possible to calculate basic enzyme kinetic parameters for the enzyme under study (*see* **Note 16**).

4.3.2.2 To Detect Acyl-CoA Synthase and Related Activities of FadD Enzymes Using Radioactivity

In this assay, acyl-CoA or acyl-AMP is detected directly. In order to do this, radiolabeled fatty acids must be used.

1. Add 2 nmol [1-^{14}C] palmitic acid (or other fatty acid) to 0.5 mL polypropylene tubes. This will be 0.1 μCi for 50 Ci/mol. Prepare duplicate tubes: one will be the test reaction tube, the other will be the control reaction tube.
2. Dry off any solvent and reconstitute in 2 μL 5 mM Triton X-100.
3. Prepare the test reaction tube by adding the following components:

   (a) 2 μL of 5 mM Triton X-114
   (b) 2 μL of 20 mM DTT in 0.5 M Tris-HCl, pH 7.4
   (c) 2 μL of 50 mM ATP (in 25 mM MgCl$_2$)
   (d) 2 μL of 20 mM coenzyme A
   (e) Milli-Q water to a final volume of 10 μL

4. Prepare the control reaction tube by adding the following components:

   (a) 2 μL of 5 mM Triton X-114
   (b) 2 μL of 20 mM DTT in 0.5 M Tris-HCl, pH 7.4
   (c) 2 μL of 50 mM ATP (in 25 mM MgCl$_2$)
   (d) Milli-Q water to a final volume of 10 μL

5. Warm to the assay temperature of 30°C.
6. Add 10 μL of the preparation with enzyme activity (final volume 20 μL). Include no more than 0.1 milli-units enzyme per assay (*see* **Note 15**).
7. Incubate for 5 min at 30°C.
8. Add 5 μL of 1 M acetic acid water to quench reaction.
9. Apply the entire mixture to the concentrating zone of a silica gel TLC plate.
10. Develop plate with butan-1-ol/acetic acid/water (15:5:8).
11. Quantify the amount of label in every spot, preferably using a phosphorimager. This will give you the nmol acyl-CoA and acyl-AMP formed (*see* **Note 15**). These have Relative Mobilities of ~0.4 and ~0.5, respectively, where the Relative Mobility of free fatty acid = 1.0 (*see* **Note 17**).

4.3.3 Assay for FadE Enzyme Activity: Fatty Acyl CoA Dehydrogenase

1. Prepare the test and control reactions containing the following components in a 1-mL cuvette (*see* **Note 15**):

   1 to 175 μL of of cell-free preparation (*see* **Note 15**)
   0.2 M potassium phosphate buffer, pH 7.0/5 mM DTT to 175 μL
   30 μL of 0.5 M potassium phosphate buffer, pH 8.0

50 μL of 6 mM EDTA
50 μL of 0.6 mM DCPIP
100 μL of 14 mM PMS
50 μL of 2 mM FAD (*see* **Note 18**)
Milli-Q water to a final volume of 900 μL (test) or 1000 μL (control)

2. Bring to the assay temperature of 25°C.
3. Allow the change in A_{600} to stabilize so that there is no or almost no difference between the two cuvettes (*see* **Note 10**).
4. Add 100 μL of 0.3 mM fatty acyl-CoA to the test cuvette.
5. Keeping the cuvettes at 25°C, measure the rates of fall in A_{600} (or the difference in rates of fall). Stop no later than when the fall in the cuvette containing fatty acyl-CoA is $A_{600} = 0.16$.
6. Calculate activity in enzyme units (*see* **Note 10** for definition), using maximum rates of $\Delta A_{600}/min$ (*see* **Note 7**):

$\Delta A_{600}/min$ with complete mix $- \Delta A_{600}/min$ with mix with fattyacyl-CoA omitted

divided by 16.1. If the volume is not 1 mL, multiply by the volume in milliliters. If the pathlength the light travels through is not 1 cm (it is in all standard cuvettes, but it will have to be measured as the depth of the assay mixture in ELISA plate wells), divide by the pathlength in centimeters. If different concentrations of substrates are used, it will be possible to calculate basic enzyme kinetic parameters for the enzyme under study (*see* **Note 16**).

4.4 Notes

1. Mid-log phase is recommended unless the experimental design requires otherwise. OD_{600} values of ~0.6 for a 1-cm light path are usually considered mid-log, but this should be determined for the conditions and strains being used. Because light scattering is measured in suspensions of bacteria, the OD values do not obey Beer's law, so readings obtained from the same suspension may vary from one instrument to another. For log-phase cultures, studies should be done before the OD_{600} is half the value at the end of log phase. After this, the proportion of bacteria ceasing to divide becomes increasingly significant, and the culture becomes heterogenous.

2. Most scintillation counters take 5-mL vials: for the inserts, 0.5-mL Eppendorf vials with press-on lids (left open) are ideal as they can be pulled out by the lids. Press-on tops for the scintillation vials must be avoided as they may cause splashing of the contents of the vial. An alternative to Buddemeyer vials is to use Becton Dickinson BACTEC technology (Becton-Dickinson, Franklin Lakes, NJ, USA).

3. A range of fatty acids should be tested to obtain the most information from conducting this assay. The assay should be performed first with palmitic acid (16:0). Then, a range of saturated acids with even numbers of carbon atoms between 4 and 24 should be tried. These are denoted 4:0 to 24:0. In this notation, the second number is the number of double bonds; 18:0 and 24:0 could be priorities as the fatty acyl groups of mycobacterial lipids are commonly this chain length, while 10:0 and 12:0 are most rapidly used by a number of enzyme systems in mycobacteria, so these should be used also. At the same time, some unsaturated acids should be tried, starting with oleic acid (18:1). Fatty acids

found in the host such as myristic acid (14:0) and even arachidonic acid (20:4) should also be considered. Later on, more unusual fatty acids such as odd chain length (15:0, 17:0, etc.) and very-long-chain fatty acids (up to 30:0, difficult to handle because they are extremely poorly water-compatible) could be tried. If potassium salts cannot be purchased, they can easily be made by 1 mL 1.2 M KOH to 1 mol fatty acid (dry; as the free acid) and heating in a PTFE capped tube at 100°C for 1 h. These can then be used directly in assays requiring fatty acids (potassium salts). Be careful with polyunsaturated fatty acids as some are susceptible to light and air, so follow the instructions that are provided by the manufacturers for labile ones. In the FadE assay, acyl-CoAs are the substrates. These are always ready for use as supplied. There is a smaller range of acyl chains available commercially for acyl-CoAs than for the free acids; the considerations for prioritizing which acyl-CoAs to use in assays will be the same as for free acids.

4. To make a cell-free preparation with FadD activity, either (a) harvest a culture of mycobacteria and prepare a cell-free extract by bead-beating or sonication [11] or (b) clone *fadD* genes so as to prepare affinity-purified his6-tagged proteins [9]; follow protocols for using pET vectors in the BL21-DE3 strain of *Escherichia coli* (Amersham Biosciences, Uppsala, Sweden). Other mechanical methods for disruption (e.g., French press) are acceptable. Cell-extracts should be clarified by centrifugation at 5000 to 27,000 × *g* for 10 min and may be filtered through a cellulose acetate filter (0.2-μm pore size glass fiber prefilter for volumes over 2 mL). Some replicates should be done with desalted extracts in case there are small molecules that give background activity or are inhibitors. To desalt preparations, take 2.5-mL portions and desalt using Pharmacia PD-10 desalting columns (Amersham Biosciences), following the manufacturer's instructions. The final buffer should be 50 mM Tris-HCl, pH 7.4. The protein eluate may desalted and concentrated using semipermeable membranes with a molecular weight cutoff of 10,000 Da.

5. Some activity is lost on freezing and thawing, although enzyme preparations can be kept for a year at –20°C. Therefore, the amount of loss of activity should be determined by performing replicates of assays done with a fresh enzyme preparation.

6. The individual enzymes for the coupling enzyme mix should be purchased as rabbit muscle enzymes supplied as a suspension in ammonium sulfate. This source of the individual enzymes is recommended as it gives the lowest background in the assays, and the enzymes are the most stable in these preparations. All are available from Sigma-Aldrich (Poole, UK). The coupled assay described here assays the production of 5'-AMP. The principle of the assay is that the myokinase in the coupling enzyme mix will convert AMP formed to ADP, the pyruvate kinase uses the ADP to convert phosphoe-*nol*pyruvate to pyruvate, then the lactate dehydrogenase in the mix converts the pyruvate thus formed and the NADH added to the assay to lactate and NAD. The rate of conversion of NADH to NAD is followed by measuring the fall in A_{340}. For more theory, and the stoichiometry of the assay, see Ref. 12. All activities are expressed in international enzyme units, referred to as "units," defined as the enzyme required to convert 1 μmol substrate/min. If there are myokinase or NADH oxidizing enzymes in the enzyme preparation (a consideration in crude extracts), there will be background activity, but the assay can still be done by subtracting the background obtained in assays with CoA and fatty acid omitted from the assay.

7. Millimolar extinction coefficients for the compounds detected by spectrophotometry in this chapter are as follows: for NADH, 6.23 at 340 nm; for DCPIP, 16.1 at 600 nm. These coefficients are values for the absorbance of light at the wavelength given (at which absorbance is maximal) of a solution at pH 7, through a light path of 1 cm. The coefficient is only valid for solutions at concentrations where absorbance is proportional to concentration: considerably lower than 1 mM.

8. NADH is unstable in acid, so should be kept in dilute alkali on ice. In this solution, the alkali is dilute enough to not affect pH in the assay.

9. Use phosphate buffer for enzyme preparations to be assayed for FadE. To make a cell-free preparation with FadD activity, either (a) harvest a culture of mycobacteria and prepare a cell-free extract by method 3.1 (bead-beating) or method 3.2 (sonication) as described in a previous volume of *Mycobacteria Protocols* [11].These extracts can be used for all the enzymes, except FadD enzymes, of β-oxidation (referenced in this article [6]). Alternatively, (b) clone *fadE* genes so as to prepare affinity-purified his6-tagged proteins [13] using pET vectors in the BL21-DE3 strain of *Escherichia coli* (Pharmacia); follow protocols for using pET vectors in the BL21-DE3 strain of *Escherichia coli* (Pharmacia). Other mechanical methods for disruption (e.g., French press) are acceptable. Cell-extracts should be clarified by centrifugation at 5000 to 27,000 × *g* for 10 min and may be filtered through a cellulose acetate filter (0.2 μm) with a glass fiber prefilter for volumes over 2 mL. Some replicates should be done with desalted extracts in case there are small molecules that give background activity or are inhibitors. To desalt preparations, take 2.5-mL portions and desalt using Pharmacia PD-10 desalting columns, following the manufacturer's instructions. The final buffer should be 0.2 M potassium phosphate buffer (pH 7.0) with 5 mM dithiothreitol. The protein eluate may desalted and concentrated using semipermeable membranes with a molecular weight cutoff of 10,000.

10. The FadE enzyme is a flavoprotein: this is assayed by its flavin reducing and bleaching the DCPIP. The intermediate electron acceptor, PMS, has to be added to allow this bleaching reaction to occur. The rate of reduction of DCPIP is measured by following the fall in A_{600}. All activities are expressed in international enzyme units, referred to as "units," defined as the enzyme required to convert 1 μmol substrate/min. There may be a background rate, especially if crude extracts are used. This should be less than the difference in rates in the cuvettes when substrate is added to one, and must be less than 0.16.

11. Triplicates are standard in metabolomic or transcriptomic experiments when statistics based on the normal distribution (usually Student's *t*-test) will be used in their analysis; Quadruplicates should be done if a nonparametric test is to be done.

12. This suspension does not have to be exactly 5 mg/mL, as in the protocol the next step is to determine the actual weight of bacteria in a sample. Using this value, the results for $^{14}CO_2$ evolved should always be calculated as specific activities (i.e., as dpm/mg). The OD_{600} given in the methods for suspensions of bacteria only applies to the Helios gamma spectrophotometer (Thermospectronic, Cambridge, UK) and should be determined for each make of spectrophotometer .

13. Hazards are both the bacteria and the radioactivity, though the $^{14}CO_2$ evolved should all have been absorbed by the NaOH. Open in a fume hood or in a class I safety cabinet.

14. Alkalis are generally quenching agents. Therefore, ensure quench correction is used and counting efficiency is determined so that dpm values are obtained to get true quantitative results. Hyamine hydroxide 1 M may be used in place of NaOH. The advantage is that it is compatible with more scintillation fluids. The disadvantage is that the solvent has to be allowed to dry, and there is the risk of fumes in the incubation vessel.

15. The number of units in the enzyme preparation will have to be worked out by trial and error, but routinely assays should be run to give a fall in A_{340} of less than 0.3 for FadD. For FadE, the fall in A_{600} should be less than 0.16. In both these continuous assays, calculations must be done from the maximum rate of reaction. For FadD, only a single time point is recorded, when no more than 25% of the labeled fatty acid should be transformed if units of activity are to be determined. The volume of 0.5 M Tris-HCl should be adjusted to give a final concentration of 50 mM if the volume of enzyme preparation is other than 100 μL. The volumes can be scaled down for an ELISA plate assay, to a total assay volume of 100 to 200 μL.

16. To calculate enzyme kinetic parameters such as K_m, V_{max}, (s^{-1}) from the data obtained from the assays described in this chapter, consult Ref. 14 for a straightforward, practical guide. In the standard assays given here, the substrates are at saturating concentration, so that, at 0.25 times the concentration, you may see very little difference in reaction rates.

Below this concentration, a little trial and error will be needed to find a range of concentrations that gives a measurably different range of reaction rates. Design the experiment bearing in mind you will be plotting graphs with the reciprocal of the substrate concentration on the x-axis, so both a linear range of concentrations or serial dilutions are not useful. Kinetic data with very hydrophobic fatty acid substrates should be treated and presented with caution, especially when comparing and contrasting kinetic data from different FadD. This is because, the longer the chain length, the less fatty acid will be in solution and the more will be complexed with the detergent included in the assay.

17. It is possible that acyl phosphates will also be detected by this system, too. HPLC may be used as an alternative to TLC to separate the products [9]: a radioactivity detector will be needed in line to quantify the proportion of label in each product. Verification of the products may be done using a mass spectrometer to analyze the products from HPLC [9].

18. Cloned acyl-CoA dehydrogenase from *M. tuberculosis* has the characteristic yellow-green of a flavoprotein [13], so an assay without added FAD should be adeuate. However, 0.1 mM FAD was included in the assay in this work [13] so is included in the protocol here. It should be possible to omit FAD from the assay and for its omission to have no effect.

Acknowledgment P.R.W. was supported through DEFRA funding.

References

1. Cole S. T., Brosch R., Parkhill J., Garnier T., Churcher C., Harris D., et al. (1998). Deciphering the biology of *Mycobacterium tuberculosis* from the complete genome sequence. Nature; 393: 537–544.

2. Siddiqi S. H., Hwangbo C. C., Silcox V., Good R. C., Snide D. E., & Middlebrook G. (1984). Rapid radiometric methods to detect and differentiate *Mycobacterium/M.bovis* from other mycobacterial species. Am Rev Respir Dis; 130: 634–640.

3. Camargo E. E., Kertcher J. A., Larson S. M., Tepper B. S. & Wagner H. N. (1982). Radiometric measurement of differential metabolism of fatty acid by mycobacteria. Int J Lepr Other Mycobact Dis; 50: 200–204.

4. Heifets L. B., Iseman M. D., Cook J. L., Levy P. J. & Drupa I. (1985). Determination of the *in vitro* susceptibility of *Mycobacterium tuberculosis* to cephalosporins by radiometric and conventional means. Antimicrob Agents Chemother; 27: 11–15.

5. Camargo E. E., & Wagner H. N. (1987). Radiometric studies on the oxidation of 1-14C fatty acids and U-14C amino acids by mycobacteria. Nucl Med Biol; 14: 43–49.

6. Wheeler P. R., Bulmer K., & Ratledge C. (1991) Fatty acid oxidation and the beta-oxidation complex in *Mycobacterium leprae* and two axenically cultivable mycobacteria that are pathogens. J Gen Microbiol; 137: 885–893.

7. Wheeler P. R., & Ratledge C. (1988). Use of carbon sources for lipid biosynthesis in *Mycobacterium leprae*: a comparison with other pathogenic mycobacteria. J Gen Microbiol; 134: 2111–2121.

8. Franzblau S. G. (1988). Oxidation of Palmitic acid by *Mycobacterium leprae* in an axenic medium. J Clin Microbiol; 26: 18–21.

9. Trivedi O. A., Arora P., Sridharan V., Tickoo R., Mohanty D., & Gokhale R. S. (2004). Enzyme activation and transfer of fatty acids as acyl-adenylates in mycobacteria. Nature; 428: 441–445.

10. Shulz R. (1974). Long chain enoyl-CoA hydatase from pig heart. J Biol Chem; 249: 2704–2709.

11. Parish T., & Wheeler P. R. (1998). Preparation of cell-free extracts from mycobacteria. Methods Mol Biol; 101: 77–89.

12. Hosaka K., Mishina M., Kamiryo T., & Numa S. (1981). Long-chain acyl-CoA synthetases I and II from *Candida lipolytica*. Methods Enzymol; 71: 325–333.
13. Mahadevan U., & Padmanaban G. (1998). Cloning and expression of an acyl-CoA dehydrogenase from *Mycobacterium tuberculosis*. Biochem Biophys Res Commun; 244: 893–897.
14. Qiu J (1993). Enzymology. In: Chambers J. A. A., & Rickwood D, eds. Biochemistry Labfax. Bios Scientific, Oxford, UK,1993: 101–144.

Chapter 5
Analysis of Lipid Biosynthesis and Location

Paul Robert Wheeler

Abstract A procedure for metabolic labeling of all cellular lipids starting with a culture of mycobacteria is described in this chapter using either a pulse-chase or a simple labeling experimental design. Three fractions are produced for subsequent lipid analysis: (1) the culture filtrate; (2) a readily released surface lipid fraction; and (3) the killed, labeled bacteria. A standardized, TLC-based method for general lipid analysis that can be used to quantify the labeling of all the mycobacterial lipids is given as well as a protocol for analyzing the fatty acyl moieties of the lipids.

Keywords acetate · fatty acid · lipid metabolism · mycobacteria · *Mycobacterium tuberculosis* · propionate

5.1 Introduction

More than 20% of the dry weight of mycobacteria is composed of lipid, and at least 10% of the genome encodes for proteins involved in lipid metabolism. Many of these genes still have broad functional annotations; for example, the *mmpL/S* genes are predicted to be involved in lipid transport, but the type of lipid transported has to be determined for every one [1, 2, 3]. Polyketide synthases are clearly involved in the biosynthesis of complex lipids, but again, the role of every *pks* and *pps* gene has to be determined; for example, by investigating the effect of their genetic manipulation on the lipid content of the bacteria [4]. In this chapter, methods are given to follow the biosynthesis of such lipids and to determine their subcellular location. A second question is what happens when genes involved in lipid metabolism are deleted or lost during evolution, as in a major U.K. field isolate of *Mycobacterium bovis* [5], or controlled by global regulation events, such as in the macrophage, where four *fadD* genes and nine *fadE* genes are upregulated in *Mycobacterium tuberculosis*

P.R. Wheeler
Tuberculosis Research Group, Veterinary Laboratories Agency (Weybridge),
New Haw, Addlestone, Surrey, KT15 3NB, UK
e-mail: p.wheeler@vla.defra.gsi.gov.uk

T. Parish, A.C. Brown (eds.), *Mycobacteria Protocols*,
doi: 10.1007/978-1-59745-207-6_5, © Humana Press, Totowa, NJ 2008

[6]. Such questions can only be addressed using biochemical approaches such as the ones outlined in this chapter. This chapter concentrates on methods that can be used for overall lipid analyses (Fig. 5.1).

The use of enzyme extracts and more specialized biochemical methods is largely outside the scope of this chapter. However, a useful method that can be used as the basis for modification for more general use is the fatty acid synthase (Fas) assay. This uses a clarified extract that can be sterile filtered. The core method is described in *Methods in Enzymology* [7, 8], with a useful modification [9] in that commercially available cyclodextrin can be used to relieve feedback inhibition by acyl-CoA where this powerful effect [10] is an issue. With modification, the biosynthesis of multimethylated fatty acids such as the mycocerosic acids [11, 12], polyketides, and even some complex esters such as phthiocerol dimycocerasate (PDIM) analogues [13] can be assayed. However, Fas-type assays cannot be used for assaying the formation of most complex lipids, including mycolates.

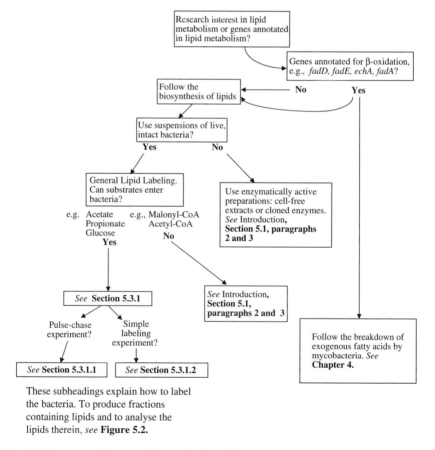

Fig. 5.1 The scope of the metabolic methods in this article

Mycolic acid biosynthesis can be assayed using a wall and membrane-based preparation called "P60" separated from intact cells on a 60% Percoll gradient [14, 15, 16]. The use of P60 has been extended to a wider range of lipid biosynthetic activities [17, 18]. One disadvantage of using P60 is the presence of culturable bacteria [19]. However, malonyl-CoA, which does not pass through the plasma membrane, was active in the P60 preparation [15] showing there was material from disrupted bacteria in P60. Given this disadvantage, the safest way to use P60, and the one recommended here, is in low water assays. In low water assays, P60 is in a minimum volume of aqueous phase, and a nonpolar phase, usually hexane, is included in the assay [15]. This has the advantages that nonpolar substrates can be both extracted and added, difficult in aqueous phase assays, and that any bacteria in the P60 are rendered permeable by the solvent so that substrates such as malonyl-CoA have access to all the enzymes in the preparation.

The protocols described here start with a culture of mycobacteria and involve following the biosynthesis of all lipids from simple carbon sources in intact mycobacteria. All of the major lipids can be labeled using one of three alternative [14C]-labeled carbon sources enabling lipid biosynthesis to be easily followed. The use of the radiolabeled substrates makes quantification easy and thus protocols to generate metabolically labeled lipids are described (Fig. 5.1) (*see* Section 5.3.1) from which lipid fractions can be generated (*see* Section 5.3.2) for analysis and quantification (*see* Sections 5.3.3 and 5.3.4; Fig. 5.2).

Before the lipids can be analyzed, metabolically labeled lipid fractions must be prepared for analysis. The main way of following the biosynthesis and turnover of lipids in growing mycobacteria is by adding a radiolabeled metabolite to an actively growing culture and tracing the incorporation of the label, over time, into lipids. Simple labeling experiments can be performed, or alternatively, a pulse-chase experiment can be used where the labeled compound is washed away and metabolites followed over time. The timings in the procedures are appropriate for slow-growing bacteria but can be adapted for fast growers, such as *Mycobacterium smegmatis*. After labeling, preparation of the bacteria for lipid analysis (Fig. 5.2) gives the option of extracting several fractions: (i) lipids right at the surface of the cell that are almost entirely the outer leaflet of the outer permeability barrier, (ii) the lipids secreted into the culture medium (probably a similar set of lipids to the fraction i), and (iii) the residue consisting of the plasma membrane lipids and mycolic acids covalently attached to the cell wall matrix of arabinogalactan-peptidoglycan. Further extraction of lipids from fraction (iii) (or analysis of the whole labeled bacteria without extraction) can also be performed to leave a final residue of cell wall with only covalently bound lipids attached (mycolic acids).

General methods to analyze the lipids are given. In practice, a general thin-layer chromatography (TLC)-based method [20, 21] will reveal all the principal lipids, and this is presented in outline. A method for quantification of lipid fractions that focuses on this TLC-based method is also described. The TLC method will reveal the lipid classes, but sometimes it is necessary to analyze fatty

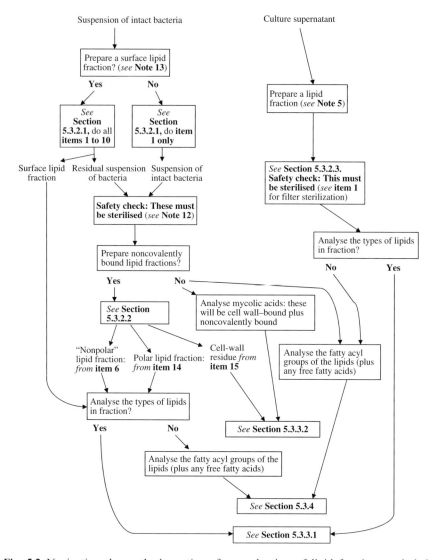

Fig. 5.2 Navigating the methods sections for production of lipid fractions and their analysis

acids that are esterified to the lipids. This can give an insight into the metabolic state of the organisms, the role of certain genes (especially those in fatty acid metabolism), and the properties of the lipids in greater detail. Therefore, a method for analyzing the constituent fatty acyl groups of the lipids is given using HPLC in combination with UV and radioactive detection [22], with a note on using gas chromatography (GC) as an alternative.

5.2 Materials

Reagents should be AnalaR grade for normal work or HPLC grade for HPLC.

5.2.1 Metabolic Labeling of Lipids

5.2.1.1 Labeling Suspensions of Bacteria: Simple Labeling Experiment

1. A suitable culture of the mycobacteria being studied (*see* **Note 1**).
2. Centrifuge that can be operated between 37°C and 4°C.
3. Sealed centrifuge buckets to contain any spills of radioactivity or bacteria.
4. Centrifuge tubes (e.g., 50-mL Falcon tubes).
5. Labeled compound: [1-^{14}C] for acetate, propionate, or other fatty acids.
6. Milli-Q water. Keep at 4°C.

5.2.1.2 Labeling and Chasing Suspensions of Bacteria: A "Pulse-Chase" Experiment

1. A suitable culture of the mycobacteria being studied (*see* **Note 1**).
2. Centrifuge that can be operated between 37°C and 4°C.
3. Sealed centrifuge buckets to contain any spills of radioactivity or bacteria.
4. Centrifuge tubes (e.g., 50-mL Falcon tubes).
5. Labeled compound: [1-^{14}C] for acetate, propionate, or other fatty acids.
6. Fresh culture medium. Warm to culture temperature before use (*only needed for pulse-chase experiments*).
7. Milli-Q water. Keep at 4°C.

5.2.2 Preparing Labeled Samples for Lipid Analysis

5.2.2.1 Preparation of a Surface Lipid Fraction

1. Milli-Q water. Keep at 4°C.
2. Polytetrafluoroethane (PTFE)-capped Pyrex tubes (16 mm diameter × 125 mm).
3. Hexane.
4. 0.05% (w/v) decylamine in hexane (*only needed if a surface lipid fraction is to be prepared*).
5. PTFE syringe filters, 0.20-μm pore size, 13-mm diameter with low retention volume (less than ~30 μL).
6. Water bath at 80°C with rack for 16-mm-diameter tubes.
7. 0.1 M HCl in Milli-Q water.
8. Scintillation vials.
9. Equipment for drying down samples under a stream of nitrogen, that is, hot block, evaporating unit with multiple nozzles.

10. Toluéne-based scintillation fluid; for example, Scintran Cocktail NA (VWR Merck, Lutterworth, UK).
11. Scintillation counter.

5.2.2.2 Preparation of Noncovalently Bound Lipid Fractions from Suspensions of Mycobacteria

1. Chloroform.
2. Methanol.
3. 0.3% (w/v) NaCl in Milli-Q water.
4. Methanol/0.3% (w/v) NaCl in Milli-Q water (10:1).
5. Hexane.
6. Rotatory mixer.
7. Scintillation vials.
8. Equipment for drying down samples under a stream of nitrogen, that is, hot block, evaporating unit with multiple nozzles.
9. Toluene-based scintillation fluid; for example, Scintran Cocktail NA (VWR Merck).
10. Scintillation counter.
11. Chloroform/methanol/0.3% (w/v) NaCl in Milli-Q water (9:10:3).
12. Chloroform/methanol/0.3% (w/v) NaCl in Milli-Q water (5:10:4).
13. Chloroform/methanol (2:1).

5.2.2.3 Preparation of a Culture Supernatant Lipid Fraction

1. Cellulose acetate or (Polyvinylidine fluoride, PVDF) syringe filters, 0.20-μm pore size, 25-mm diameter with glass fiber prefilter.
2. Freeze dryer.
3. 0.3% (w/v) NaCl in Milli-Q water.
4. Chloroform.
5. Methanol.
6. Separating funnel capable of holding 50 mL comfortably.
7. Scintillation vials.
8. PTFE-capped Pyrex tubes (16 mm diameter × 125 mm).
9. Equipment for drying down samples under a stream of nitrogen, that is, hot block, evaporating unit with multiple nozzles.
10. Scintillation counter.

5.2.2.4 Preparation of the Wall-Bound Mycolic Acids

1. PTFE-capped Pyrex tubes (16 mm diameter × 125 mm).
2. Tetrabutylammonium hydroxide (TBAH): supplied as a 40% solution. Make a 15% (w/v) solution in Milli-Q water.
3. Dichloromethane.
4. Iodomethane.
5. Rotatory mixer.

6. 0.1 M HCl in Milli-Q water.
7. Milli-Q water.
8. Scintillation vials.
9. Equipment for drying down samples under a stream of nitrogen, that is, hot block, evaporating unit with multiple nozzles.
10. Chloroform/methanol (2:1).
11. Hexane.

5.2.3 Analysis and Quantification of the Labeled Lipids

5.2.3.1 Analysis of the Noncovalently Bound Lipids by TLC

1. TLC plates: aluminum-backed silica gel 60 F254 plates, 20 cm × 20 cm (Merck 1.05554). Cut each plate into nine equal squares (6.6 cm × 6.6 cm).
2. Glass capillaries with marks at 5 μL or 10 μL.
3. TLC chambers.
4. Hairdryer.
5. Hexane.
6. Ethyl acetate.
7. Acetone.
8. Toluene.
9. Chloroform.
10. Methanol.
11. Milli-Q water.
12. Acetic acid (glacial).
13. Solvent systems by volume (Table 5.1).

   (a) Hexane/ethyl acetate (98:2) and hexane/acetone (98:2)
   (b) Hexane/acetone (92:8) and toluene/acetone (95:5)
   (c) Chloroform/methanol (96:4) and toluene/acetone (80:20)
   (d) Chloroform/methanol/water (250:75:4) and chloroform/acetone/methanol/water (100:120:5:6)
   (e) Chloroform/methanol/water (10:5:1) and chloroform/acetic acid/methanol/water (40:25:3:6)

14. PhosphorImager (examples are PMI, Bio-Rad Laboratories, Hercules, CA, USA or FLA or BAS Series, Fujifilms, Bedford, UK) (or autoradiography equipment).

5.2.3.2 Analysis of Mycolic Acids

1. TLC plates: glass-backed silica gel 60 F254 plates, 10 cm × 10 cm with concentrating zone (Merck 13748).
2. Glass capillaries with marks at 5 μL or 10 μL.
3. TLC chambers.
4. Hairdryer.

Table 5.1 Solvent Systems for TLC Analysis of Noncovalently Bound Lipids

System	Solvent mixtures (prepare immediately before use)		Lipids developed
	First direction	Second direction (run once)	
A	Hexane/ethyl acetate (98:2). Run ×3.	Hexane/acetone (98:2)	PDIMs, triacylglycerol (TAG), menaquinone (MK), cholesterol esters (CE)
B	Hexane/acetone (92:8). Run ×3.	Toluene/acetone (95:5)	Trehalose mycolipenates (TTMP), free fatty acids (FFA)
C	Chloroform/methanol (96:4). Run once.	Toluene/acetone (80:20)	Phenolic glycolipids (PGL), free fatty acids, glucose monomycolate
D	Chloroform/methanol/water (250:75:4). Run once.	Chloroform/acetone/methanol/water (100:120:5:6)	*Nonpolar fraction:* Sulfolipid (SL), mono- and diacylated trehalose, including trehalose mycolates (TMM, TDM). *Polar fraction:* Monoacylglycerol (MAG), unknown glycolipids (?, g)
E	Chloroform/methanol/water (10:5:1). Run once.	Chloroform/acetic acid/methanol/water (40:25:3:6)	Phospholipids (DPG, PE, PI) Phosphatidylinositolmannosides (PIM) with up to 6 mannose residues

The patterns of lipids on each plate are shown in Figure 5.3. It is safe to use the sprays used to visualize lipids described in Refs. 20, 21 with [14C] lipids and possible to detect their radioactivity after spraying.

5. Hexane/ethyl acetate (95:5).
6. PhosphorImager (or autoradiography equipment).

5.2.3.3 Quantification of the Lipids on TLC Plates

1. PhosphorImager (or autoradiography equipment).

5.2.4 Analysis and Quantification of the Fatty Acyl Chains of the Lipids

1. Equipment for drying down samples under a stream of nitrogen, that is, hot block, evaporating unit with multiple nozzles.
2. Tetrabutylammonium hydroxide: supplied as a 40% solution; make a 15% (w/v) solution in Milli-Q water.
3. Dichloromethane.
4. Pentafluorobenzylbromide.
5. Rotatory mixer.
6. Diethyl ether.
7. 0.1 M HCl in Milli-Q water.
8. PTFE syringe filters, 0.20-μm pore size, 4 mm diameter.
9. Scintillation vials.

10. PTFE-capped Pyrex tubes (16 mm diameter × 125 mm).
11. 1,4-Dioxan, HPLC grade, filtered and degassed.
12. Acetonitrile, HPLC grade, filtered and degassed.
13. Acetonitrile/1,4-dioxan (7:3).
14. Waters (Milford, MA, USA) μBondapak C18 10-μm reversed phase column (3.9 mm dia-meter ×150 mm) or similar.
15. Scintillation fluid to mix with eluate from HPLC; for example, Packard Ultima-Flo M (PerkinElmer, Waltham, MA, USA).
16. Toluene-based scintillation fluid; for example, Scintran Cocktail NA (VWR Merck).
17. Fatty acid standards. A useful range is 16:0, 18:0, 20:0, 24:0, and 30:0 (all saturated) plus 18:1 and 24:1 (monounsaturated), all available commercially. At least one [^{14}C]-labeled fatty acid standard, corresponding with unlabeled standard(s), should also be used: 18:0, 18:1, or 24:0 are the most useful.

5.3 Methods

5.3.1 Metabolic Labeling of Lipids

5.3.1.1 Labeling Suspensions of Bacteria: Simple Labeling Experiment

1. Prewarm centrifuge and all media to be used in the experiment to the temperature used for growth and incubation.
2. Add one 25 μCi radiolabeled compound per 100 mL of culture of mycobacteria (*see* **Notes 1, 2, and 3**). Keep the cultures at the same temperature as that used for growth.
3. Place cultures back in the incubator for 2 h to 24 h depending on the amount of label needed for analysis (*see* **Notes 2 and 4**).
4. At the end of the incubation period, harvest the mycobacteria by centrifugation at 4000 × g for 10 min in the prewarmed centrifuge. Keep the culture supernatant if required for lipid analysis (*see* Section 5.3.2.3 and **Note 5**).
5. Wash the pellet in Milli-Q water at 4°C (in the same volume as the sample volume). Spin at 4000 × g for 10 min. Wash once if no detergent was used in the growth medium, twice if detergent was used in the growth medium (*see* **Note 6**).
6. For preparing samples for lipid analysis from the pellet, go to Section 5.3.2.1.

5.3.1.2 Labeling and Chasing Suspensions of Bacteria: A "Pulse-Chase" Experiment

1. Prewarm centrifuge and all media to be used in the experiment to the temperature used for growth and incubation.
2. Add 50 μCi of radiolabeled compound directly to each 100 mL of culture. Keep the cultures at the same temperature as that used for growth (*see* **Notes 1, 2, and 3**).

3. Place cultures back in the incubator for 2 h (*see* **Notes 2 and 4**). This is the **pulse** stage.

4. At the end of the pulse stage, pellet the mycobacteria by centrifugation at $4000 \times g$ for 10 min in the prewarmed centrifuge. Keep the culture supernatant if required to for lipid analysis (*see* Section 5.3.2.3 and **Note 5**).

5. Wash the pellet in fresh, prewarmed growth medium ($\sim$100 mL) by resuspension and centrifugation at $4000 \times g$ for 10 min in a prewarmed centrifuge.

6. Resuspend the washed pellet in 100 mL fresh, prewarmed growth medium. Take a 10-mL sample, label as "**pulsed but not chased.**"

7. Centrifuge the remaining 90 mL at $4000 \times g$ for 10 min in the prewarmed centrifuge.

8. Resuspend the pellet in 90 mL fresh prewarmed growth. Put this suspension back in the incubator—this is the **chase** part of the experiment.

9. Pellet the 10 mL sample—**pulsed but not chased** (from step 6). Wash the pellet at $4000 \times g$ for 10 min in 40 mL Milli-Q water at 4°C. Wash once if no detergent was used in the growth medium, twice if detergent was used in the growth mcdium (*see* **Note 6**).

10. For preparing samples for lipid analysis from the pellet, go to Section 5.3.2.1.

11. Take samples from the **chase** part of the experiment at required intervals for analysis (*see* **Note 7**). From each sample, harvest the mycobacteria by centrifugation at $4000 \times g$ for 10 min in the prewarmed centrifuge.

12. Wash each pellet from the **chase** part of the experiment (in the same volume as the sample volume) at $4000 \times g$ for 10 min in Milli-Q water at 4°C. Wash once if no detergent was used in the growth medium, twice if detergent was used in the growth medium (*see* **Note 6**).

13. For preparing samples for lipid analysis from each pellet, go to Section 5.3.2.1 (*see* **Note 8**).

5.3.2 Preparing Labeled Samples for Lipid Analysis

The following procedures can be carried out on the pellets and culture filtrates derived from the labeling experiments generated in Section 5.3.1.1 and/or Section 5.3.1.2.

5.3.2.1 Preparation of a Surface Lipid Fraction (*see* Note 9)

1. Resuspend each of the washed pellets (from Section 5.3.1.1, item 5, Section 5.3.1.2, items 9 and 12) in 1 to 2 mL Milli-Q water in PTFE-capped Pyrex tubes (16 mm diameter $\times$ 125 mm). The tubes should be preweighed so that the weight of bacteria harvested can be determined.

2. To obtain surface lipids, add 2 vol 0.05% decylamine in hexane to each suspension. Briskly mix by inverting 4 times (*see* **Note 10**).

3. Allow layers to settle or centrifuge at $250 \times g$ for 2 min to separate the layers (*see* **Note 11**).

4. Take the upper, hexane-rich layer and sterilize by passing through an 0.20-μm pore size PTFE filter (13 mm diameter) into a clean Pyrex tube.

5. Sterilize the remaining lower, aqueous layer including the bacteria in the capped tube by heating at 80°C for 30 min (see **Note 12**). The preparation of lipid fractions from this residual suspension is described in Section 5.3.2.2.

6. Wash the upper layer (to remove the decylamine) with 2 vol 0.1 M HCl; mix for 5 min using a rotatory mixer. Centrifuge at 250 × g for 2 min and remove the lower layer. Repeat, but this time transfer the upper layer into a clean PTFE-capped Pyrex tube.

7. Discard all lower-layer material.

8. Measure the volume of the upper layer and take a sample (5%) into a scintillation vial.

9. Dry the bulk of the material in the Pyrex tube under a stream of nitrogen at 40°C. Reconstitute in 50 μL chloroform/methanol (2:1) prior to TLC analysis described in Section 5.3.3.1.

10. Dry the sample in the scintillation vial under a stream of nitrogen at 40°C. Add scintillation fluid and perform scintillation counting on the vial (see **Note 13**).

5.3.2.2 Preparation of Noncovalently Bound Lipid Fractions from Suspensions of Mycobacteria

The following procedures should be carried out on the residual suspension after extraction of the surface lipid fraction (from Section 5.3.1.3, step 5) or directly on suspensions of mycobacteria (see **Note 9**).

1. Freeze-dry the suspensions still in the PTFE-capped Pyrex tubes, with the caps loose.

2. Weigh the dried bacteria (or bacterial residue) (see **Note 14**).

3. Add 2 mL methanol/0.3% NaCl (10:1) and 1 mL hexane to each tube. Mix on a rotary mixer for 15 min. Centrifuge at 250 × g for 2 min.

4. Remove the upper layer into a clean PTFE-capped Pyrex tube marked with "NP."

5. Add 1 mL hexane to the residual lower layer. Mix on a rotary mixer for 15 min. Centrifuge at 250 × g for 2 min.

6. Remove the upper layer, and combine it with the corresponding upper layer already in each tube marked "NP." This is the **nonpolar fraction** (see **Note 15**). Keep the lower layer, which also has a solid residue, to extract polar lipids (continues from item 10) after items 7 to 9 have been completed.

7. Measure the volume of the **nonpolar fraction** and take a sample (5%) into a scintillation vial (see **Note 13**).

8. Dry the sample in the scintillation vial under a stream of nitrogen at 40°C. Add scintillation fluid and perform scintillation counting on the vial (see **Note 13**).

9. Dry the bulk of the **nonpolar fraction** in the Pyrex tube under a stream of nitrogen at 40°C. Reconstitute in 50 μL chloroform and put aside for TLC analysis described in Section 5.3.3.1.

10. Return to the tube containing the lower layer and solid residue from which the upper layer was removed in item 6. With the cap on the tube, heat at 100°C for 5 min, allow to cool to 40°C, loosen the lid (care: may be under pressure) to allow any hexane remaining to escape, then allow to cool to room temperature.

11. Add 2.3 mL chloroform/methanol/0.3% NaCl (9:10:3) to the residual layer. Mix on a rotary mixer for 60 min. Centrifuge at $500 \times g$ for 5 min. Transfer the solvent to a clean PTFE-capped Pyrex tube marked with the sample "P."

12. Add 0.75 mL chloroform/methanol/0.3% NaCl (5:10:4). Mix on a rotary mixer for 30 min. Centrifuge at $500 \times g$ for 5 min. Transfer the solvent and combine it with the corresponding upper layer already in each tube marked "P." Repeat this step.

13. Add 1.3 mL chloroform and 1.3 mL 0.3% NaCl to each tube marked "P" (*see* **Note 16**). Centrifuge at $250 \times g$ for 5 min.

14. Remove the lower layer into a clean PTFE-capped Pyrex tube marked with the sample i/d and "P." This is the **polar fraction** (*see* **Note 15**).

15. Discard the upper layer but keep the solid residue for mycolic acid analysis (Section 5.3.2.4).

16. Measure the volume of the **polar fraction** and take a sample (5%) into a scintillation vial (*see* **Note 13**). Dry the sample in the scintillation vial under a stream of nitrogen at 40°C. Add scintillation fluid and perform scintillation counting on the vial.

17. Dry the bulk of the **polar fraction** in the Pyrex tube under a stream of nitrogen at 40°C. Reconstitute in 50 μL chloroform/methanol (2:1) prior to TLC analysis described in Section 5.3.3.1.

5.3.2.3 Preparation of a Culture Supernatant Lipid Fraction (*see* Note 5)

1. Sterilize the culture supernatant by filtering through two 0.20-μm pore size filters.

2. Freeze-dry and reconstitute by making a slurry in 4.1 mL 0.3% NaCl (*see* **Note 2**).

3. Add 18.2 mL methanol and 6.7 mL chloroform.

4. Transfer the slurry/solution into a separating funnel.

5. Wash out the original vessel with a mixture of 6.5 mL chloroform and 6.5 mL methanol: add this to the separating funnel, mix by inversion 2 or 3 times, allow to settle, take off the lower layer. Discard the upper layer.

6. Measure the volume of the retained lower layer.

7. Place a sample (10%) into a scintillation vial. Dry under a stream of nitrogen at 40°C (*see* **Note 13**). Add scintillation fluid and perform scintillation counting on the vial.

8. Transfer the bulk of the retained lower layer in a clean PTFE-capped Pyrex tube (16 mm diameter × 125 mm). Dry under a stream of nitrogen at 40°C. Reconstitute in 50 μL of chloroform/methanol (2:1 by vol.) prior to TLC analysis described in Section 5.3.3.1.

5.3.2.4 Preparation of the Wall-Bound Mycolic Acids

1. To the solid residue from Section 5.3.2.2, item 15, add 1 mL 15% TBAH. Seal tubes with the PTFE-lined caps. Heat at 105°C for 18 h. Allow to cool to room temperature.
2. Add 1 mL water and 2 mL dichloromethane. Add 100 μL iodomethane. Mix on a rotary mixer for 30 min. Centrifuge at $250 \times g$ for 5 min.
3. Remove the upper layer and wash the lower, organic layer with 4 mL 0.1 M HCl. Mix on a rotary mixer for 15 min. Centrifuge at $250 \times g$ for 5 min.
4. Remove the upper layer and wash the lower, organic layer with 4 mL Milli-Q water. Mix on a rotary mixer for 15 min. Centrifuge at $250 \times g$ for 5 min.
5. Transfer the lower layer into a clean PTFE-capped Pyrex tube. Discard the upper layer.
6. Measure the volume of the lower layer and take a sample (5%) into a scintillation vial (*see* **Note 13**). Add scintillation fluid and perform scintillation counting on the vial.
7. Dry the bulk of the material in the Pyrex tube under a stream of nitrogen at 40°C. Reconstitute in 50 μL chloroform/methanol (2:1) prior to TLC analysis. It is probable that there will be too much material to reconstitute in 50 μL. If so, carry out items 8 to 10 below (*see* **Note 17**):
8. Add 2 mL 0.1 M HCl and 1 mL hexane to the tube. Mix on a rotary mixer for 5 min. Centrifuge at $250 \times g$ for 5 min.
9. Remove the lower layer and wash the upper, organic layer again with 2 mL of Milli-Q water. Mix on a rotary mixer for 5 min. Centrifuge at $250 \times g$ for 5 min.
10. Transfer the upper layer into a clean PTFE-capped Pyrex tube. Discard the lower layer. Reconstitute in 50 μL chloroform/methanol (2:1) prior to TLC analysis described in Section 5.3.3.2.

5.3.3 Analysis and Quantification of the Labeled Lipids

5.3.3.1 Analysis of the Noncovalently Bound Lipids by TLC

The methods required for the analysis of the labeled material depend very much on the products of interest to the researcher. The classic methods for analysis of all classes of mycobacterial lipids use two-dimensional TLC and are given in detail, with diagrams of the TLC plates, in a chapter in a previous volume of *Mycobacteria Protocols* [20]. Those TLC solvent systems used are summarized here.

The following fractions (*see* **Note 18**) may be analyzed by these methods: nonpolar fraction, polar fraction (both from Section 5.3.2.2), culture supernatant (from Section 5.3.2.3), and surface lipids (from Section 5.3.2.1). The nonpolar lipid fraction should be developed in solvent systems A to D (Table 5.1). The polar lipids should be developed in systems D and E. The other fractions may be developed in all solvent systems.

1. Apply 25,000 dpm to 100,000 dpm per TLC plate using a capillary tube (5 or 10 μL capacity). Apply as a spot 1 cm from the left-hand corner of a 6.6 cm × 6.6 cm TLC plate for each of the solvent systems in Table 5.1.
2. Run in the first direction to within a few millimeters of the right-hand edge. Dry using a stream of warm air from a hair dryer. Repeat 3 times (*see* **Note 19**) for systems A and B (Table 5.1) only. Dry the plate well before running in the second direction.
3. Turn the plate through 90 degrees to the left so the lipids streaked out by the first solvent are on the bottom of the plate (Fig. 5.3). Then place the plate in the second solvent and run to within a few millimeters of the top of the plate.
4. Use autoradiography or a PhosphorImager to analyze the distribution and quantity of the lipids on each place (described in Section 5.3.3.3).
5. If the developed plate is badly smeared (*see* **Note 20** and Figure 5.3), repeat, but with less material.
6. Identify individual lipids by reference to Figure 5.3.

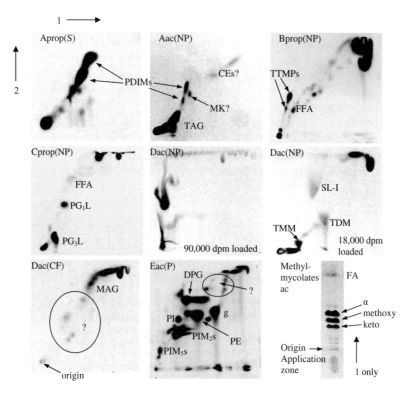

Fig. 5.3 Autoradiographs of TLC plates. A to E refer to the solvent systems in Table 5.1. The fractions used were NP, nonpolar; P, polar; S, surface lipids; CF, culture filtrate. The [^{14}C] substrates used were ac, acetate; prop, propionate. The abbreviations for the lipids marked on the plates correspond with those used in Table 5.1. Subscript numbers refer to the number of saccharide units. The lipids were from members of the *M. tuberculosis* complex chosen best to illustrate each solvent system. Numbered arrows denote the direction in which solvent was run

5.3.3.2 Analysis of Mycolic Acids

1. Apply 25,000 dpm to 100,000 dpm per TLC plate using a capillary (5 or 10 μL capacity): apply as a streak in the concentrating zone of a glass-backed silica gel plate. About eight samples can be applied to a single plate.
2. Run in the solvent system (*see* **Note 17**): hexane/ethyl acetate (95:5) 3 times (*see* **Note 19**).
3. Use autoradiography or a PhosphorImager to analyze the distribution and quantity of the lipids on each plate (described under Section 5.3.3.3).
4. If the developed plate is badly smeared (*see* **Note 20**), repeat, but with less material.

5.3.3.3 Quantification of the Lipids on TLC Plates

1. Using the values of dpm in the samples taken above, calculate total dpm in each fraction (i.e., polar, nonpolar, cell wall–bound mycolate, culture filtrate, and surface lipid fractions). Referring back to the dry weight of the bacteria, calculate dpm per mg (dry weight bacteria) (*see* **Note 21**).
2. For TLC, quantify the label in each lipid separated by TLC (Fig. 5.3), using a PhosphorImager. Follow the manufacturer's instructions to expose the TLC plates, taking care not to contaminate the PhosphorImager screen with radioactivity. Read out the relative amounts of label in every spot on each TLC plate *including those at the origin and solvent front*. The absolute amounts of label in each lipid can then be calculated in item 1 from the radioactivity in each of the above fractions. The identification of individual lipids is from their position on the plates (Fig. 5.3) [20, 21].
3. For autoradiography: Autoradiographs can be made using a suitable film (e.g., Kodak Biomax MR Eastman Kodak, Rochester, NY, USA for [^{14}C]). Expose TLC plates for 2 days for 100,000 dpm; 5 days for 25,000 dpm. The intensity of the spots can be analyzed (*see* **Note 22**) by a densitometer or PhosphorImager software (e.g., Bio-Rad Quantity One Bio-Rad Laboratories, Hercules, CA, USA).

5.3.4 Analysis and Quantification of the Fatty Acyl Chains of the Lipids (see Notes 23, 24, and 25)

The protocol described here is our preferred HPLC-based method because volatile, radioactive material does not have to be handled. (*see* **Note 25**).

1. Dry all the samples and standards under a stream of nitrogen in Pyrex tubes (*see* **Note 24**). Add 1 mL 15% (w/v). tetrabutylammonium (TBA) hydroxide. Ensure tubes are sealed with the PTFE-lined caps.
2. Incubate at 105°C for 18 h. The TBA salts thus prepared are stable for months at 4°C.
3. Cool the TBA salt solutions to 25°C.

4. Add 2 mL dichloromethane and 50 μL pentafluorobenzylbromide (PFB). Mix well for 18 h using a rotatory mixer to obtain PFB esters of all the fatty acids (*see* **Note 25**).

5. Allow to settle or centrifuge at $250 \times g$ for 2 min to separate the layers.

6. Remove the lower layer into a clean tube and dry down under a stream of nitrogen at 40°C. Discard the upper layer.

7. Dissolve the material in 1 mL diethyl ether.

8. Wash twice to remove salts and side-products by adding 2 mL 0.1 M HCl then mixing for 5 min using a rotatory mixer. Centrifuge at $250 \times g$ for 2 min and remove the lower layer.

9. After the second wash, transfer the upper layer into a polypropylene syringe fitted with a 4-mm-diameter PTFE HPLC syringe filter (*see* **Note 26**). Filter into a clean PTFE-capped Pyrex tube.

10. Measure the volume of the filtrate.

11. Place a sample (5%) into a scintillation vial. Dry the sample in the scintillation vial under a stream of nitrogen at 40°C. Add scintillation fluid and perform scintillation counting on the vial (*see* **Note 13**).

12. Dry the bulk of the filtrate in the PTFE-capped Pyrex tube under a stream of nitrogen at 40°C.

13. Reconstitute in 100 μL acetonitrile/1,4-dioxan (7:3) (*see* **Note 27**).

14. Inject 20-μL portions onto the HPLC column. Adjust these volumes so each radioactive peak contains ~10,000 dpm or more. Complex mixtures will need at least 100,000 dpm per injection, whereas for simpler mixtures, 40,000 dpm per injection is sufficient. This is likely to be an iterative process so follow the order of HPLC injections in items 15 to 18 below:

15. First, determine the retention times of peaks (*see* **Note 28**) detected by UV-absorbance for the HPLC system for each of the fatty acid–PFB standards (*see* **Note 27**).

16. Second, use the radioactive standard(s) to determine the retention times of peaks detected by the radioactivity detector. Determine a correction factor if the retention times differ for the same fatty acyl-PFB ester through the two detectors (*see* **Note 27**).

17. Use the radioactive standard(s) to calibrate the peak area determined by the radioactivity detector against the quantity of radioactivity injected.

18. Finally, inject the samples, together with radioactive standards, and determine the radioactivity in each peak. Identify each peak from its retention time, using the correction factor determined in item 16 above, if necessary. It will probably be necessary to go back to item 14 and readjust volumes, then repeat item 18 as an iterative process.

5.4 Notes

1. Mid-log phase is recommended unless the experimental design requires otherwise. Optical density at 600 nm wavelength (OD_{600}) values of ~0.6 for a 1-cm light path are usually considered mid-log, but this should be determined for the conditions and strains being

used. Because light scattering is measured in suspensions of bacteria, the OD values do not obey Beer's law, so readings obtained from the same suspension may vary from one instrument to another. For log-phase cultures, studies should be done before the OD_{600} is half the value at the end of log phase. After this, the proportion of bacteria ceasing to divide becomes increasingly significant and the culture becomes heterogenous, with respect to the metabolic state of the bacteria.

2. The culture volume of 100 mL gives good labeling in this protocol, with specific radio-activities high enough for TLC and HPLC analysis. There will usually be enough for 4 to 12 analyses. For long incubations (e.g., 24 h), it may be possible to use less radiolabeled material (5 to 10 µCi). For pulse-chase experiments, 2 h is the maximum pulse time for slow-growing mycobacteria. Shorter times are acceptable if sufficient label is incorporated. It is important to ensure specific radioactivities are high so as not to have to load too much mass of material to achieve the required loading of radioactivity. This problem is likely to arise if all the material in a fraction has to be loaded on a single TLC plate or HPLC injection. In this case, it would also be difficult to dissolve all the material. If smaller volumes than 100 mL are desired to be handled, the best alternative is to pellet the bacteria and resuspend them in a small volume of medium or buffer. However, pelleting and resuspension do perturb the metabolism of the bacteria.

3. For efficient labeling of methyl branched fatty acids, and lipids that contain them (e.g., phthiocerol dimycocerosate, glycosylphenolphthiocerol dimycocerosate, trehalose mycolipenates and related tetra-acylated trehaloses and sulfolipid), use [^{14}C]-propionate. For efficient labeling of straight-chain fatty acids and mycolic acids, and lipids that contain them (e.g. acylglycerols, free fatty acids, trehalose mycolates including cord factor, and phospholipids and their mannosides), use [^{14}C]-acetate. [^{14}C]-Glucose is useful for labeling glycolipids, but avoid [1-^{14}C] glucose as the labelled carbon is easily lost when pentoses (e.g., arabinose) are formed.

4. Airtight culture vessels should be used and opened in a fume hood or, if they contain pathogens, a safety cabinet. On opening, vent any $^{14}CO_2$ formed. If the vessels are not airtight, a CO_2 trap should be included: either a hyamine hydroxide–soaked filter in the cap arrangement or 1 M NaOH in a side arm if a flask is used. If, after monitoring, it is clear that $^{14}CO_2$ is not released, it may be permissible to relax these precautions. If biohazard waste (from pathogens) is mixed with radioactivity, solid waste should be autoclaved as nonradioactive waste, then disposed of by specially trained staff: liquid waste should be decanted into glass, safe-break bottles, double wrapped in autoclave bags, both lightly sealed, autoclaved, then disposed of by specially trained staff.

5. Is it worth analyzing culture supernatant? This issue was raised by Converse et al. [2]. It is true that there is little radioactivity in the culture filtrate using this protocol. If the fraction only represents shed surface lipids, it may not be worth the trouble. My suggestion is to analyze the culture filtrate the first time you use this protocol and decide, on the basis of your own research interest, whether to do it in future.

6. A decision has to be taken about whether to use detergent in the culture medium. If the culture filtrate is to be analyzed, it should not be used, as it will prove almost impossible to separate the large amount of detergent from the small amount of lipid. Consider using tyloxapol (0.025% w/v is recommended) in place of Tween detergents, which are fatty acyl esters and can be used as a carbon source.

7. This protocol provides enough material for two to six samples from the chase part of the experiment: if every lipid is to be analyzed, take only two or three samples; if the focus is one or two lipids, there may be enough lipid labeled to take four to six samples. Take the last sample around the generation time: ~20 h for *M. tuberculosis*. The pulse stage should be proportional to the mean generation time so that most of the metabolism subsequent to accumulation of the substrate and its initial incorporation occurs during the chase phase. As the pulse should not exceed 2 h in *M. tuberculosis*, it follows that it should be 10 to 30 min for *M. smegmatis* and other fast-growing mycobacteria.

8. Cell wall components other than lipids will be labeled in this protocol, notably lipo-glycans such as lipoarabinomannan and cell wall glycans (*see* Chapter 3).

9. In designing the experiment, think if it is really necessary to prepare a surface lipid fraction. The location of mycobacterial lipids has been substantially worked out [23, 24], so, in general, it is only necessary to investigate their location if the mechanism of location is under study. If surface lipids are not to be analyzed separately, do not add the decylamine in hexane. Instead, sterilize the suspension of bacteria in Milli-Q water in PTFE-capped Pyrex tubes directly, by heating at 80°C for 30 min and proceed directly to Section 5.3.2.2.

10. Decylamine is added to ensure efficient release of sulfolipid [25] and has been used recently to show the requirement for the *mmpL8* gene for transfer of this lipid to surface layers [2]. The author has validated this method (1) using a *mmpL7* mutant and (2) in typical strains showing phthiocerol dimycocerosate (PDIM) in the surface lipid whereas menaquinone is not shown. So no, or no more than a trace of menaquinone should appear in the surface lipid fraction prepared this way. If it does, you are being too vigorous. The opposite applies with PDIM—this method should be extracting 90% or more: if it is not, be more vigorous with the mixing.

11. Important safety issue: Only use sealed centrifuge buckets capable of holding flammable liquid and live bacteria.

12. *Mycobacterium bovis* and *M. tuberculosis* can be killed by heating in a water bath at 80°C for 30 min or 95°C for 5 min. You should establish the efficacy of this yourself—especially if working with other mycobacteria. Killing the bacteria first is critical because the sterile material will be freeze-dried to determine its dry weight, a process that could generate aerosols. Preweigh the tubes empty so the dried bacteria do not have to be handled.

13. The lipid solutions in scintillation vials should be dried because chloroform is a powerful quenching agent. It is best to use a water-intolerant scintillation fluid unless water is used. Ensure quench correction is used and counting efficiency is determined so that dpm values are obtained to get true quantitative results.

14. The volumes given are for 50 mg dry weight bacteria. Keep these if the dry weight is less, but scale up if the dry weight is more than 50 mg.

15. If this fraction is very hazy, filter through a 13-mm PTFE filter to clarify it before drying it down.

16. The addition of the chloroform and 0.3% NaCl solution at this step causes the material to separate into two phases for the first time.

17. At this point, the method described here is different from that described in Ref. 20, which could be used as an alternative.

18. None of this material contains any cell wall–bound mycolate.

19. Dry plates to be developed more than once with a stream of warm air from a hair dryer between each solvent run: use fresh solvent for each run.

20. Smearing is more of a problem with the more polar systems in Table 5.1, so it can be possible to develop a satisfactory TLC plate for PDIMs (waxes, apolar molecules) but obtain an uninterpretable smear for the more polar trehalose mycolates and sulfolipids (Fig. 5.3) with exactly the same loading. It is possible to load as little as 10,000 dpm to a TLC plate and analyze the labeling pattern by increasing the exposure time (e.g., 7 to 10 days using Kodak Biomax MR film). A PhosphorImager would be much faster.

21. Dpm per culture or relative dpm into each lipid are limited in that they cannot be compared between cultures. Thus, dpm/mg (dry weight) bacteria is recommended. Alternatively, dpm per OD_{600} unit is acceptable but requires OD_{600} to be proportional to the weight of bacteria to be precisely quantitative. Dpm and μCi have been used throughout as the measure of radioactivity. This is convenient, but Bq could be used instead: 1 Bq = 1 dps (i.e., 60 dpm); 1 μCi = 2,200,000 dpm = 37 kBq (37,000 dps).

22. The dynamic range using autoradiographs is so low ($\sim$100) that the radioactivity in dense spots is likely to be underestimated. Software designed for a PhosphorImager

cannot be used satisfactorily to identify saturated spots in autoradiographs. In this case, scrape spots into a vial and perform scintillation counting, at least on the densest spots. For the less dense spots, readouts are proportional to the radioactivity.

23. The fatty acyl groups must first be released from the starting material, which can be individual lipids, lipid fractions, or even labeled bacteria (Fig. 5.2).

24. The entire procedure should also be carried out with a series of unlabeled standards of representative chain length. A useful range is 16:0, 18:0, 20:0, 24:0, 30:0 (saturated) plus 18:1 and 24:1 (monounsaturated fatty acids), which are available commercially. At least one [14C]-labeled fatty acid standard must also be used: 18:0, 18:1, or 24:0 are the most useful.

25. If GC is used [12, 15, 26], substitute 100 µL methyl iodide instead of the pentafluorobenzyl bromide to make fatty acyl methyl esters. If a mass spectrometer is used [12, 26] to detect the fatty acyl derivative ions, ask the operator which derivative to make—either is usable, but the sensitivity is very high for PFB esters. The method described here will also produce PFB esters of mycolic acids: these are best washed off the HPLC column with 1,4-dioxan.

26. The filtration step, essentially to protect the HPLC microbore tubing from particulates being injected into it, is done now. As the volume of the sample is high at this stage, any losses of sample are minimized.

27. Routinely, materials for HPLC analysis are dissolved in the starting mobile phase. However, the starting mobile phase in this method is acetonitrile, which does not completely dissolve the material described here, whereas the acetonitrile/dioxan mixture does. However, there is no increase in pressure when the dissolved material is applied, showing that precipitation does not occur, at least to the extent of interfering with the analysis. A gradient of 0 to 70% 1,4-dioxan in acetonitrile should be used with a flow rate of 2.2 mL/min. This can be achieved by filtering and degassing each of the solvents and mixing them in an HPLC gradient mixer. For a narrow range of fatty acyl esters (16:0 to 24:0), acetonitrile alone, as an isocratic system, may be used with a flow rate of 1.4 mL/min. Set the UV detector at 254 nm to detect the PBF esters. To detect radioactivity, mix with scintillation fluid pumped at twice the rate of the solvent and run through an in-line scintillation detector. At the end of each gradient, or for the isocratic system at the end of each day (or earlier if there is an increase in pressure), wash out the HPLC system in 1,4-dioxan. We use a Waters µBondapak C18 10-µm reversed phase column (3.9 mm diameter × 150 mm) with these solvent systems to resolve fatty acyl esters for fatty acids of 16 carbons or longer [22]. However, other reversed phase columns of the same dimensions could be used instead so long as the gradient and flow rate are optimized.

28. Standards are not generally available for fatty acids of more than 32 carbons or complex structures (e.g., the multimethylated mycocerosates), so these very-long-chain fatty acids are best detected by mass spectrometry [26, 27].

Acknowledgment P.R.W. was supported through DEFRA funding.

References

1. Camacho LR, Constant P, Raynaud C, Laneelle MA, Triccas JA, Gicquel B, Daffe M & Guilhot C (2001). Analysis of the phthiocerol dimycocerosate locus of *Mycobacterium tuberculosis*: evidence that this lipid is involved in the cell wall permeability barrier. J Biol Chem; 276: 19845–19854.
2. Converse SE, Mougous JD, Leavell MD, Leary JA, Bertozzi CR & Cox JS (2003). MmpL8 is required for sulfolipid-1 biosynthesis and *Mycobacterium tuberculosis* virulence. Proc Nat Acad Sci USA; 100: 6121–6126.

3. Cox JS, Chen B, McNeil M & Jacobs WR Jr (1999). Complex lipid determines tissue-specific replication of *Mycobacterium tuberculosis* in mice. Nature; 402: 79–83.
4. Minnikin DE, Kremer L, Dover LG & Besra GS (2002). The methyl-branched fortifications of *Mycobacterium tuberculosis*. Chem Biol; 9: 545–553.
5. Mostowy S, Inwald J, Gordon S, Martin C, Warren R, Kremer K, Cousins D & Behr MA (2005). Revisiting the evolution of *Mycobacterium bovis*. J Bacteriol; 187: 6386–6395.
6. Schnappinger D, Ehrt S, Voskuil MI, Liu Y, Mangan JA, Monahan I, Dolganov G, Efron B, Butcher PD, Nathan C & Schoolnik GK (2003). Transcriptional adaptation of *Mycobacterium tuberculosis* within macrophages: insights into the phagosomal environment. J Exp Med; 198: 693–704.
7. Bloch K (1975). Fatty acid synthases from *Mycobacterium phlei*. Methods Enzymol; 35: 84–90.
8. Wood WI & Peterson DO (1981). Fatty acid synthase from *Mycobacterium smegmatis*. Methods Enzymol; 71: 110–116.
9. Wheeler PR, Bulmer K & Ratledge C (1990). Enzymes for biosynthesis de novo and elongation of fatty acids in mycobacteria grown in host cells: is *Mycobacterium leprae* competent in fatty acid biosynthesis? J Gen Microbiol; 136: 211–217.
10. Bloch K (1977). Control mechanisms for fatty acid synthesis in *Mycobacterium smegmatis*. Adv Enzymol Relat Areas Mol Biol; 45: 1–84.
11. Fernandes ND & Kolattukudy PE (1998). A newly identified methyl-branched chain fatty acid synthesizing enzyme from *Mycobacterium tuberculosis* var. bovis BCG. J Biol Chem; 273: 2820–2828.
12. Rainwater DL & Kolattukudy PE (1983). Synthesis of mycocerosic acids from methyl-malonyl coenzyme A by cell-free extracts of *Mycobacterium tuberculosis* var. bovis BCG. J Biol Chem; 258: 2979–2985.
13. Trivedi OA, Arora P, Vats A, Tickoo R, Sridharan V, Mohanty D & Gokhale RS (2005). Dissecting the mechanism and assembly of a complex virulence mycobacterial lipid. Mol Cell; 17: 631–643.
14. Slayden RA, Lee RE, Armour JW, Cooper AM, Orme IM, Brennan PJ & Besra GS (1996). Antimycobacterial action of thiolactomycin: an inhibitor of fatty acid and mycolic acid synthesis. Antimicrob Agents Chemother; 40: 2813–2819.
15. Wheeler PR, Besra GS, Minnikin DE & Ratledge C (1993). Stimulation of mycolic acid biosynthesis by incorporation of cis-tetracos-5-enoic acid in a cell-wall preparation from *Mycobacterium smegmatis*. Biochim Biophys Acta; 1167: 182–188.
16. Lacave C, Quemard A & Laneelle G (1990). Cell-free synthesis of mycolic acids in *Mycobacterium aurum*: radioactivity distribution in newly synthesized acids and presence of cell wall in the system. Biochim Biophys Acta; 1045: 58–68.
17. Salman M, Lonsdale JT, Besra GS & Brennan PJ (1999). Phosphatidylinositol synthesis in mycobacteria. Biochim Biophys Acta; 1436: 437–450.
18. Banerjee A, Dubnau E, Quemard A, Balasubramanian V, Um KS, Wilson T, Collins D, de Lisle G & Jacobs WR Jr. (1994). InhA, a gene encoding a target for isoniazid and ethionamide in *Mycobacterium tuberculosis*. Science; 263: 227–200.
19. Salman M, Brennan PJ & Lonsdale, JT (1999). Synthesis of mycolic acids of mycobacteria: an assessment of the cell-free system in light of the whole genome. Biochim Biophys Acta; 1437: 325–332.
20. Besra GS (1998). Preparation of cell-wall fractions from mycobacteria. Parish T & Stoker N, eds. Mycobacteria protocols, vol. 101. Humana Press, Totowa, NJ: 91–107.
21. Dobson G, Minnikin DE, Minnikin SM, Parlett JH & Goodfellow M (1985). Systematic analysis of complex mycobacterial lipids. Goodfellow M & Minnikin DE, eds. Chemical methods in bacterial systematics. Academic Press, London: 207–265.
22. Wheeler PR & Anderson PM (1996). Determination of the primary target for isoniazid in mycobacterial mycolic acid biosynthesis with *Mycobacterium aurum* A+. Biochem J; 318: 451–457.

23. Kremer L & Besra GS (2005). A waxy tale, by *Mycobacterium tuberculosis*. Cole ST, Eisenach KD, McMurray DN & Jacobs WR, eds. Tuberculosis and the tubercle bacillus. ASM Press, Washington, DC: 287–307.
24. Draper P & Daffe M (2005). The cell envelope of *Mycobacterium tuberculosis* with special reference to the capsule and outer permeability barrier. Cole ST, Eisenach KD, McMurray DN & Jacobs WR, eds. Tuberculosis and the tubercle bacillus. ASM Press, Washington, DC: 261–274.
25. Goren MB (1970). Sulfolipid I of *Mycobacterium tuberculosis* strain H37Rv. Purification and properties. Biochim Biophys Acta; 210: 116–126.
26. Sirakova TD, Thirumala AK, Dubey VS, Sprecher H & Kolattukudy PE (2001). The *Mycobacterium tuberculosis* pks2 gene encodes the synthase for the hepta- and octa-methyl-branched fatty acids required for sulfolipid synthesis. J Biol Chem; 276: 16833–16839.
27. Qureshi N, Takayama K & Schnoes HK (1980). Purification of C30-56 fatty acids from *Mycobacterium tuberculosis* H37Ra. J Biol Chem; 255: 182–189.

Chapter 6
Whole Genome Analysis Using Microarrays

S.J. Waddell, J. Hinds and P.D. Butcher

Abstract The development of microarray technology has allowed the genomes of mycobacteria to be directly compared to identify DNA regions that differ between strains due to deletion, insertion, or sequence divergence. The use of microarrays in comparative genomics has proved to be a valuable tool for comparing both mycobacterial species and strains. We describe here the methodology for comparing two mycobacterial DNA samples by microarray hybridization, from labeling and slide preparation, to DNA microarray analysis options. Further developments in microarray design and methodology promise to ensure that microarrays remain an important resource for comparative genomic studies in the future.

Keywords comparative genomics · deletion analysis · hybridization · microarray · *Mycobacterium*

6.1 Introduction

DNA microarrays have become an essential tool in the study of mycobacterial comparative genomics. We describe here the preparation of labeled DNA samples and the hybridization, image acquisition, and simple analysis of microarrays to identify differences in mycobacterial genome content. The sequencing projects of *Mycobacterium tuberculosis* H37Rv [1] and more recently *M. tuberculosis* CDC1551 [2] and *Mycobacterium bovis* [3] have allowed differences in mycobacterial DNA to be examined at the nucleotide level on a genome-wide scale. The application of PCR product and oligonucleotide based microarrays [4] has enabled this whole genome approach to be extended to compare multiple genomes without genome sequencing. The underlying principle is that genomic differences may be identified by hybridizing DNA from an unknown

S.J. Waddell

Medical Microbiology, Division of Cellular & Molecular Medicine, St. George's University of London, Cranmer Terrace, Tooting, London, SW17 0RE, UK

e-mail: swaddell@sgul.ac.uk

T. Parish, A.C. Brown (eds.), *Mycobacteria Protocols,*
doi: 10.1007/978-1-59745-207-6_6, © Humana Press, Totowa, NJ 2008

mycobacterial strain to a microarray representing the DNA content of a sequenced reference strain. DNA sequences on the array that are not hybridized with labeled genomic DNA from the unknown sample are hypothesized to be absent or highly divergent in the unknown mycobacterial genome. To perform a direct comparison and build in an internal control, this analysis is commonly performed in a competitive reaction by cohybridizing labeled DNA from the unknown strain with differently labeled DNA from the reference strain.

The microarray platform used for comparative genomics directly determines the information gained from these studies. The genomic content represented on the microarray is important, as clearly presence/absence calls can only be made for genes that are represented on the array. Therefore, approaches to ensure the array represents multiple mycobacterial strains/ species are useful to interrogate a gene pool that ideally includes all genes, in all strains, so that all genomic differences can be detected. The *M. tuberculosis* complex microarray used at St. George's, University of London (ArrayExpress Accession: A-BUGS-23), was designed and generated to represent the sequenced genomes of *M. tuberculosis* strains H37Rv and CDC1551 and *M. bovis* 2122/97 [5, 6]. The precise nature of the genomic differences observed from microarray analysis is also dependent on the microarray platform with regard to the type and number of reporter elements used. Single nucleotide sequence differences are more likely to be detected by short oligonucleotide reporters ($\sim$20 bp) than by long oligonucleotides (50 to 70 bp) or PCR products (100 to 1000 bp), which are more permissive of an increasing degree of sequence divergence. Likewise, determining the presence/absence of short regions of DNA will be improved by a higher resolution of reporters tiled along the genome compared with reporters every few hundred bases or one reporter per gene as is typical for most array platforms designed for gene expression analysis. The protocol detailed here is based on the analysis of mycobacterial DNA on microarrays generated by printing gene-specific PCR products (or oligonucleotides) onto poly-lysine or amino-silane coated glass slides. For more information on printing and postprocessing PCR-based microarrays for *M. tuberculosis*, see Wilson et al. [7].

Microarray analysis has been used as a tool in mycobacterial molecular epidemiology, comparing genomic deletions acquired in *M. tuberculosis* clinical strains [8] and investigating outbreak strains [9]. Comparison of mycobacterial genomes by microarray has also identified genes that may influence the pathogenicity or host-specificity of mycobacteria, for example the RD regions missing in *M. bovis* BCG v *M. tuberculosis* H37Rv [10] or *Mycobacterium microti* v *M. tuberculosis* H37Rv [11]. The identification of differences in genome sequence may also aid in the discovery of novel diagnostics for *M. tuberculosis* by distinguishing unique patterns of DNA deletion [12, 13]. Indeed, investigation of a series of genomic deletions has led to a revised evolutionary scenario for the *M. tuberculosis* complex [14].

6.2 Materials

6.2.1 DNA Labeling and Purification

1. Random primers (3 μg/μL) (Invitrogen, Carlsbad, USA).
2. dNTPs (5 mM dATP/dGTP/dTTP, 2 mM dCTP) (Invitrogen).
3. Cy3-dCTP 25 nmol (GE Healthcare Chalfont St. Giles, UK) light sensitive. Store at –20°C.
4. Cy5-dCTP 25 nmol (GE Healthcare) light sensitive. Store at –20°C.
5. 10x REact 2 buffer (Invitrogen).
6. DNA polymerase I Large Fragment (Klenow) (3 to 9 U/μL) (Invitrogen).
7. DNase-free water (Sigma, Saint Louis, USA).
8. Mycobacterial DNA (RNase-treated, phenol/chloroform free) 1 to 5 μg test DNA/hybridization.
9. MinElute PCR purification kit (Qiagen, Hilden, Germany).
10. 20x SSC (saline sodium citrate) (Sigma), 0.2 μm filter sterilized.
11. 20% (w/v) SDS (sodium dodecyl sulfate) (Flowgen Bioscience, Wilford, UK), filter sterilized.

6.2.2 Prehybridization of Microarray Slide

1. 0.2-μm filtered 20% (w/v) SDS (sodium dodecyl sulfate).
2. 0.2-μm filtered 20x SSC (saline sodium citrate).
3. Bovine serum albumin (BSA) fraction V 96% to 99%: 100 mg/mL, 0.2 μm filter sterilized (Sigma).
4. Prehybridization buffer: 3.5x SSC, 0.1% SDS, 10 mg/mL BSA.
5. Coplin staining jar (Fisher Scientific, Pittsburgh, USA).
6. Slide staining troughs (Raymond A. Lamb, London, UK).
7. Slide staining racks (Raymond A. Lamb).
8. 50-mL Falcon tubes (Fisher Scientific).
9. Propan-2-ol (VWR, West Chester, USA).

6.2.3 Hybridization and Slide Washing

1. Hybridization chamber II (Corning, Corning, USA), Hybridization cassette (TeleChem International, Sunnyvale, USA), or suitable alternative.
2. DNase-free water, 0.2 μm filter sterilized.
3. 22 × 22 mm LifterSlip (Erie Scientific, Portsmouth, USA).
4. Water bath set at 65°C.
5. Wash A: 1x SSC, 0.05% SDS.
6. Wash B: 0.06x SSC.

6.2.4 Microarray Scanning, Data Acquisition, and Analysis

1. Dual laser microarray scanner such as Affymetrix 418/428 (MWG Biotech, Ebersberg, Germany), Axon GenePix (Molecular Devices, Sunnyvale, USA), Scanarray (PerkinElmer, Waltham, USA).
2. Image analysis software such as ImaGene (BioDiscovery, El Segundo, USA), GenePix Pro (Molecular Devices), Quantarray (PerkinElmer), Blue-Fuse for Microarrays (BlueGnome, Cambridge, UK).
3. Microarray analysis packages such as GeneSpring GX (Agilent Technologies, Santa Clara, USA), GeneSight (BioDiscovery), Rosetta Resolver (Agilent Technologies), BlueFuse for Microarrays (BlueGnome), GACK [15].

6.3 Methods

The microarray strategy adopted is largely dependent on the aims and scale of the project. We would recommend using a common reference DNA in one of the Cy3/5 channels if multiple DNA samples are to be compared (*see* **Note 1**). This provides a more flexible experimental design strategy because samples can be added to the analysis easily and results may be compared across experiments using the same reference source. Additionally, the analysis of multiple mycobacterial DNA samples is simpler to interpret when compared with the same reference DNA. This strategy, however, may require additional microarray hybridizations compared with a direct hybridization strategy and a supply of suitable reference DNA. We would recommend repeating each hybridization at least 3 times if significance testing is to be applied to the microarray data set; this may not be necessary, however, for large sample comparisons or if confirmatory PCR or sequencing is performed.

6.3.1 DNA Labeling and Purification

1. Prepare one Cy3 and one Cy5 sample per microarray hybridization. Add 3 μg random primers to 1 to 5 μg DNA in a final volume of 41.5 μL (made up to the correct volume with DNase-free water) (*see* **Notes 2, 3** and **4**).
2. Heat to 95°C for 5 min, snap cool on ice, and centrifuge briefly.
3. Add 8.5 μL master mix containing 5 μL 10x REact 2 buffer, 1 μL dNTPs, 1.5 μL Cy3-dCTP or Cy5-dCTP, and 1 μL DNA polymerase I large fragment (Klenow) (*see* **Note 5**).
4. Mix and incubate at 37°C for 90 min (*see* **Note 6**).
5. Centrifuge briefly, combine Cy3 and Cy5 labeled DNA samples in a single tube (*see* **Note 7**). Add 500 μL Buffer PB (from MinElute kit) to each sample.
6. Mix and apply to MinElute column and centrifuge at 13,000 × *g* for 1 min (*see* **Note 8**).

7. Discard flow-through and place column back into the same collection tube.
8. To wash, add 500 μL Buffer PE to MinElute column and centrifuge for 1 min (at 13,000 × g), discard flow-through, and place column back into the same collection tube.
9. Add 250 μL Buffer PE to the column and centrifuge at 13,000 × g for 1 min, discard flow-through, and place MinElute column back into the same collection tube.
10. Centrifuge at 13,000 × g for an additional minute to remove residual wash buffer, and place MinElute column into a fresh 1.5-mL tube
11. Add 30.2 μL DNase-free water to the column membrane; incubate at room temperature for 1 min, then centrifuge at 13,000 × g for 1 min (see **Note 9**).
12. Add 9 μL 20x SSC and 6.8 μL 2% SDS to the labeled DNA samples (final concentration of 4x SSC and 0.3% SDS), mix thoroughly (see **Notes 9** and **10**).
13. Heat samples at 95°C for 2 min. Allow to cool slowly (approximately 2 to 5 min, do not place on ice), centrifuge briefly, and store in the dark while the hybridization chambers are set up (see **Note 6**).

6.3.2 Prehybridization of Microarray Slide

1. Mix 50 mL prehybridization solution in a Coplin jar and incubate at 65°C for 20 to 30 min to equilibrate (this can be performed while the DNA samples are labeling).
2. Incubate the microarray slide in the preheated prehybridization buffer at 65°C for 20 min.
3. Place slide in slide rack and rinse in 500 mL water in a staining trough for 1 min with agitation.
4. Rinse slide in 500 mL propan-2-ol for 1 min with agitation (see **Note 11**).
5. Place slide into 50-mL Falcon tube and centrifuge at 1500 × g for 5 min to dry (see **Note 12**).
6. Store in dark, dust free box until hybridization.

6.3.3 Hybridization and Slide Washing

1. Place the prehybridized microarray slides (from Section 6.3.2) into the hybridization cassettes; add two 15-μL aliquots of 0.2 μm filtered DNase-free water to the wells in each cassette (see **Note 13**).
2. Carefully place LifterSlip over the printed area of the microarray slide, ensuring that the LifterSlip is the correct way up and is not scratched or dusty (see **Note 14**).
3. Slowly and steadily pipette the hybridization sample (from Section 6.3.1) underneath the LifterSlip, the solution should be drawn evenly under the

LifterSlip by capillary action; pipette any excess to top and bottom of LifterSlip edges (*see* **Note 9**).

4. Seal the hybridization cassette, submerge immediately in water bath at 65°C, incubate overnight (16 to 20 h) (*see* **Note 15**).
5. Preheat 500 mL Wash A buffer and staining trough to 65°C (*see* **Note 16**).
6. Remove slide from hybridization cassette, immediately dip slide into Wash A, allow the coverslip to fall off in the buffer, then place slide into a slide rack submerged in Wash A in the staining trough (*see* **Notes 17** and **18**).
7. Agitate in Wash A for 2 min (*see* **Notes 17** and **18**).
8. Agitate in 500 mL Wash B (at room temperature) for 2 min. Transfer into a second staining trough of 500 mL Wash B, and agitate for a further 2 min (*see* **Notes 17** and **18**).
9. Centrifuge in 50-mL Falcon tubes at 1500 × *g* for 5 min (*see* **Note 12**).
10. Carefully place slides into a dust-free slide box, store in the dark, scan immediately (*see* **Note 6**).

6.3.4 Microarray Scanning, Data Acquisition, and Analysis

1. Scan microarray sequentially at 532 nm and 635 nm corresponding with Cy3 and Cy5 excitation maxima using a dual laser microarray scanner. Microarrays should be scanned to achieve the best dynamic range (*see* **Notes 18** and **19**).
2. Derive comparative spot intensities and apply flagging algorithms using image analysis software such as ImaGene, GenePix, Quantarray, or BlueFuse. Perform analysis to discriminate between present or absent DNA sequences using this software or import into further analysis packages (*see* **Note 20**).
3. The primary aim of the analysis methods employed for comparative genomics data is to identify genes that are present or absent in the strain of interest compared with the reference strain. The test strain signal intensities for each gene are divided by the reference strain signal intensities to give a ratio that when normalized will be equal to 1 for genes present in both test and reference strains, greater than 1 for genes present in the test strain only, and less than 1 for genes present in the reference strain only (*see* **Note 21** and Figure 6.1).

   Analysis approaches therefore apply cutoffs to the ratio data to identify genes present in both strains, genes present in the test strain but absent or highly divergent in the reference strain, and genes present in the reference strain but absent or highly divergent in the test strain. These cutoffs may be determined and applied using a number of methods (*see* **Note 22**); however, an analysis strategy that considers the distribution of data to set cutoffs and that also incorporates genome position is likely to be most informative. The dynamic fold-change analysis strategy uses the spread of the data distribution measured on a subset of genes considered present in all strains to

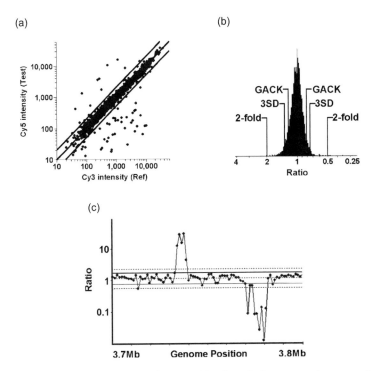

Fig. 6.1 Typical distribution of data from an *M. tuberculosis* comparative genomics study shown as **(a)** scatterplot of signal intensities for each channel with twofold cutoff lines detailed on the graph; **(b)** frequency distribution of ratios showing cutoffs for different analysis approaches; and **(c)** location analysis of the same *M. tuberculosis* comparative genomics data to demonstrate increased confidence in regions of difference between strains based on chromosome position

determine ratio cutoffs, typically set at 3 SD ± the median. This may be combined with location analysis where chromosome position is integrated with ratio-based cutoffs to improve the confidence in calling presence/absence of adjacent genes located in the same genomic region, as shown in Figure 6.1c (*see* **Note 21**).

6.4 Notes

1. In many microarray studies where multiple samples are to be compared, a common reference is used and the samples compared indirectly (Cy3 reference vs. Cy5 DNA1 or DNA2) rather than performing multiple direct comparisons (Cy3 DNA1 vs. Cy5 DNA2). The common reference should hybridize to all (or at least most) of the elements on the microarray (so accurate Cy3/5 ratios can be calculated). Thus a common reference may be genomic DNA derived from a reference strain such as H37Rv, the combined PCR products printed onto the slide (in the case of a PCR-based microarray), combined

genomic DNA from multiple strains, or amplified DNA. We have found that simply using *M. tuberculosis* H37Rv genomic DNA is acceptable in most instances. Stocks of *M. tuberculosis* genomic DNA may be requested from Colorado State University, TB Vaccine Testing and Research Materials (HHSN266200400091C).

2. Cy3 or Cy5 channels may be used as the reference channel (if using); we commonly use Cy3 for the DNA reference. A dye swap experiment (when the same samples are labeled with the alternative Cy-dye) may also be useful to ensure that any differences identified are not related to dye-dependent effects.

3. We regularly use 2 to 5 μg mycobacterial genomic DNA for each labeling reaction. DNA amplification may be required where genomic DNA is limited or the same common reference DNA is required for many microarray analyses (e.g., the REPLI-g system; Qiagen). We have not found it necessary to fragment/shear or digest genomic DNA before labeling, however this may be an option if hybridization signals are consistently weak. DNA quality may be assessed by gel electrophoresis and/or optical density 260:280 ratio. DNA samples should be RNA-free and not contain excessive salt or phenol:chloroform contaminants.

4. We generally use random primers to prime the labeling reactions; however, genome-directed primers designed against the mycobacterial genomes of interest is an alternative [16].

5. This protocol describes the direct labeling of DNA with Cy3/5-dCTP; dCTP was chosen as mycobacterial genomes are particularly GC rich. Indirect labeling methods have not been tested for use in comparative genomics in our laboratory. Labeling efficiency may be checked by spectrophotometer if problems arise (e.g., NanoDrop ND-1000; NanoDrop Technologies, Wilmington, USA).

6. The Cy-dyes are sensitive to light; therefore incubate and store tubes or hybridized slides in the dark. We also recommend using amber-colored tubes to help reduce this problem.

7. If using a common reference, combine the reference labeling reactions before redistributing into Cy3/5 reaction mixes; this will help to reduce array:array variation.

8. We use Qiagen MinElute columns for purification of the labeled samples; however, a number of similar commercially available columns are also suitable. Ethanol precipitation is also effective at removing unincorporated Cy-dye.

9. The hybridization volumes detailed in this protocol are for use with a 22 × 50 mm LifterSlip area; if using 22 × 22 mm LifterSlips, a final hybridization volume of around 23 μL is recommended (elute in 15.9 μL DNase-free water, add 3.5 μL SDS, 4.6 μL SSC). This volume should be adjusted down if flat coverslips are used. We have found that hybridizations using LifterSlips are more even, and there is less array:array and user:user variation compared with that of conventional flat coverslips.

10. Do not create a SSC/SDS hybridization master mix, as the SDS will begin to precipitate out. Pipette the SSC and SDS separately; mixing thoroughly between additions.

11. It is possible to recycle the propan-2-ol used to rinse the prehybridized slides; change every few months depending on frequency of use. Discard water used to wash slides after 8 slides, or each day.

12. Slides should be centrifuged immediately after washing to prevent any drying of the wash buffers onto the microarray surface. Slides should be placed microarray side down and label side down in the Falcon tubes for centrifugation to help prevent scratching of the array surface.

13. Water is added to the hybridization cassettes to retain the humidity and to prevent the hybridization solution from drying onto the slides.

14. Compressed air may be used to remove dust from slides, coverslips, and slide boxes. Take care not to expel vapor from the pressurized canister.

15. The hybridization temperature using this method is 65°C. Lower temperatures may however be required for microarray formats using a formamide-based system.

16. Prepare Wash A buffer the previous day and incubate overnight at 65°C with staining trough.

17. Performing the first wash at 65°C helps to remove unwanted background from the microarrays and retains the hybridization stringency through the first wash. When washing, repeatedly raise and lower slide rack in wash buffer keeping the slides immersed in buffer as far as possible. Be careful not to splash Wash A into Wash B containers, and clean or change gloves between washes to limit the transfer of SDS from Wash A to Wash B. If washing multiple microarray slides, leave the slides in the slide rack in Wash A while others are unpacked. We wash 4 microarrays at a time and recommend changing the Wash B buffers every 8 slides. Discard, ensuring that the coverslips left in the Wash A staining trough are disposed of appropriately.

18. High background in the Cy3 channel is likely to be due to excess salt or to drying of the slides between wash steps; high background in the Cy5 channel may be due to SDS coming though the washes. If Cy3 or Cy5 background remains consistently high, review washing process changing gloves and/or slide racks between washes, and/or adding an additional Wash B step.

19. A pre-scan may often be helpful to check correct slide orientation and define scan area. When scanning, adjust PMT or laser settings to scan just below saturation in each channel to maximize the dynamic range. Multiple scans using different PMT/laser settings may be useful if image analysis packages such as MAVI Pro.2.6.0 (MWG Biotech) are to be used. Avoid multiple test scans if possible as photobleaching may occur.

20. Microarray data should also be deposited into a MIAME-compliant (minimum information about a microarray experiment) database such as ArrayExpress (http://www.ebi.ac.uk/arrayexpress), GEO (http://www.ncbi.nlm.nih.gov/geo/), or CIBEX (http://cibex.nig.ac.jp/) soon after analysis for data sharing, further meta-analysis, and publication.

21. Figure 6.1 illustrates a typical distribution of *M. tuberculosis* comparative genomic data and the various cutoffs applied to determine presence/absence of genes in each strain. In the scatterplot, Figure 6.1a, genes highly divergent or absent in the reference strain compared with the test strain are located above the twofold cutoff diagonal line; this corresponds with the genes on the left side of Figure 6.1b and the first peak in Figure 6.1c. Genes absent or highly divergent in the test strain compared with the reference are situated below the twofold cutoff diagonal line in Figure 6.1a, to the right in Figure 6.1b, and in the second peak in Figure 6.1c.

22. The simplest analysis strategy uses a fixed fold-change cutoff; an arbitrary ratio cutoff, typically at twofold (ratio >2 or ratio <0.5), is applied to the microarray data. This method takes no account of the variance in the distribution of data. A second analysis method, probability of presence, uses GACK software [15] to model the data distribution to determine the expected probability of presence (EPP), which is then used as a cutoff; the default is binary output with EPP threshold at 50% (as shown in Figure 6.1).

Acknowledgments S.J.W. was funded by an EU STREP Sixth Framework Programme Priority (LHP-CT-2004-012187). S.J.W., J.H., and P.D.B. would like to thank Kate Gould for her critical reading of the manuscript and acknowledge the Wellcome Trust and its Functional Genomics Resources Initiative for funding the multicollaborative microbial pathogen microarray facility at St. George's (BµG@S).

References

1. Cole, S. T., Brosch, R., Parkhill, J., Garnier, T., Churcher, C., Harris, D., Gordon, S. V., Eiglmeier, K., Gas, S., Barry, C. E. III, Tekaia, F., Badcock, K., Basham, D., Brown, D., Chillingworth, T., Connor, R., Davies, R., Devlin, K., Feltwell, T., Gentles, S., Hamlin, N., Holroyd, S., Hornsby, T., Jagels, K., Krogh, A., McLean, J., Moule, S., Murphy, L., Oliver, K., Osborne, J., Quail, M. A., Rajandream, M. A., Rogers, J., Rutter, S.,

Seeger, K., Skelton, J., Squares, R., Squares, S., Sulston, J. E., Taylor, K., Whitehead, S., and Barrell, B. G. (1998) Deciphering the biology of *Mycobacterium tuberculosis* from the complete genome sequence. *Nature.* **393**, 537–544.

2. Fleischmann, R. D., Alland, D., Eisen, J. A., Carpenter, L., White, O., Peterson, J., DeBoy, R., Dodson, R., Gwinn, M., Haft, D., Hickey, E., Kolonay, J. F., Nelson, W. C., Umayam, L. A., Ermolaeva, M., Salzberg, S. L., Delcher, A., Utterback, T., Weidman, J., Khouri, H., Gill, J., Mikula, A., Bishai, W., Jacobs, W. R. Jr., Venter, J. C., and Fraser, C. M. (2002) Whole-genome comparison of *Mycobacterium tuberculosis* clinical and laboratory strains. *J Bacteriol.* **184**, 5479–5490.

3. Garnier, T., Eiglmeier, K., Camus, J. C., Medina, N., Mansoor, H., Pryor, M., Duthoy, S., Grondin, S., Lacroix, C., Monsempe, C., Simon, S., Harris, B., Atkin, R., Doggett, J., Mayes, R., Keating, L., Wheeler, P. R., Parkhill, J., Barrell, B. G., Cole, S. T., Gordon, S. V., and Hewinson, R. G. (2003) .The complete genome sequence of *Mycobacterium bovis*. *Proc Natl Acad Sci USA.* **100**, 7877–7882.

4. Schena, M., Shalon, D., Davis, R. W., and Brown, P. O. (1995) Quantitative monitoring of gene expression patterns with a complementary DNA microarray. *Science.* **270**, 467–470.

5. Hinds, J., Witney, A. A., and Vass, J. K. (2002) Microarray design for bacterial genomes, *Methods in Microbiology: Functional Microbial Genomics* (Wren, B. W. and Dorrell, N. eds.), Academic Press, London, pp. 67–82.

6. Hinds, J., Laing, K. G., Mangan, J. A., and Butcher, P. D. (2002) Glass slide microarrays for bacterial genomes, *Methods in Microbiology: Functional Microbial Genomics* (Wren, B. W. and Dorrell, N. eds.), Academic Press, London, pp. 83–99.

7. Wilson, M., Voskuil, M., Schnappinger, D., and Schoolnik, G. K. (2001) Functional genomics of *Mycobacterium tuberculosis* using DNA microarrays, *Methods in Molecular Medicine, vol. 54: Mycobacterium tuberculosis Protocols* (Parish, T. and Stoker, N. G. eds.), Humana Press Inc., Totowa, NJ, pp. 335–357.

8. Tsolaki, A. G., Hirsh, A. E., DeRiemer, K., Enciso, J. A., Wong, M. Z., Hannan, M., Goguet de la Salmoniere, Y. O., Aman, K., Kato-Maeda, M., and Small, P. M. (2004) Functional and evolutionary genomics of *Mycobacterium tuberculosis*: insights from genomic deletions in 100 strains. *Proc Natl Acad Sci USA.* **101**, 4865–4870.

9. Rajakumar, K., Shafi, J., Smith, R. J., Stabler, R. A., Andrew, P. W., Modha, D., Bryant, G., Monk, P., Hinds, J., Butcher, P. D., and Barer, M. R. (2004) Use of genome level-informed PCR as a new investigational approach for analysis of outbreak-associated *Mycobacterium tuberculosis* isolates. *J Clin Microbiol.* **42**, 1890–1896.

10. Behr, M. A., Wilson, M. A., Gill, W. P., Salamon, H., Schoolnik, G. K., Rane, S., and Small, P. M. (1999) Comparative genomics of BCG vaccines by whole-genome DNA microarray. *Science.* **284**, 1520–1523.

11. Garcia-Pelayo, M. C., Caimi, K. C., Inwald, J. K., Hinds, J., Bigi, F., Romano, M. I., van Soolingen, D., Hewinson, R. G., Cataldi, A., and Gordon, S. V. (2004) Microarray analysis of *Mycobacterium microti* reveals deletion of genes encoding PE-PPE proteins and ESAT-6 family antigens. *Tuberculosis (Edinb).* **84**, 159–166.

12. Cockle, P. J., Gordon, S. V., Lalvani, A., Buddle, B. M., Hewinson, R. G., and Vordermeier, H. M. (2002) Identification of novel *Mycobacterium tuberculosis* antigens with potential as diagnostic reagents or subunit vaccine candidates by comparative genomics. *Infect Immun.* **70**, 6996–7003.

13. Warren, R. M., van Pittius, N. C., Barnard, M., Hesseling, A., Engelke, E., de Kock, M., Gutierrez, M. C., Chege, G. K., Victor, T. C., Hoal, E. G., and van Helden, P. D. (2006) Differentiation of *Mycobacterium tuberculosis* complex by PCR amplification of genomic regions of difference. *Int J Tuberc Lung Dis.* **10**, 818–822.

14. Brosch, R., Gordon, S. V., Marmiesse, M., Brodin, P., Buchrieser, C., Eiglmeier, K., Garnier, T., Gutierrez, C., Hewinson, G., Kremer, K., Parsons, L. M., Pym, A. S.,

Samper, S., van Soolingen, D., and Cole, S. T. (2002) A new evolutionary scenario for the *Mycobacterium tuberculosis* complex. *Proc Natl Acad Sci USA*. **99**, 3684–3689.

15. Kim, C. C., Joyce, E. A., Chan, K., and Falkow, S. (2002) Improved analytical methods for microarray-based genome-composition analysis. *Genome Biol*. **3**, RESEARCH0065.

16. Talaat, A. M., Hunter, P., and Johnston, S. A. (2000) Genome-directed primers for selective labeling of bacterial transcripts for DNA microarray analysis. *Nat Biotechnol*. **18**, 679–682.

Chapter 7
Use of DNA Microarrays to Study Global Patterns of Gene Expression

Roberta Provvedi, Giorgio Palù and Riccardo Manganelli

Abstract DNA microarray technology represents an extremely powerful tool to understand the biology of *Myobacterium tuberculosis* and its interaction with the host. It opens up the possibility of monitoring the expression level of thousands of genes in parallel, thus the ability to test the effect on global transcription of different experimental conditions. Whole genome microarrays consist either of PCR amplicons or oligonucleotides representing every open reading frame in a genome printed on a slide in a high-density matrix. The gene identity and position of each spot is known and can be tracked.

Transcription profiling experiments are designed to compare gene expression in bacteria exposed to two different conditions. The RNA from the two different cultures is extracted and reverse transcribed to obtain differentially labeled cDNA by incorporating dUTP or dCTP conjugated with either Cy5 or Cy3, two fluorophores able to emit fluorescence of two different wavelengths. Equal amounts of the two differentially labeled cDNA are mixed, applied to the array surface, and allowed to hybridize to the corresponding gene-specific target. The microarray is finally scanned to obtain two overlapping images each relative to the fluorescence emitted from each label. The images obtained are then analyzed by several software packages to identify and quantify the spots corresponding with the gene-specific probes. After image processing, the data are normalized and then analyzed to determine those genes whose differential expression between the two samples is statistically significant. However, the statistical analysis of microarray data alone is not usually considered enough to confirm differential expression of a gene, and validation with an independent technique, such as quantitative RT-PCR, is required.

Keywords DNA microarrays · global regulation of gene expression · transcriptomics

R. Manganelli
Department of Histology, Microbiology and Medical Biotechnologies, University of Padova, Via Gabelli 63, 35121 Padova, Italy
e-mail: riccardo.manganelli@unipd.it

T. Parish, A.C. Brown (eds.), *Mycobacteria Protocols*,
doi: 10.1007/978-1-59745-207-6_7, © Humana Press, Totowa, NJ 2008

7.1 Introduction

The genome of *Mycobacterium tuberculosis* measures 4.4 Mb and is the largest, known to date, among obligate human pathogens and intracellular bacteria [1]. It encodes approximately 190 regulatory proteins, including 13 sigma factors, 11 two-component systems, five unpaired response regulators, two unpaired histidine kinases, 11 protein kinases, and more than 140 other putative transcription regulators [2]. Moreover, the ratio between the number of alternative sigma factors and genome size is higher in *M. tuberculosis* than in the other obligate human pathogens, suggesting that gene expression modulation plays an important role in the biology of this bacterium [3].

DNA microarray technology opens up the possibility of monitoring the expression level of thousands of genes in parallel, thus the ability to test the effect on global transcription of different experimental conditions (e.g., those mimicking host environment or drug treatment) or to help elucidate the role of transcriptional regulators on bacterial physiology, and represents an extremely powerful tool to understand the biology of *M. tuberculosis* and its interaction with the host [4].

7.1.1 DNA Microarray Technology

Whole genome microarrays consist either of PCR amplicons or oligonucleotides representing every open reading frame in a genome, printed on a slide in a high-density matrix in a way that the identity and position of each spot is known and can be tracked.

Transcription profiling experiments are designed to compare the transcription profile in bacteria exposed to two different conditions, for instance bacterial cultures exposed to a drug or not exposed to it. The RNA from the two different cultures is extracted and reverse transcribed to obtain differentially labeled cDNA. Differential labeling is usually obtained incorporating dUTP or dCTP conjugated with either Cy5 or Cy3, two fluorophores able to emit fluorescence of two different wavelengths. Equal amounts of the two differentially labeled cDNA are mixed, applied to the array surface, and allowed to hybridize to the corresponding gene-specific target, which acts as a hybridization capture probe. Finally, the microarray is scanned to obtain two overlapping images each relative to the fluorescence emitted from each label [5, 6].

Usually, experiments are carried out analyzing at least three independent biological sample sets. Each sample is hybridized twice through reverse labeling of the respective cDNAs.

7.1.2 Normalization and Data Analysis

The images obtained scanning the microarrays must be analyzed to identify and quantify the spots corresponding with the gene-specific probes. Several software packages, either provided by commercial microarray scanner manufacturers or available on the World Wide Web, can be used for image processing.

Usually, the first step is normalization of the data. This is necessary to correct systematic errors introduced by differences in the initial amount of mRNA in the samples or by different efficiencies in the labeling process or in the detection of the fluorescence of the two different labels. After normalization, data are ready for quantitative and statistical analysis. Statistical analysis is of extreme importance as it allows the determination of the significance of the data and the identification of those genes whose differential expression between the two samples is statistically significant. Also in this case, several approaches are possible, and both commercial and public software is available. The result of this analysis is a "statistically significant genes list." This often includes several genes that show small changes in expression ratio and is usually refined eliminating the genes falling below an arbitrary twofold (induction or repression) cutoff [7].

It is worth mentioning that the statistical analysis of microarray data alone is not usually considered enough to confirm differential expression of a gene, and validation with an independent technique is requested, at least for those genes that are considered more important for the biological implications of the data and for those that will be targets of following studies based on the microarray data. Usually, quantitative RT-PCR is the method of choice [8].

7.1.3 Public Access Databases for Microarray Data

Microarray experiments produce an incredible amount of data, forcing the investigators to focus on a few main issues. The consequence is that most of the data is underutilized. In order to make the complete microarray data sets available to the scientific community, many scientific journals require the raw data to be placed in a public access database prior to publication. The most widely used databases are ArrayExpress at the European Bioinformatics Institute (EBI) (http://www.ebi.ac.uk/arrayexpress/) and the Gene Expression Omnibus at the National Center for Biotechnology Information (NCBI) (http://www.ncbi.nlm.nih.gov/geo/). To facilitate the interpretation of the data, microarray databases require the use of standardized descriptions using a defined format (e.g., Minimum Information About a Microarray Experiment; MIAME).

7.1.4 Experimental Design

DNA microarrays are often used to answer two main types of question: the first is what happens when bacteria are subjected to a particular environment, and the second is what is the physiologic role of a specific protein (usually a global transcription regulator). In the first case, the transcription profiles of wild-type bacteria subjected to different environmental conditions are compared, and in the second case the comparison is performed between two isogenic strains usually the mutant lacking the regulator and parental strain; subjected to the same conditions [4].

7.1.4.1 Comparing the Transcription Profile of Wild-Type Bacteria Subjected to Different Environmental Conditions

This type of experiment can be used to study the variation of the transcription profile during the different phases of growth or to study the transcriptional response of bacteria to adverse environmental conditions, such as exposure to antimycobacterial drugs or to environmental conditions mimicking the intra-phagosomal environment (e.g., acid stress, oxidative stress, surface stress, starvation) [9]. When performing these experiments, investigators always face the problem of deciding the degree of stress to apply to the culture (e.g., drug concentration and the time of the exposure before collecting the samples). Even if no rules can be formulated, it is a good idea to use at least two different degrees of stress (inhibitory and subinhibitory) and different time points. Recently, we studied the transcriptional response of *M. tuberculosis* to vanco-mycin at two different drug concentrations; one able to slightly reduce the bacterial division rate, and one (MIC × 10) totally inhibiting cell division. In the first experiments, bacteria were grown in the presence of a subinhibitory concentration and harvested at mid-log phase. For the second experiment, vancomycin was added to a mid-exponential culture and bacteria were harvested after 1 or 4 h (in these conditions, vancomycin treatment was bacteriostatic but not bactericidal). This allowed us to recognize two different patterns of transcriptional response each typical of the two drug concentrations (Provvedi R et al., manuscript in preparation).

Recently, the development of more sensitive techniques of RNA extraction and labeling has allowed the study of the transcriptional response to the intraphagosomal environment [10] and the comparison of the transcription profiles of bacteria grown *in vitro* with those of bacteria growing in infected tissues [11, 12]. Although these techniques are still labor intensive, they provide a vast resource to the understanding of *M. tuberculosis* physiology and interaction with the host.

7.1.4.2 Comparing the Transcription Profile of Isogenic Mutants Subjected to the Same Environmental Conditions

Microarray experiments can be designed to understand the role of specific proteins, often transcriptional regulators, in bacterial physiology. In this case, the transcription profile of a mutant unable to synthesize the protein of interest is compared with that of the wild-type parental strain. Ideally, this should be performed under the conditions inducing the activity of the transcription regulators. In this case, it is possible to compare the transcriptional response of mutant and wild type to these inducing conditions [13]. However, sometimes the *in vitro* inducing conditions are unknown, and this represents a serious obstacle for the identification of the specific regulons. Investigators have attempted to circumvent this problem by comparing the trancriptomes of mutant and parental strains in different phases of growth in the absence of specific inducing conditions. However, the relevance of this data is sometimes difficult to assess as only a relatively small fraction of the differentially expressed genes are likely to be directly regulated by the protein of interest [3]. Alternative methodologies such as overexpression of the protein of interest may constitute an interesting option and have been used successfully in case of sigma factors [14, 15].

7.2 Materials

7.2.1 Growth and Collection of Bacteria

1. Albumin-dextrose complex supplement (ADN): 25 g bovine serum albumin, fraction V (BSA), 10 g D-glucose, 4.25 g NaCl in 500 mL water. Filter sterilize with 0.2 μm pore size filter.
2. 20% (v/v) Tween 80: mix 20 mL Tween 80 with 80 mL distilled water, filter sterilize.
3. 7H9-Tw medium: Dissolve 4.7 g Middlebrook 7H9 broth base (BD-Difco, Detroit, MI) in 900 mL deionized water and add 2 mL glycerol, mix well and filter sterilize. Add 5 mL 20% (v/v) Tween 80 (final concentration of 0.05% v/v). Add 100 mL ADN immediately before use (*see* **Note 1**).
4. 7H10-Tw plates: Dissolve 19 g Middlebrook 7H10 medium (Difco) in 900 mL deionized water. Add 8.4 mL 60% glycerol (v/v) end 5 mL 20% (v/v) Tween 80 (final concentration of 0.05%). Autoclave. When media is cooled to ~50°C, add 100 mL ADN enrichment. Mix well and pour in standard plastic Petri dishes.
5. 225 mL polypropylene conical tubes (Becton Dickinson Labware, Franklin Lakes, NJ).
6. Incubator with rolling apparatus.
7. Centrifuge.
8. Dry ice.

7.2.2 Bacterial Lysis and RNA Extraction

1. Trizol Reagent (Invitrogen, Carlsbad, CA).
2. 0.1-mm sterile zirconia/silica beads (Biospec Products, Bartlesville, OK).
3. 2-mL screw-cap microcentrifuge tubes with O-rings (Fisher Scientific, Pittsburgh, PA).

4. Mini-BeadBeater-8 (Biospec Products).
5. Heavy Phase Lock Gel tubes (Eppendorf, Hamburg, Germany).
6. Chloroform/isoamyl alcohol (24:1).
7. Isopropanol.
8. RNase-free distilled water.
9. High Salt Solution: 0.8 M sodium citrate, 1.2 M NaCl in RNase-free distilled water.
10. 75% ethanol v/v (made with RNase-free distilled water).
11. DNAse I RNase-free (Ambion, Inc., Austin, TX).
12. 10X buffer DNase I (Ambion).
13. RNeasy Mini Kit (Qiagen GmbH, Hilden, Germany).

7.2.3 Reverse Transcription, cDNA Labeling, and Purification

1. 0.2-mL sterile PCR tubes.
2. RNase-free distilled water.
3. Random Primers (3μg/μL, Invitrogen).
4. Superscript II RNaseH⁻ reverse transcriptase (Invitrogen).
5. 5X First Strand reaction buffer: 250 mM Tris-HCl, pH 8.3, 375 mM KCl, 15 mM $MgCl_2$. (Invitrogen).
6. 100 mM DTT (Invitrogen).
7. Low dTdNTP mix (5 mM dATP, 5 mM dGTP, 5 mM dCTP, and 0.2 mM dTTP).
8. Cy3-dUTP and Cy5-dUTP (GE Healthcare, Piscataway, NJ) (*see* **Note 2**).
9. PCR thermal cycler.
10. CyScribe GFX purification kit (GE Healthcare) (*see* **Note 3**).
11. Microcon YM-30 (Millipore).
12. TE solution (10 mM Tris/HCl, pH 8.0, 1 mM EDTA).
13. Sterile Milli-Q water.

7.2.4 DNA Microarrays Prehybridization and Hybridization

1. Microarray slides (*see* **Notes 4, 5, and 6**).
2. 10% w/v BSA solution.
3. 10% w/v SDS solution.
4. Prehybridization Solution: 76 mL distilled water, 30 mL 10% w/v BSA, 1.2 mL 10% w/v SDS (*see* **Note 7**).
5. Glass slide staining jar.
6. Waterbath.
7. 50-mL conical tubes.
8. Milli-Q water (*see* **Note 8**).
9. Isopropanol.

10. 10 mg/mL tRNA (Invitrogen).
11. 20X SSC: 3 M NaCl, 0.3 M sodium citrate, pH 7.0.
12. Formamide.
13. 1% SDS solution.
14. LifterSlip coverslips 22 × 22 (Erie Scientific, Portsmouth, NH) (*see* **Note 9**).
15. Microarray chamber (ABgene, Epsom, UK).
16. Wash solution A: 2X SSC, 0.1% SDS.
17. Wash solution B: 1X SSC, 0.05% SDS.
18. Wash solution C: 0.06X SSC.

7.2.5 Image Acquisition

1. Affymetrix 428 array scanner.
2. Affimetrix 428 Reader software (version 1.1) (*see* **Note 10**).

7.2.6 Software for Data Analysis

1. Imagene (BioDiscovery, Inc., El Segundo, CA).
2. DNMAD: Diagnosis and Normalization for MicroArray Data (http://dnmad.bioinfo.cipf.es/cgi-bin/dnmad.cgi).
3. TIGR Multiple Experiment Viewer (TMEV) (http://www.tm4.org/mev.html).

7.3 Methods

7.3.1 Growth and Collection of Bacteria

1. Streak 10 to 15 μL of *M. tuberculosis* frozen cultures onto 7H10-Tw plate and incubate at 37°C for approximately 6 days.
2. Resuspend bacteria in 4 mL 7H9-Tw liquid medium to get an optical density (OD$_{540}$) of approximately 0.2 to 0.3.
3. Inoculate 31.5 mL 7H9-Tw medium in a 225-mL polypropylene conical tube with 3.5 mL of the preculture and roll at 40 to 50 rpm at 37°C until an OD$_{540}$ of 0.2 to 0.3 (approximately 3 to 4 days) (*see* **Note 11**).
4. Collect cells by centrifugation for 5 min at 1500 × *g* at room temperature.
5. Pipette off the supernatant and freeze pellets immediately in dry ice.

7.3.2 Bacterial Lysis and RNA Extraction

1. Resuspend the frozen pellets in 1 mL Trizol reagent by vortex and pipetting.
2. Transfer the suspension to a 2-mL screw-cap tube containing 0.5 mL 0.1-mm zirconia/silica beads.

3. Place the tube with the sample in the BeadBeater and subject it to 3×30-s pulses with a 1-min rest on ice between each session (*see* **Note 12**).

4. Incubate sample for 5 min at room temperature inverting periodically.

5. Centrifuge 45 s in a microcentrifuge at $14,000 \times g$ and transfer the suspension to a 2-mL tube with Heavy Phase Lock Gel.

6. Add 300 μL chloroform/isoamyl alcohol (24:1). Invert rapidly for 15 s and continue inverting periodically for 2 min.

7. Centrifuge 10 min at $14,000 \times g$, transfer the aqueous phase (approximately 540 μL) to a tube containing 270 μL isopropanol.

8. Add 270 μL High Salt Solution.

9. Invert several times and precipitate over night at $4°C$.

10. Centrifuge 10 min at $14,000 \times g$ in the cold, aspirate the isopropanol, and wash the pellet with 1 mL 75% ethanol.

11. Centrifuge again.

12. Remove supernatant and spin quickly to eliminate the last drops.

13. Air dry and resuspend the pellet in 90μL RNase-free water (*see* **Note 13**).

14. Add 10 μL 10X DNase I buffer and 2 μL DNase I.

15. Incubate 1 h at $37°C$.

16. Add 350 μL RLT buffer provided by RNeasy Mini Kit (add 10 μL β-mercaptoethanol to RLT before using).

17. Vortex, add 265 μL 95% ethanol, vortex again.

18. Transfer the sample to a RNeasy column and continue RNA purification according to the instructions of the manufacturer (Qiagen).

7.3.3 Reverse Transcription, cDNA Labeling, and Purification

1. Bring 0.5 to 5 μg of RNA to a volume of 11.5 μL in an 0.2-mL PCR tube with RNase-free water and add 1.47 μL random primers (3 μg/μL) (*see* **Note 14**).

2. Pulse spin to collect contents.

3. Heat 2 min at $98°C$, snap cool on ice.

4. Add 11.1 μL of the following reaction mix: 5 μL 5X First-Strand buffer, 2.5 μL 100 mM DTT, 2.3 μL low dT dNTP mix, 1.5 μL Cy3-dUTP or Cy5-dUTP (*see* **Note 15**).

5. Add 1.2 μL Superscript II, vortex to mix, and spin to collect contents.

6. Incubate 10 min at $25°C$ followed by 90 min at $42°C$ in PCR thermal cycler (*see* **Note 16**).

7. Mix together both reactions and purify labeled cDNA using GFX columns according to the instructions provided by the manufacturer (GE Healthcare).

8. Prime a Microcon YM-30 with 100 μL TE and spin for 3 min in a microcentrifuge at $14,000 \times g$ at room temperature. Empty flow-throw in the collection tube and return it to primed column.

9. Bring the volume of the purified labeled cDNA to 100 μL with Milli-Q water and add it to column. Centrifuge 10 min as above.

10. Add 7.5 μL Milli-Q water into the column and incubate 2 min.

11. Recover cDNA by inverting the Microcon into a new collection 1.5-mL tube and centrifuge for 2 min as above.

7.3.4 DNA Microarrays Prehybridization and Hybridization

1. Add Prehybridization Solution to a glass slide-staining jar and place post-processed slides array side up into solution (*see* **Notes 17 and 18**).

2. Place the staining jar containing the Prehybridization Solution and slides into a waterbath and hybridize for 1 h underwater at 42°C (*see* **Note 19**).

3. Take the prehybridized slides out of the prehybridization solution and place them into a 50-mL conical tube containing Milli-Q water. Wash slides for 2 min with constant shaking (*see* **Note 20**).

4. Transfer slides to 50-mL conical tubes containing 40 mL isopropanol and wash for 2 min with constant shaking.

5. Remove slides from the isopropanol bath, wipe one of the edges to remove excess liquid, and air dry by keeping them tilted in one hand (*see* **Notes 21 and 22**).

6. Add the following to the 7.5-μL labeled sample: 0.75 μL 10 mg/mL tRNA, 1.5 μL 20X SSC, 3.75 μL formamide, 1.5 μL 1% SDS.

7. Heat to 98°C for 2 min. Spin in centrifuge for 1 min at $14,000 \times g$. Let cool for 5 min (*see* **Note 23**).

8. Apply 22×22 LifterSlip coverslip and pipette cooled 15 μL hybridization solution on microarray (*see* **Note 24**).

9. Hybridize in a microarray chamber with 40 μL water under each slide (20 μL under both ends of each slide) overnight in 50°C waterbath for 16 h (*see* **Notes 25 and 26**).

10. Remove the chamber from the waterbath and blot off external water. Disassemble the chamber and transfer slides to a tray containing wash solution A.

11. Remove coverslip while array is submerged in wash solution A (*see* **Note 27**).

12. Transfer to a 50-mL conical tube and rinse for 1 min in 40 mL wash solution B with shaking.

13. Rinse in wash solution C with shaking. Transfer to a new 50-mL conical tube and wash 2 min in 40 mL fresh wash solution C with shaking.

14. Take slides in the wash solution C bath to the centrifuge to keep the slides wet until you can dry them rapidly. Transfer in an empty 50-mL conical tube and spin at $68 \times g$ for 3 min to dry (array edge up).

15. Place the slide in a light-proof box. Scan as soon as possible (*see* **Note 28**).

7.3.5 Image Acquisition

1. Put the slide to be scanned into the slide holder of the scanner and follow the instructions provided by the manufacturer.
2. Perform a low-resolution scan of the slide and adjust the overall brightness of the emission on a slide by the gain tool (*see* **Notes 29 and 30**).
3. Once the gain setting has been established, proceed with a high-resolution scanning of the array. Store the row data files as 16-bit TIFF images (*see* **Note 31**).
4. Open the TIFF images using the microarray image analysis program Imagene.
5. Load the file containing gene list or Gene IDs to associate gene information with spot location (*see* **Note 32**).
6. Identify and remove features that are flawed technically and might yield misleading data and extract the numerical data from the spot intensities (*see* **Note 33**).
7. After quantification is complete, save the data in a database that correlate the spot positions (numbers) with their identification and descriptive information.

7.3.6 Data Analyses

7.3.6.1 Diagnosis and Normalization

We use a Web-based tool, Diagnosis and Normalization of spotted cDNA MicroArrayData (DNMAD), available at http://dnmad.bioinfo.cnio.es [16] (*see* **Note 34**).

Program input:

1. The server accepts GenePix files as input, therefore, if GenePix Pro software has not been used to scan the arrays, it is necessary to create customized files containing the following columns and in this same order (*see* **Note 35**): Block, Column, Row, Name, ID, F635 Mean, B635 Median, F532 Mean, B532 Median, and Flags (F635 and F532 indicate the spot foreground intensity values from samples labeled with Cy5 and Cy3, respectively; B635 and B532 indicate the spot background intensity values from samples labeled with Cy5 and Cy3, respectively).
2. Introduce the layout of the array into the interface along with the data files (*see* **Note 36**).
3. You can choose options such as the use of flags to exclude certain spots from further analysis or background correction (*see* **Note 37**).
4. The default normalization method is "print-tip loess," a robust local regression for each print-tip group of the array. Alternatively, the "global loess" option can be used, where the local robust regression is carried out over the whole array instead of for each print-tip group (*see* **Notes 38 and 39**).

Program output:

5. The result file contains the normalized $\log_2$ ratios of expression for every spot (*see* **Note 40**).
6. Output data also include (1) *boxplots* for all the arrays and for each array showing individual print-tip groups, to asses the need for normalization for a particular array; (2) *MA plots* with the regression curve for each print-tip group in order to observe the efficiency of normalization; (3) *diagnostic plots* such as (i) histograms of the row pixel intensities ($\log_2$) of the red and green mean foreground that help to identify problems in the scanner settings or in the hybridizations; (ii) density plots of the $\log_2$ intensities of both channels to give an idea about the signal distribution across a chip; (iii) images of the arrays including the red and green background and the unnormalized and normalized ratio values, which should help to identify spot damaged arrays or spatial patterns (*see* **Notes 41 and 42**).
7. The server accepts multiple arrays, which can show differences between their scales, therefore slide scale normalization is provided in order to reduce differences in the scales of the arrays that are being normalized.

7.3.6.2 Identification of Differentially Expressed Genes

The overall reproducibility of the microarray experiments can be evaluated using Significance Analysis of Microarray (SAM) [17] (*see* **Note 43**).

1. The input data to SAM must have a general format of one gene per row and one sample per column.
2. The cutoff for significantly regulated genes is determined by a tuning parameter delta chosen by the user based on the false-positive rate (*see* **Note 44**).

7.4 Notes

1. Middlebrook 7H9 liquid medium supplemented with ADN is the most commonly used medium to grow *M. tuberculosis* cultures in microarray analysis experiments. However, depending on the test conditions, some authors report the use of different media such as Minimal Medium [MM; 0.5% (w/v) asparagine, 0.5% (w/v) KH_2PO_4, 2% (v/v) glycerol, 0.5 mg/L $ZnCl_2$, 0.1 mg/L $MnSO_4$, and 40 mg/L $MgSO_4$, supplemented with 0.05% (v/v) Tween 80 and 10% (v/v) ADN]; Sauton's medium [4 g/L asparagine, 2 g/L sodium citrate, 0.5 g/L $K_2HPO_4 \cdot 3 H_2O$, 0.5 g/L $MgSO_4 \cdot 7 H_2O$, 0.05 g/L ferric ammonium citrate, 60 g/L glycerol, supplemented with 10% (v/v) ADN]; or modified Dubos Albumin broth (Difco) supplemented with 0.3% (v/v) glycerol.
2. Cy3 ($excitation_{max}$ = 550 nm; $emission_{max}$ = 570 nm) and Cy5 ($excitation_{max}$ = 649 nm; $emission_{max}$ = 670 nm) are available as dCTP or dUTP analogs. When labeling *M. tuberculosis* cDNA, it is necessary to use dUTP instead of dCTP because of the high GC content in its genome. If too many fluorophores are incorporated close together on the cDNA, a quenching reaction will occur, causing a loss of signal.
3. Alternatively, Qiagen MinElute Reaction Cleanup Kit has been reported to work as well.
4. Institutes producing *M. tuberculosis* DNA microarrays:

Microarray Core Facility, The University of Texas, Medical Center at Dallas, Texas (http://microarray.swmed.edu)

TB Vaccine Testing and Research Materials Contract, Colorado State University, Colorado (http://www.cvmbs.colostate.edu/microbiology/tb/materials.htm)

Microarray Facility of the Albert Einstein College of Medicine (http://microarray1 k.aecom. yu.edu)

Affymetrix GeneChip Arrays (*see* **Note 5**) (http://www.affymetrix.com/index.affx)

TIGR, The Institute for Genome Research, Pathogen Functional Genomics Resource Center (http://pfgrc.tigr.org/programs/programs_microarray.shtml)

The Center for Applied Genomics, The Public Health Research Institute, Newark, New Jersey (http://www.cag.icph.org/)

Bacterial Microarray Group, Medical Microbiology, Department of Cellular and Molecular Medicine, St George's Hospital Medical School, United Kingdom (http://bugs.sghms. ac.uk/)

Stanford: Functional Genomics Facility, Stanford, California (http://www.microarray. org/sfgf/)

5. Affymetrix GeneChip oligonucleotide microarrays consist of small DNA fragments (referred to as probes), synthesized at specific locations on a coated quartz surface. Every probe on an Affymetrix microarray has 25-nucleotide length, and a set of 11 to 20 probes are used to represent each gene. These probes are called PM (Perfect Match) probes. For each PM probe, Affymetrix also builds probes containing a single substitution at the middle base. These probes, called MM (Mismatch) probes, are designed to measure nonspecific hybridization. The Affymetrix technology requires special procedures for cDNA labeling, image processing, and data analysis that are not described in this chapter. A *Genechip Expression Analysis Technical Manual* describing the protocols for an Affymetrix experiment is available at http://www.affymetrix.com/technology/ge_analysis/index.affx.

6. We use slides from The Center for Applied Genomics (http://www.cag. icph.org/). The array layout is composed of 16 subarrays (4 × 4) containing 4295 70-mer oligonucleotides representing 3924 open reading frames from *M. tuberculosis* strain H37Rv and 371 unique open reading frames from strain CDC1551 that are not present in the H37Rv strain's annotated gene complement.

7. This volume is based on a glass slide staining jar with a base measurement of 11 × 8 cm. This dish allows for a prehybridization of 1 to 4 slides. Adjust volume accordingly based on size of dish and/or number of slides.

8. Make sure the Milli-Q system is well maintained; it has been reported that water from poorly maintained systems can damage Cy5.

9. LifterSlip coverslips present raised edges designed to provide separation between the DNA microarray and coverslip, allowing for even dispersal of hybridization solutions. They offer a simple solution to the problem of hybridization nonuniformity.

10. The Affymetrix 428 array scanner is currently out of production. Many authors use GenePix scanners (Axon Instruments) supplied with the GenePix Pro software to collect, analyze, and normalize fluorescent intensity data. GenePix Pro can be also acquired separately.

11. 225-mL polypropylene conical tubes can be incubated in a rolling apparatus and are quite convenient as they can be directly centrifuged to collect cells when required. Moreover, after removing the culture, they can be easily disposed of. Clearly, it is possible to use different kinds of vessels, including glass flasks, although special sterilizing programs may be required to ensure the complete killing of pathogenic bacteria before reusing them.

12. If lysing many samples, treat them singly for the first pulse and then proceed with the second and third pulse with all the samples together.

13. It may need 10 min at 55°C to 60°C to dissolve.

14. We typically use 2 μg RNA. Amounts of RNA higher than 5 μg sometimes can increase background.

15. Never mix unincorporated Cy3 or Cy5 dUTP because a quenching reaction will occur.

16. After this step, samples can be frozen or left at 4°C overnight if needed.

17. Prepare Prehybridization Solution fresh each time.

18. Make sure there are no bubbles on the surface of the slides.

19. We keep the glass slide–staining jar with the cover lid edge about 1 cm out from water.

20. Alternatively, the prehybridized slides can be placed into a slide holder submerged in a slide washing tray containing Milli-Q water. We prefer to use 50-mL conical tubes and shake vigorously by hand to reduce background.

21. Alternatively, slides can be dried by centrifugation spinning at $68 \times g$ in a 50-mL conical tube for 3 min. In this case, it is important to keep the slides wet (leaving them into the isopropanol) until they are transferred to an empty tube and then loaded into the centrifuge.

22. Complete this step within 1 h of hybridization.

23. Do not apply hot samples to the array because this would cause a circular pattern of high background forms where the droplets first touched the array. Moreover, do not put samples on ice after heating because this would cause SDS precipitation.

24. If using a regular coverslip, 10 μL hybridization solution is enough. In this case, elute labeled cDNA from Microcon YM-30 with 5 μL Milli-Q water and scale down the hybridization solution reagents. If bubbles appear under the coverslip, they will slowly migrate to the edge provided that the volume is accurate. Alternatively, apply a thin wall of rubber cement around the four corners of the coverslips to avoid movement. This step is recommended if the waterbath is not perfectly leveled. Never tap down on the cover glass once it is laid on the array.

25. Water is added to slides to supply a source of humidity and prevent evaporation of the sample during hybridization.

26. Do not put the microarray chamber directly in contact with the bottom of the waterbath; many waterbaths contain a heating element at the bottom and this can alter the temperature inside the chamber.

27. Do this step very carefully to avoid the coverslip scratching the array.

28. Cy5/Cy3 photoactivity is light sensitive.

29. Image acquisition settings should be adjusted to obtain equal fluorescence intensities from the Cy3 and Cy5 channels. Most array scanner software programs provide signal intensity histograms to observe the relative intensities of both channels. For this purpose, we look at *sigA* intensity (as *sigA* is frequently used as internal standard in RT-PCR experiments due to its expression remaining unaffected under different physiologic conditions) [18]. It is worthwhile mentioning that even if the two channels are not perfectly balanced, this does not results in a significant error because minor imbalances are corrected by the normalization procedure.

30. Sometimes the default settings produce a slide image with some saturated spots or pixels. Saturated pixels do not contain complete data. Therefore, if there are saturated pixels in spots important to the experiment, you can improve the image by adjusting the gain setting. Alternatively, you can acquire two different images and carry out your analysis with both of them.

31. Choose gain settings that give the highest signal-to-noise ratio. Decide on gain setting quickly and avoid rescanning the slide multiple times because photobleaching will reduce the signal with each subsequent scan. Cy5 dye is more vulnerable to bleaching and fading than is Cy3.

32. This is usually achieved by creating a grid and adjusting it to fit the array spots properly. Imagene combines the grid to a gene list or Gene IDs to create a template, which is required to easily quantify and name the various features on the array.

33. The act of quantification converts the images' pixel intensities into numerical values to be used later in expression analysis.

34. DNMAD represent an easy-to-use tool to normalize Affymetrix and two-color micro-array experiments and perform other preprocessing options. DNMAD offers the advantage of using files coming directly from the scanner and the possibility of entering more than one slide at a time. Moreover, it is completely integrated within the Gene Expression Profile Analysis Suite (GEPAS), a Web-based pipeline that includes tools for functional annotation of the experiment, different possibilities for clustering, gene selection, class prediction, and array comparative genomics hybridization management [19].

35. This can be easily achieved by saving as text an Excel file where data extracted from quantification analysis is pasted under the corresponding column. These columns constitute the minimal set of columns required to perform the normalization. If the microarray analysis program assigns different values from GenePix Pro to characterize features (i.e., good spots, bad spots, etc.), remember to convert them into the GenePix format. This can be easily done using the Excel function "IF."

36. The layout refers to the number of rows and columns in the main grid and the number of rows and columns within each subgrid (each of the squares or rectangles defined by the rows and columns of the main grid). Your array facility should tell you how the array is structured. It is very important not trying to do any guesswork about the layout because if you make a mistake here, all print-tip based methods and all image plots will be wrong.

37. If you choose the "background subtraction" option, where the background median for each spot is subtracted from the foreground mean of each spot, you can end up with many spots with negative values, and the log of a negative value is not defined. Using the "half" method, any intensity less than 0.5 after subtracting the background is reset equal to 0.5 (and thus you end up with a very tiny value that will not lead to a missing value when taking logs).

38. Small variations in the many steps that produce an array image can make comparison across arrays problematic. Sources of variations can be due to differences in sample preparation, hybridization, labeling efficiencies, chemical properties of different dyes, pin tips, slide batches, or scanner settings. Any of these variations can be corrected by normalization. No one normalization method will correct all type of variation. However, print-tip loess normalization has been considered one of the most effective methods to remove the systematic sources of variation associated with microarray experiments [7].

39. Print-tip loess method can only be used if the following assumptions are met: (i) the number of points to normalize must be large in each print tip group; (ii) very few genes should be differentially expressed; and (iii) there should be an approximately equal number of upregulated and downregulated genes in each print-tip group. If these assumptions cannot be met for each print-tip group, you should choose global loess [7].

40. It is usual in microarray data analysis to use logarithms to base 2. Log_2 ratios are chosen for intuitive convenience. The reason is that the ratio of the raw Cy5 and Cy3 intensities is transformed into the difference between the logs of the intensities of the Cy5 and Cy3 channels. Therefore, twofold upregulated genes correspond with a log ratio of +1, and twofold downregulated genes correspond with a log ratio of –1. Genes that are not differentially expressed have a log ratio of 0. These log ratios have a natural symmetry, which reflects the biology and is not present in the raw fold difference.

41. A boxplot is a very useful graph that shows the shape of the distribution of the data as well as its central value (median) and variability. Boxplots of data distributions across microarray chips before and after normalization can help investigators decide if the normalizations applied have had the intended effects and can help to identify problematic chips. Boxplots can highlight problems with the arrays or with the normalization that have been applied to the data. If, after normalization, the median values are not similar across the arrays and/or the distributions across the replicates are not similar, your data may be problematic, or the normalizations you have applied may not have been appropriate [7].

42. MA plots show the relationship of signal ratios to signal intensities, where values are usually reported on the log2 scale. For a large chip where most data is not expressed at

different levels across the treatments, after normalization you expect data to fall approximately along a straight horizontal line along 0. Before normalization, you may see that the line along which the majority of the data cluster is curved reflects the dye bias of some genes. One aim of normalization is to "straighten out" this line [7].

43. SAM is a statistical method for finding significant values derived from t-statistics but using a resampling scheme to determine if the expression of any gene is significantly related to the response [17]. It is available as an Excel plug-in at http://www-stat. stanford.edu/~tibs/SAM/ or as part of microarray and analysis applications suits such as TIGR MultiExperiment Viewer (TMeV) developed at The Institute for Genomic Research (TIGR) and which can be downloaded at www.tigr.org/software/tm4/.

44. SAM key steps in selecting significant genes consist in (i) calculate a t-statistic for each gene; these are called experimentally observed t-statistics; (ii) perform permutation and calculate average t-statistics for each gene based on permuted sample groups; these are called calculated t-statistics; (iii) graph experimentally observed t-statistics against calculated t-statistics (this is the essence of SAM: the whole idea is to compare these two sets of statistics and identify really changed genes by selecting those with observed t-statistics being greater than calculated t-statistics); (iv) allow a user to define a delta value that defines how much bigger (or smaller for downregulated genes) the observed t-statistics need to be, compared with the calculated t-statistics, in order for a gene to be called significant; (v) calculate false discovery rate (FDR) based on the experimental data (original data), the permuted data, and the user-defined delta value. The final result is a list of significant genes with their corresponding FDRs (reported as q-value) and observed t-statistics [reported as score (d); a large t-statistic is equivalent to a small p-value in a t-test]. Fold changes are also calculated so significant genes can be selected for those with a fold change greater than a certain value. Beyond SAM, other methods can be used to identify significant differentially expressed genes; for example, when more than two experimental conditions must be analyzed, analysis of variance (ANOVA) is often used.

Acknowledgment The authors wish to thank Issar Smith for valuable discussion and for carefully reading the manuscript.

References

1. Cole, S. T., Brosch, R., Parkhill, J., Garnier, T., Churcher, C., Harris, D., Gordon, S. V., Eiglmeier, K., Gas, S., Barry, C. E. III, Tekaia, F., Badcock, K., Basham, D., Brown, D., Chillingworth, T., Connor, R., Davies, R., Devlin, K., Feltwell, T., Gentles, S., Hamlin, N., Holroyd, S., Hornsby, T., Jagels, K., Krogh A., McLean J., Moule S., Murphy L., Oliver K., Osborne J., Quail M. A., Rajandream M.-A., Rogers J., Rutter S., Seeger K., Skelton J., Squares R., Squares S., Sulston J. E., Taylor K., Whitehead S., and Barrell, B. G. (1998). Deciphering the biology of *Mycobacterium tuberculosis* from the complete genome sequence. *Nature* **393**, 537–544.
2. Tekaia, F., Gordon, S. V., Garnier, T., Brosch, R., Barrell, B. G. and Cole, S. T. (1999). Analysis of the proteome of *Mycobacterium tuberculosis in silico*. *Tuber Lung Dis* **79**, 329–342.
3. Rodrigue, S., Provvedi, R., Jacques, P. E., Gaudreau, L. and Manganelli, R. (2006). The sigma factors of *Mycobacterium tuberculosis*. *FEMS Microbiol. Rev.* **30**, 926–941.
4. Butcher, P. D. (2004). Microarrays for *Mycobacterium tuberculosis*. *Tuberculosis* **84**, 131–137.
5. Duggan, D. J., Bittner, M., Chen, Y., Meltzer, P. and Trent, J. M. (1999). Expression profiling using cDNA microarrays. *Nat. Genet.* **21**, 10–14.

6. Wilson, M., Voskuil, M., Schnappinger, D. and Schoolnik, G. (2001). Functional genomics of *Mycobacterium tuberculosis* using DNA microarrays. In *Mycobacterium tuberculosis Protocols* (Parish, T. and Stoker, N. G. eds.) Humana Press, Totowa, NJ, pp. 335–357.

7. Stekel, D. (2003). *Microarray Bioinformatics*, Cambridge University Press, Cambridge, UK.

8. Manganelli, R., Tyagi, S. and Smith, I. (2001). Real-time PCR using molecular beacons: a new tool to identify point mutations and to analyze gene expression in *Mycobacterium tuberculosis*. In *Mycobacterium tuberculosis Protocols* (Parish, T. and Stoker, N. G. eds.) Humana Press, Totowa, NJ, pp. 295–310.

9. Kendall, S. L., Rison, S. C., Movahedzadeh, F., Frita, R. and Stoker, N. G. (2004). What do microarrays really tell us about *M. tuberculosis*? *Trends Microbiol.* **12**, 537–544.

10. Schnappinger, D., Ehrt, S., Voskuil, M. I., Liu, Y., Mangan, J. A., Monahan, I. M., Dolganov, G., Efron, B., Butcher, P. D., Nathan, C. and Schoolnik, G. K. (2003). Transcriptional adaptation of *Mycobacterium tuberculosis* within macrophages: insights into the phagosomal environment. *J. Exp. Med.* **198**, 693–704.

11. Cappelli, G., Volpe, E., Grassi, M., Liseo, B., Colizzi, V. and Mariani, F. (2006). Profiling of *Mycobacterium tuberculosis* gene expression during human macrophage infection: Upregulation of the alternative sigma factor G, a group of transcriptional regulators, and proteins with unknown function. *Res. Microbiol.* **157**, 445–455.

12. Talaat, A. M., Lyons, R., Howard, S. T. and Johnston, S. A. (2004). The temporal expression profile of *Mycobacterium tuberculosis* infection in mice. *Proc. Natl. Acad. Sci. U. S. A.* **101**, 4602–4607.

13. Manganelli, R., Fattorini, L., Tan, D., Iona, E., Orefici, G., Altavilla, G., Cusatelli, P. and Smith, I. (2004). The extra cytoplasmic function sigma factor SigE is essential for *Mycobacterium tuberculosis* virulence in mice. *Infect. Immun.* **72**, 3038–3041.

14. Dainese, E., Rodrigue, S., Delogu, G., Provvedi, R., Laflamme, L., Brzezinski, R., Fadda, G., Smith, I., Gaudreau, L., Palu, G. and Manganelli, R. (2006). Posttranslational regulation of *Mycobacterium tuberculosis* extracytoplasmic-function sigma factor SigL and roles in virulence and in global regulation of gene expression. *Infect. Immun.* **74**, 2457–2461.

15. Hahn, M. Y., Raman, S., Anaya, M. and Husson, R. N. (2005). The *Mycobacterium tuberculosis* extracytoplasmic-function sigma factor SigL regulates polyketide synthases and secreted or membrane proteins and is required for virulence. *J. Bacteriol.* **187**, 7062–7071.

16. Vaquerizas, J. M., Conde, L., Yankilevich, P., Cabezon, A., Minguez, P., Diaz-Uriarte, R., Al-Shahrour, F., Herrero, J. and Dopazo, J. (2005). GEPAS, an experiment-oriented pipeline for the analysis of microarray gene expression data. *Nucleic Acids Res.* **33**, W616–620.

17. Tusher, V. G., Tibshirani, R. and Chu, G. (2001). Significance analysis of microarrays applied to the ionizing radiation response. *Proc. Natl. Acad. Sci. U. S. A.* **98**, 5116–5121.

18. Manganelli, R., Dubnau, E., Tyagi, S., Kramer, F. R. and Smith, I. (1999). Differential expression of 10 sigma factor genes in *Mycobacterium tuberculosis*. *Mol. Microbiol.* **31**, 715–724.

19. Montaner, D., Tarraga, J., Huerta-Cepas, J., Burguet, J., Vaquerizas, J. M., Conde, L., Minguez, P., Vera, J., Mukherjee, S., Valls, J., Pujana, M. A., Alloza, E., Herrero, J., Al-Shahrour, F. and Dopazo, J. (2006). Next station in microarray data analysis: GEPAS. *Nucleic Acids Res.* **34**, W486–491.

Chapter 8
Two-Dimensional Gel Electrophoresis-Based Proteomics of Mycobacteria

Jens Mattow, Frank Siejak, Kristine Hagens, Julia Kreuzeder, Stefan H.E. Kaufmann and Ulrich E. Schaible

Abstract Two-dimensional gel electrophoresis (2-DE) in combination with mass spectrometry (MS) is the classic proteomics approach used to monitor the dynamics of protein abundance and posttranslational modifications in biological systems. In this chapter, we provide detailed protocols for 2-DE–based proteomics of mycobacteria. Adequate standard operating procedures for mycobacterial culture, subcellular fractionation, and selective enrichment of proteins are indispensable prerequisites for targeted proteome analyses. Therefore, we also provide approved protocols for selective and efficient extraction of cytosolic, secreted, and hydrophobic plasma membrane proteins of mycobacteria, as well as for isolation of mycobacteria from infected macrophages.

Keywords mass spectrometry · mycobacteria · *Mycobacterium tuberculosis* · proteomics · tandem mass spectrometry · two-dimensional electrophoresis

8.1 Introduction

8.1.1 From Genome to Proteome

The availability of the complete genome sequences of *Mycobacterium tuberculosis* H37Rv [1] and other mycobacterial strains/species including *M. tuberculosis* CDC1551 [2], *Mycobacterium leprae* TN [3], and *Mycobacterium bovis* AF2122/97 [4] has paved the way for functional genomics (i.e., global transcriptome, proteome, and metabolome analyses). The term *proteome* was originally coined to describe the entire set of proteins encoded by a genome [5]. It more commonly refers to the protein complement of a biological system at a given time point under defined conditions. Unlike the genome, but similar to the transcriptome,

J. Mattow
Department of Immunology, Max Planck Institute for Infection Biology, Charitéplatz 1, D-10117 Berlin, Germany
e-mail: mattow@mpiib-berlin.mpg.de

T. Parish, A.C. Brown (eds.), *Mycobacteria Protocols*,
doi: 10.1007/978-1-59745-207-6_8, © Humana Press, Totowa, NJ 2008

the proteome is not static but varies according to developmental and functional state and in response to physiologic and environmental conditions. In contrast with the transcriptome, the proteome provides information on cotranslational and posttranslational protein modifications, which determine protein function and topology. In addition, several studies demonstrated a poor correlation between cellular mRNA and protein levels, possibly due to cotranslational and posttranslational protein modification/processing, posttranscriptional control of protein translation, and/or different turnover rates of mRNA versus protein [6, 7, 8]. Hence, proteomics is the only approach to describe the actual protein inventory of a biological system as the additive outcome of both gene expression and posttranscriptional events. However, proteomics may not only be employed for protein profiling but also for numerous other purposes including analysis of the spatial distribution and temporal dynamics of proteins [9, 10] as well as characterization of protein complexes and cellular networks [11, 12]. Therefore, proteomics represents an important tool for functional cell analysis and systems biology.

8.1.2 Proteomics Approaches

Proteomics involves the separation and analysis of complex protein mixtures derived from biological samples. The classic proteomics approach is based on separation of proteins by two-dimensional gel electrophoresis (2-DE) followed by 2-DE image analysis and identification of relevant 2-DE spots by mass spectrometry (MS) techniques [13, 14]. Until recently, peptide mass fingerprinting (PMF) by matrix-assisted laser desorption/ionization mass spectrometry (MALDI-MS) was the most widely used protein identification technique in 2-DE–based proteomics. PMF involves the generation of proteolytic peptides by enzymatic or chemical in-gel protein digestion (by means of residue-specific proteolytic enzymes or chemical reagents) and the determination of their masses by MS. The fragmentation pattern of a protein depends on the specificity of the proteolytic enzyme/chemical reagent used but is characteristic of that protein. This means that as long as the protein cleavage is complete, it will generate a unique set of characteristic peptides. For protein identification, the experimental peptide masses of the protein in question are searched against a nonredundant protein database using an appropriate search program to create a list of likely protein identifications [15, 16]. Proteins from organisms with fully sequenced genomes can often be identified by PMF alone. However, PMF has repeatedly been questioned as the method of choice for reliable protein identification. This notion is based on the observation that the success rate for PMF is much lower when applied to organisms with incompletely sequenced genomes. In addition, PMF often fails to reliably identify low-molecular-mass proteins and protein mixtures, particularly at low protein amounts [17, 18, 19]. In this context, it should also be noted that even 2-DE spots often contain more than one protein component [20, 21]. Therefore, it is advantageous to complement PMF by tandem mass spectrometry (MS/MS) applications to allow unambiguous identification of 2-DE spots.

Protein identification by MS/MS is accomplished by acquiring fragment ion mass spectra on individual peptides of the protein in question. MS/MS involves the fragmentation of peptides into ions and measuring the mass of each ion. MS/MS measurements can be performed by means of both MALDI and electrospray-ionization (ESI) mass spectrometers. For protein identification, the database search algorithm mimics the MS/MS experiment by calculating the possible peptide-fragment masses for each entry in a protein database taking into account the specificity of the enzyme used. The experimental fragment masses are compared with the calculated fragment masses, and a score is calculated for the comparison. The optimal choice of ionization mode (MALDI or ESI) and sample introduction (infusion, HPLC, solid-phase desorption) depends on the nature of the sample to be analyzed and on the experimental information desired [22, 23]. It should be noted that the utility of MS and MS/MS for protein identification depends on the possibility to compare experimental mass data with theoretical peptide/fragment ion masses, generated through *in silico* digestion and/or fragmentation of protein sequences contained within public protein databases. Therefore, it is advantageous if the genome of the organism investigated has been fully sequenced. If this is not the case, proteins may be identified by interspecies comparison and/or by comparing experimental mass data against theoretical peptide/fragment ion mass data deduced from translated cDNAs and/or expressed sequence tags.

Despite the wide use of 2-DE–based proteomics, limitations to this approach include restrictions regarding the analysis of hydrophobic or low-abundance proteins and proteins with extreme isoelectric points or molecular weights [14, 24, 25, 26]. To overcome these limitations, complementary non–gel-based (shotgun) and hybrid proteomic workflows have been developed. The term *shotgun* proteomics refers to the direct analysis of proteolytic peptides derived from digestion of complex protein mixtures by combined use of one-dimensional (1-D) or multidimensional liquid chromatography (LC) for peptide separation and MS/MS applications for protein identification [27, 28, 29]. Recently, the use of 1-D SDS-PAGE in combination with LC-MS/MS (either online LC-ESI-MS/MS or offline LC-MALDI-MS/MS) for proteomic profiling has become popular. The term *Gel-C-MS/MS* describes this hybrid proteomics approach. Gel-C-MS/MS does not face the above-mentioned limitations of 2-DE and is relatively simple to apply [23, 30].

In 2-DE–based proteomics, relative protein quantification relies on comparative image analysis of two or more sets of protein patterns derived from distinct biological states. Prior to image analysis, the gel-separated proteins are detected by chromophoric staining with visible dyes such as Coomassie Brilliant Blue (CBB) or silver nitrate [14, 31]. An alternative strategy, the two-dimensional fluorescence difference gel electrophoresis (2-D DIGE) technique [32], uses fluorescent dyes to differently label protein samples prior to 2-DE analysis. This allows two samples to be coseparated and visualized on a single 2-DE gel. Complementary to this, several techniques for differential metabolic, enzymatic, or chemical labeling of proteins/peptides with stable isotopes have been developed to allow protein quantification in LC-MS/MS–based and Gel-C-MS/MS–based proteomics applications. These include the use of isotope coded affinity tags (ICAT) [33], stable isotope labeling of amino acids in cell culture (SILAC) [34], global internal standard technology (GIST) [35], isobaric tags (iTRAQ) [36], isotope-coded protein labels (ICPL) [37], and $_{18}O$ enzymatic labeling of peptides [38]. In addition, shotgun proteomics may also be employed for label-free relative quantification of proteins [30]. See reviews for more information [22, 23, 30].

Because of the development of novel approaches, the "old lady" 2-DE has been pronounced dead on many occasions, but without doubt 2-DE is still, and will remain, an important key technology in proteomics. It should be noted that 2-DE–based proteomics is the approach of choice for parallel and independent analysis of structurally distinct protein variants/species resulting from alternative mRNA splicing and/or differential cotranslational/posttranslational protein modification/processing [39, 40]. This relates to the fact that 2-DE is employed to separate intact proteins. This is in contrast with LC-MS/MS–based proteomics applications, where complex protein samples are digested into peptides, which are subsequently separated by LC and analyzed by MS/MS. Because of its peptide-centric nature, shotgun proteomics does not distinguish between peptides of identical amino acid sequence that are derived from multiple different proteins or from different protein species of a single protein. This is also one of the major reasons why we preferably employ 2-DE–based proteomics for proteome analysis of mycobacteria. In addition, 2-DE–based proteomics is still the only approach that can be routinely used for parallel comparative profiling of large sets of complex protein samples. In contrast with LC-MS/MS or Gel-C-MS/MS applications, 2-DE–based proteomics allows simultaneous analysis of multiple biological and/or technical replicates to control quality of, and increase confidence in, protein quantification. However, it is expected that LC-MS/MS and Gel-C-MS/MS applications for relative protein quantification of large sample sets will be developed in the future. The basis for this development was already established with the development of iTRAQ reagents. These are stable isotope reagents, which currently can be used to differentially label peptides in up to eight different samples in parallel [36].

8.1.3 The Use of Proteomics in Mycobacterial Research

Proteomics has greatly contributed to a better understanding of the biology of *M. tuberculosis* and other microbial pathogens [13, 41]. It has been used for comparative proteomic profiling of virulent and attenuated mycobacterial strains [42, 43, 44, 45, 46], *M. tuberculosis* strains/clinical isolates varying in virulence [47, 48, 49, 50], and mycobacteria grown under different physiologic/environmental conditions (e.g., mycobacteria grown under aerobic and anaerobic/hypoxic conditions) [51, 52, 53, 54]. This has led to identification of vaccine candidates as well as putative diagnostic markers, virulence factors, and drug targets. Proteomics has also been employed for identification of secreted proteins and T-cell antigens of mycobacteria [43, 50, 55, 56, 57, 58, 59, 60, 61], mycobacterial proteins targeted by reactive nitrogen intermediates produced in the host [62], and characterization of functional protein networks in mycobacteria [63, 64, 65]. We have used both 2-DE–based and LC-MS/MS–based proteomics to compare virulent *M. tuberculosis* and attenuated *M. bovis* BCG strains and *M. tuberculosis* clinical isolates/strains of varying virulence [42, 43, 44, 45, 66]. In addition, targeted deletion mutants of *M. tuberculosis* have been analyzed for downstream regulatory effects on global protein composition (unpublished data). Finally, we have also employed proteomics to monitor differences in protein composition of *M. tuberculosis* grown under different environmental conditions (e.g., bacilli grown in broth vs. inside murine macrophages) [67]. Our analyses were primarily aimed at the identification of proteins that are important or essential for the intracellular survival, virulence, and persistence of mycobacteria and revealed numerous vaccine candidates and putative virulence factors.

8.1.4 *The Completion of the* **M. tuberculosis** *Proteome, a Vision for the Future*

The following sections describe methods we have used for 2-DE–based proteome analysis of mycobacteria, including protocols for cultivation of mycobacteria, selective extraction, and purification of soluble cytosolic, secreted, and hydrophobic plasma membrane proteins, as well as for isolation of intracellular mycobacteria. In our experience, controlled and consistent standard operating procedures for mycobacterial culture, sample preparation, and prefractionation are required to realize the full potential of proteomics, which is a challenge when working with mycobacteria, which have a rigid cell wall. One of the major goals of proteomics is to define all the proteins encoded in the genome of an organism. However, because of the high complexity, temporal dynamics, and spatial variability of an organism's proteome, it is almost impossible to achieve this goal. The description of the complete proteome of an organism calls for analysis of all developmental stages in response to the entire environmental and physiologic conditions. The multitude of responses to physiologic and environmental stimuli (most of them not even known yet), which generate distinct protein patterns, makes it hard to imagine that we might ever establish the complete proteome of any complex biological system. An intracellular pathogen, such as *M. tuberculosis*, is even less tractable to proteomics. When analyzing the proteome of intracellularly grown *M. tuberculosis*, we observed that isolation of mycobacteria from phagosomes of infected host cells bears the burden of strong contamination with host cell proteins, which limits access to less abundant mycobacterial proteins [67]. Even more importantly, mycobacterial proteins secreted into the phagosomal lumen are lost for analysis due to the prefractionation strategy used (i.e., the washing step to strip off the phagosomal membrane). However, these proteins are of particular interest due to their potential as T-cell antigens or virulence factors [50, 56]. It is reasonable to assume that the level of contamination will be even higher upon purification of mycobacteria from host tissue. Thus, it is not possible to comprehensively analyze the proteome of mycobacteria in tissues from hosts with active or latent infection with current technologies; although such analyses hold essential keys to our understanding of pathogenesis, persistence, and drug targeting. However, adequate culture models can be used to analyze mycobacteria under conditions that mimic the *in vivo* situation. For example, Wayne and Hayes established a culture model in which progressive oxygen depletion causes mycobacteria to shift into a nonreplicating state [68]. Proteome analysis of such cultures has provided novel insights into the molecular mechanisms underlying mycobacterial dormancy in latent infection [52, 53, 69].

Standard proteomic applications only allow analysis of abundant proteins, whereas those less abundant, such as regulators and signaling proteins, often escape detection. Therefore, the parallel analysis of subproteomes with reduced complexity provides a practical approach toward more comprehensive proteome analysis. Here, the main challenge is to selectively purify proteins because of cellular location or physicochemical properties [70]. Subcellular fractionation can be used to reduce sample complexity and to get information on location, processing, and trafficking of proteins. Other protein features can also be exploited to reduce sample complexity and get hold of certain protein classes even when underrepresented in the whole proteome. These include secondary protein modifications such as phosphorylation, glycosylation, methylation, acylation, and biotinylation. Whereas many protein modifications allow protein purification and detection by specific protein tagging and/or affinity chromatography [70, 71], detergent-based extraction methods can be used to selectively enrich lipoproteins and hydrophobic membrane proteins [25, 72, 73]. Here, it becomes evident that a global, not to say complete, proteome analysis calls for creativity, imagination, and development of a wide range of techniques. A major challenge for future proteome studies will be to provide improved methods to selectively access low-abundance proteins. This will be an important step forward toward the description of the "entire" proteome. We hope that this chapter will broaden the understanding and appreciation of proteomics and foster further investigations in this exciting field of tuberculosis research.

8.2 Materials

8.2.1 Cultivation of Mycobacteria and Purification of Proteins

8.2.1.1 Purification of Soluble Proteins from Mycobacterial Whole Cell Lysates

1. 7H9-ADC medium: Dissolve 4.7 g Middlebrook 7H9 broth base (Difco, Detroit, MI) in 900 mL deionized water and add 2 mL glycerol, mix well,

and filter sterilize. Add 5 mL 20% (v/v) Tween 80 (final concentration of 0.05% v/v). Add 100 mL albumin dextrose catalase (ADC) immediately before use.

2. PBS-TW; PBS, 0.05% (v/v) Tween 80.

3. Proteinase inhibitor cocktail: Make stock solutions of 25 mg/mL E64, 50 mg/mL leupeptin, 50 mg/mL pepstatin, and 100 mg/mL TLCK (Sigma-Aldrich, Taufkirchen, Germany) in DMSO. Store the proteinase inhibitor stock solutions in 10-mL aliquots at –80°C. To prepare ready-to-use proteinase inhibitor cocktail, mix equal volumes of the stock solutions.

4. Sonifier 450 (Branson, Cincinnati, OH) equipped with a cup horn "High Intensity" (Heinemann, Schwäbisch-Gmünd, Germany) with metal cooling jacket connected with a FRYKA Multistar cooling device (Fryka, Esslingen, Germany).

8.2.1.2 Purification of Secreted Proteins from Mycobacterial Short-Term Culture Supernatants

1 7H9-ADC medium (see Section 8.2.1.1).

2. PBS-TW.

3. Sauton minimal medium. To prepare 1 L Sauton minimal medium, add 4 g L-asparagine to 250 mL deionized water. Swirl the solution at 80°C until the asparagine is dissolved. Add 2 g citric acid monohydrate, 0.5 g $MgSO_4 \cdot 7 H_2O$, 0.5 g K_2HPO_4, 0.05 g ferric ammonium (III)-citrate, 4.82 g pyruvic acid (sodium salt), 4.82 g D (+)-glucose, and 60 mL glycerol. Add deionized water to a final volume of 900 mL. Dissolve the chemicals, adjust the pH to 6.8 by addition of 32% (v/v) ammonia, add deionized water to a final volume of 1 L, and autoclave prior to use.

4. GP Express Plus membrane filters (pore size 0.22 μm; Millipore, Bedford, MA).

5. 0.15% (w/v) sodium deoxycholate (DOC).

6. 75% (w/v) trichloroacetic acid (TCA).

7. Acetone.

8. 0.02 M Tris, pH 9.0.

9. Bandelin Sonorex Ultrasonic Bath RK 100 H (Schalltec, Mörfelden, Germany).

8.2.1.3 Alternative Procedure for Purification of Secreted Proteins from Mycobacterial Short-Term Culture Supernatants

1. 7H9-OADC medium: Dissolve 4.7 g Middlebrook 7H9 broth base (Difco, Detroit, MI) in 900 mL deionized water and add 2 mL glycerol, mix well, and filter sterilize. Add 5 mL 20% (v/v) Tween 80 (final concentration of 0.05% v/v). Add 100 mL oleic albumin dextrose catalase (OADC) immediately before use.

2. Sauton minimal medium (see Section 8.2.1.2) supplemented with 0.05% (v/v) Tween 80.

3. N salt medium: 100 mM Bis/Tris HCl, pH 7.0, supplemented with 5 mM KCl, 7.5 mM $(NH_4)_2SO_4$, 0.5 mM $KHSO_4$, 1 mM KH_2PO_4, 10 mM $MgCl_2$, and 38 mM glycerol.
4. PBS-TW.
5. GP Express Plus membrane filters (pore size 0.22 μm; Millipore, Bedford, MA).
6. Vivacell centrifugal filter units (molecular weight cut off (MWCO) 5 kDa; Sartorius, Göttingen, Germany).
7. 0.15% (w/v) DOC.
8. 75% (w/v) TCA.
9. Acetone.
10. 0.02 M Tris, pH 9.0.
11. Bandelin Sonorex Ultrasonic Bath RK 100 H (Schalltec, Mörfelden, Germany).

8.2.1.4 Purification of Hydrophobic Proteins from Crude Plasma Membranes of Mycobacteria

1. 7H9-ADC medium (*see* Section 8.2.1.1).
2. PBS-TW.
3. Proteinase inhibitor cocktail (*see* Section 8.2.1.1).
4. Lysis buffer: 50 mM Tris, pH 7.4, 10 mM magnesium chloride, 0.02% (w/v) sodium azide.
5. Sonifier 450 (Branson, Cincinnati, OH) equipped with a cup horn "High Intensity" (Heinemann, Schwäbisch-Gmünd, Germany) with metal cooling jacket connected with a FRYKA Multistar cooling device (Fryka, Esslingen, Germany).
6. 100 mM sodium carbonate, pH 11.4.
7. Triton X-114 (TX-114; Sigma-Aldrich, Taufkirchen, Germany).
8. 0.15% (w/v) DOC.
9. 75% (w/v) TCA.
10. Acetone.

8.2.1.5 Isolation of Mycobacteria from Phagosomes of Infected Murine Bone Marrow–Derived Macrophages

1. Macrophages differentiated from bone marrow–derived precursor cells of C57BL/6 mice.
2. D10/HS medium: Dulbecco's Modified Eagle Medium (DMEM; high glucose, sodium carbonate buffered) supplemented with 10 mM HEPES, 10 mM pyruvate, 10 mM glutamine, 10% (v/v) fetal calve serum (FCS), and 5% (v/v) horse serum (HS).
3. D10 medium: DMEM supplemented with 10 mM HEPES, 10 mM pyruvate, 10 mM glutamine, and 10% (v/v) FCS.
4. 28-gauge needles & syringes (Braun, Melsungen, Germany).

5. Proteinase inhibitor cocktail (*see* Section 8.2.1.1). All buffers/solutions listed below were supplemented with 4 μL proteinase inhibitor cocktail per milliliter.
6. Lysis buffer: 20 mM HEPES, pH 6.5, 8.55% (w/v) sucrose.
7. 20 mM HEPES, pH 6.5, 0.05% (v/v) Tween 80.
8. 20 mM HEPES, pH 6.5, 0.05% (v/v) NP40.
9. 20 mM HEPES, pH 6.5, 0.005% (v/v) NP40.
10. 0.6 M potassium chloride, 0.05% (v/v) NP40.
11. 20 mM HEPES, pH 6.5, 12% (w/v) sucrose.
12. 20 mM HEPES, pH 6.5, 50% (w/v) sucrose.

8.2.2 Pretreatment of 2-DE Samples

1. 2-DE sample buffer: 50 mM Tris, pH 7.1, 50 mM potassium chloride.
2. Urea.
3. 1.4 M dithiothreitol (DTT; Biomol, Hamburg, Germany).
4. 40% (v/v) Servalytes 2–4 (Serva, Heidelberg, Germany).
5. Triton X-100 (TX-100; Sigma-Aldrich, Taufkirchen, Germany).
6. ASB-C8Φ (Calbiochem, Darmstadt, Germany).
7. Ultrafree-MC Durapore PVDF centrifugal filter devices (pore size 0.22 μm; Millipore, Bedford, MA).

8.2.3 Protein Separation by 2-DE (see Note 1)

1. 2D gel apparatus: WITA (Teltow, Germany).
2. Rod gels for IEF: containing 9 M urea, 3.5% (w/v) acrylamide, 0.3% (w/v) piperazine diacrylamide, 4% (v/v) Servalytes pH 2–11 (Serva, Heidelberg, Germany) and 2% (v/v) TX-100.
3. Second dimension gel: 23 × 30 cm gel, 0.75 to 1.5 mm thick, 15% acrylamide.

8.2.4 Visualization of Proteins in 2-DE Gels

8.2.4.1 Silver Staining

1. Fixative: 50% (v/v) ethanol, 10% (v/v) acetic acid.
2. Incubation solution: 30% (v/v) ethanol, 0.5% (v/v) glutaraldehyde, 4.1% (w/v) sodium acetate, 0.2% (w/v) sodium thiosulfate.
3. Dye solution: 0.1% (w/v) silver nitrate, 0.01% (v/v) formaldehyde.
4. Washing solution: 2.5% (w/v) sodium carbonate.
5. Developing solution: 2.5% (w/v) sodium carbonate, 0.01% (v/v) formaldehyde.
6. Stop solution: 0.05 M EDTA, 0.02% (w/v) thimerosal.

8.2.4.2 CBB G-250 Staining

1. Fixative: 50% (v/v) methanol, 2% (v/v) ortho-phosphoric acid.
2. Incubation solution: 34% (v/v) methanol, 2% (v/v) ortho-phosphoric acid, 17% (w/v) ammonium sulfate.
3. Dye: CBB G-250 (BioRad, Hercules, CA).
4. 25% (v/v) methanol.

8.2.4.3 Imidazole-Zinc Reverse Staining

1. Sensitizer solution: 0.2 M imidazole, 0.1% (w/v) SDS.
2. Developing solution: 0.2 M zinc sulfate.
3. 50 mM ammonium bicarbonate, pH 7.8, 100 mM DTT.

8.2.5 Preparation of Samples for MALDI-MS Analysis

8.2.5.1 Enzymatic In-Gel Digestion of Gel-Separated Proteins

Trypsin Digestion

1. Lyophilized porcine sequencing grade modified trypsin (Promega, Madison, WI).
2. Trypsin solution: reconstitute lyophilized trypsin in 50 mM acetic acid to a final concentration of 0.1 µg/µL. Store the resultant stock solution in 10-µL aliquots at –20°C. To prepare ready-to-use trypsin solution, mix 1 volume trypsin stock solution with 24 volumes digestion buffer.
3. Acetonitrile (ACN).
4. Trifluoroacetic acid (TFA).
5. Destaining buffer: 200 mM ammonium bicarbonate, pH 7.8, 50% (v/v) ACN.
6. Digestion buffer: 50 mM ammonium bicarbonate, pH 7.8, 5% (v/v) ACN.
7. Stop solution: 60% (v/v) ACN, 0.5% (v/v) TFA.
8. Vacuum centrifuge: Concentrator 5301 (Eppendorf, Hamburg, Germany).

Endoproteinase Glu-C Digestion

1. Lyophilized endoproteinase Glu-C sequencing grade (Roche, Penzberg, Germany).
2. Glu-C solution: reconstitute lyophilized Glu-C in deionized water to a final concentration of 0.1 µg/µL. Mix 1 volume of the resultant solution with 24 volumes of digestion buffer to prepare ready-to-use Glu-C solution (*see* **Note 2**).
3. ACN.
4. TFA.

5. Destaining buffer: 200 mM ammonium bicarbonate, pH 7.8, 50% (v/v) ACN.
6. Digestion buffer: 50 mM ammonium bicarbonate, pH 7.8, 5% (v/v) ACN.
7. Stop solution: 60% (v/v) ACN, 0.5% (v/v) TFA.
8. Vacuum centrifuge: Concentrator 5301 (Eppendorf, Hamburg, Germany).

8.2.5.2 Treatment of Proteolytic Peptides Prior to MALDI-MS Analysis

1. ZipTip$_{\mu\text{-}C18}$ pipette tips (Millipore, Bedford, MA).
2. ACN.
3. TFA.
4. Equilibration/washing buffer: 0.1% (v/v) TFA.
5. Elution buffer: 60% (v/v) ACN, 0.3% (v/v) TFA.
6. 2% (w/v) alpha-cyano-4-hydroxycinnamic acid (CHCA; Sigma-Aldrich, Taufkirchen, Germany) in 50% (v/v) ACN, 0.3% (v/v) TFA.
7. 4% (w/v) 2,5-dihydroxy benzoic acid (DHB; Bruker Daltonics, Bremen, Germany) in 33% (v/v) ACN, 0.5% (v/v) TFA.
8. Stainless steel MALDI sample plates (Applied Biosystems, Framingham, MA).

8.2.6 Protein Identification by MALDI-MS

1. 4700 Proteomics Analyzer (Applied Biosystems, Framingham, MA).
2. 4700 Explorer Software Version 3.0 (Applied Biosystems).
3. MASCOT Database Search Algorithm Version 2.0 (Matrix Science, London, UK).
4. MASCOT Daemon Tool Version 2.0 (Matrix Science).

8.3 Methods

8.3.1 Cultivation of Mycobacteria and Purification of Proteins

This section provides information for cultivation of mycobacteria, extraction of soluble proteins from mycobacterial whole cell lysates (Section 8.3.1.1), purification of secreted proteins from mycobacterial short-term culture supernatants Sections 8.3.1.2 and 8.3.1.3), selective enrichment of hydrophobic proteins from crude mycobacterial plasma membranes (Section 8.3.1.4), and isolation of mycobacteria from phagosomes of infected murine bone marrow–derived macrophages (Section 8.3.1.5) (*see* **Notes 3 and 4**).

8.3.1.1 Purification of Soluble Proteins from Mycobacterial Cell Lysates (*see* Note 5)

To allow replicate 2-DE analyses, prepare a mid log-phase mycobacteria suspension with a volume of 500 mL and a cell density of 1×10^8 to 2×10^8 per mL. This should produce a bacterial cell pellet with a wet weight of 0.5 to 0.75 g.

1. Inoculate 50 mL 7H9-ADC medium with 2×10^8 mycobacteria from seed lot in a 100-mL Erlenmeyer flask.
2. Incubate the mycobacteria using an appropriate shaking incubator under gentle agitation for 6 to 8 days at 37°C to prepare a mid log-phase culture with a cell density of 1×10^8 to 2×10^8 per mL.
3. Transfer the 50 mL culture into a 500-mL Erlenmeyer flask containing 150 mL fresh 7H9-ADC medium and incubate the bacteria for 4 to 5 days to a cell density of 1×10^8 to 2×10^8 per mL.
4. Transfer the 200 mL culture into a 1-L Erlenmeyer flask filled with 300 mL fresh 7H9-ADC medium and incubate the bacteria for another 4 to 5 days to a cell density of 1×10^8 to 2×10^8 per mL.
5. Centrifuge at $4000 \times g$ for 15 min at 4°C to sediment the mycobacteria.
6. Wash the mycobacteria twice with 100 mL chilled PBS-TW.
7. Resuspend the mycobacteria in 1 mL chilled PBS-TW and transfer them into a 2-mL Eppendorf reaction tube.
8. Centrifuge at $4000 \times g$ for 5 min at 4°C to sediment the mycobacteria and decant the supernatant.
9. Add 4 µL proteinase inhibitor cocktail to the sedimented cells.
10. Disrupt and homogenize the mycobacteria by sonication at 4°C (*see* Note 6).
11. Collect the sonicate and store it at –80°C.
12. Continue with sample preparation for 2-DE.

8.3.1.2 Purification of Secreted Proteins from Mycobacterial Short-Term Culture Supernatants (*see* Note 7)

1. Grow mycobacteria as described in steps 1 to 4 of Section 8.3.1.1 in 7H9-ADC medium to prepare a mid log-phase bacteria culture with a volume of 500 mL and a cell density of 1×10^8 to 2×10^8 per mL.
2. Centrifuge at $4000 \times g$ for 15 min at 4°C to sediment the mycobacteria.
3. Wash the mycobacteria 3 times with 100 mL chilled PBS-TW to deplete the large quantity of albumin resulting from the 7H9-ADC medium, which may confound analysis of secreted mycobacterial proteins.
4. Resuspend the mycobacteria in 5 to 10 mL Sauton minimal medium and transfer them into a 1-L Erlenmeyer flask containing 490 mL Sauton minimal medium.
5. Culture the mycobacteria under gentle agitation for 12 to 24 h at 37°C to allow release of secreted proteins to the extracellular milieu (*see* Note 8).

6. Collect the culture supernatant and filter it through a GP Express Plus membrane filter (pore size 0.22 μm).
7. Add 0.1 vol 0.15% (w/v) DOC and incubate for 10 min at RT with shaking.
8. Add 0.1 vol 75% (w/v) TCA and incubate overnight at 4°C with shaking.
9. Centrifuge at 4000 × g for 15 min at 4°C to sediment the precipitated proteins.
10. Wash the proteins with 100 mL chilled acetone.
11. Centrifuge at 4000 × g for 15 min at 4°C.
12. Resuspend the proteins in 500 μL 0.02 M Tris, pH 9.0.
13. Homogenize the sample by incubating it for 2 to 3 min in a Bandelin Sonorex Ultrasonic Bath RK 100 H (Schalltec, Mörfelden, Germany) filled with chilled water.
14. Collect the sonicate and store it at –80°C.
15. Continue with sample preparation for 2-DE.

8.3.1.3 Alternative Procedure for Purification of Secreted Proteins from Mycobacterial Short-Term Culture Supernatants (*see* Note 9)

1. For preparation of a preculture, inoculate 10 mL 7H9-OADC medium with 2×10^8 mycobacteria from seed lot.
2. Culture the mycobacteria under gentle agitation for 6 to 8 days at 37°C in a 250-mL Erlenmeyer flask to a cell density of 1×10^8 to 2×10^8 per mL.
3. Transfer 2 mL of the seed culture into a 1000-mL roller bottle containing 200 mL Sauton medium supplemented with 0.05% (v/v) Tween 80 to achieve an OD_{600} of 0.01 to 0.02.
4. Culture the mycobacteria for 4 to 5 days at 37°C on an appropriate roller apparatus until the culture reaches an OD_{600} of 0.25 to 0.4 (*see* Note 10).
5. Centrifuge at 4000 × g for 10 min at 4°C to sediment the mycobacteria.
6. Wash the sedimented mycobacteria 3 times with 100 mL chilled PBS-TW.
7. Resuspend the mycobacteria in 5 to 10 mL Sauton minimal medium supplemented with 0.05% (v/v) Tween 80 or N salt medium and transfer them into a 1000-mL roller bottle containing 200 mL of the appropriate medium.
8. Culture the mycobacteria for 24 h at 37°C.
9. Collect the culture supernatant and filter it twice through a GP Express Plus membrane filter (pore size 0.22 μm).
10. Concentrate the proteins on spin filters with a 5-kDa MWCO or using TCA and acetone precipitation as described in steps 7 to 15 of Section 8.3.1.2 (*see* Note 11).
11. Collect the concentrated proteins and store them at –80°C.
12. Continue with sample preparation for 2-DE.

8.3.1.4 Purification of Hydrophobic Proteins from Crude Plasma Membranes of Mycobacteria (*see* Note 12)

1. Grow mycobacteria as described in steps 1 to 4 of Section 8.3.1.1 in 7H9-ADC medium to prepare a mid log-phase bacteria culture with a volume of 500 mL and a cell density of 1×0^8 to 2×10^8 per mL.

2. Centrifuge at 4000 x g for 15 min at 4°C to sediment the mycobacteria.
3. Wash the mycobacteria twice with 100 mL chilled **PBS-TW**.
4. Add 4 μL proteinase inhibitor cocktail to the mycobacterial pellet and gently stir the sample for approximately 10 s.
5. Add 1.5 to 2 mL lysis buffer and gently stir the sample for approximately 10 s.
6. Disrupt and homogenize the mycobacteria by sonication as described in Section 8.3.1.1.
7. To remove unbroken cells and debris, centrifuge the sonicate twice, first at 4000 × g for 10 min at 4°C and then at 23,000 × g for 30 min at 4°C. Discard the sediments.
8. To recover a crude membrane fraction, spin the supernatant at 150,000 × g for 1 h at 4°C. Decant the supernatant, remove mycobacterial lipids sticking to the inner surface of the centrifugation tube by rinsing with lysis buffer, and collect the sediment.
9. To strip off peripheral and hydrophilic proteins, resuspend the sedimented crude membranes in 100 mM sodium carbonate, pH 11.4, to a final protein concentration of 1 mg/mL.
10. Stir the resulting solution for 30 min at 4°C.
11. Centrifuge at 150,000 × g for 1 h at 4°C and decant the supernatant.
12. For extraction of hydrophobic proteins, subject the sedimented membranes to TX-114 phase partitioning as described below.
13. Resuspend the sedimented membranes in lysis buffer to a final protein concentration of 10 mg/mL.
14. Add Triton TX-114 to a final concentration of 2% (v/v). Pretreat the TX-114 by mixing 1 vol TX-114 with 9 vol deionized water, incubate the resultant suspension for 10 min at 37°C, centrifuge at 1000 × g for 10 min at 21°C to allow phase partitioning, and finally remove the upper aqueous phase.
15. Add 4 μL proteinase inhibitor cocktail and stir the resulting solution for 1 h at 4°C.
16. Centrifuge at 150,000 × g for 1 h at 4°C to sediment insoluble proteins.
17. Collect the supernatant and transfer it into a fresh reaction tube.
18. Incubate for 10 min at 37°C to allow phase partitioning.
19. Centrifuge at 1000 × g for 10 min at room temperature (RT).
20. Carefully remove the upper aqueous phase containing hydrophilic proteins.
21. Collect the lower detergent (TX-114) phase enriched for hydrophobic proteins.
22. Wash the detergent to minimize contamination by soluble proteins. Add a volume of lysis buffer corresponding with the volume of the aqueous phase removed in step 20. Incubate for 10 min at 37°C. Centrifuge at 1000 × g for 10 min at 21°C to allow phase partitioning. Carefully remove the aqueous supernatant. Repeat.
23. Precipitate the proteins in the detergent phase by adding 5 vol chilled acetone.
24. Incubate overnight at –20°C.
25. Centrifuge at 14,000 × g for 30 min at 4°C to sediment the proteins.

26. Collect and air-dry the sediment.
27. Store the extracted proteins at –80°C.
28. Continue with sample preparation for 2-DE.

8.3.1.5 Isolation of Mycobacteria from Phagosomes of Infected Murine Bone Marrow–Derived Macrophages (*see* Note 13)

1. Plate macrophages differentiated from bone marrow–derived precursor cells of C57BL/6 mice [74] into T75 cell culture flasks. Add 10^7 macrophages and 10 mL D10/HS medium per flask. Prepare ~40 culture flasks to achieve a protein amount sufficient for replicate 2-DE analyses.
2. Infect the macrophages at a multiplicity of infection (MOI) of 5:1 with *M. tuberculosis* H37Rv and incubate them for 2 h at 37°C in a 7% carbon dioxide atmosphere.
3. Replace the D10/HS medium by 10 mL D10 medium to remove extracellular mycobacteria and further incubate the macrophages at 37°C.
4. Because *M. tuberculosis* infection drives macrophages into cell death at days 2 to 3 p.i. (post infection), add 10^7 uninfected macrophages per flask at day 3 p.i.
5. Incubate the macrophages for another 3 days at 37°C (*see* Note 14).
6. Scrape off the macrophages in 2 mL cold lysis buffer per flask. Unless otherwise indicated, the following steps are performed at 4°C.
7. Centrifuge at $470 \times g$ for 5 min at 4°C to sediment the macrophages.
8. Resuspend the macrophages in 5 to 10 mL cold lysis buffer to achieve a cell density of 10^9 per mL.
9. For cell disruption, pass the macrophages approximately 20 times through a 28-gauge syringe. Control the degree of lysis by microscopy and stop the procedure at a nuclei/intact cell ratio of 9:1.
10. Centrifuge the resulting homogenate at $120 \times g$ for 8 min at 4°C to remove debris and nuclei.
11. Collect supernatant and pellet separately.
12. Resuspend the pellet in 2 mL lysis buffer, pass the resulting suspension 10 times through a 28-gauge syringe, and then centrifuge at $120 \times g$ for 8 min at 4°C to sediment debris and nuclei.
13. Collect the supernatant and pool it with the supernatant from step 11.
14. Centrifuge the pooled supernatants twice at $120 \times g$ for 8 min at 4°C to remove undisrupted cells and nuclei.
15. Collect the supernatant (approximately 8 to 10 mL).
16. For purification of phagosomes, overlay 2-mL aliquots of the supernatant on sucrose step gradients containing 2 mL 20 mM HEPES, pH 6.5, 50% (w/v) sucrose under 2 mL 20 mM HEPES, pH 6.5, 12% (w/v) sucrose.
17. Centrifuge the sucrose step gradients at $850 \times g$ for 45 min at 4°C without brake.
18. Recover the phagosome-containing sucrose interfaces and control the purity of the collected fractions by microscopy.
19. Wash the phagosomes with 35 vol 20 mM HEPES, pH 6.5, 0.05% (v/v) Tween 80.

20. Centrifuge at 3500 × g for 15 min at 4°C to sediment the phagosomes.
21. To remove the phagosomal membranes, resuspend the phagosomes in 1 mL 20 mM HEPES, pH 6.5, 0.05% (v/v) NP40.
22. Centrifuge at 19,000 × g for 5 min at 4°C to sediment the mycobacteria.
23. Wash the cell pellet with 1 mL 0.6 M KCl, 0.05% (v/v) NP40.
24. Incubate the sample for 5 min on ice.
25. Centrifuge at 19,000 × g for 5 min at 4°C and collect the pellet.
26. Repeat steps 22 to 25.
27. Resuspend the cell pellet in 1 mL 20 mM HEPES, pH 6.5, 0.005% (v/v) NP40 and incubate the sample for 5 min on ice.
28. Centrifuge at 19,000 × g for 5 min at 4°C and collect the sedimented mycobacteria.
29. Store the isolated mycobacteria at –80°C.
30. Continue with protein purification.

8.3.2 Pretreatment of 2-DE Samples

1. Determine the volumes of the samples resulting from protein purification (Sections 8.3.1.1 to 8.3.1.4).
2. If necessary, add 2-DE sample buffer to adjust to a final protein concentration of approximately 20 mg/mL.
3. Add in order: 1.08 g urea, 100 µL 1.4 M DTT, and 100 µL 40% (v/v) Servalytes 2–4 per mL of sample to denature and reduce the proteins (see **Note 15**).
4. For solubilization of hydrophilic proteins (purified from mycobacterial cell lysates or culture supernatants), add TX-100 to a final concentration of 2% (v/v). For solubilization of hydrophobic plasma membrane proteins, additionally add ASB-C8Φ to a final concentration of 2% (v/v) (see **Note 16**).
5. Incubate the sample under gentle agitation for 30 min at RT.
6. Centrifuge at 100,000 × g for 30 min at RT. Samples derived from mycobacterial cell lysates should be centrifuged at 23,000 × g for 30 min at RT to remove debris, and sterile-filtered using Ultrafree-MC Durapore PVDF centrifugal filter devices (pore size 0.22 µm) prior to this ultracentrifugation step.
7. Collect the cleared supernatant and subject it to 2-DE analysis.

8.3.3 Protein Separation by 2-DE (see Notes 1 and 17 to 21)

1. Apply 60 to 800 µg of protein samples to the anodic side of the IEF gels (see **Note 18**).
2. Focus under nonequilibrium pH gradient electrophoresis (NEPHGE) conditions (1800 Vh (volts hours)).
3. Use the IEF gel as a stacking gel and run the second dimension on a 15% acrylamide gel.

8.3.4 Visualization of Proteins in 2-DE Gels (see Note 22)

To obtain best staining results, prepare all solutions immediately before use. Perform all incubation steps at RT under gentle shaking, unless otherwise stated. Cover the staining trays during all steps of the staining procedures to avoid evaporation of solvents and/or gel contamination by exogenous keratins.

8.3.4.1 Silver Staining (see Note 23)

1. Soak the gel for 2 h in 1 L fixative.
2. Store the gel overnight at 4°C in fixative without shaking.
3. Soak the gel for 2 h in 1 L incubation solution.
4. Incubate the gel twice for 20 min in 4 L deionized water.
5. Incubate the gel for 30 min in 1 L dye solution.
6. Discard the dye solution and quickly rinse the gel for 15 s in 1 L deionized water.
7. Incubate the gel for 1 min in 1 L 2.5% (w/v) sodium carbonate.
8. For protein visualization, incubate the gel for 4 to 6 min in 1 L developing solution.
9. To stop the developing process, incubate the gel for 20 min in 1 L stop solution.
10. Store the stained gel at 4°C and dry with a vacuum dryer.

8.3.4.2 CBB G-250 Staining (see Note 24)

1. Soak the gel overnight in 1 L fixative.
2. Incubate the gel 3 times in 2 L tap water for 30 min.
3. Soak the gel for 1 h in 1 L incubation solution.
4. For protein staining, add undissolved CBB G-250 dye powder to a final concentration of 0.066% (w/v) and incubate the gel for a total of 5 days.
5. Wash the gel for 1 min in 1 L 25% (v/v) methanol.
6. Wash the gel for 30 s in 1 L deionized water.
7. Shrink-wrap the stained gel and store it at 4°C.

8.3.4.3 Imidazole-Zinc Reverse Staining (see Note 25)

1. Wash the gel for 30 s in 1 L deionized water.
2. Soak the gel for 15 min in 1 L sensitizer solution.
3. Incubate the gel for approximately 30 s in 1 L developing solution until the gel background becomes white leaving transparent protein spots.
4. Wash the gel 3 times for 30 s in 1 L deionized water.
5. Shrink-wrap the stained gel and store it at 4°C.
6. Prior to enzymatic in-gel digestion, mobilize the proteins by soaking the excised gel material for 10 to 20 min until the gel material is completely transparent in 0.5 mL 50 mM ammonium bicarbonate, pH 7.8, 100 mM DTT.

8.3.5 Preparation of Samples for MALDI-MS Analysis (see Note 26)

This section describes the preparation of peptide samples for MALDI-MS analysis. Gel-separated proteins are digested into proteolytic peptides using sequence-specific proteases or chemical reagents. These are desalted and concentrated and subsequently spotted onto MALDI sample plates.

8.3.5.1 Enzymatic In-Gel Digestion of 2-DE–Separated Proteins

Trypsin Digestion (*see* **Notes 27 to 28**)

1. Wash fresh 500 μL reaction tube twice with 200 μL 100% ACN.
2. Rinse the gel for 10 min with distilled water to remove any particulate matter.
3. Excise protein spot of interest with a clean scalpel and transfer it into the reaction tube. Cut as close to the edge of the protein spot as possible to reduce the amount of "background" gel.
4. Chop the excised gel material into pieces of about 1×1 mm using stainless steel tweezers.
5. Add 500 μL destaining buffer and incubate under agitation for 30 min at 30°C to wash and destain the gel material.
6. Discard the supernatant.
7. Equilibrate the gel material by adding 500 μL digestion buffer.
8. Incubate under agitation for 30 min at 30°C.
9. Discard the supernatant.
10. To destain strongly dyed gel material, repeat the steps 7 to 9.
11. Prior to addition of trypsin solution, dry the gel material by means of a vacuum centrifuge to ensure that the endoproteinase will be able to diffuse well into the gel matrix.
12. Add 25 μL trypsin solution (containing 0.1 μg trypsin).
13. After 15 to 20 min, check the sample. If all liquid has been absorbed by the gel material, add an appropriate volume of digestion buffer to cover the gel pieces.
14. Incubate the sample under agitation overnight at 37°C to allow protein digestion.
15. Centrifuge the sample at $1000 \times g$ for 10 s at RT to spin down condensation water sticking to the inner surface of the reaction tube.
16. Collect the peptide-containing supernatant and transfer it into a fresh 200-μL reaction tube.
17. Add 30 μL stop solution to the gel material and incubate for 10 min at RT to terminate the protein digestion and to shrink the material.
18. Collect the supernatant and pool it with the supernatant of step 16.
19. Add 30 μL 100% ACN to the gel material and incubate for 10 min at RT.

20. Collect the supernatant and pool it with the supernatant of the previous extraction.
21. Dry the resulting peptide solution using a vacuum centrifuge (*see* **Note 29**).
22. Store the dried peptides at –80°C.

Endoproteinase Glu-C Digestion (*see* **Notes 30 to 31**)

1. Excise protein spot of interest and treat the gel material as described in steps 1 to 11 of Section 8.3.5.1.1.
2. Add 25 µL Glu-C solution (containing 0.1 µg Glu-C).
3. Incubate the sample under agitation for 4 to 16 h at 25°C to allow protein digestion.
4. Treat the sample as described in steps 15 to 22 of Section 8.3.5.1.1.

8.3.5.2 Treatment of Proteolytic Peptides Prior to MALDI-MS Analysis (*see* Note 32)

1. Wash ZipTip$_{\mu-C18}$ 3 times with 10 µL elution buffer.
2. Equilibrate ZipTip$_{\mu-C18}$ 3 times with 10 µL equilibration/washing buffer.
3. Dissolve the dried peptides resulting from enzymatic protein cleavage (from Section 8.3.5.1) in 20 µL equilibration/washing buffer.
4. Load ZipTip$_{\mu-C18}$ with peptides by fully aspirating and dispensing the resulting peptide solution approximately 20 times through the chromatography material.
5. Wash the loaded ZipTip$_{\mu-C18}$ 3 times with 10 µL equilibration buffer.
6. Wash a fresh 500 µL reaction tube twice with 100 µL elution buffer.
7. Add 1 to 2 µL elution buffer into the reaction tube.
8. To elute the peptides, carefully aspirate and dispense the elution buffer 3 times through the chromatography material of the loaded ZipTip$_{\mu-C18}$ without introducing air, and then collect the eluate (*see* **Note 33**).
9. For manual spotting of concentrated peptides onto MALDI sample plates, we routinely apply the "dried droplet technique." Mix equal volumes (0.2 to 1 µL) of concentrated peptide solution and MALDI-MS matrix solution and load 0.2 to 1 µL of the resultant mixture onto the sample plate.
10. Allow samples to air-dry before introducing the MALDI sample plate into the mass spectrometer (*see* **Notes 34 and 35**).

8.3.6 Protein Identification by MALDI-MS *(see Note 36)*

The following instructions assume the use of a 4700 Proteomics Analyzer (Applied Biosystems, Framingham, MA).

1. Employ PMF in combination with MS/MS fragmentation analysis of a minimum of three highly abundant peptides to allow unambiguous identification of 2-DE–separated proteins.

2. Acquire the MS spectra in the reflectron mode of the 4700 Proteomics Analyzer and calibrate the spectra using peptides occurring because of autolysis of trypsin or matrix-specific mass peaks as internal markers to provide high mass accuracy [21].

3. Acquire the MS/MS spectra without applying collision-induced dissociation.

4. Process the acquired MS and MS/MS spectra using the 4700 Explorer Software Version 3.0 (Applied Biosystems).

5. Create the peak lists of the MS and MS/MS spectra using the Peak-to-MASCOT Script of the 4700 Explorer Software. To this end, we routinely apply the following filter settings: mass range: from 500 Da to 4000 Da (MS); 60 Da to precursor mass –20 Da (MS/MS); peak density: $\leq$10 peaks per 200 Da (MS), $\leq$20 peaks per 100 Da (MS/MS); minimal signal-to-noise-ratio: 20 (MS), 3 (MS/MS); minimal peak area: 1000 (MS), 0 (MS/MS); maximal number of peaks per spectrum: 100 (MS, MS/MS). We do not apply smoothing, and the peaks are not de-isotoped.

6. For protein identification, search the peaklists of the MS and MS/MS spectra against an appropriate nonredundant protein database. For evaluation of mass data derived from analysis of *M. tuberculosis* proteins a protein database comprising all 3998 currently known *M. tuberculosis* H37Rv proteins can be downloaded from the TubercuList server of the Pasteur Institute (http://genolist.pasteur.fr/TubercuList/; Release R7). For database searching, we routinely use the search algorithm MASCOT version 2.0 (Matrix Science, London, UK) in conjunction with the MASCOT Daemon Tool version 2.0. Note, however, that there are several other search programs that may also be used for this purpose including SEQUEST and MS-FIT [15, 16]. Our standard search parameters are as follows: search mode: MS/MS ions search; enzyme: trypsin/P; protein mass: unrestricted; peptide ion mass tolerance: $\pm$30 ppm; fragment mass tolerance: $\pm$0.3 Da; maximum number of missed cleavage sites: 1; variable protein/peptide modifications: acetylation of N-termini of proteins, modification of cysteines by acrylamide (propionamide), and oxidation of methionines. The primary identification criterion is a significant ($p < 0.05$) MASCOT search result. To increase the confidence in protein identification, all search results are manually validated, particularly MS/MS identifications based on few peptide assignments.

8.4 Notes

1. See Klose and Kobalz [31] and Zimny-Arndt et al. [75] for detailed information. For comparative 2-DE–based proteome analysis of mycobacteria, we routinely use high-resolution gels with a size of 23 × 30 cm and a thickness of 0.75 mm (analytical gels) or 1.5 mm (preparative gels). These have a very high loading capacity and allow separation

of ~1800 protein spots from mycobacterial cell lysates [42, 44]. The complete equipment required for this 2-DE technique can be purchased from WITA (Teltow, Germany). Minigels (size: 6.6 × 8 cm) are only used for preliminary analyses or when gel resolution is not a problem.

2. Lyophilized Glu-C is quite stable at 2°C to 8°C. However, a decrease in activity will occur if stored in solution. A solution in deionized water can be used for 1 to 2 days at maximum, if stored at 2°C to 8°C. Thus, whenever possible, use freshly reconstituted Glu-C to obtain best results.

3. All protocols have successfully been used in practice. Figures 8.1 to 8.3 depict 2-DE patterns of mycobacterial proteins that were purified using the described prefractionation methods. The application of our routine method for isolation of secreted proteins from mycobacterial short-term culture supernatants even as short as 24 h (Section 8.3.1.2) resulted in identification of both mycobacterial proteins harboring cleavable N-terminal signal sequences and proteins lacking such sequences including many cytosolic marker proteins such as chaperones (Fig. 8.2) [43]. It can be assumed that the extracellular presence of the latter proteins was due to bacterial autolysis in combination with high levels of protein abundance and stability rather than specific export. Several conditions could cause autolysis such as high cell densities, extended culture time, agitation during culture, centrifugation, and sudden change of culture conditions. Hence, selective enrichment of secreted mycobacterial proteins calls for optimization of growth conditions and efficient methods to purify secreted proteins from short-term culture supernatants. For this reason, we also include an alternative protocol for purification of secreted proteins from mycobacterial short-term culture supernatants (Section 8.3.1.3). There is evidence that the application of this protocol minimizes the degree of mycobacterial autolysis [58]. Alternatively, mycobacteria may also be grown as surface pellicle on Sauton minimal medium to minimize autolysis and unspecific release of cytosolic proteins [76].

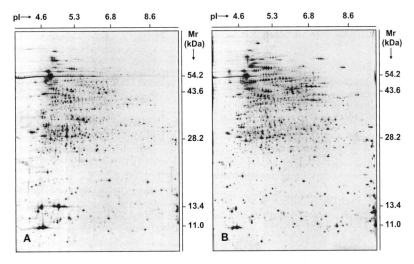

Fig. 8.1 Two-dimensional gel electrophoresis patterns of soluble proteins purified from whole cell lysates of broth-cultured and intracellularly grown *M. tuberculosis* H37Rv. Soluble proteins were purified from whole cell lysates of **(A)** broth-cultured and **(B)** intracellularly grown *M. tuberculosis* H37Rv (Sections 8.3.1.1 and 8.3.1.5), separated by 2-DE (15% polyacrylamide gels; size: 23 × 30 cm), and visualized by silver staining

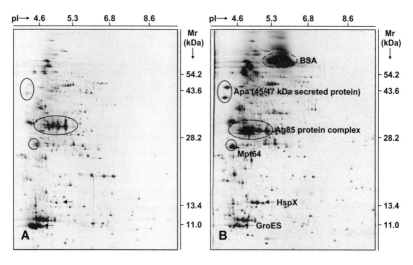

Fig. 8.2 Two-dimensional gel electrophoresis patterns of proteins purified from stationary-phase and short-term culture supernatants of broth-cultured *M. tuberculosis* H37Rv. *M. tuberculosis* H37Rv was grown in 7H9-ADC medium to prepare a mid log-phase pre-culture and subsequently n Sauton minimal medium to allow protein secretion to the extracellular milieu (Section 8.3.1.2). After **(A)** 21 days or **(B)** 24 h of culture in Sauton medium, supernatants were harvested. Proteins purified from these stationary-phase and short-term culture supernatants were subjected to 2-DE (15% polyacrylamide gels; size: 23 × 30 cm) followed by silver staining. An in-depth analysis of the 2-DE–separated proteins by MS and Edman degradation revealed many known secreted mycobacterial proteins of high abundance such as members of the antigen 85 complex and Mpt64 (circles) [21, 43]. However, it also identified several abundant mycobacterial proteins lacking a cleavable N-terminal signal sequence that are commonly considered as cytosolic proteins such as chaperones HspX and GroES (arrows) at both time points investigated

4. *M. tuberculosis* and other pathogenic mycobacteria are facultative intracellular pathogens that survive and proliferate inside host macrophages. This hallmark of mycobacterial virulence is accompanied by changes in environmental conditions and bacterial metabolism, which lead to changes in mycobacterial mRNA and protein levels [67, 77, 78, 79, 80 81]. The protocol provided in Section 8.3.1.5 has been developed for isolation of *M. tuberculosis* H37Rv from murine bone marrow–derived macrophages. This protocol may, however, also be employed to isolate mycobacteria from other cell lines such as human macrophage lines J774, Raw-B, and THP-1. It should be noted that the isolation procedure of intraphagosomal mycobacteria given here is not perfect. We observed a substantial number of host cell proteins that copurified with mycobacteria, and the mycobacterial protein yield derived from even large macrophage cultures was still limited. It was not possible to increase the yield of mycobacterial proteins without simultaneously enhancing the amount of contaminating host cell proteins. Hence, improved protocols for purification of mycobacteria from host cells are required.

5. This protocol is adapted from Jungblut et al. [42]. 2-DE patterns of soluble proteins extracted from cell lysates of broth-cultured and intracellularly grown *M. tuberculosis* H37Rv are shown in Figure 8.1.

6. For disruption and homogenization of mycobacterial cells, we routinely use a Sonifier 450 (Branson, Cincinnati, OH) equipped with a cup horn "High Intensity" (Heinemann, Schwäbisch-Gmünd, Germany) with metal cooling jacket connected with a FRYKA

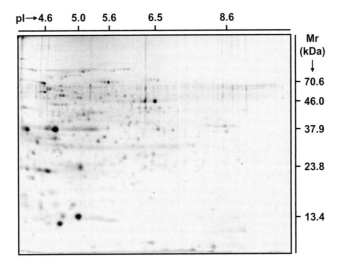

Fig. 8.3 2-DE pattern of hydrophobic proteins isolated from crude plasma membranes of broth-cultured *M. bovis* BCG Danish strain 1331. Hydrophobic proteins were extracted from crude plasma membranes of broth-cultured *M. bovis* BCG Danish strain 1331 (Section 8.3.1.4), separated by 2-DE (15% polyacrylamide gel; size: 6.6 × 8 cm), and visualized by silver staining. To enhance the number of hydrophobic proteins analyzable by 2-DE, IEF (first dimension) was performed in the presence of 9 M urea, 2% (v/v) TX-100, and 2% (v/v) ASB-C8F

Multistar cooling device (Fryka, Esslingen, Germany). This equipment allows one to sonicate pathogenic samples with very small volumes (even below 100 μL) in sealed reaction tubes indirectly (i.e., without immersion of the sonifier tip into the sample). For sonification, we routinely apply the following parameters: temperature: 4°C; time = 10 to 20 min; output = 100%; interval = 50%.

7. This protocol is adapted from Mattow et al. [43]. Mycobacteria are first grown in 7H9-ADC medium and then in Sauton minimal medium. 2-DE patterns of proteins purified from stationary-phase and short-term culture supernatants of broth-cultured *M. tuberculosis* H37Rv are depicted in Figure 8.2.

8. To increase the protein yield, the incubation time may be prolonged. This will, however, also lead to an increase of unspecific protein release as a consequence of enhanced autolysis.

9. This protocol was kindly provided by Dr. Sarah Fortune (Harvard School of Public Health, Department of Immunology and Infectious Diseases, Boston, MA). Mycobacteria are first grown in 7H9-OADC medium and then in Sauton minimal medium or other defined media without protein supplement or detergent such as N salt media. Note that you get about 100 μg of protein per 100 mL of collection.

10. When secreted proteins from different strains will be compared, the goal is to have the cultures matched for growth phase (within 0.1 OD unit). Mycobacteria tend to clump in Sauton medium, even in the presence of 0.05% (v/v) Tween 80, which may cause errors in the OD readings. The extent of the error can be roughly judged by comparing pellet volumes when the bacterial culture filtrates are harvested.

11. If comparing secreted proteins from different strains, filter concentration can be more reproducible. If using filter concentration, some hydrophobic proteins may stick to the filter. These can be retrieved by rinsing the filter with a small volume of 6 M guanidinium HCL, 1% *N*-octyl beta-D-glucopyranoside if proteins will be

analyzed directly by MS or 50 mM Tris, pH 7.1, 50 mM potassium chloride if proteins will be subjected to 2-DE.

12. This protocol is adapted from Mattow et al. [82]. A 2-DE pattern of hydrophobic proteins extracted from crude plasma membranes of broth-cultured *M. bovis* BCG is shown in Figure 8.3.

13. This protocol is adapted from Mattow et al. [67]. A 2-DE pattern of soluble proteins purified from a cell lysate of intracellularly grown *M. tuberculosis* H37Rv is shown in Figure 8.1B.

14. The time point for macrophage lysis was chosen to recover a sufficient amount of protein material for 2-DE–based proteome analysis. The specified culture conditions (MOI; duration of infection) assume the use of our in-house *M. tuberculosis* H37Rv strain, which is less virulent than other H37Rv variants. The use of other H37Rv variants or mycobacterial strains will require adaptation of the culture conditions.

15. These additions will double the sample volume resulting in final concentrations of approximately 9 M urea, 70 mM DTT, and 2% (v/v) Servalytes 2–4.

16. We have tested several nonionic and zwitterionic detergents for their ability to efficiently solubilize hydrophobic proteins in 2-DE samples (unpublished data). These included TX-100 Sigma-Aldrich, Taufkirchen, Germany), ASB-C8Φ and ASB-14 (Calbiochem, Bad Soden, Germany), as well as Chaps (Serva, Heidelberg, Germany). In our hands, the combined use of 2% (v/v) TX-100 and 2% (v/v) ASB-C8Φ resulted in the best 2-DE resolution and reproducibility. Substitution of 9 M urea by 7 M urea and 2 M thiourea did not significantly improve the 2-DE gel quality.

17. 2-DE as a combination of isoelectric focusing (IEF) and SDS-PAGE has been developed by Klose [83] and O'Farrell [84] and is used to separate proteins from complex biological protein mixtures based on their charge (isoelectric point) and size (mass). Many laboratories employ the IPG-Dalt technique by Görg et al. [26, 85] for 2-DE separation of proteins. In this approach, IEF is performed on immobilized pH gradient (IPG) strips, which are commercially available. Despite its wide use and recent methodological improvements, the IPG-Dalt technique suffers from certain limitations including relatively low resolution [14]. To evade these restrictions, we use a complementary high-resolution 2-DE technique for separation of complex protein mixtures, which combines nonequilibrium pH gradient electrophoresis (NEPHGE) carrier ampholyte IEF and SDS-PAGE [31, 83]. This technique offers a very high gel-to-gel reproducibility and allows to separate up to 10,000 protein spots from complex protein mixtures in a single 2-DE gel [31].

18. For standard applications, IEF (first dimension) is performed in rod gels. Gels used for IEF separation of hydrophobic proteins should additionally comprise 2% (v/v) ASB-C8Φ (Calbiochem, Darmstadt, Germany), a zwitterionic detergent. For preparation of high-resolution 2-DE gels (size: 23 × 30 cm), 60 µg (analytical gels) or 400 to 800 µg (preparative gels) of protein samples should be loaded. In the case of minigels, we apply 10 to 15 µg (analytical gels) or 40 to 50 µg (preparative gels) of protein sample. SDS-PAGE (second dimension) is routinely performed in 15% (w/v) polyacrylamide gels using the IEF gels as stacking gels. Under these conditions, the applied 2-DE technique allows one to separate proteins with a pI ranging from about 4 to 11.5 and a mass between 6 and 140 kDa [43, 86].

19. Upon 2-DE separation, proteins are routinely visualized by either silver staining (analytical gels) or CBB G-250 staining (preparative gels), and 2-DE gels are calibrated as described [87].

20. In comparative proteomic profiling experiments, we analyze at least three independent samples (biological replicates) per mycobacterial strain and/or biological state. In addition, individual samples are analyzed in replicate to provide a rough estimate of technical variation. For quantitative evaluation of 2-DE protein patterns, we use the image analysis software tool PDQuest Version 7.1.0 (BioRad, Hercules, CA). Note, however, that there are several other software programs that may also be used for 2-DE image

analysis such as Melanie (Genebio, Geneva, Switzerland), Phoretix 2D (Nonlinear Dynamics, Newcastle upon Tyne, UK), Delta2D (Decodon, Greifswald, Germany), ProteomeWeaver (Definiens, Munich, Germany), and Z3 (Compugen, Tel Aviv, Israel) [88, 89].

21. During 2-DE, staining of preparative 2-DE gels, enzymatic in-gel digestion of proteins, and preparation of MS samples, special caution must be taken to minimize any sample contamination by keratin and other impurities. To avoid sample contamination, wear a laboratory coat and gloves during all these processes, exclusively use sequencing or LC grade chemicals, shrink-wrap preparative polyacrylamide gels immediately upon protein staining, and carry out in-gel digestion of proteins in a sterile hood.

22. Proteins in polyacrylamide gels can be visualized by several different staining procedures. However, because of the presence of chemical cross-linkers and/or oxidizing/reducing reagents, not all staining procedures are compatible with enzymatic in-gel digestion of proteins and MS [90, 91, 92]. Because of its high sensitivity, we routinely use the silver staining technique described in Section 8.3.4.1 to visualize proteins in analytical poly-acrylamide gels. Note, however, that this silver staining technique is incompatible with MS. In contrast, colloidal CBB G-250 (Section 8.3.4.2) and imidazole-zinc reverse staining (Section 8.3.4.3) are compatible with MS and can therefore be used for staining of preparative polyacrylamide gels. The provided protocols have been optimized for polyacrylamide gels with a size of 20 cm × 30 cm and a thickness of 0.75 mm (analytical gels) or 1.5 mm (preparative gels). Other gel formats may require the application of slightly different experimental conditions (solution volumes; incubation times).

23. This protocol is adapted from Jungblut and Seifert [93].

24. This staining method by Doherty et al. [94] is more sensitive than CBB R-250 staining and requires no gel destaining.

25. This staining method by Fernandez-Patron et al. [95] is highly sensitive and may be used to visualize proteins that are not detectable by CBB dyes. Note that imidazole-zinc reverse staining and CBB R-250 staining may also be used in tandem [95, 96].

26. The methods described in this section are adapted from Refs. 21, 42, 43, 44, 67, 97, 98.

27. Trypsin is the most commonly used protease in proteome analysis. It selectively cleaves peptide bonds C-terminal to lysine and arginine residues [99]. However, at Lys-Pro and Arg-Pro bonds, it cleaves at a much slower rate than at other amino acid residues [100]. Arginine and lysine are usually fairly common amino acid residues. However, certain types of proteins including mycobacterial integral plasma membrane proteins show a relatively low arginine and lysine content. Other proteins such as many PE/PPE proteins of *M. tuberculosis* are almost free from arginine and lysine residues. These proteins may escape from identification by MS, when tryptic protein digestion is employed for gen-eration of proteolytic peptides. Accordingly, the proteome analysis of these proteins may benefit from using alternative identification strategies and/or complementary proteolytic enzymes such as Glu-C.

28. The type of trypsin that we use for protein digestion has been treated with TPCK to inactivate chymotryptic activity and modified by acetylation of the ε-amino groups of lysine residues to prevent autolysis. It is also free of impurities that might interfere with LC and/or MS applications.

29. Use of an ammonium bicarbonate buffer system means that no residue is left on drying. Instead, you just get ammonia and carbon dioxide gases.

30. Under certain conditions, it is advantageous to use complementary endoproteinases instead of or in parallel with trypsin for specific protein cleavage, in particular for the analysis of proteins with a low arginine and lysine content or when the aim is to achieve a very high MS sequence coverage. There are a variety of endoproteinases that may be used for this purpose such as chymotrypsin, Asp-N, Lys-C, and Glu-C. In addition, proteo-lytic peptides may also be generated by chemical cleavage of proteins with cyanogen bromide specifically cleaving proteins at the C-terminus of methionyl residues [101, 102].

In our hands, protein digestion by endoproteinase Glu-C (*Staphylococcus aureus* protease V8) proved to be a valuable addition to tryptic protein digestion. Glu-C selectively cleaves peptide bonds C-terminal to glutamic acid and aspartic acid residues [103]. It prefers cleavage at glutamic acid residues over aspartic acid residues at a 100-fold greater rate in all buffers and cleaves Glu-Pro and Asp-Pro bonds very slowly [104].

31. The type of Glu-C that we use for protein digestion is free of impurities that might interfere with LC and/or MS applications. The protocol for Glu-C in-gel digestion of proteins is very similar to the trypsin digestion protocol, and we use the same buffers for tryptic and Glu-C digestion of proteins.

32. It is advantageous or even necessary (for direct ESI-MS/MS applications) to desalt and concentrate peptides resulting from enzymatic protein cleavage prior to MS analysis. However, in our experience, peptide mixtures resulting from enzymatic digestion of 2-DE–separated proteins can usually be analyzed by MALDI-MS without desalting when the protein digestion is performed in an ammonium bicarbonate buffer system.

33. Sample recovery may be improved at the expense of peptide concentration by increasing the elution volume. ACN is volatile, and evaporation can occur rapidly. If this occurs, add more elution buffer to recover peptides. ZipTip pipette tips may be used repeatedly. To avoid cross-contamination of consecutive peptide samples, wash $ZipTip_{\mu\text{-}C18}$ twice with 10 μL elution buffer, twice with 10 μL 100% ACN, and 3 times with 10 μL equilibration/washing buffer after use.

34. It is also possible to elute peptides directly from ZipTip pipette tips onto MALDI sample plates. This approach is very useful for spotting of samples containing only trace amounts of peptides. For direct sample spotting in matrix, elute peptides with 0.2 to 0.5 μL 60% (v/v) ACN, 0.3% (v/v) TFA supplemented with 2% (w/v) CHCA.

35. As described, peptide mixtures resulting from enzymatic in-gel digestion of 2-DE–separated proteins do not need to be desalted prior to MALDI-MS analysis when the digestion is performed using an ammonium bicarbonate buffer system. For direct MALDI-MS analysis, dissolve dried peptides resulting from enzymatic protein cleavage in 0.5 to 2 μL 33% ACN, 0.3% TFA prior to mixing them with matrix solution for sample spotting. When necessary, peptide solutions can be concentrated or even completely dried using a vacuum centrifuge and then resuspended in an appropriate volume of 33% (v/v) ACN, 0.3% (v/v) TFA prior to MALDI-MS. Note, however, that evaporation is a critical procedure, which may cause loss of sample.

36. We routinely employ MALDI-MS PMF in combination with MALDI-MS/MS analysis of three highly abundant peptides for identification of 2-DE–separated proteins. Peptides derived from in-gel digestion of 2-DE–separated proteins may, however, also be analyzed by combination of LC and MS/MS. Most laboratories use HPLC directly coupled with a quadrupole time-of-flight (TOF) mass spectrometer to allow this type of analysis. In our experience, online LC-MS/MS performed with a quadrupole TOF mass spectrometer is a valuable addition to MALDI-MS and MS/MS to increase the confidence in protein identification, particularly for analysis of complex peptide mixtures. We have, for example, recently used MALDI-MS and MS/MS as well as online-LC-MS/MS to analyze mycobacterial plasma membrane proteins [82]. MALDI-MS/MS may also be employed in combination with LC to allow in-depth analysis of complex peptide mixtures. In this approach, the acquisition of the MS/MS data is decoupled from the LC separation of peptides. Eluting peptides are collected onto MALDI sample plates using a fraction-collecting robot and then analyzed by MALDI-MS/MS without the time constraints of an online method. As a corollary, data acquisition decisions can be made more efficiently and can be repeated, thereby increasing the number of peptides that can be analyzed from complex peptide mixtures [23, 30].

Acknowledgments The authors thank Dr. Peter Jungblut (MPIIB), Dr. Frank Schmidt (University of Oslo, Biotechnology Centre of Oslo, Oslo, Norway), and Dr. Achim Treumann (Royal College of Surgeons, Department of Clinical Pharmacology, Dublin, Ireland) for their fruitful cooperation, Dr. Sarah Fortune (Harvard School of Public Health, Department of Immunology and Infectious Diseases, Boston, MA) for providing us with the protocol described in Section 8.3.1.3, Dr. Stephen Reece (MPIIB) for critical reading of the manuscript, and the Bundesministerium für Bildung und Forschung (Competence Network "Neue Methoden zur Erfassung des Gesamtproteoms von Bakterien" J.M., S.H.E.K., U.E.S.; Competence network "PathoGenoMik-Plus"; J.M., S.H.E.K.), the Deutsche Forschungsgemeinschaft (DFG Priority Programme SPP1131; U.E.S.), the Royal Society Wolfson Research Merit Award, UK (U.E.S.), and the European Community (Project "Structural and Functional Genomics of *M. tuberculosis*"; S.H.E.K.) for financial support.

References

1. Cole ST, Brosch R, Parkhill J et al. Deciphering the biology of *Mycobacterium tuberculosis* from the complete genome sequence. Nature 1998; 396:190–198.
2. Fleischmann RD, Alland D, Eisen JA et al. Whole-genome comparison of *Mycobacterium tuberculosis* clinical and laboratory strains. J Bacteriol 2002; 184:5479–5490.
3. Cole ST, Eiglmeier K, Parkhill J et al. Massive gene decay in the leprosy bacillus. Nature 2001; 409:1007–1011.
4. Garnier T, Eiglmeier K, Camus JC et al. The complete genome sequence of *Mycobacterium bovis*. Proc Natl Acad Sci USA 2003; 100:7877–7882.
5. Wilkins MR, Pasquali C, Appel RD et al. From proteins to proteomes: large scale protein identification by two-dimensional electrophoresis and amino acid analysis. Biotechnology (NY) 1996; 14:61–65.
6. Gygi SP, Rochon Y, Franza BR, Aebersold R. Correlation between protein and mRNA abundance in yeast. Mol Cell Biol 1999; 19:1720–1730.
7. Greenbaum D, Colangelo C, Williams K, Gerstein M. Comparing protein abundance and mRNA expression levels on a genomic scale. Genome Biol 2003; 4:117.
8. Anderson L, Seilhamer J. A comparison of selected mRNA and protein abundances in human liver. Electrophoresis 1997; 18:533–537.
9. Stoeckli M, Chaurand P, Hallahan DE, Caprioli RM. Imaging mass spectrometry: a new technology for the analysis of protein expression in mammalian tissues. Nat Med 2001; 7:493–496.
10. Chaurand P, Stoeckli M, Caprioli RM. Direct profiling of proteins in biological tissue sections by MALDI mass spectrometry. Anal Chem 1999; 71:5263–5270.
11. Ho Y, Gruhler A, Heilbut A et al. Systematic identification of protein complexes in Saccharomyces cerevisiae by mass spectrometry. Nature 2002; 415: 180–183.
12. Gavin AC, Bosche M, Krause R et al. Functional organization of the yeast proteome by systematic analysis of protein complexes. Nature 2002; 415: 141–147.
13. Jungblut PR. Proteome analysis of bacterial pathogens. Microbes Infect 2001; 3: 831–840.
14. Wittmann-Liebold B, Graack HR, Pohl T. Two-dimensional gel electrophoresis as tool for proteomics studies in combination with protein identification by mass spectrometry. Proteomics 2006; 6: 4688–4703.
15. Sadygov RG, Cociorva D, Yates JR III. Large-scale database searching using tandem mass spectra: looking up the answer in the back of the book. Nat Methods 2004; 1: 195–202.
16. Shadforth I, Crowther D, Bessant C. Protein and peptide identification algorithms using MS for use in high-throughput, automated pipelines. Proteomics 2005; 5: 4082–4095.

17. Mattow J, Jungblut PR, Müller EC, Kaufmann SH. Identification of acidic, low molecular mass proteins of *Mycobacterium tuberculosis* strain H37Rv by matrix-assisted laser desorption/ionization and electrospray ionization mass spectrometry. Proteomics 2001; 1: 494–507.

18. Gevaert K, Vandekerckhove J. Protein identification methods in proteomics. Electrophoresis 2000; 21: 1145–1154.

19. Lahm HW, Langen H. Mass spectrometry: a tool for the identification of proteins separated by gels. Electrophoresis 2000; 21: 2105–2114.

20. Schmidt F, Schmid M, Jungblut PR, Mattow J, Facius A, Pleissner KP. Iterative data analysis is the key for exhaustive analysis of peptide mass fingerprints from proteins separated by two-dimensional electrophoresis. J Am Soc Mass Spectrom 2003; 14: 943–956.

21. Mattow J, Schmidt F, Höhenwarter W, Siejak F, Schaible UE, Kaufmann SH. Protein identification and tracking in two-dimensional electrophoretic gels by minimal protein identifiers. Proteomics 2004; 4: 2927–2941.

22. Aebersold R, Mann M. Mass spectrometry-based proteomics. Nature 2003; 422: 198–207.

23. Ong SE, Mann M. Mass spectrometry-based proteomics turns quantitative. Nat Chem Biol 2005; 1: 252–262.

24. Rabilloud T. Membrane proteins ride shotgun. Nat Biotechnol 2003; 21: 508–510.

25. Santoni V, Molloy M, Rabilloud T. Membrane proteins and proteomics: un amour impossible? Electrophoresis 2000; 21: 1054–1070.

26. Görg A, Weiss W, Dunn MJ. Current two-dimensional electrophoresis technology for proteomics. Proteomics 2004; 4: 3665–3685.

27. Delahunty C, Yates JR III. Protein identification using 2D-LC-MS/MS. Methods 2005; 35: 248–255.

28. Washburn MP, Wolters D, Yates JR III. Large-scale analysis of the yeast proteome by multidimensional protein identification technology. Nat Biotechnol 2001; 19: 242–247.

29. Wolters DA, Washburn MP, Yates JR III. An automated multidimensional protein identification technology for shotgun proteomics. Anal Chem 2001; 73: 5683–5690.

30. Domon B, Aebersold R. Mass spectrometry and protein analysis. Science 2006; 312: 212–217.

31. Klose J, Kobalz U. Two-dimensional electrophoresis of proteins: an updated protocol and implications for a functional analysis of the genome. Electrophoresis 1995; 16:1034–1059.

32. Unlu M, Morgan ME, Minden JS. Difference gel electrophoresis: a single gel method for detecting changes in protein extracts. Electrophoresis 1997; 18:2071–2077.

33. Gygi SP, Rist B, Gerber SA, Turecek F, Gelb MH, Aebersold R. Quantitative analysis of complex protein mixtures using isotope-coded affinity tags. Nat Biotechnol 1999; 17:994–999.

34. Ong SE, Blagoev B, Kratchmarova I et al. Stable isotope labeling by amino acids in cell culture, SILAC, as a simple and accurate approach to expression proteomics. Mol Cell Proteomics 2002; 1:376–386.

35. Chakraborty A, Regnier FE. Global internal standard technology for comparative proteomics. J Chromatogr A 2002; 949:173–184.

36. Ross PL, Huang YN, Marchese JN et al. Multiplexed protein quantitation in Saccharomyces cerevisiae using amine-reactivre isobaric tagging reagents. Mol Cell Proteomics 2004; 3:1154-1169

37. Schmidt A, Kellermann J, Lottspeich F. A novel strategy for quantitative proteomics using isotope-coded protein labels. Proteomics 2005; 5:4–15.

38. Mirgorodskaya OA, Kozmin YP, Titov MI, Korner R, Sonksen CP, Roepstorff P. Quantitation of peptides and proteins by matrix-assisted laser desorption/ionization

mass spectrometry using (18)O-labeled internal standards. Rapid Commun Mass Spectrom 2000; 14:1226–1232.

39. Jungblut P, Thiede B, Zimny-Arndt U et al. Resolution power of two-dimensional electrophoresis and identification of proteins from gels. Electrophoresis 1996; 17:839–847.

40. Jungblut P, Thiede B. Protein identification from 2-DE gels by MALDI mass spectrometry. Mass Spectrom Rev 1997; 16:145–162.

41. Jungblut PR, Hecker M. Proteomics of microbial pathogens. Proteomics 2004; 4:2829–2830.

42. Jungblut PR, Schaible UE, Mollenkopf HJ et al. Comparative proteome analysis of *Mycobacterium tuberculosis* and *Mycobacterium bovis* BCG strains: towards functional genomics of microbial pathogens. Mol Microbiol 1999; 33:1103–1117.

43. Mattow J, Schaible UE, Schmidt F et al. Comparative proteome analysis of culture supernatant proteins from virulent *Mycobacterium tuberculosis* H37Rv and attenuated *Mycobacterium bovis* BCG Copenhagen. Electrophoresis 2003; 24:3405–3420.

44. Mattow J, Jungblut PR, Schaible UE et al. Identification of proteins from *Mycobacterium tuberculosis* missing in attenuated *Mycobacterium bovis* BCG strains. Electrophoresis 2001; 22:2936–2946.

45. Schmidt F, Donahoe S, Hagens K et al. Complementary analysis of the *Mycobacterium tuberculosis* proteome by two-dimensional electrophoresis and isotope-coded affinity tag technology. Mol Cell Proteomics 2004; 3:24–42.

46. Sinha S, Arora S, Kosalai K, Namane A, Pym AS, Cole ST. Proteome analysis of the plasma membrane of *Mycobacterium tuberculosis*. Comp Funct Genom 2002; 3:470–483.

47. Pheiffer C, Betts JC, Flynn HR, Lukey PT, van Helden P. Protein expression by a Beijing strain differs from that of another clinical isolate and *Mycobacterium tuberculosis* H37Rv. Microbiology 2005; 151:1139–1150.

48. He XY, Zhuang YH, Zhang XG, Li GL. Comparative proteome analysis of culture supernatant proteins of *Mycobacterium tuberculosis* H37Rv and H37Ra. Microbes Infect 2003; 5:851–856.

49. Betts JC, Dodson P, Quan S et al. Comparison of the proteome of *Mycobacterium tuberculosis* strain H37Rv with clinical isolate CDC 1551. Microbiology 2000; 146:3205–3216.

50. Bahk YY, Kim SA, Kim JS et al. Antigens secreted from *Mycobacterium tuberculosis*: identification by proteomics approach and test for diagnostic marker. Proteomics 2004; 4:3299–3307.

51. Starck J, Kallenius G, Marklund BI, Andersson DI, Akerlund T. Comparative proteome analysis of *Mycobacterium tuberculosis* grown under aerobic and anaerobic conditions. Microbiology 2004; 150:3821–3829.

52. Rosenkrands I, Slayden RA, Crawford J, Aagaard C, Barry CE, III, Andersen P. Hypoxic response of *Mycobacterium tuberculosis* studied by metabolic labeling and proteome analysis of cellular and extracellular proteins. J Bacteriol 2002; 184:3485–3491.

53. Boon C, Li R, Qi R, Dick T. Proteins of *Mycobacterium bovis* BCG induced in the Wayne dormancy model. J Bacteriol 2001; 183:2672–2676.

54. Betts JC, Lukey PT, Robb LC, McAdam RA, Duncan K. Evaluation of a nutrient starvation model of *Mycobacterium tuberculosis* persistence by gene and protein expression profiling. Mol Microbiol 2002; 43:717–731.

55. Andersen P, Askgaard D, Ljungqvist L, Bennedsen J, Heron I. Proteins released from *Mycobacterium tuberculosis* during growth. Infect Immun 1991; 59:1905–1910.

56. Covert BA, Spencer JS, Orme IM, Belisle JT. The application of proteomics in defining the T cell antigens of *Mycobacterium tuberculosis*. Proteomics 2001; 1:574–586.

57. Dobos KM, Spencer JS, Orme IM, Belisle JT. Proteomic approaches to antigen discovery. Methods Mol Med 2004; 94:3–17.
58. Fortune SM, Jaeger A, Sarracino DA et al. Mutually dependent secretion of proteins required for mycobacterial virulence. Proc Natl Acad Sci U S A 2005; 102:10676–10681.
59. Rosenkrands I, Weldingh K, Jacobsen S et al. Mapping and identification of *Mycobacterium tuberculosis* proteins by two-dimensional gel electrophoresis, microsequencing and immunodetection. Electrophoresis 2000; 21:935–948.
60. Sinha S, Kosalai K, Arora S et al. Immunogenic membrane-associated proteins of *Mycobacterium tuberculosis* revealed by proteomics. Microbiology 2005; 151:2411–2419.
61. Weldingh K, Rosenkrands I, Jacobsen S, Rasmussen PB, Elhay MJ, Andersen P. Two-dimensional electrophoresis for analysis of *Mycobacterium tuberculosis* culture filtrate and purification and characterization of six novel proteins. Infect Immun 1998; 66:3492–3500.
62. Rhee KY, Erdjument-Bromage H, Tempst P, Nathan CF. S-nitroso proteome of *Mycobacterium tuberculosis*: enzymes of intermediary metabolism and antioxidant defense. Proc Natl Acad Sci USA 2005; 102:467–472.
63. Strong M, Graeber TG, Beeby M et al. Visualization and interpretation of protein networks in *Mycobacterium tuberculosis* based on hierarchical clustering of genome-wide functional linkage maps. Nucleic Acids Res 2003; 31:7099–7109.
64. Strong M, Mallick P, Pellegrini M, Thompson MJ, Eisenberg D. Inference of protein function and protein linkages in *Mycobacterium tuberculosis* based on prokaryotic genome organization: a combined computational approach. Genome Biol 2003; 4:R59.
65. Mawuenyega KG, Forst CV, Dobos KM et al. *Mycobacterium tuberculosis* functional network analysis by global subcellular protein profiling. Mol Biol Cell 2005; 16:396–404.
66. Mollenkopf HJ, Grode L, Mattow J et al. Application of mycobacterial proteomics to vaccine design: Improved protection by *Mycobacterium bovis* BCG prime-Rv3407 DNA boost vaccination against tuberculosis. Infect Immun 2004; 72:6471–6479.
67. Mattow J, Siejak F, Hagens K et al. Proteins unique to intraphagosomally grown *Mycobacterium tuberculosis*. Proteomics 2006; 6:2485–2494.
68. Wayne LG, Hayes LG. An in vitro model for sequential study of shiftdown of *Mycobacterium tuberculosis* through two stages of nonreplicating persistence. Infect Immun 1996; 64:2062–2069.
69. Boon C, Dick T. *Mycobacterium bovis* BCG response regulator essential for hypoxic dormancy. J Bacteriol 2002; 184:6760–6767.
70. Stasyk T, Huber LA. Zooming in: fractionation strategies in proteomics. Proteomics 2004; 4:3704–3716.
71. Lescuyer P, Hochstrasser DF, Sanchez JC. Comprehensive proteome analysis by chromatographic protein prefractionation. Electrophoresis 2004; 25:1125–1135.
72. Bordier C. Phase separation of integral membrane proteins in Triton X-114 solution. J Biol Chem 1981; 256:1604–1607.
73. Fujiki Y, Hubbard AL, Fowler S, Lazarow PB. Isolation of intracellular membranes by means of sodium carbonate treatment: application to endoplasmic reticulum. J Cell Biol 1982; 93:97–102.
74. Schaible UE, Kaufmann SHE. Studying trafficing of intracellular pathogens in antigen-presenting cells. In: Sansonetti P, Zychlinsky A, eds. Methods in Microbiology, Volume 31. Academic Press, New York, 2002, p. 343–360.
75. Zimny-Arndt U, Schmid M, Ackermann R, Jungblut PR. Classical proteomics: two-dimensional electrophoresis/MALDI mass spectrometry. In: Pasatolic L, ed. Mass Spectrometry of Proteins and Peptides. Humana Press, Totowa, NJ, 2007.
76. Malen H, Berven FS, Fladmark KE, Wiker HG. Comprehensive analysis of exported proteins from *Mycobacterium tuberculosis* H37Rv. Proteomics 2007; 7:1702–1718.

77. Dubnau E, Fontan P, Manganelli R, Soares-Appel S, Smith I. *Mycobacterium tuberculosis* genes induced during infection of human macrophages. Infect Immun 2002; 70:2787–2795.

78. Monahan IM, Betts J, Banerjee DK, Butcher PD. Differential expression of mycobacterial proteins following phagocytosis by macrophages. Microbiology 2001; 147:459–471.

79. Rachman H, Strong M, Ulrichs T et al. Unique transcriptome signature of *Mycobacterium tuberculosis* in pulmonary tuberculosis. Infect Immun 2006; 74:1233–1242.

80. Schnappinger D, Ehrt S, Voskuil MI et al. Transcriptional adaptation of *Mycobacterium tuberculosis* within macrophages: Insights into the phagosomal environment. J Exp Med 2003; 198:693–704.

81. Sturgill-Koszycki S, Haddix PL, Russell DG. The interaction between *Mycobacterium* and the macrophage analyzed by two-dimensional polyacrylamide gel electrophoresis. Electrophoresis 1997; 18:2558–2565.

82. Mattow J, Siejak F, Hagens K et al. An improved strategy for selective and efficient enrichment of integral plasma membrane proteins of mycobacteria. Proteomics 2007; 7:1687–1701.

83. Klose J. Protein mapping by combined isoelectric focusing and electrophoresis of mouse tissues. A novel approach to testing for induced point mutations in mammals. Humangenetik 1975; 26:231–243.

84. O'Farrell PH. High resolution two-dimensional electrophoresis of proteins. J Biol Chem 1975; 250:4007–4021.

85. Görg A, Postel W, Westermeier R, Gianazza E, Righetti PG. Gel gradient electrophoresis, isoelectric focusing and two-dimensional techniques in horizontal, ultrathin polyacrylamide layers. J Biochem Biophys Methods 1980; 3:273–284.

86. Jungblut PR, Bumann D, Haas G et al. Comparative proteome analysis of *Helicobacter pylori*. Mol Microbiol 2000; 36:710–725.

87. Aksu S, Scheler C, Focks N et al. An iterative calibration method with prediction of post-translational modifications for the construction of a two-dimensional electrophoresis database of mouse mammary gland proteins. Proteomics 2002; 2:1452–1463.

88. Dowsey AW, Dunn MJ, Yang GZ. The role of bioinformatics in two-dimensional gel electrophoresis. Proteomics 2003; 3:1567–1596.

89. Raman B, Cheung A, Marten MR. Quantitative comparison and evaluation of two commercially available, two-dimensional electrophoresis image analysis software packages, Z3 and Melanie. Electrophoresis 2002; 23:2194–2202.

90. Westermeier R, Marouga R. Protein detection methods in proteomics research. Biosci Rep 2005; 25:19–32.

91. Westermeier R. Sensitive, quantitative, and fast modifications for Coomassie Blue staining of polyacrylamide gels. Proteomics 2006; 6:61–64.

92. Miller I, Crawford J, Gianazza E. Protein stains for proteomic applications: which, when, why? Proteomics 2006; 6:5385–5408.

93. Jungblut PR, Seifert R. Analysis by high-resolution two-dimensional electrophoresis of differentiation-dependent alterations in cytosolic protein pattern of HL-60 leukemic cells. J Biochem Biophys Methods 1990; 21:47–58.

94. Doherty NS, Littman BH, Reilly K, Swindell AC, Buss JM, Anderson NL. Analysis of changes in acute-phase plasma proteins in an acute inflammatory response and in rheumatoid arthritis using two-dimensional gel electrophoresis. Electrophoresis 1998; 19:355–363.

95. Fernandez-Patron C, Hardy E, Sosa A, Seoane J, Castellanos L. Double staining of coomassie blue-stained polyacrylamide gels by imidazole-sodium dodecyl sulfate-zinc reverse staining: sensitive detection of coomassie blue-undetected proteins. Anal Biochem 1995; 224:263–269.

96. Fernandez-Patron C, Castellanos-Serra L, Hardy E et al. Understanding the mechanism of the zinc-ion stains of biomacromolecules in electrophoresis gels: generalization of the reverse-staining technique. Electrophoresis 1998; 19:2398–2406.
97. Mollenkopf HJ, Mattow J, Schaible UE, Grode L, Kaufmann SH, Jungblut PR. Mycobacterial proteomes. Methods Enzymol 2002; 358:242–256.
98. Thiede B, Höhenwarter W, Krah A et al. Peptide mass fingerprinting. Methods 2005; 35:237–247.
99. Northrop JH, Kunitz M. Isolation of protein crystals possessing tryptic activity. Science 1931; 73:262–263.
100. Perona JJ, Craik CS. Structural basis of substrate specificity in the serine proteases. Protein Sci 1995; 4: 337–360.
101. Gross E, Witkop B. Nonenzymatic cleavage of peptide bonds: the methionine residues in bovine pancreatic ribonuclease. J Biol Chem 1962; 237:1856–1860.
102. Cordoba OL, Linskens SB, Dacci E, Santome JA. 'In gel' cleavage with cyanogen bromide for protein internal sequencing. J Biochem Biophys Methods 1997; 35:1–10.
103. Drapeau GR, Boily Y, Houmard J. Purification and properties of an extracellular protease of *Staphylococcus aureus*. J Biol Chem 1972; 247:6720–6726.
104. Birktoft JJ, Breddam K. Glutamyl endopeptidases. Methods Enzymol 1994; 244:114–126.

Chapter 9
Transport Assays and Permeability in Pathogenic Mycobacteria

Marie-Antoinette Lanéelle and Mamadou Daffé

Abstract Mycobacteria produce an effective permeability layer that consists of a mycolic acid–containing cell wall. This protection confers a natural resistance to many chemical agents and results in a low permeability toward both hydrophilic and lipophilic agents. The permeability of cells is classically measured using methods that generally need cell suspensions and are hazardous with pathogens (e.g., nutrient and antibiotic uptake). A major problem encountered with mycobacteria is their propensity to form aggregates; the addition of detergent to the cell suspension is not recommended as this disorganizes the cell envelope, rendering it more permeable to antibiotics. To circumvent this problem, growing cells are uniformly labeled with [^{3}H]-uracil, allowing a quantification of the aliquots; then, the uptake of [^{14}C]-labeled probes is followed during the first minutes. To avoid the generation of aerosols associated with the commonly used filtration methods, centrifugation through an oil mixture is the preferred alternative technique for use with *Mycobacterium tuberculosis*.

Keywords chenodeoxycholate uptake · glycerol uptake · *Mycobacterium tuberculosis* · permeability barrier · transport assays

9.1 Introduction

Many of the unique properties of mycobacteria, such as their acid-fastness, their slow growth rate, and their natural resistance to antibiotics, are at least partly related to the poor penetration of solutes across the cell wall. The mycobacterial cell wall constitutes a very efficient outer permeability barrier and is composed of the covalently linked peptidoglycan-arabinogalactan-mycolic acids (the cell wall core) and the outermost layer (also called *capsule*

M. Daffé
Department of 'Molecular Mechanisms of Mycobacterial Infections',
Institut de Pharmacologie et Biologie Structurale du Centre National
de la Recherche Scientifique and Université Paul Sabatier, 205 Route de Narbonne,
31077 Toulouse cedex 04, France
e-mail: daffe@ipbs.fr

T. Parish, A.C. Brown (eds.), *Mycobacteria Protocols*,
doi: 10.1007/978-1-59745-207-6_9, © Humana Press, Totowa, NJ 2008

in the case of pathogenic species) [1]. The cell wall has a high lipid content, constituting up to 40% of the cell wall dry mass [2]. In all the current models proposed to date, the outer permeability barrier consists of an asymmetric bilayer composed of mycolic acids covalently linked to arabinogalacan and various extractable lipids such as trehalose dimycolate and waxes [3, 4, 5]. In the case of pathogenic, slow-growing mycobacterial species (including the "non-cultivable" *Mycobacterium leprae*), this barrier is surrounded by a capsule-like coat of polysaccharides and proteins [5].

The presence of the low-permeability barrier in the mycobacterial cell wall leads to a restriction of the availability of water-soluble molecules, which include common nutrients, and could partly explain the slow growth rate of these organisms. This barrier is also likely to contribute to the intrinsic resistance of mycobacteria to many antimicrobial agents. Small hydrophilic substances can pass through the pore-forming proteins (porins), whereas larger hydrophilic molecules are unable to gain access to the inside of the mycobacterial cells. The influx of small hydrophilic agents is, however, very slow, presumably because the cell wall contains only a small number of porins exhibiting channels with small diameters compared with Gram-negative bacteria. For instance, the *Mycobacterium chelonae* porins have been shown to possess permeability far lower than that of *Escherichia coli* porins [6, 7, 8]. Lipophilic molecules are able to diffuse across the outer permeability layer of mycobacteria through the lipid domain, but this diffusion is slowed down by the lipid bilayer, which has an unusually low fluidity and abnormally high thickness. This layer is composed of a highly ordered organization of a long chain of mycolic acids aligned in a direction perpendicular to the surface creating a nearly crystalline-like structure [9, 10, 11, 12].

One key question when referring to the permeability of mycobacteria is the difficulty in distinguishing between the binding of molecules to cell wall components from their true entry into the cell and, as a consequence, to precisely measure the permeability of the mycobacterial cell wall to drugs. Many antibiotic agents are quite hydrophobic, and it is impossible to distinguish partition into the lipid interior of the walls and membranes from true entry across these barriers. When the agents possess sites of protonation, they show an unequal distribution across the plasma membrane because of the proton-motive force across this region. Moreover, the methods used to measure permeability imply that cells exist as unicellular dispersions in contact with the medium, a property not commonly encountered in mycobacteria. The addition of detergent, such as Tween 80, used for culture growth to give a more homogenous cell distribution, is not recommended as it alters the cell wall organization [13, 14].

Quantitative measurements of the permeability of the cell wall began with the pioneering work of Nikaido and colleagues [15]. The first data were obtained by measuring the hydrolysis of β-lactams by intact mycobacterial cells: cell wall permeability was calculated by assuming that drug molecules first diffuse through the cell wall and are then hydrolyzed by β-lactamases presumably localized in the hypothetical periplasmic space. Cell wall permeability to cephalosporins was shown to be 10 times lower than the permeability of the notoriously impermeable *Pseudomonas aeruginosa* outer membrane [15]. Another approach was the study of the permeability coefficient of nutrients by the estimation of their kinetics uptake. Nutrient molecules are taken up by active transport systems of high specificity (amino acids) or by facilitated diffusion (glycerol) and penetrate through the mycobacterial cell wall more slowly than through the outer membrane of *E. coli* [15].

To confirm the implication of the outer layer constituents in the impermeability of mycobacteria, radiolabeled nutrients or antibiotics have been used. Assuming that the hydrophobic compound chenodeoxycholate would diffuse through lipid domains, the permeability of the lipid pathway was measured and shown to be the major penetration pathway for hydrophobic molecules. Accordingly, this anionic agent can be used as an indicator of diffusion rates through the lipid pathway [11]. Mycobacterial mutants in which genes encoding proteins involved in wall lipid metabolism are deleted have a higher initial uptake rate of chenodeoxycholate (Fig. 9.1), indicating that cell wall mycolic acids are responsible for producing a permeability barrier for the diffusion of both hydrophobic and hydrophilic molecules [16, 17, 18]. The structures of mycolic acids have also been shown to play a crucial role in determining the fluidity of the hydrocarbon domain by the presence of *trans* cyclopropyl groups in the long chains [19, 20]. A mutant devoid of oxygenated mycolic acids exhibited a very low initial rate of uptake and accumulation of the hydrophobic chenodeoxycholate compared with the wild-type strain of *Mycobacterium tuberculosis*, suggesting an increased rigidity of the lipid layer [21].

Covalently linked mycolic acids and extractable lipids (e.g., phthiocerol dimycocerosates) present in the *M. tuberculosis* cell wall are also involved in the permeability barrier as probes have been shown to enter the cell more rapidly in lipid mutants than in the wild-type strain [14]. However, the lipid layer does not represent the only barrier, as observed with a mutant in which the gene encoding for the major porin of *Mycobacterium smegmatis*, MspA, was inactivated; the mutant presented a reduced rate of uptake of the hydrophobic probe, supporting the hypothesis that the loss of MspA indirectly reduces the permeability through the lipid pathway [22]. For the hydrophilic pathway, the uptake of glycerol, a small organic molecule that is the preferential carbon source for mycobacterial growth, was chosen [23]. As a consequence of its small size and neutral charge, both passive and facilitated diffusions have been proposed for its uptake. To inhibit protein-dependent transport, the uptake experiments were performed in melting ice; the data obtained showed that transport was abolished, indicating that this molecule enters by facilitated diffusion [23]. However, the glycerol uptake rate of a mutant affected in its mycolic acid content compared with that of the parental strain (Fig. 9.1) indicated that glycerol transport occurred at least partly through the lipid domains [16].

This chapter describes protocols adapted for use with pathogenic mycobacteria in order to determine the uptake rates of both hydrophobic and hydrophilic probes (i.e., chenodeoxycholate and glycerol, respectively).

Fig. 9.1 Uptakes of (A)
chenodeoxycholate and (B)
glycerol by *Mycobacterium
tuberculosis*, wild type (filled
symbols) and mycolate-
deficient mutant (open
squares). (Reprinted from
Jackson, M., Raynaud, C.,
Lanéelle, M.-A., Guilhot, C.,
Laurent-Winter, C.,
Enserguiex, D., Gicquel, B.,
and Daffé, M. [1999]
Inactivation of the antigen
85C gene profoundly affects
and alters the permeability
of the *Mycobacterium
tuberculosis* cell envelope.
Mol. Microbiol. 31,
1573–1587, with
permission.)

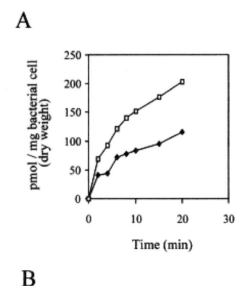

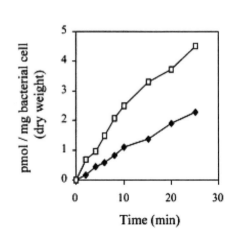

9.2 Materials

9.2.1 Quantitation of Biomass Using [^{3}H]-Uracil Labeling

1. [5, 6-^{3}H]-uracil: 2×10^{-5} M, 1.85 TBq/mmol (Amersham, Little Chalfont, UK).
2. Middlebrook 7H9 medium (Difco, Detroit, MI) supplemented with 2 mL glycerol per liter.
3. Transport buffer: 10 mM phosphate-buffered saline (PBS), pH 7.2 to 7.4: 100 mL 1.5 M NaCl, 10 mL 1 M phosphate buffer pH 7.4 (500 mL 1 M K_2HPO_4, 100 mL 1 M KH_2PO_4), and adjust to 1 L with distilled water.

4. Erlenmeyer flasks: 250 mL and 50 mL.
5. Screw-capped glass tubes, 140 mm length, 15 mm diameter.
6. Sterile glass beads, 3 mm diameter: aliquot 10 g per glass tube.
7. 5-mL and 10-mL plastic pipettes with filters.
8. Aerosol-resistant tips (e.g., Rainin Instrument Co. Inc., Oakland, CA).
9. Counting vials.
10. Centrifuge Sorvall RC5B, Waltham, MA or equivalent equipped with an aerosol containment rotor.
11. New Brunswick incubator G24, Edison, NJ or equivalent.
12. Mettler Balance; Viroflay, France.
13. Packard Tricarb 1900 TR counter: Loveland, CO equipped with a ^{3}H /^{14}C program.

9.2.2 Accumulation Assays

1. [^{14}C]-glycerol: 6.5×10^{-6} M, 5.66 GBq/mmol (Amersham).
2. [^{14}C]-chenodesoxycholate: 2×10^{-5} M, 1.85 GBq/mmol (American Radiolabeled Chemicals Inc., St. Louis, MO).
3. Plastic test tubes, 10 mL with caps.
4. 5-mL plastic pipettes with filters.
5. Aerosol-resistant tips (e.g., Rainin Instrument Co. Inc.).
6. Eppendorf conic microtubes 1.5 mL (Treff AG, Degersheim, Switzerland).
7. Micro tube cutter (Scienware, Bel-Art, Pequannock, NJ)
8. Silicone oil $d_{25°C} = 1.049$ or $d_{25°C} = 1.07$.
9. Codex paraffin oil (Gifrer, Décines, France).
10. Oils mixture: mix 1 vol silicone oil and 0.2 vol paraffin oil (carefully). Distribute 250 μL in aliquots into microtubes. Store at room temperature.
11. Scintillation liquid (Aqualuma Lumac, Groningen, The Netherlands).
12. Counting vials.
13. Microcentrifuge.
14. Waterbath shaker for 10-mL test tubes.
15. Sonication bath.

9.2.3 Measuring Radiolabels and Data Analysis

1. Packard Tricarb 1900 TR counter equipped with a ^{3}H/^{14}C program.

9.3 Methods

9.3.1 Quantitation of Biomass Using [^{3}H]-Uracil Labeling

1. Grow mycobacterial cells in 100 mL Middlebook 7H9 medium plus glycerol in a 250-mL Erlenmeyer flask. Harvest cells from a well-grown culture

(estimated by the thickness of the veil at the surface of the medium) by discarding the medium. Alternatively, centrifuge the culture.

2. Add 10 g sterile 3-mm glass beads to the wet cell pellet. Shake gently for 30 to 60 by hand to disperse the cells until a creamy paste is obtained.
3. Add 30 mL fresh Middlebrook 7H9 culture medium and decant the aggregates (around 5 min) to obtain a suspension devoid of aggregates.
4. Transfer 25 mL of the cell suspension into a 50-mL sterile Erlenmeyer flask.
5. Add 10 μL [^{3}H]-uracil.
6. Incubate overnight at 37°C with shaking at 125 rpm.
7. Collect cells by centrifugation at 3000 × g for 15 min at 4°C in 10-mL closed plastic tubes in a centrifuge equipped with an aerosol-free rotor.
8. Wash the pellet with 20 mL transport buffer.
9. Centrifuge for 10 min at 3000 × g at 4°C.
10. Resuspend the pellet in the transport buffer and adjust the ^{3}H-labeled cell suspension with transport buffer as in step 3. This is the [^{3}H]-labeled cell suspension to be used for accumulation assays.
11. Transfer 2 × 500-μL aliquots of the suspension of [^{3}H]-labeled cells into two preweighed counting vials.
12. Allow the samples to dry overnight at room temperature in a safety cabinet. Weigh each in vials.
13. Add 5 mL of the scintillation liquid and count using a scintillation counter equipped with a ^{3}H/^{14}C program.
14. Correlate [^{3}H]-labeling with dry weight. One milliliter of the suspension will contain around 40 mg of [^{3}H]-labeled cells (*see* **Note 1**).

9.3.2 Accumulation Assays

Accumulation assays should be carried out with continuous agitation to avoid cell sedimentation and at room temperature. At least three independent uptake experiments should be carried out.

1. Dispense 1 mL of the [^{3}H]-radiolabeled cell suspension (from Section 9.3.1) into 10-mL plastic test tubes. Place the tubes on the shaker.
2. Add 10 μL [^{14}C]-labeled tracer (glycerol or chenodeoxycholate) (*see* **Note 2**) to each tube.
3. Immediately remove 100 μL of the suspension and place onto the surface of the oil mixture in the microtube (*see* **Note 3**).
4. Centrifuge at 13,000 × g for 1 min and check for the formation of cell pellet, at least for the first tube, which depends both on the quality of the oil mixture and the density of the suspension. In the case of absence of such a pellet, either centrifuge for 2 min or use a more dense suspension.
5. Place on dry ice. This sample represents time 0.
6. Repeat steps 3 to 5 at intervals time of 2, 4, 6, 8, 10, 15, and 20 min.
7. Once all samples have been collected, cut off the bottom of each tube (containing the cell pellet) and drop into a counting vial.

8. Add 5 mL scintillation solution. Leave the closed vials overnight to kill the cells (*see* **Note 4**).
9. Sonicate the vials for 30 min in a waterbath until the cell pellet is dispersed in the scintillation solution.
10. Count using a scintillation counter equipped with a $^3H/^{14}C$ program.

9.3.3 Measuring Radiolabels and Data Analysis

Counting is carried out using a Packard Tricarb 1900 TR equipped with $^3H/^{14}C$ program (Sections 9.3.1 and 9.3.2). An illustration of the glycerol and the chenodeoxycholate uptakes by *M. tuberculosis* is shown in Figure. 9.1.

1. Count the incorporation of $[^{14}C]$-tracer (in disintegrations per minutes) for each vial and correct for the residual radioactivity at time 0.
2. Weigh the aliquots of $[^3H]$-labeled cells, add 5 mL of liquid scintillation fluid, and count the radioactivity to correlate $[^3H]$-labeling to cell dry weight.
3. Calculate, the dpm of $[^{14}C]$-labeled tracer taken up per milligram of dry cells (dpm per mg dry weight) for each time point
4. Transform dpm into dps (disintegrations per seconds) to obtain the values in Becquerel (1 Bq = 1 dps).
5. Given the specific activity of the $[^{14}C]$-labeled tracer (Bq/mol), express the results as the number of picomoles of tracer accumulated by 1 mg of bacterial cells (picomoles per mg dry weight).

9.4 Notes

1. Because mycobacteria have the propensity to grow as aggregates, it is not always possible to correlate the optical density of the suspension to the weight of cells; as a consequence, the quantification of the biomass in aliquots is difficult. This problem is solved by labeling the cells by $[^3H]$-uracil to calculate biomass. The validity of the method has been shown using proline transport in *Mycobacterium aurum* [23], which tends to grow without aggregation. Measurement of proline uptake rates were the same whether they were standardized to dry weights (calculated from either optical density measurements) or $[^3H]$-uracil counting [23]. Uracil incorporation in the cells increased for the first 10 min, after which labeling remained constant. An alternative solution to aggregation is the use of detergents, such as Tween 80, in small amounts. However, Tween 80, even at the low concentrations (0.01% to 0.05%) necessary to obtain cell suspensions, alters the cell surface properties and can modify nutrient uptake rates; consequently, its use for permeability studies is not recommended [13, 14].
2. Two tracers are suggested here. The uptake of glycerol can be used as a reference for small organic molecules. Glycerol is the preferential substrate for mycobacterial growth and has been used to evaluate the role of the cell wall mycolates in the permeability of hydrophilic molecules [16]. Chenodeoxycholate is a negatively charged hydrophobic probe that diffuses through the lipid domain and has been used to evaluate the role of the lipid fluidity on diffusion [11, 16, 17,20].

3. The uptake rates of hydrophilic molecules such as glycerol are about 10- to 20-fold slower than that of chenodeoxycholate. Small neutral molecules such as glycerol are not trapped by a charge inside bacterial cells. Therefore, it is not possible to use filtration methods to separate cells from the medium, as a large proportion of the label would be lost during washing the filter. Centrifugation through an oil mixture prevents this problem and, in addition, is advantageous when dealing with pathogens. An alternative method to separate the labeled cells from the transport medium uses 0.5 M sucrose in place of oil [24]. However, this method is applicable only in the case of active transport (e.g., amino acids); in the case of molecules like glycerol that freely diffuse and bear no charge, a large part of the labeling could be lost during the centrifugation through the sucrose solution.
4. This protocol has been used with *Mycobacterium tuberculosis*. In this case, all the assays are performed in a containment laboratory using safety cabinets equipped with HEPA filters. The materials listed are appropriate for these types of experiments (e.g., aerosol-free rotors for centrifugation). The separation of labeled bacteria from the transport medium by centrifuging cells through silicone oil is much safer than filtration methods that generates aerosols. In addition, the scintillation liquid killed the cells after overnight contact, as no growth on Loewenstein-Jensen medium, the selective medium for mycobacteria, was observed.

References

1. Daffé, M., and Draper, P. (1998) The envelope layers of mycobacteria with reference to their pathogenicity. *Adv. Microbiol. Physiol.* 39, 131–203.
2. Goren, M. B. and Brennan P. J. (1979) Mycobacterial lipids: chemistry and biological activities, in *Tuberculosis* (Youmans, G. P., ed.) W.B. Saunders, Philadelphia, pp 63–193.
3. Brennan, P., and Nikaido, H. (1995) The envelope of mycobacteria. *Annu. Rev. Biochem.* 64, 29–63.
4. Draper, P. (1998) The outer parts of the mycobacterial envelope as permeability barriers. *Frontiers Biosci.* 3:d1253–d1261.
5. Draper, P. and Daffé, M. (2005) The cell envelope of *M. tuberculosis* with special reference to the capsule and the outer permeability barrier, in *Tuberculosis* (Cole, S. T. et al., eds.) ASM Press, Washington, DC, pp 261–273.
6. Jarlier, V., Gutman L. and Nikaido, H. (1991) Interplay of cell wall barrier and beta-lactamase activity determines high resistance to beta-lactam antibiotics in *Mycobacterium chelonae. Antimicrob. Agents Chemother.* 35, 1937–1939.
7. Trias, J., Jarlier, V., Benz, R. (1992) Porins in the cell wall of *Mycobacterium chelonae. Science* 258, 1479–1481.
8. Niederweis M. (2003) Mycobacterial porins—new channel proteins in unique outer membranes. *Mol. Microbiol.* 49, 1167–1177.
9. Nikaido, H., Kim S. H., and Rosenberg E.Y. (1993) Physical organization of lipids in the cell wall of *Mycobacterium chelonae. Mol. Microbiol.* 8, 1025–1030.
10. Jarlier, V. and Nikaido, H. (1994) Mycobacterial cell wall: structure and role in natural resistance to antibiotics. *FEMS Microbiol. Lett.* 123, 11–18.
11. Liu, J., Barry C. E. III, Besra, G. S., and Nikaido, H. (1996) Mycolic acid structure determines the fluidity of the mycobacterial cell wall. *J. Biol. Chem.* 271, 29545–29551.
12. Nikaido, H., (2001) Preventing drug access to targets: cell surface permeability barriers and active efflux in bacteria. *Semin. Cell Dev. Biol.* 12, 215–223.
13. Ortalo-Magné, A., Lemassu, A., Lanéelle, M.-A., Bardou, F., Silve, G., Gounon P., Marchal, G., and Daffé, M. (1996) Identification of the surface-exposed lipids on the cell envelopes of *Mycobacterium tuberculosis* and other mycobacterial species. *J. Bacteriol.* 178, 456–461.

14. Camacho, L. R., Constant, P., Raynaud, C., Lanéelle, M. A., Triccas, J., A., Gicquel, B., Daffé, M., and Guilhot, C. (2001) Analysis of the phthiocerol dimycocerosate locus of *Mycobacterium tuberculosis. J. Biol. Chem.* 276, 19845–19854.
15. Jarlier, V. and Nikaido, H. (1990) Permeability barrier to hydrophilic solutes in *Mycobacterium chelonei. J. Bacteriol.* 172, 1418–1423.
16. Jackson, M., Raynaud, C., Lanéelle, M.-A., Guilhot, C., Laurent-Winter, C., Ensergueix, D., Gicquel, B., and Daffé, M. (1999) Inactivation of the antigen 85C gene profoundly affects and alters the permeability of the *Mycobacterium tuberculosis* cell envelope. *Mol. Microbiol.* 31, 1573–1587.
17. Liu, J., and Nikaido, H. (1999) A mutant of *Mycobacterium smegmatis* defective in the biosynthesis of mycolic acids accumulates meromycolates. *Proc. Natl. Acad. Sci. U. S. A.* 96, 4011–4016.
18. Wang, L., Slayden, R. A., Barry, C. E. III and Liu, J. (2000) Cell wall structure of a mutant of *Mycobacterium smegmatis* defective in the biosynthesis of mycolic acids. *J. Biol. Chem.* 10, 7224–7229.
19. Liu, J., Rosenberg, Y. and Nikaido, H. (1995) Fluidity of the lipid domain of cell wall from *Mycobacterium chelonae. Proc. Natl. Acad. Sci. U. S. A.* 92, 11254–11258.
20. Yuan, Y., Crane D. C., Musser, J. M., Sreevatsan, S. and Barry, C. E. III (1997) MMAS-1, the branch point between *cis*- and *trans*- cyclopropane-containing oxygenated mycolates in *Mycobacterium tuberculosis. J. Biol. Chem.* 272, 10041–10049.
21. Dubnau, E., Chan, J., Raynaud, C., Mohan, V. P., Lanéelle, M.A., Yu, K., Quémard, A., Smith, I., and Daffé, M. (2000) Oxygenated mycolic acids are necessary for virulence of *Mycobacterium tuberculosis* in mice. *Mol. Microbiol.* 36, 630–637.
22. Stephan, J., Mailaender, C., Etienne, G., Daffé, M. and Niederweis, M. (2004) Multidrug resistance of a porin deletion mutant of *Mycobacterium smegmatis. Antimicrob. Agents Chemother.* 48, 4163–4170.
23. Bardou, F., Raynaud, C., Ramos, C., Lanéelle, M.A., and Lanéelle, G. (1998) Mechanism of isoniazid uptake in *Mycobacterium tuberculosis. Microbiology* 144, 2539–2544.
24. Raynaud, C., Papavinasasundaram, K.G., Speight, R.A., Springer, B., Sander, P., Böttger, E.C., Colston, J.C. and Draper, P. (2002) The function of OmpATb, a pore-forming protein of *Mycobacterium tuberculosis. Mol. Microbiol.* 46, 191–201.

Chapter 10
Continuous Culture of Mycobacteria

Joanna Bacon and Kim A. Hatch

Abstract Batch cultures have predominately been used for the study of physiology and gene expression in mycobacteria. This chapter describes the assembly of chemostats and the methodology that is being used for growing *Mycobacterium tuberculosis* in continuous culture, which provides the greatest control over experimental conditions. It is difficult to determine the underlying genetic changes that enable *M. tuberculosis* to adapt to the host environment, but *in vitro* experiments aid the interpretation of gene expression profiles of the bacillus *in vivo*. Selecting relevant host conditions for study presents a major challenge. Oxygen availability has been identified as an important environmental stimulus and is a simple parameter to adjust and monitor. Described here are continuous culture methods to determine the response of *M. tuberculosis* to low oxygen environments.

Keywords chemostat · continuous culture · gene expression · mycobacteria

10.1 Introduction

10.1.1 Growth Models

Cultures of mycobacterial species have been established in static cultures, stirred/shaken batch cultures (which could either be roller bottles or flasks), fed-batch, continuous culture, and modified fermenters for controlled stationary phase studies. The greatest control over experimental conditions is provided by continuous culture [1, 2, 3], but access to the specialist expertise and facilities required to perform chemostat cultures of mycobacterial spp., particularly pathogenic strains, is rare. This chapter describes the assembly of chemostats and the methodology for growing *M. tuberculosis* in continuous culture.

J. Bacon
TB Research, Health Protection Agency, CEPR, Porton Down, Salisbury,
SP4 OJG, UK
e-mail: Joanna.bacon@hpa.org.uk

T. Parish, A.C. Brown (eds.), *Mycobacteria Protocols*,
doi: 10.1007/978-1-59745-207-6_10, © Humana Press, Totowa, NJ 2008

However, some of the equipment and methodology described could be applied to batch systems to develop models that are better defined and controlled or monitored more effectively.

Batch cultures of mycobacterial spp. are used routinely in the research laboratory. These are closed systems with a finite supply of every necessary nutrient for growth. The organism will undergo a lag phase as cells prepare biochemically for growth, followed by a period of exponential growth. Cells then enter stationary phase at the point at which an essential nutrient starts to run out or environmental conditions become unfavorable (e.g., due to a change in pH). A fed-batch culture is essentially a batch culture that is supplied with fresh nutrients to extend the logarithmic (exponential) phase of growth. Fed-batch also has other advantages by offering a means of (a) using substrates given in a controlled manner, which would otherwise inhibit growth if present in higher concentrations, (b) increasing the production of biomass, (c) producing metabolites associated with stationary phase, or (d) overcoming catabolite repression.

Mycobacteria can use a wide range of carbon sources. Glycerol is often included in media because it can be used as the sole carbon source by most of the culturable mycobacteria. These organisms can also use a variety of nitrogen sources such as amino acids and ammonium. The type of media chosen for the isolation and identification of mycobacterial species in a clinical setting is likely to be different from the choice of medium used to define the transcriptional response and metabolic adaptation to fatty acids, which are likely to be carbon sources available to bacilli *in vivo*.

Considerations to be made when setting up a culture include whether the organism is a slow or fast grower, which carbon source is the most relevant, and whether the medium needs to be minimal or defined. For example, not all mycobacterial spp. will grow on glucose, but most will grow on glycerol [4]. Some species, such as *Mycobacterium leprae,* will not grow *in vitro* and need to be cultured *in vivo*. The composition of the medium, the volume of the culture vessel, and the degree of aeration should all be relevant to the experiment and an integral part of the experimental design.

10.1.2 Growth of Mycobacterial Species in Chemostats for Bacterial Physiology and Gene Expression Studies

It is difficult to determine the underlying genetic changes that enable *M. tuberculosis* to adapt to the host environment solely by studying gene expression of the bacillus *in vivo* because of the low number of cells that can be recovered from clinical material and the inherent environmental heterogeneity within each sample. In order to aid interpretation of *in vivo* gene expression studies and to dissect the signals that are responsible for the control of subsets of genes, more focused *in vitro* experiments are necessary [5, 6, 7]. Selecting relevant host conditions for study *in vitro* presents a major challenge. Key parameters that affect the growth of a pathogen in the host include nutrient status [8, 9, 10], environmental pH [11], nitric oxide [12, 13], and atmospheric conditions [3, 14, 15, 16, 17, 18]. As tuberculosis is predominately a respiratory disease, oxygen availability has been an obvious choice for investigation and is also a simple parameter to adjust and monitor. However, deciding on an appropriate oxygen level that mimics the availability of oxygen *in vivo* presents another hurdle as the oxygen levels that *M. tuberculosis* is exposed to in different niches of the body will be fluctuating throughout development of the disease. In general, extremes of a particular parameter are selected as being representative of the spectrum of conditions that may be encountered by the pathogen in the host (e.g., comparisons can be made after growth under aerobic versus micro-aerophilic conditions).

Continuous culture is a method for prolonging the exponential phase of the organism at a predetermined fixed doubling time under defined and controlled conditions. The apparatus in which cells are grown continuously is called a chemostat, and this has been used for the continuous culture of *M. tuberculosis* [1, 2, 3], *Mycobacterium bovis,* and *Mycobacterium bovis BCG* [19, 20]. Cells from these cultures have been used as a defined and reproducible inoculum in aerosol infection studies [21, 22], vaccine trials [23, 24], physiology and gene expression studies [2, 3, 5], and in macromolecular profiling [19].

In the chemostat, the feed medium is designed to contain all required nutrients in excess apart from one. Organisms are grown in a medium in which the concentration of an essential selected nutrient is reduced so that it will be depleted first, thereby limiting growth. Fresh medium is added via a peristaltic pump, while effluent is removed by a second peristaltic pump (or by an overflow device). Growth is then proportional to the rate of addition of the fresh medium containing the growth rate–limiting nutrient. In this way, the growth rate can be controlled precisely by varying the rate of addition of the medium, and the influence of particular carbon and nitrogen sources can be determined by modifying the medium composition. The chemostat also provides independent control of pH (by the automatic addition of acid or alkali to maintain the pH at a predetermined level), temperature, and dissolved oxygen (by altering gas flow, or altering stirring speed). Microorganisms growing in a chemostat will eventually reach a "steady-state" (usually after approximately ten mean generation times) where the chemical composition and metabolic activity of the cells are in balance with the environment. The properties of cells are highly reproducible over time in steady-state, and there is good reproducibility in gene expression profiles in cells grown under identical conditions in replicate experiments [3, 5]. In this way, the chemostat permits the study of true cause-and-effect relationships, because single parameters (such as pH, growth rate, or oxygen concentration) can be varied independently while keeping all other factors constant. Likewise, cells will be lost ("washed-out") if they cannot grow under a set of imposed conditions.

10.1.3 Some Simple Calculations Required for Establishing Chemostat Culture

The flow of medium into the vessel (F) is related to the culture volume (V) by the dilution rate (D) where $D = F/V$. The volume is expressed in mL, the flow rate is expressed in mL h^{-1}, so that the dilution rate is therefore expressed as h^{-1}. Under steady-state conditions, the biomass remains constant, therefore the specific growth rate (μ) must equal the dilution rate (i.e., $\mu = D$). The dilution rate is related to the doubling time (T_d) by the equation $T_d = \log_e 2/D$. The equation ($X = Y_{x/s}(S_o - S)$) has been derived from the material balance equation of the limiting nutrient across the system [25, 26]. The relationship between the yield of cells using substrate, $Y_{x/s}$ (grams of biomass per gram of substrate), and the limiting substrate can be calculated using $X = Y_{x/s}(S_o - S)$ where X is the biomass at steady-state (g L^{-1}) and S_o and S are the concentrations of the limiting substrate in the feed and residual substrate in the outflow, respectively.

10.1.4 Adapted Chemostat Models for Studying Stationary Phase and Persistence

Growth phase and sampling time can be modified in order to produce cells that are rapidly growing, in stationary phase, or in a persistent state. Chemostat vessels can be adapted to establish controlled batch models when there is a requirement for studying cells that are in stationary phase through depletion of nutrients independently of oxygen fluctuations or pH [9]. Methods to determine whether growth has reached stationary phase or if cells are in steady-state include optical density measurement, total viable count measurements (retrospective), and analysis of the CO_2 content of the off-gas from a culture.

10.1.5 Mycobacterial Species and Strains

Mycobacterial species can be divided into two main groups: the fast growers and the slow growers (*see* **Note 1**). However, species vary enormously in the rate at which they grow, and there is huge diversity within each group. The slow-growing mycobacteria tend to be those that are pathogenic and include *M. bovis*, *M. tuberculosis*, *Mycobacterium microti*, *Mycobacterium kansasii*, *Mycobacterium scrofulaceum*, *Mycobacterium gordonae*, and *Mycobacterium avium*. These organisms produce visible colonies on solid medium within 10 to 28 days. Fast growers include *Mycobacterium fortuitum*, *Mycobacterium chelonae*, *Mycobacterium smegmatis*, and *Mycobacterium phlei* and can be observed on solid agar after 5 days of incubation. Despite being fast growers, they can have a much lower growth rate compared with other bacteria.

10.2 Materials

10.2.1 Assembling the Chemostat

1. 1-L glass vessel (D.J. Lee & Co., Ferndown, UK; made to order).
2. Stopper, silicone, SZ4 (Cole-Parmer, Hanwell, London, UK; 06298-08).
3. Stainless steel A4 form B washers (RS Components Ltd, Corby, UK; 183-9051).
4. pH probe (220 mm) plus leads (Mettler Toledo, gel filled) (Brighton Systems, New Haven, UK; 104054481).
5. Phoenix Dissolved Oxygen (DO) probe (220) mm plus leads (Brighton Systems; 027 NG15.220.DL).
6. Phoenix membrane kit (for DO probe) (Brighton Systems; M05000T1-K5).
7. Phoenix filling solution (DO probe electrolyte) (Brighton Systems; 692-542).
8. Silicone rubber compound, flowable fluid (RS Components Ltd; R001069).
9. Dual 2-wire PFA sheathed stainless steel RT probe plus leads (temperature probe) (Brighton Systems; made to order).
10. Titanium chemostat top plate plus collars and screw fittings (CEPR, HPA, Porton Down, UK; made to order) *or* PTFE (polytetrafluoroethylene) chemostat top plate plus collars and screw fittings (Radleys, Saffron Walden, UK; made to order).
11. Silicone rubber gasket (CEPR; made to order).
12. Non-melting silicon grease (RS Components Ltd; 6026).
13. Esco silicone tubing, 4 × 1.6 mm (bore × wall) (Scientific Laboratory Supplies Ltd. [SLS], Nottingham, UK; TUB7042).
14. Portexsil silicone tubing, 6 × 2 mm (bore × wall) (SLS; 435464-1).
15. Esco silicone tubing, 1 × 2 mm (bore × wall) (SLS; TUB7012).
16. Ty-Fast cable ties, 186 mm (RS Components Ltd; 178-484).
17. Ty-Fast cable ties, 141 mm (RS Components Ltd; TY125-40-100).
18. 6-mL glass tubing (D.J. Lee & Co.; made to order).
19. Tubing connectors, Y-shaped for 4- to 5-mm tubing ID (VWR International, Lutterworth, UK; 229-3442).
20. Tubing connectors, Y-shaped for 6- to 7-mm tubing ID (VWR International; 229-3444).
21. Tubing connectors, T-shaped for 6- to 7-mm tubing ID (VWR International; 229-3424).
22. Glass media addition anti grow-back device (D.J. Lee & Co.; made to order).
23. Glass side arms (D.J. Lee & Co.; made to order).
24. Quickfit PTFE washer (QW 18/7) (Fischer Scientific, Loughborough, UK; QAK-235-v).
25. Quickfit silicone rubber ring (QR 18/7) (Fischer Scientific; QAK-175-B).
26. Quickfit plastic screw cap (QC 18/11) (Fischer Scientific; QAK115-U).
27. Titanium sample port connector (20 mm ID) (CEPR; made to order).

28. In-line connectors (fitting, ¼ inch) (Cole-Palmer; 06360-90).
29. Acro 37 TF vent device with 0.2-μm PTFE membrane (Pall Corporation, Ann Arbor, Michigan US; 4464).
30. Nalgene 2-L, 2126 Heavy Duty Vacuum Bottles, polypropylene (Jencons pls, Leighton Buzzard UK; 72-363).
31. 10-L glass media duran bottle (SLS; BOT5210).
32. Anglicon magnetic stirrer unit (Brighton Systems; MS01).
33. Watson Marlow Bredel 101U/R auto/manual control variable speed pump (0.06 to 2 rpm) (Watson Marlow Limited, Falmouth, UK; 010.4002.00U).
34. Watson Marlow Bredel 101U/R auto/manual control variable speed pump (1.0 to 32 rpm) (Watson Marlow Limited; 010.4202.00U).
35. 1-L glass media duran bottles (SLS; BOT5216).
36. Stirring bars, PTFE, wheel, 45 mm (VWR International, Lutterworth, UK; 442-0080).
37. Stirring bars, PTFE, wheel, 57 mm (VWR International; 442-0081).
38. Tape heater 5-inch diameter (24 v/50 w) (Brighton Systems; made to order).
39. Anglicon microlab fermenter control panels plus leads (Brighton Systems; made to order).
40. Soloview Data logging software package (Brighton Systems; made to order).
41. Buffer, reference standard, pH 4.0 ± 0.01 at 25°C.
42. Buffer, reference standard, pH 7.0 ± 0.01 at 25°C.
43. Hydrochloric acid.
44. Sodium hydroxide pellets.
45. Glass universals for sampling (20 mm ID) (SLS; BOT 5006).
46. Titanium tubing connectors (CEPR; made to order).
47. Digital Thermometer (Tempcon Instrumentation limited, Ford, UK; 2046T).
48. Clips, tubing, 30 mm (VWR International; 229-0592).
49. Clips, tubing, 40 mm (VWR International; 229-0593).
50. Keck ramp clamp tubing clamps, 3/8 inch (Cole-Palmer; KH-06835-07).
51. Keck ramp clamp tubing clamps, 1/4 inch (Cole-Palmer; KH-06835-03).
52. Bulb hand-pump (blowing ball with reservoir) (VWR International; 612-9953).
53. Nitrogen gas (BOC Medical, Worsley, UK).
54. Air pump (Brighton Systems; made to order).
55. Middlebrook 7H9 broth:

Na_2HPO_4	2.5 g
KH_2PO_4	1.0 g
Monosodium glutamate	0.5 g
$(NH_4)_2SO_4$	0.5 g
Sodium citrate	0.1 g

$MgSO_4 \cdot 7\,H_2O$	0.05 g
Ferric ammonium citrate	0.04 g
$CuSO_4 \cdot 5\,H_2O$	1.0 mg
Pyridoxine	1.0 mg
$ZnSO_4 \cdot 7\,H_2O$	1.0 mg
Biotin	0.5 mg
$CaCl_2 \cdot 2\,H_2O$	0.5 mg
ADC/OADC enrichment	100.0 mL
(pH 6.6 ± 0.2 at 25°C)	

10.2.1.1 Preparation

Add the glycerol to 900 mL distilled water and add the remaining components, except for the ADC or OADC enrichment. Mix thoroughly. Gently heat and bring to the boil. Autoclave the mixture for 15 min at 121°C. Cool to 50°C and aseptically add 100 mL OADC or ADC enrichment.

56. Middlebrook 7H10 agar.

   Composition of 7H10 agar per liter:

Agar	15.0 g
Na_2HPO_4	1.5 g
KH_2PO_4	1.5 g
$(NH4)_2SO_4$	0.5 g
L-glutamic acid	0.5 g
Sodium citrate	0.4 g
Ferric ammonium citrate	0.04 g
$MgSO_4 \cdot 7H_2O$	0.025 g
$ZnSO_4 \cdot 7H_2O$	1.0 mg
Pyridoxine	1.0 mg
Biotin	0.5 mg
$CaCl_2 \cdot 2H_2O$	0.5 mg
Malachite Green	0.25 mg
OADC enrichment	100.0 mL
Glycerol	5.0 mL

10.2.1.2 Preparation

Add the glycerol to 900 mL distilled water and add the remaining components, except for the OADC enrichment. Mix thoroughly. Gently heat and bring to the boil. Autoclave the mixture for 15 min at 121°C. Cool to 50°C and aseptically add 100 mL OADC enrichment. Mix thoroughly and pour into Petri dishes.

57. ADC enrichment.

Composition of ADC enrichment per 100 mL:

Bovine serum albumin fraction V	5.0 g
Glucose	2.0 g
Catalase	0.003 g
Distilled water	100.0 mL

10.2.1.3 Preparation

Add the components to the distilled water and bring the volume to 100 mL. Mix thoroughly and then filter sterilze the solution. ADC enrichment is available as a premixed powder from BBL Microbiology Systems and Difco Laboratories.

58. OADC enrichment.

Composition of OADC enrichment per 100 mL:

Bovine serum albumin fraction V	5.0 g
Glucose	2.0 g
NaCl	0.85 g
Oleic acid	0.05 g
Catalase	4.0 mg

10.2.1.4 Preparation

Add the components to the distilled water and bring the volume to 100 mL. Mix thoroughly and then filter sterilze the solution. OADC enrichment is available as a premixed powder from BBL Microbiology Systems and Difco Laboratories.

59. CAMR Mycobacterium Medium (CMM).

Composition of CMM per liter:

ACES buffer	10.0 g
KH_2PO_4	0.22 g
Distilled water	500.0 mL
CMYCO solution 1	10.0 mL
CMYCO solution 2	10.0 mL
CMYCO solution 3	100.0 mL
CMYCO solution 4	10.0 mL
Biotin (10 µg/mL solution)	10.0 mL

NaHCO$_3$	0.042 g
CMYCO solution 5	10.0 mL
Tween 80	2.0 mL

10.2.1.5 Preparation

Add the first two ingredients to the first volume of distilled water (500 mL). Add the remaining ingredients and solutions in the order listed. Stir the solution to dissolve all the ingredients. Adjust the pH to 6.5 with 20% KOH. Filter sterilize the medium using a 0.2-μm filter. Store the medium between 2°C to 8°C, in the dark and use within 2 months of the production date.

60. CMYCO solution 1.

Composition of CMYCO solution 1 per liter:

CaCl$_2$·2H$_2$O	0.055 g
MgSO$_4$·7H$_2$O	21.40 g
ZnSO$_4$·7H$_2$O	2.88 g
Distilled water	1.0 L

10.2.1.6 Preparation

To make up CMYCO solution 1, add the ingredients to the water and stir to dissolve. Store it at 2°C to 8°C. Use it within a 6-month period from the date of production.

61. CMYCO solution 2.

Composition of CMYCO solution 2 per liter:

CoCl$_2$·6H$_2$O	0.048 g
CuSO$_4$·5H$_2$O	0.0025 g
MnCl·4H$_2$O	0.002 g
Conc. HCL	0.5 mL
Distilled water	1.0 L

10.2.1.7 Preparation

To make up CMYCO solution 2, add the ingredients to the water and stir to dissolve. Store it at 2°C to 8°C. Use it within a 6-month period from the date of production.

62. CMYCO solution 3.

Composition of CMYCO solution 3 per liter:

| L-serine | 1.0 g |
| L-alanine | 1.0 g |

L-arginine	1.0 g
L-asparagine	20.0 g
L-aspartic acid	1.0 g
L-glycine	1.0 g
L-glutamic acid	1.0 g
L-isoleucine	1.0 g
L-leucine	1.0 g
Distilled water	1.0 L

10.2.1.8 Preparation

To make up CMYCO solution 3, add the ingredients to the water and stir to dissolve. Store it at 2°C to 8°C. Use it within a week of production.

63. CMYCO solution 4.

Composition of CMYCO solution 4 per liter:

Pyruvic acid sodium salt	100.0 g
Distilled water	1.0 L

10.2.1.9 Preparation

Make CMYCO solution 4 up fresh every time CMM is prepared. Add the pyruvic acid to the water and stir to dissolve the ingredients.

64. CYMCO solution 5.

Composition of CMYCO solution 5 per liter:

$FeSO_4 \cdot 7H_2O$	1.0 g
Conc. HCL	0.5 mL
Distilled water	1.0 L

10.2.1.10 Preparation

Make CMYCO solution 5 up fresh every time CMM is made. Add the ingredients to the water and stir to dissolve the ingredients.

10.2.2 Establishing Steady-State Growth

1. Chemostat (set up as in Section 10.3.1).
2. CMM or Middlebrook 7H9 medium supplemented with OADC/ADC (*see* Section 10.2.1 for the recipe).

10.2.3 Alteration of DOT Levels in Chemostat Culture

1. Chemostat (set up as in Section 10.3.1).
2. Spare 1-L glass vessel: (D.J. Lee & Co.; made to order).
3. CMM or Middlebrook 7H9 medium supplemented with OADC/ADC (*see* Section 10.2.1 for the recipe).

10.2.4 Monitoring Chemostat Parameters

1. CMM or Middlebrook 7H9 medium supplemented with OADC/ADC (*see* Section 10.2.1 for the recipe).
2. 40% v/v formaldehyde.
3. Columbia blood agar plates (Biomerieux, Basingstoke, UK; 43050).
4. Middlebrook 7H10 agar plates with OADC enrichment (*see* Section 10.2.1 for the recipe).
5. Hand-held pH meter.

10.3 Methods

10.3.1 Assembling the Chemostat

Assemble the chemostat as shown in Figure 10.1.

1. Rinse the 1-L glass chemostat vessel with distilled water. Check that the probes are intact and that the oxygen probe contains electrolyte (*see* **Note 2**).
2. Insert the probes through the ports in the top plate (Fig. 10.2). If a titanium top plate is used, this is achieved by pushing the probes through foam bungs, which are firmly screwed into the ports in the top plate. If a PTFE top plate is used, this is supplied with fittings for the probes.
3. Assemble effluent and medium lines using rubber tubing and connect them to the vessel (*see* **Note 3**).
4. Assemble and connect the air inlet, the off-gas condensate bottle (duran bottle), and sample port to the vessel. Add vent filters (0.2 μm) to air inlets, air outlets, sample port side arms, waste bottles, collection bottles, and medium bottles to maintain sterility and to prevent the build up of pressure in the vessel (*see* **Note 4**).
5. Place essential equipment in the safety cabinet, which will house the chemostat. This includes the electronic stirrer, two peristaltic pumps (medium and effluent), two autoclaved waste pots, and the medium source.
6. Add a 45-mm magnetic flea and 600 mL water to the chemostat vessel.
7. Place the vessel on top of the stirrer and on the heat pad.
8. Connect the probes to the correct controller, which will maintain growth parameters at set values within the culture vessel.

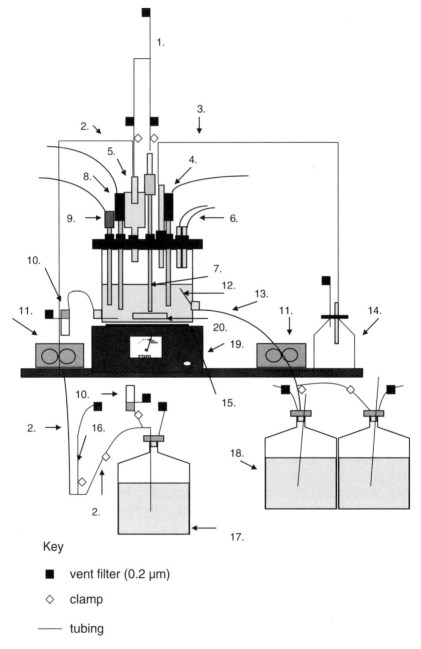

Key

■ vent filter (0.2 µm)

◇ clamp

—— tubing

Fig. 10.1 The chemostat system developed at The Health Protection Agency, CEPR (Porton Down, UK). 1, air inlet; 2, medium supply; 3, air outlet/off-gas; 4, pH probe; 5, anti–grow-back device as part of medium inlet; 6, spare ports for addition of acid and alkali; 7, air inlet tube; 8, temperature probe; 9, oxygen probe; 10, sample port; 11, peristaltic pumps; 12, effluent port; 13, effluent line; 14, off-gas condensate; 15, heat pad; 16, glass pipette for flow-rate measurement; 17, medium supply; 18, effluent waste pots or cell collection pots; 19, stirrer; 20, magnetic flea

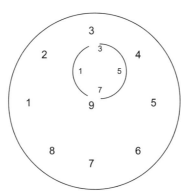

Fig. 10.2 Port designation for chemostats used at CEPR (Porton Down, UK). 1, Air outlet/off-gas; 2, Sample port (an alternative sampling port can be via a side arm, which is 2 cm from the bottom of the vessel); 3, temperature probe (dual two-wire PFA (perfluoroakoxy plastic)-sheathed stainless steel RT (resistance thermometer) probe); 4, DOT electrode; 5, spare port for acid addition; 6, spare port for alkali addition; 7, pH electrode

9. Check the oxygen probe has a signal and a rough span between 0% dissolved air saturation (via nitrogen addition) and 100% dissolved air saturation (via air addition) (*see* **Note 5**).

10. Calibrate the pH probe with pH 4 and pH 7 buffers. If the pH is to be controlled, acid (1 M HCl) and alkali (1 M NaOH) should be delivered in response to a fluctuation in acidity levels via pumps, which will also respond to the controller (*see* **Note 6**).

11. Heat the water in the vessel to 37°C using the heat pad. To calibrate the temperature probe, use a Tempcon hand-held digital thermometer to monitor the temperature in the vessel until it reaches 37°C. Set the temperature reading on the controller to 37°C (*see* **Note 7**).

12. Apply grease to the rubber gasket and seal the vessel with the top plate. Pressure-test the vessel prior to autoclaving to check for air leaks. Submerge the vessel in a sink of water with the probe fittings just under the water level. Vent filters should be above the water line so that they do not get wet (*see* **Note 4**). Leave all clamps in place apart from the air inlet and air outlet clamps, which should be removed. Extend the air outlet tubing past the vent filter using an additional length of tubing, and place the end of the tubing below the water surface. Place a bulb hand-pump onto the air inlet line and pump air in. Air bubbles should be seen coming out of the air outlet. Clamp the air outlet (between the vessel and the vent filter) and pump gently a couple of times. Observe whether air bubbles rise out of the vessel, particularly from the probe fittings on the top plate, side arms, and so forth. If any leaks appear, repair and re-test. Release the clamp on the air outlet and re-clamp the air inlets before autoclaving.

13. Autoclave the vessel at 121°C for 30 min to achieve sterility.
14. Calibrate the oxygen probe by warming up the vessel to 37°C while stirring, and pump in nitrogen and air alternately until calibrated between 0% and 100% dissolved air saturation, which is equivalent to between 0% and 20% dissolved oxygen tension (DOT) (*see* **Note 8**).
15. Ensure waste and medium bottles are connected to the chemostat.
16. Switch off heater. Drain the water from the chemostat.
17. Fill the vessel with 400 mL medium via the medium line. Warm the medium to 37°C.
18. Make up the inoculum by taking approximately five confluent plates of mycobacterial colonies, which have been incubated for 3 weeks (*see* **Note 9**), and scrape them into 10 mL of autoclaved, distilled water. Add the inoculum through the sample port.
19. Ensure that the stirrer unit is responding to the dissolved air saturation setting on the controller by seeing an increase in stirrer speed as the DOT level decreases (*see* **Note 10**).

10.3.2 Establishing Steady-State Growth

1. Leave in batch mode for approximately 48 h. For the first 24 h, the air inlet should be closed at the cabinet connection to allow the DOT set point of 10% DOT to be reached (*see* **Note 11**). After this, open the air inlet as the culture will now require additional oxygen.
2. Switch the culture to fed-batch mode by starting the media pump at a flow rate of 5 mL h^{-1}. Set medium pump by calibrating it to the required flow rate. Keep the culture at 5 mL h^{-1} for 24 h to increase the culture volume to 500 mL (*see* **Note 12**).
3. Start the culture in continuous mode at 5 mL h^{-1} by switching on the effluent pump at a speed that is higher than the medium pump in order to maintain the culture volume at 500 mL. This flow rate is maintained for 2 days to establish the culture in continuous mode prior to an increase in flow rate (*see* **Note 13**).
4. Increase the flow rate to 10 mL h^{-1} for 2 days.
5. Increase the flow rate to 15 mL h^{-1} and monitor the culture daily (*see* Notes 14 and 15.
6. Establish the culture in steady-state. The DOT level and optical density (OD) should be stable for at least 7 to 8 days.
7. Sample the culture for 5 to 7 days in steady-state. Alternatively, collect the whole culture in one sample depending on the needs of the experiment.
8. Alter the level of a particular nutrient by exchanging the existing medium bottles for those containing the modified medium (*see* **Note 16**). Alter oxygen levels by changing the set point on the controller and re-establish steady-state if required (the focus of this method, *see* Section 10.3.3). Alter pH levels by the addition of acid/alkali.

10.3.3 Alteration of DOT Levels in Chemostat Culture

1. Establish steady-state cultures at 10% DOT (*see* Section 10.3.2).
2. Remove samples from the chemostat if required.
3. Change the vessel at this stage if wall growth has started to appear (*see* **Note 17**).
4. Set the DOT to 4% on the controller and allow the culture to adjust over a 3-day period (*see* **Note 18**).
5. Set the DOT level to 2%, and let the culture adjust over 24 h.
6. Repeat this again to 1% DOT, then 0.4% DOT, and finally to 0.2% DOT over a period of 3 days.
7. Re-establish steady-state and sample.

10.3.4 Monitoring Chemostat Parameters

10.3.4.1 Daily

1. Check that the volume of liquid in the chemostat vessel is constant.
2. Check medium is entering the vessel and effluent is going into the waste pot.
3. Check that medium and waste volumes are at the expected levels and that pumps, stirrer, and magnetic flea are all working correctly.
4. Check the waste level and swap the waste to an empty pot containing neat disinfectant if required.
5. Fill in the chemostat run sheet for temperature level, pH level, DOT level, and stirrer speed (rpm) (*see* **Note 15**).
6. Sample 5 to 10 mL of the culture for optical density. Kill the cells by the addition of 1/10 volume of 40% formaldehyde (v/v). Shake the sample vigorously and leave for 16 h before the sample can be measured for optical density (*see* **Note 19**). Dilute each sample five-fold in sterile, distilled water and place the resulting cell suspension in a plastic cuvette. Read the optical density at 540 nm against water (these readings are important for determining when the culture has passed into mid-logarithmic growth and then when it reaches steady-state).

10.3.4.2 Weekly

1. Monday: Take a sample for independent pH measurement using a hand-held pH probe. Carry out purity checks on agar (2 × blood agar and 2 × Middlebrook agar plates) and optical density.
2. Friday: Check the waste levels and if necessary divert the waste line to an empty waste bottle. Check that there is sufficient medium supply available for the culture to use over the weekend.
3. Weekend: Check the chemostat once during the weekend either on a Saturday evening or Sunday morning to make sure that all parameters are correct, all equipment is working correctly, and that there are no leaks.

10.3.4.3 Occasionally

1. Move the tubing through the pumps approximately every 2 weeks in order to maintain elasticity of the tubing and to reduce the likelihood of splits developing.
2. Check flow rate (*see* **Note 13**) and temperature as required.

10.4 Notes

1. Strains of mycobacterial species (mentioned in Section 10.1.5) are available from several national collections. The National Collection of Type Cultures (NCTC) supplies strains on application to: The curator, NCTC, Health Protection Agency, 61 Colindale Ave, London, NW9 5HT, UK. Cultures may also be supplied by ATCC upon application to The American Type Culture Collection, 12301 Parklawn Drive, Rockville, MD. For a complete list of culture collections worldwide refer to: http://www.kumc.edu/instruction/medicine/microbio/lutkenhaus/colection.htm.
2. Store the pH probes in 3 M KCl and rinse with distilled water before use. Oxygen probes are stored dry. Change the membrane on the oxygen probe and refill with fresh electrolyte before every culture. Leak-test oxygen probes in a beaker of water 24h after changing the membrane. If leaks occur, change the membrane and re-test.
3. There are three thicknesses of tubing. To achieve a slow flow rate, the internal diameter of the tubing that runs through the medium pump head should have a bore width of no more than 1 mm. However, thicker tubing with a wider bore width of 4mm is used for the medium and effluent lines, and the thickest tubing with a bore width of 6mm is used for all gaseous inlets and outlets.
4. Dead ends (some of which will be used for drawing off media or cell samples), such as air outlets, sample port side arms, waste bottles, collection bottles and media bottles, flow rate measurement devices, also need to be fitted with vent filters to ensure that gas can be released continuously from the culture during growth and pressure does not build up in the vessel. The air outlet/off-gas line can be fed into a CO_2 analyzer for measurement of CO_2 levels. If for any reason these filters become wet, they will be blocked, a vacuum will build up, and the culture will be sucked up through the air inlet.
5. Do not calibrate oxygen probes before autoclaving because the electrolyte is affected by the heat. Oxygen is only transferred evenly throughout the culture if it is stirred or shaken effectively. Standing cultures of *M. tuberculosis* will result in microenvironments in which the oxygen levels will be very low reaching micro-aerophilic or anaerobic levels.
6. There is a pH/temperature compensation mode on the controller to compensate for temperature differences because pH calibration is done at room temperature. Fittings that are exposed to the acid or alkali should be made of inert metal such as titanium to prevent corrosion caused particularly by the acid.
7. The heat pad is electronically controlled by the controller unit to maintain temperature at 37°C.
8. The dissolved oxygen tension (DOT) is maintained by an immediate response of the controller to a drop in oxygen level, which in turn alters the stirrer speed to draw more air into the medium. The controller unit automatically controls the extent to which the culture is stirred via a magnetic stirring device and a flea. A DOT of 10% is equivalent to 50% dissolved air saturation. The controller displays the dissolved air saturation and not the DOT.

9. It is not advisable to use plates that are more than 4 weeks old because growth in the chemostat will be slow, and cells will be more clumped in culture.

10. 10% DOT is used in this protocol and is considered to be standard aerobic conditions [3].

11. The DOT level in the culture could be above 10% DOT and will need to drop to 10% DOT as soon as possible. It is also important for the stirrer speed to increase to disperse the cells. Once the DOT has dropped to the set point (10%), the controller will inform the stirrer to increase its speed to maintain a DOT of 10%.

12. DOT levels may fluctuate during a transition from batch to fed-batch.

13. A high flow rate too early on in continuous mode may lead to culture "washout." The flow rate is measured using a device that is constructed using a glass pipette that has been inserted into the tubing between the medium bottle and the medium pump via a connector with a T junction in it. The pipette is capped with a piece of tubing and a vent filter (Fig. 10.1). The bottom of the pipette is normally clamped off. The clamp is removed, and medium is drawn up into the pipette using a syringe attached to the vent filter at the top of the pipette. The medium bottle is then clamped off so that the culture subsequently draws the medium from the pipette and not from the medium bottle. The speed at which the culture uses the media from the pipette is then measured. The flow rate and dilution rates can then be calculated (*see* Section 10.1.3). Remember to remove the clamp from the medium bottle and replace the clamp at the bottom of the pipette once flow rate determinations have been completed.

14. Using the calculation in Section 10.1.3 a flow rate of 15 mL h^{-1} will give a dilution rate of 0.03 h^{-1} and a mean generation time of 23 h. This mean generation time has previously been used for gene expression studies, however slower growth rates have also been used for macromolecular profiling [19].

15. Run sheets and Soloview software are used to record parameters on a daily basis.

16. To obtain the correct level of limiting nutrient, preliminary cultures should be set up with the limiting nutrient absent from the culture. The culture may not grow at all or there may be a low level of growth followed by culture washout. A pulse of a small amount of the limiting nutrient is added to the vessel and subsequently the medium source to rescue the culture. This will be accompanied by a rise in the optical density. Each time a small amount of nutrient is added, the optical density is recorded. An alternative method is to set up shaking batch cultures containing varying amounts of the limiting nutrient and determine the lowest level of nutrient that will support growth. To confirm that a particular nutrient is limiting growth during continuous culture, a small amount of the limiting nutrient is added to the chemostat vessel. The culture will respond by an increase in optical density.

17. Under certain growth conditions, mycobacteria will adhere to the walls of the vessels and the probes. Once this starts to occur, the optical density is likely to fall and the DOT levels will fluctuate. The culture will no longer be in steady-state and will need to be transferred to another vessel.

18. Oxygen levels should not be dropped rapidly as the culture will wash out.

19. For guidance only, all methods associated with safe working with pathogenic mycobacterial spp. should be validated in accordance with health and safety practices within a specific workplace. A formal containment level 3 (CL3) training course should be attended, and the regulations on lab access, code of practice, CL3 lab procedure (including accident, fire, etc.) should be understood. Risk assessments should be in place. Decontamination procedures (fumigation of cabinets, waste, and microbial decontamination) should all be understood and in place. CL3 cabinets should be appropriately designed, validated, and serviced. Users should be trained in chemostat set-up, operation of chemostats at CL3, and the use of controller units. Users should also be trained in the maintenance of stock cultures, total viable counts, OD and pH measurements, purity, flow-rate checks, harvesting of mycobacterial cultures, and centrifugation/filtration.

Acknowledgments This was funded by the Department of Health and the Health Protection Agency, UK. The views expressed in this chapter are those of the authors and not necessarily those of the Department of Health or Health Protection Agency. The authors acknowledge Dr. Brian James for the huge contribution he has made to the development of chemostat models for the growth of *M. tuberculosis*. The authors also express their gratitude to Jon Allnutt for technical information and to Prof. Philip Marsh for his constructive comments.

References

1. James, B. W.,Williams, A. Marsh, P. D. (2000). The physiology and pathogenicity of *Mycobacterium tuberculosis* grown under controlled conditions in a defined medium. *J Appl Microbiol* **88**, 669–677.
2. Bacon, J., Dover, L. G., Hatch, K. A., Zhang, Y., Gomes, J.M., Kendall, S. L., Wernisch, L., Stoker, N. G., Butcher, P. D., Besra, G. S. and Marsh, P. D. (2007). The lipid composition and transcriptional response of *Mycobacterium tuberculosis* grown under iron-limitation in continuous culture: identification of a novel wax ester. *Microbiology* **153** (5), 1435–1444.
3. Bacon, J., James, B. W., Wernisch, L., Williams, A., Morley, K. A., Hatch, G. J., Mangan, J. A., Hinds, J., Stoker, N. G., Butcher, P. D. and Marsh, P. D. (2004). The influence of reduced oxygen availability on pathogenicity and gene expression in *Mycobacterium tuberculosis*. *Tuberculosis* **84**, 205–217.
4. Keating, L. A., Wheeler, P. R., Mansoor, H., Inwald, J. K., Dale, J., Hewinson, R. G. and Gordon, S. V. (2005). The pyruvate requirement of some members of the *Mycobacterium tuberculosis* complex is due to an inactive pyruvate kinase: implications for *in vivo* growth. *Mol Microbiol* **56**, 163–174.
5. James, B. W., Bacon, J., Hampshire, T., Morley, K. A. and Marsh, P. D. (2002). *In vitro* gene expression dissected: chemostat surgery for *Mycobacterium tuberculosis*. *Comparative and Functional Genomics* **3**, 345–347.
6. Kendall, S. L., Rison, S. C., Movahedzadeh, F., Frita, R. and Stoker, N. G. (2004). What do microarrays really tell us about *M. tuberculosis*? *Trends Microbiol* **12**, 537–544.
7. Bacon, J. and Marsh, P. D. (2007) Transcriptional responses of *Mycobacterium tuberculosis* exposed to adverse conditions *in vitro*. *Curr Mol Med* **7**, 277–286.
8. Voskuil, M. I., Visconti, K. C. and Schoolnik, G. K. (2004). *Mycobacterium tuberculosis* gene expression during adaptation to stationary phase and low-oxygen dormancy. *Tuberculosis* **84**, 218–227.
9. Hampshire, T., Soneji, S., Bacon, J., James, B. W., Hinds, J., Laing, K., Stabler, R. A., Marsh, P. D. and Butcher, P. D. (2004). Stationary phase gene expression of *Mycobacterium tuberculosis* following a progressive nutrient depletion: a model for persistent organisms? *Tuberculosis* **84**, 228–238.
10. Betts, J. C., Lukey, P. T., Robb, L. C., McAdam, R. A. and Duncan, K. (2002). Evaluation of a nutrient starvation model of *Mycobacterium tuberculosis* persistence by gene and protein expression profiling. *Mol Microbiol* **43**, 717–731.
11. Fisher, M. A., Plikaytis, B. B. andShinnick, T. M. (2002). Microarray analysis of the *Mycobacterium tuberculosis* transcriptional response to the acidic conditions found in phagosomes. *J Bacteriol* **184**, 4025–4032.
12. Ohno, H., Zhu, G., Mohan, V. P., Chu, D., Kohno, S., Jacobs, W. R., Jr. and Chan, J. (2003). The effects of reactive nitrogen intermediates on gene expression in *Mycobacterium tuberculosis*. *Cell Microbiol* **5**, 637–648.
13. Voskuil, M. I., Schnappinger, D., Visconti, K. C., Harrell, M. I., Dolganov, G. M., Sherman, D. R. and G. K., S. (2003). Inhibition of respiration by nitric oxide induces a *Mycobacterium tuberculosis* dormancy program. *J Exp Med* **198**, 705–713.

14. Voskuil, M. I. (2004). *Mycobacterium tuberculosis* gene expression during environmental conditions associated with latency. *Tuberculosis* **84**, 138–143.

15. Sherman, D. R., Voskuil, M., Schnappinger, D., Liao, R., Harrell, M. I. and Schoolnik, G. K. (2001). Regulation of the *Mycobacterium tuberculosis* hypoxic response gene encoding alpha-crystallin. *Proc Natl Acad Sci U S A* **98**, 7534–7539.

16. Muttucumaru, D. G., Roberts, G., Hinds, J., Stabler, R. A. and Parish, T. (2004). Gene expression profile of *Mycobacterium tuberculosis* in a non-replicating state. *Tuberculosis (Edinb)* **84**, 239–246.

17. Wayne, L. G. and Lin, K. Y. (1982). Glyoxylate metabolism and adaptation of *Mycobacterium tuberculosis* to survival under anaerobic conditions. *Infect Immun* **37**, 1042–1049.

18. Wayne, L. G. and Hayes, L. G. (1996). An *in vitro* model for sequential study of shiftdown of *Mycobacterium tuberculosis* through two stages of nonreplicating persistence. *Infect Immun* **64**, 2062–2069.

19. Beste, D. J., Peters, J., Hooper, T., Avignone-Rossa, C., Bushell, M. E. and McFadden, J. J. (2005). Compiling a molecular inventory for *Mycobacterium bovis BCG* at two growth rates: evidence for growth rate-mediated regulation of ribosome biosynthesis and lipid metabolism. *J Bacteriol* **187**, 1677–1684.

20. Beste D.J., Laing E., Bonde B., Avignone-Rossa C, Bushell M.E., McFadden J.J. Transcriptome analysis identifies growth rate modulation as a component of the adaptation of the mycobacteria to survival inside the macrophage. *J Bacteriol* **189**, 3969–3976.

21. Williams, A., avies, A., Marsh, P. D., Chambers, M. A. and Hewinson, R. G. (2000). Comparison of the protective efficacy of bacille calmette-Guerin vaccination against aerosol challenge with *Mycobacterium tuberculosis* and *Mycobacterium bovis*. *Clin Infect Dis* **30**(Suppl 3), S299–301.

22. Williams, A., James, B. W., Bacon, J., Hatch, K. A., Hatch, G. J., Hall, G. A. and Marsh, P. D. (2005). An assay to compare the infectivity of *Mycobacterium tuberculosis* isolates based on aerosol infection of guinea pigs and assaessment of bacteriology. *Tuberculosis* **85**, 177–184.

23. Williams, A., Hatch, G. J., Clark, S. O., Gooch, K. E., Hatch, K. A., Hall, G. A., Huygen, K., Ottenhoff, T. H., Franken, K. L., Andersen, P., Doherty, T. M., Kaufmann, S. H., Grode, L., Seiler, P., Martin, C., Gicquel, B., Cole, S. T., Brodin, P., Pym, A. S., Dalemans, W., Cohen, J., Lobet, Y., Goonetilleke, N., McShane, H., Hill, A., Parish, T., Smith, D., Stoker, N. G., Lowrie, D. B., Kallenius, G., Svenson, S., Pawlowski, A., Blake, K. and Marsh, P. D. (2005). Evaluation of vaccines in the EU TB Vaccine Cluster using a guinea pig aerosol infection model of tuberculosis. *Tuberculosis* **85**, 29–38.

24. Vipond, J.,Vipond, R., Allen-Vercoe, E., Clark, S. O., Hatch, G. J., Gooch, K. E., Bacon, J., Hampshire, T., Shuttleworth, H., N.P., M., Blake, K., Williams, A. and Marsh, P. D. (2006). Selection of novel TB vaccine candidates and their evaluation as DNA vaccines against aerosol challenge. *Vaccine* **24**, 6340–6350.

25. Herbert, D., Elsworth, R. and Telling, R. C. (1956). The continuous culture of bacteria; a theoretical and experimental study. *J Gen Microbiol* **14**, 601–622.

26. Herbert, D. (1976). *Continuous Culture 6 Applications and New Fields* (Dean, A. C. R., Ellwood, D. C., Evans, C. G. T. and J, M., eds.), **6**, Ellis Horwood Ltd, Chichester.

Chapter 11
Measuring Minimum Inhibitory Concentrations in Mycobacteria

Frederick A. Sirgel, Ian J.F. Wiid and Paul D. van Helden

Abstract An agar dilution method for measuring minimum inhibitory concentrations (MICs) of *Mycobacterium tuberculosis*, based on the method of proportion, is described. *Mycobacterium* strains are grown on Middlebrook 7H10 (or 7H11) agar medium with twofold serially diluted drug concentrations in order to determine specific inhibitory values. The proportion of bacilli resistant to a given drug is determined by comparing the number of colony-forming units (CFU) on a drug-free control with those growing in the presence of drug within a specific concentration range. The MIC is defined as the lowest drug concentration that inhibits growth of more than 99% of a bacterial population of *M. tuberculosis* on solid Middlebrook medium within 21 days of incubation at 37°C. The proportion method, the absolute concentration method, and the resistant ratio method have traditionally been used as standard procedures for antimycobacterial drug-susceptibility testing (DST), and reference data are mainly based on these methods. DST concepts and alternative procedures that have been adopted for DST are also briefly discussed.

Keywords agar proportion · antimicrobial agents · MICs · *Mycobacterium tuberculosis* · susceptibility testing

11.1 Introduction

Several quantitative and qualitative drug-susceptibility testing (DST) methods have been described that incorporate antimicrobial agents at specific concentrations into solid or liquid media [1, 2, 3]. Conventional microbiological procedures differentiate between drug-susceptible and drug-resistant *Mycobacterium* strains by the observation of visible bacterial growth or metabolic inhibition in drug-containing media. DST results are influenced by methodology such as

P.D. van Helden
Division of Molecular Biology and Human Genetics, Stellenbosch University,
P.O. Box 19063, Tygerberg 7505, South Africa
e-mail: pvh@sun.ac.za

T. Parish, A.C. Brown (eds.), *Mycobacteria Protocols*,
doi: 10.1007/978-1-59745-207-6_11, © Humana Press, Totowa, NJ 2008

the composition of the medium, inoculum size, pH, and incubation conditions [1]. Phospholipids, large-molecular-weight proteins in egg media, and certain amino acids may also reduce drug activity [1]. Diverse minimum inhibitory concentration values are therefore anticipated for the same drug against a specific organism when DST is performed on different media. Susceptibility methods and the interpretation of results vary greatly, and standardized methods are therefore recommended to minimize interpretation errors and to ensure accurate and reproducible results. Minimum inhibitory concentration data of antituberculosis drugs are mainly based on methods that have been performed on either solid media (agar or Löwenstein-Jensen media) or with radiometric BACTEC 460 broth. Results obtained by the latter methods provide reference information that is still regarded as the gold standard for DST, despite many other techniques that have been developed.

A combination of four drugs isoniazid (H), rifampicin (R), pyrazinamide (Z), and ethambutol (E)—is recommended for the treatment of active tuberculosis if caused by drug-susceptible *M. tuberculosis*. Therapy consists of two phases: an initial 2-month intensive phase with HRZE, followed by an additional 4-month continuation phase with HR [4]. The lengthy dosing regimen with multiple drugs increases the risk of nonadherence and therefore clinical failure, which contributes to the development of drug-resistant strains and multidrug-resistant tuberculosis.

Various DST procedures, both quantitative and qualitative, have been described to measure the *in vitro* susceptibility of mycobacteria to antimicrobial agents [2]. Quantitative methods measure specific inhibitory (bacteriostatic) and bactericidal (killing) values of drugs, and these are respectively expressed as the minimum inhibitory concentration (MIC) and the minimum bactericidal concentration (MBC) [2, 3]. The MIC is the lowest concentration of a drug that prevents visible growth of bacteria, or at least 99% of the bacterial population, under defined *in vitro* conditions [2, 3]. In contrast with the inhibition effect (MIC), the MBC quantitates the bactericidal activity of an antimicrobial agent and is defined as the lowest drug concentration that kills at least 99.9% of a bacterial population [2]. Qualitative methods on the other hand are designed to provide only a suggested interpretation of susceptibility without estimating specific inhibitory values [5]. A qualitative result depends on a single breakpoint, although additional concentrations can be included if required [3]. Breakpoints have been used as a standard to distinguish between susceptible and resistant strains [3].

Mycobacterium strains that lack previous exposure to antituberculosis agents, the so-called wild strains, show little variation in their susceptibility to first-line drugs [5, 6, 7]. Clinical isolates with susceptibility levels that are significantly greater than the uniform MIC distribution of wild strains are therefore categorized as drug-resistant strains [6]. *M. tuberculosis* typically displays a bimodal distribution of MICs with a highly susceptible population with low MICs and a group with relatively high MICs that has some mechanism(s) of resistance [8]. Microbiological parameters such as the MIC distribution for a collection of clinical isolates in association with clinical and pharmacologic factors (achievable blood and tissue drug levels) are therefore used to determine appropriate breakpoint concentrations and effective drug dosages [9, 10]. Breakpoint concentrations should be more than the highest MIC found among the wild-type strains but significantly less than the lowest MIC for isolates that are regarded as resistant; it should also be less than the achievable drug plasma level at a standard or tolerable drug dosage [6]. MIC values are often determined to confirm borderline qualitative susceptibility test results or to measure the antimicrobial potential of a new compound. The interpretation of DST results for *M. tuberculosis* is generally based on the proportion method, the resistance ratio method, or the absolute concentration method [6, 7, 11].

The proportional method allows an exact estimation of the number of mutants in a bacterial population that is resistant to a specific drug [3], and this proportion is expressed as a percentage. The proportion of resistance is determined by comparing the number of colony-forming units (CFU) that develop on drug-containing medium to those growing on drug-free controls. The absolute concentration method is similar to the proportional method with the exception that results are interpreted differently. The absolute concentration method defines the MIC as the lowest drug concentration that inhibits visible growth of a bacterial culture when exposed to several dilution concentrations of a drug. Bacterial growth of less than 20 CFU at a specific drug concentration is allowed, whereas 20 or more CFU confirms resistance. Standardization of the drug concentrations and inoculum size is critical in this method, as the main cause of error is variation in inoculum size. The resistance ratio method determines and expresses resistance as the ratio of the MIC for the test strain to the MIC of a standard susceptible reference strain such as *M. tuberculosis* H37Rv or wild-type strains. A ratio of 2 or less distinguishes fully susceptible isolates from resistant ones, whereas a ratio of 8 or more indicates that the test isolate is highly resistant [12]. This method is not suitable to measure intermediate- and low-level resistance. The respective use of solid and liquid media in DST is briefly discussed in **Note 1**.

11.2 Materials

11.2.1 Preparation of Middlebrook 7H9 Inoculum Medium

1. Middlebrook 7H9 broth base (Becton Dickinson Sparks, MD, USA).
2. Double-distilled water.
3. 10% w/v Tween 80 (Polysorbate 80) (Sigma-Aldrich St Louis, MO, USA).
4. Albumin-dextrose-catalase (ADC) enrichment (Becton Dickinson).
5. 5- to 10 mL sterile screw-cap tubes.

Table 11.1 Preparation of McFarland Standards

McFarland scale	Reagents	
	1% BaCl$_2$ (mL)	1% H$_2$SO$_4$(mL)
0.5	0.05	9.95
1	0.1	9.9
2	0.2	9.8
3	0.3	9.7
4	0.4	9.6
5	0.5	9.5
6	0.6	9.4
7	0.7	9.3
8	0.8	9.2
9	0.9	9.1
10	1.0	9.0

11.2.2 Preparation of Inoculum from a Middlebrook 7H9 Broth Culture

1. 7H9 inoculum medium.
2. *M. tuberculosis* test strains.
3. Control strains: *M. tuberculosis* H$_{37}$Rv (ATCC 27294, susceptible to all standard antituberculosis agents and, e.g., *M. tuberculosis* strain ATCC 35822 that is resistant to isoniazid).
4. McFarland standards: No. 0.05 and No. 1 (*see* **Note 2** and Table 11.1).
5. Sterile glass or plastic beads with a diameter of ca. 6 mm.
6. Sterile double-distilled water.

11.2.3 Preparation of Antimicrobial Stock Solutions

1. Antimicrobial agents (Sigma-Aldrich) (Table 11.2).
2. Solvents (Table 11.2).
3. Double-distilled water.

Table 11.2 Solvents Appropriate for Dissolving Antimicrobial Compounds

Antimicrobials	Solvent
Capreomycin; cycloserine; ethambutol hydrochloride; isoniazid; kanamycin	Sterile double-distilled water
amikacin; streptomycin sulfate; *p*-aminosalicylic acid; pyrazinamide	
Ethionamide	Dimethyl sulfoxide or ethylene glycol
Rifampicin; rifapentine; rifabutin	Dimethyl sulfoxide or Methyl alcohol
Clofazimine	Dimethyl sulfoxide
Ciprofloxacin; moxifloxacin; ofloxacin	0.1 N NaOH

4. Calibrated analytic balance.
5. Millex-GV syringe-driven filter units with a membrane pore size of 0.22 (Millipore Corporation Bedford, MA, USA).
6. Millex-LG 0.20-μm hydrophilic PTFE (polytetrafluoroethylene) filters (Millipore Corporation).
7. Screw-cap polypropylene or polyethylene cryovials with O-rings.

11.2.4 Preparation of the Antimicrobial Dilution Range

1. Antimicrobial stock solutions.
2. 5- to 10- mL sterile screw-cap tubes.
3. Sterile diluent (double-distilled water or alternative if required).

11.2.5 Middlebrook 7H10 Agar Dilution Plates

1. Middlebrook 7H10 agar base (Becton Dickinson).
2. Double-distilled water.
3. Glycerol (reagent grade).
4. Oleic acid-albumin-dextrose-catalase (OADC) enrichment (Becton Dickinson).
5. Water bath.
6. Standard 90-mm-diameter one-compartment (or quadrant) Petri dishes.

11.2.6 Inoculation of 7H10 Agar Plates

1. Inoculum.
2. Middlebrook 7H10 agar dilution plates.
3. Sterile plastic loops.
4. L-shaped spreaders.

11.3 Methods

A modified agar dilution proportion method for measuring specific inhibitory values of antituberculosis drugs is described (Fig. 11.1). Appropriate practices and procedures, including containment equipment and facilities, are required when working with infectious agents like *M. tuberculosis*.

11.3.1 Preparation of Middlebrook 7H9 Inoculum Medium

1. Dissolve 2.4 g 7H9 in 450 mL double-distilled water. Add 2.5 mL 10% w/v Tween 80 for a final concentration of 0.05 % (*see* **Note 3**). Autoclave for 10 min at 121°C.
2. Cool to 45°C or less and add 50 mL ADC enrichment.
3. Dispense required volumes (5 to 10 mL) aseptically into sterile screw-cap test tubes.

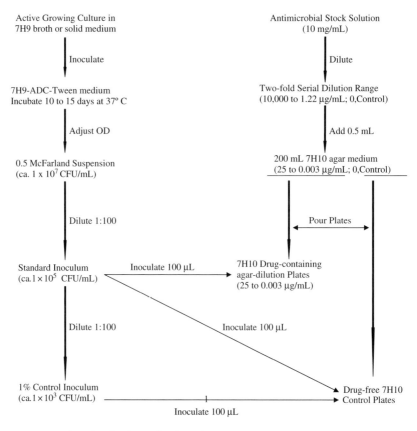

Fig. 11.1 Procedure for the determination of MICs
Incubate plates at 37°C for 21 days.

11.3.2 Preparation of Inoculum from a Middlebrook 7H9 Broth Culture (see Note 4)

1. Inoculate 5 to 10 mL 7H9-ADC-Tween inoculum medium with the bacterial strains of interest from actively growing cultures. Include at least one control strain that is susceptible to all standard antituberculosis drugs, such as *M. tuberculosis* H$_{37}$Rv (ATCC 27294). A strain that has monoresistance to the test drug may also be required (e.g., *M. tuberculosis* strain ATCC 35822, which is resistant to isoniazid).
2. Incubate the culture at 37°C (*see* **Note 5**) for about 10 to 15 days until an optical density (OD) of an 0.5 to 1 McFarland standard is reached.
3. When an adequate turbidity is reached, vortex the culture for 2 to 3 min in the presence of 5 to 8 glass beads.
4. Let the suspension stand for 20 min to allow large clumps to settle before transferring the supernatant to a sterile tube.

5. Adjust the bacterial suspensions to a turbidity equivalent to an 0.5 McFarland standard (1×10^7 CFU/mL) with sterile double-distilled water (*see* **Note 6**).
6. Dilute the suspension 10^{-2}-fold and 10^{-4}-fold to obtain (i) a standard inoculum with ca. 1×10^5 CFU/mL and (ii) a control inoculum that contains ca. 1×10^3 CFU/mL (1/100 of the standard inoculum).

11.3.3 Preparation of Antimicrobial Stock Solutions

1. Use the following formula to calculate the weight of an antimicrobial agent (*see* **Note 7**) that is required to prepare 10 mL of a 10 mg/mL stock solution (*see* **Note 8**):

$$\text{Weight(mg)} = \frac{\text{Volume required(mL)} \times \text{Desired drug concentration(mg/mL)}}{\text{Potency(mg/mg)}}$$

For example, a compound with a potency of 882 mg/g $= 0.882$ mg/mg $(= 88.2\%)$

$$\text{Weight(mg)} = \frac{10 \times 10}{0.882} = 113.37 \, \text{mg}.$$

2. Dissolve the calculated amount of drug in less than 10 mL of appropriate solvent (Table 11.2) and mix until dissolved.
3. Add more diluent (or double-distilled water) (*see* **Note 9**) to a final volume of 10 mL to obtain a drug concentration of 10 mg/mL. Further 10- or 100-fold dilutions can be made from a 10 mg/mL stock solution if required.
4. Sterilize stock solutions by filtration through a syringe filter that is compatible with the solvent (*see* **Note 10**).
5. Dispense small aliquots (2 mL) of stock solutions into sterile screw-cap polypropylene or polyethylene cryovials, and store at –70°C to –80°C (*see* **Note 11**).

11.3.4 Preparation of the Antimicrobial Dilution Range (see Note 12)

1. Prepare a twofold serial dilution from an antibiotic stock solution of 10 mg/mL as follows: dispense 2 mL of sterile diluent into each of a series of 5- to 10-mL sterile capped test tubes.
2. Add 2 mL of the antibiotic stock solution (10 mg/mL) to the first tube that contains an equal volume of diluent for a 1:2 dilution and mix thoroughly.

3. Transfer 2 mL from the first tube to the second tube and mix (1:4 dilution) (*see* **Note 13**).
4. Repeat the process for the preparation of the required dilutions.
5. No antibiotic is added to the tube labeled 0, which is used as a drug-free control.

11.3.5 Preparation of Antibiotic-Containing 7H10 Agar Dilution Plates

1. Dissolve 3.8 g 7H10 in 178.5 mL double-distilled water. Add 1 mL glycerol. Swirl ingredients into suspension and autoclave at 121°C for 10 min. Prepare one bottle of medium for each antibiotic dilution.
2. Let the medium cool to 52°C to 56°C in a water bath. Add 20 mL OADC enrichment (prewarmed to room temperature) (*see* **Note 14**).
3. Add 0.5 mL antibiotic to one bottle of medium (*see* Table 11.3) and mix well (*see* **Note 15**). Pour 20 mL quantities in 90-mm one-compartment Petri dishes and allow agar to set at room temperature without exposure to ultraviolet light (*see* **Note 16**).
4. Prepare drug-free control medium by substituting drug with 0.5 mL sterile diluent.

11.3.6 Inoculation of 7H10 Agar Plates

1. Plate 100 µL of the standard inoculum (*see* Section 11.3.2) onto both antibiotic-containing and drug-free control 7H10 agar plates in duplicate (Fig. 11.1) and spread with a sterile plastic loop or L-shaped spreader.

Table 11.3 Serial Twofold Drug Dilutions in Solvent and 7H10 Agar

Dilution number	(A) Twofold drug dilutions prepared from a 10,000 µg/mL stock solution in solvent(µg/mL)	(B) Final drug concentration in 7H10 agar(µg/mL)
0	0 (drug-free solvent)	0 (drug-free control)
1	10,000	25
2	5,000	12.50
3	2500	6.25
4	1250	3.13
5	625	1.56
6	312.50	0.78
7	156.25	0.390
8	78.13	0.195
9	39.06	0.098
10	19.53	0.049
11	9.77	0.024
12	4.88	0.012
13	2.44	0.006
14	1.22	0.003

*Mix 500 µL of solutions (A) (*see* section 11.3.4), respectively, into 200 mL 7H10 agar medium (B) (*see* Section 11.3.5)for a desired drug concentration range (Fig. 11.1).

2. Inoculate 100 µL of the 1% control inoculum only on drug-free control plates (Fig. 11.1).
3. Seal plates in plastic bags and incubate at 37°C for 21 days (*see* **Note 17**).
4. Monitor growth on 7H10 plates after 14 and 21 days of incubation. Aliquots of 100 µL from the standard and 1% control inocula should yield ca. 1 x 10^4 and 1×10^2 CFU on drug-free medium, respectively (*see* **Note 18**).

11.3.7 *Reading and Interpretation of Results*

1. Confirm that all control cultures have grown on the drug-free medium.
2. Count the number of CFU that develop on both the antibiotic-containing and antibiotic-free medium after 14 and 21 days (*see* **Note 19**).
3. Record the number of colony forming units as follows:

No colonies	=	0 or negative
Less than 200 colonies	=	Actual number of colonies
200 to 500 colonies	=	1+
500 to 1000 colonies	=	2+
Confluent growth (>1000 colonies)	=	3+
Confluent growth noticeably more than 3+	=	4+

4. Calculate the proportion and express as the percentage of resistant bacteria in the viable population to determine the MIC (*see* **Notes 20 and 21**).

11.4 Notes

1. DST on solid media such as Middlebrook 7H10 or 7H11 agar, or Löwenstein-Jensen (LJ) egg-based media, has universally been used for MIC determinations. Methods that use solid media are more easily standardized and are therefore accepted as the gold standard. MICs obtained on egg-based media are likely to differ from those on agar media, but different results may also be obtained between Middlebrook 7H10 and 7H11 agar media [1]. Many laboratories prefer to use Middlebrook 7H10 agar as a solid medium and as an alternative to egg-based media. Agar media are more expensive than egg media, but they are easy to prepare and solidify by agar rather than heat. Egg-based media require incorporation of drug before inspissation at 80°C to 85°C for 45 min to coagulate. The concentrations of certain drugs must therefore be adjusted in egg-based medium to account for their loss by heating or interaction with specific egg components. Automated liquid medium DST systems and alternative microbiological methods have also been described, but these are primarily designed for qualitative susceptibility testing. DST in liquid media has substantial advantages, in particular with regard to reduced turnaround times. Antimicrobials are added to liquid media at room temperature, and a relative short incubation period of 4 to 12 days is required to observe DST results. Various automated liquid medium DST systems have been developed with the objective to provide rapid, reproducible, and accurate results [13]. Three of these, the radiometric BACTEC-460, (BD, Sparks, MD) [11], the nonradiometric MGIT-960 (BD) [14], and the

ESP II (Difco Laboratories, Detroit, MI) [15, 16] have been recommended by the U.S. Food and Drug Administration [3]. Other automated systems that provide rapid results are the MB Redox (Biotest, Dreieich, Germany) [17] and the MB/BacT (Organon Teknika) [13, 18]. Alternative microbiological methods for DST that have been developed, mainly for qualitative purposes [6, 19], are the slide culture method [20], microcolony method on Middlebrook 7H11 agar [21, 22], microscopic observation drug susceptibility (MODS) assay [23], ETest (AB BIODISK, Solna, Sweden) for determining MICs [24], detection of ATP by bioluminescence [25], luciferase reporter phage assay [26], FASTPlaque TB-RIF test (Bio Tec Laboratories, Ipswich, UK) [27], and microplate assays [6, 28, 29].

2. McFarland Latex Standards (Hardy Diagnostics Catalogue) are more stable than barium sulfate standards due to reduced light sensitivity and have a significantly longer shelf-life. Both products are commercially available from various manufacturers. However, McFarland standards can be prepared at different optical densities in the laboratory by mixing specified amounts of 1% (w/v) $BaCl_2$ and a 1% (v/v) solution of H_2SO_4 together in proportions listed in Table 11.1. Barium sulfate standards should be stable for 6 months if stored in the dark at 20°C to 25°C and if it is tightly sealed to prevent evaporation.

3. Mycobacteria tend to clump in liquid medium and Tween 80, a non-ionic surface-active detergent, is added instead of glycerol to reduce cell clumping and to obtain homogeneous cell suspensions. Tween 80 has a high viscosity, and it is therefore difficult to aliquot small volumes. The use of a 10% (v/v) working solution is recommended to facilitate accurate measuring and transfer of the detergent. To prepare a 1:10 (v/v) dilution, add 1.0 mL (or 1.0 g) of Tween 80 with a motorized Gilson pipette to 9.0 mL of double-distilled water and mix.

4. An inoculum can also be prepared from solid media: Scrape several freshly grown colonies with a sterile loop directly from the surface of drug-free solid media (7 H10, 7 H11, or Löwenstein-Jensen media). Avoid using bacterial growth older than 2 to 3 weeks and try not to scrape off any medium. Emulsify the bacterial mass along the side of a 16 × 125 mm screw-cap tube containing five to eight glass or plastic beads (ca. 6 mm) and 5 mL 7H9-ADC-Tween medium and vortex for 2 to 3 min. The suspension should match or exceed the density of a number 0.5 McFarland standard.

5. The optimum incubation temperature for *M. tuberculosis* is 37°C, but this varies widely for different species of mycobacteria and ranges from 25°C to 45°C.

6. Saline (0.85% w/v NaCl) or 0.067 M phosphate buffer pH 7 (61.1 mL 0.067 M Na_2HPO_4, 38.9 mL 0.067 M KH_2PO_4) can also be used to suspend and dilute *M. tuberculosis* cultures.

7. Antimicrobial agents should be obtained from the manufacturer (pharmaceutical company) and not from a pharmacy. The supplier should provide the physical and chemical properties of the substance, including generic name(s); chemical structure; colour; shelf-life; storage conditions; expiry date; potency (usually expressed as mg/g of drug); solubility; and stability as a powder and in solution.

8. The potency of an antibiotic powder is expressed in units of mg/g (or μg/mg) (w/w) or as a percentage and reflects the true weight of the active drug in milligram per gram total weight of the compound [3]. The following measures of purity are normally indicated by the manufacturer: HPLC (high-performance liquid chromatography) assay, water content determined by Karl Fischer analysis or weight loss on drying, and salt/counter-ion fraction if the active fraction of a compound is supplied as a salt instead of free acid or base [3]. The potency of an antibiotic is calculated according to the following formula:

$$Potency = Assay\ purity \times Active\ fraction \times (1 - Water\ content).$$

For example, a substance with 98.5% purity (by HPLC assay), water content of 10.5% (w/w), and an active fraction of 100%, (when supplied as free acid, not as a salt) is

$$0.985 \times 1.0 \times (1 - 0.105) \times 100 = 88.2\% \text{ or } 882\text{mg/g}.$$

9. Ideally, antimicrobials should be dissolved in sterile double-distilled water. However, certain compounds require alternative solvents (Table 11.2). Water-insoluble antibiotics should be dissolved in the minimum volume of solvent required and diluted to the desired concentration in sterile distilled water. Solvents such as methyl alcohol and dimethyl sulfoxide (DMSO) can interfere with bacterial growth and should therefore be adequately diluted.

10. Millex-GV syringe-driven filter units with a membrane pore size of 0.22 μm can be used for sterilization of aqueous solutions, and Millex-LG 0.20-μm hydrophilic PTFE filters (Millipore Corporation) are recommended for the sterilization of DMSO.

11. Store drug stock solutions in vials that are suitable for storage at low temperatures, preferably at –70°C to –80°C for 12 months, provided that it is in accordance with the manufacturer's guidelines. Frozen stock solutions should be thawed only once and then discarded.

12. The concentration and span width of an antibiotic dilution range depends on the distribution of susceptibilities (MICs) among the test strains to the specific compound. A working twofold antimicrobial dilution range of 10,000 to 1.22 μg/mL in appropriate diluent is suggested. This will give a testing range of 25 to 0.003 μg/mL. Test strains that have susceptibilities distributed over a wide range of MIC values or those with limited pharmacodynamic data regarding the specific drug-pathogen combinations should be set within extended MIC dilution ranges.

13. Middlebrook 7H11 agar medium can be used as an alternative to 7H10 medium, as it is particularly suitable for the recovery of isoniazid-resistant strains of the *M. tuberculosis* complex [3, 12]. Certain antimicrobials, such as ethambutol, ethionamide, kanamycin, cycloserine, and *p*-aminosalicylic acid, should be added at different concentrations to 7H11 than that required when using 7H10 [2, 3].

14. Batch differences for the OADC supplement have been reported. This could be a reason for irregular results in susceptibility testing, even when the same procedure and medium are used [6]. Middlebrook agar base should not be stored in a refrigerator for future use after it has been autoclaved, because remelting of Middlebrook agar medium will reduce its quality.

15. Add each dilution of the drug to separate aliquots of media in turn. This ensures that the different concentrations can be poured separately minimizing the time that the microbial agents are kept at 52°C to 56°C to prevent heat inactivation.

16. Quadrant Petri dishes (diameter of 90 mm) can be used as an economical alternative to standard plates. Quadrant plates are prepared by dispensing 5 mL of drug-containing 7H10 agar at different concentrations, respectively into three labeled quadrants, while the fourth quadrant in each plate is reserved for the antibiotic-free growth control. Let the agar solidify at room temperature without exposure to daylight. Preferably use freshly prepared plates, but the media can be stored in sealed plastic bags at 4°C to 8°C for up to 10 days, provided that the antibiotic-containing agar is stable under storage conditions. Inoculate plates by spreading 100 μL of the standard inoculum onto each antibiotic quadrant and 100 μL of the 1% control inoculum onto a drug-free control quadrant. Seal plates in plastic bags and incubate at 37°C for 21 days. The inoculum volume per plate is 4 × 100 μL, and additional time is therefore required for the plates to dry before incubation. Counting large numbers of CFU is difficult and so bacterial growth should be reported as follows [3]: no colonies = 0 or negative; less than 50 colonies = actual number of CFU; 50 to 100 colonies = 1+; 100 to 200 colonies = 2+; almost confluent (200 to 500 colonies) = 3+; and confluent growth (>500 colonies) = 4+.

17. Plates should be sealed in CO_2-permeable polyethylene bags if an incubator that provides 5% to 10% CO_2 is required. An atmosphere of 10% CO_2 and 90% air is recommended for enhanced growth of *M. tuberculosis* when cultured on Middlebrook media [3].

18. A drug-susceptible reference strain with a known MIC must be included with every batch of MIC determinations to standardize results and for quality control. *M. tuberculosis* $H_{37}Rv$ (ATCC 27294) is well documented and could be used as a susceptible control organism. Inclusion of a strain that is resistant to the test drug, preferably a monoresistant strain, may also be required.

19. Growth is monitored after 14 and 21 days, but a final MIC susceptibility result should be based on a 21-day result. Adequate growth may be present after 14 days on the drug-free control medium, but drug-resistant strains may grow more slowly, and these colonies are likely to become visible later. An incubation period that extends beyond 21 days may also give unreliable results, because the drug may become partially inactivated in the medium, allowing susceptible bacteria to grow. Heat-labile drugs with a bacteriostatic rather than a bactericidal effect are particularly vulnerable to this.

20. The 1% control inoculum represents 1/100 (1%) of the standard inoculum, which should yield countable colonies on drug-free medium. This number is then used to calculate the number of viable organisms in the standard inoculum, which is needed to determine the proportion of resistant bacilli at a specific antimicrobial concentration.

21. An example is given in Table 11.4. The standard inoculum yielded confluent growth (4+) on drug-free 7H10 control medium and 100 CFU with the 1% control inoculum. Thus the standard inoculum contains 1×10^4 CFU/100 μL. A decrease in CFU was observed when the standard inoculum was exposed to drug-containing medium at doubling concentrations. Only 120, 10, 1, and 0 CFU developed on medium containing 0.003, 0.006, 0.012, and 0.024 μg/mL of drug, respectively. The percentage resistant organisms for each dilution can then be calculated according to the formula:

$$\% \text{ Resistant bacteria} = \frac{\text{Number of colonies on drug} - \text{containing medium} \times 100}{\text{Number of colonies on drug} - \text{free control}}.$$

Therefore, at the concentration of 0.003 μg/mL, the percentage is

$$\frac{120 \times 100}{1 \times 10^4} = 1.2\%.$$

Table 11.4 Estimation of the Percent Resistant Bacteria According to the Proportion Method

Drug concentration in 7H10 medium	Bacterial growth (CFU)		% Resistance
	Standard inoculum ($10^{-2} \times 0.5$ McFarland)	1% Control inoculum ($10^{-4} \times 0.5$ McFarland)	
Drug-free control	4+ = (1×10^4)	100	—
0.003 μg/mL	120	NI	1.2 (>1%)
0.006 μg/mL	10	NI	0.1
0.012 μg/mL	1	NI	0.01
0.024 μg/mL	0	NI	0

NI, Not inoculated.

Similarly at 0.006 µg/mL, the percentage was

$$\frac{10 \times 100}{1 \times 10^4} = 0.1\%.$$

Because the concentration of 0.003 µg/mL had more than 1% growth, whereas only 0.1% of the bacterial population grew at 0.006 µg/mL, the MIC is 0.006 µg/mL.

References

1. Hawkins JE. Drug susceptibility testing. In Kubica GP, Wayne LG, eds. The Mycobacteria. A Sourcebook. Part A. Marcel Dekker Inc, New York and Basel, 1984:177–193.
2. Heifets LB. Introduction: Drug susceptibility tests and the clinical outcome of chemotherapy. In: Heifets LB, ed. Drug Susceptibility in the Chemotherapy of Mycobacterial Infections. CRC Press, Inc., Boca Raton, FL, 1991:1–11.
3. National Committee for Clinical Laboratory Standards. Susceptibility testing of Mycobacteria, Norcardiae, and Other Aerobic Actinomycetes; Approved Standard. NCCLS Document M24-A. NCCLS, Wayne, PA, 2003;23(18):1–71.
4. Chan ED, Iseman MD. Current medical treatment for tuberculosis. BMJ 2002;325:1282–1286.
5. Heginbothom ML. The relationship between the in vitro drug susceptibility of opportunist mycobacteria and their in vivo response to treatment. Int J Tuberc Lung Dis 2001;5(6):538–545.
6. Heifets L, Desmond E. Clinical mycobacteriology (tuberculosis) laboratory: services and methods in tuberculosis and the tubercle bacillus. In: Cole ST, Eisenach KD, McMurray DN, Jacobs (Jr) eds. Tuberculosis and the Tubercle Bacillus. ASM Press, Washington, DC, 2005:49–69.
7. Mitchison DA. Drug resistance in tuberculosis. Eur Respir J 2005;25:376–379.
8. Dickinson JM, Mitchison DA. *In vitro* activity of new rifamycins against rifampin-resistant *M. tuberculosis* and MAIS-complex mycobacteria. Tubercle 1987;68: 177–182.
9. MacGowan AP, Wise R. Establishing MIC breakpoints and the interpretation of *in vitro* susceptibility tests. J Antimicrob Chemother 2001;48(Suppl S1):17–28.
10. Mueller M, de la Pena A, Derendorf H. Issues in pharmacokinetics and pharmacodynamics of anti-infective agents: Kill curves versus MIC. Antimicrob Agents Chemother 2004;48(2):369–377.
11. Siddiqi SH. BACTEC 460 TB system. Product and procedure manual, revision D. Becton Dickinson Microbiology Systems, Sparks, MD, 1995.
12. Heifets LB. Drug susceptibility tests in the management of chemotherapy of tuberculosis. In: Heifets LB, ed. Drug Susceptibility in the Chemotherapy of Mycobacterial Infections. CRC Press, Inc., Boca Raton, FL, 1991:89–121.
13. Bemer P, Bodmer T, Munzinger J, Perrin M, Vincent V, Drugeon H. Multicenter evaluation of the MB/BACT System for susceptibility testing of *Mycobacterium tuberculosis*. J Clin Microbiol 2004;42(3):1030–1034.
14. Chew WK, Lasaitis RM, Schio FA, Gilbert GL. Clinical evaluation of the Mycobacteria Growth Indicator Tube (MGIT) compared with radiometric (Bactec) and solid media for isolation of *Mycobacterium* species. J Med Microbiol 1998;47:821–827.
15. Bergmann JS, Woods GL. Evaluation of the ESP Culture System II for testing susceptibilities of *Mycobacterium tuberculosis* isolates to four primary antituberculous drugs. J Clin Microbiol 1998;36(10):2940–2943.
16. Woods GL, Fish G, Plaunt M, Murphy T. Clinical evaluation of Difco ESP culture system II for growth and detection of mycobacteria. J Clin Microbiol 1997;35(1):121–124.

17. Somoskövi A, Magyar P. Comparison of the Mycobacteria Growth Indicator Tube with MB Redox, Löwenstein-Jensen, and Middlebrook 7H11 media for recovery of mycobacteria in clinical specimens. J Clin Microbiol 1999;37(5):1366–1369.

18. Brunello F, Fontana R. Reliability of the MB/BACT system for testing susceptibility of *Mycobacterium tuberculosis* complex strains to anti-tuberculosis drugs. J Clin Microbiol 2000;38:872–873.

19. Heifets, LB. Drug susceptibility testing. In: Heifets LB (guest ed.), Clauser G ed. Clinics in Laboratory Medicine, vol 16 (3). W.B. Saunders, Philadelphia, 1996:641–656.

20. Dickinson JM, Allen BW, Mitchison DA. Slide culture sensitivity tests. Tubercle 1989;70:115–121.

21. Welch DF, Guruswamy AP, Sides SJ, Shaw CH, Gilchrist MJ. Timely culture for mycobacteria which utilizes a microcolony method. J Clin Microbiol 1993;31(8):2178–2184.

22. Mejia GI, Castrillon L, Trujillo H, Robledo JA. Microcolony detection in 7H11 thin layer culture is an alternative for rapid diagnosis of *Mycobacterium tuberculosis* infection. Int J Tuberc Lung Dis 1999;3(2):138–142.

23. Park WG, Bishai R, Chaisson E, Dorman SE. Performance of the microscopic observation drug susceptibility assay in drug susceptibility testing for *Mycobacterium tuberculosis*. J Clin Microbiol 2002;40(12):4750–4752.

24. Wanger A, Mills K. Testing of *Mycobacterium tuberculosis* susceptibility to ethambutol, isoniazid, rifampicin, and streptomycin by using Etest. J Clin Microbiol 1996;34(7):1672–1676.

25. Nilsson LE, Hoffner SE, Ansehehn S. Rapid susceptibility testing of *M. tuberculosis* by bioluminescence assay of mycobacterial ATP. Antimicrob Agents Chemother 1988;32:1208–1212.

26. Jacobs WR Jr, Barletta RG, Udani R, Chan J, Kalkut G, Sosne G, Kieser T, Sarkis GJ, Hatfull GF, Bloom BR. Rapid assessment of drug susceptibilities of *Mycobacterium tuberculosis* by means of luciferase reporter phages. Science 1993;260:819–822.

27. Albert H, Trollip A, Seaman T, Mole RJ. Simple, phage based (*FastPlaque*) technology to determine rifampicin resistance of *Mycobacterium tuberculosis* directly from sputum. Int J Tuberc Lung Dis 2004;8(9):1114–1119.

28. Franzblau SG, Witzig RS, McLaughlin JC, Torres P, Madico G, Hernandez A, Degnan MT, Cook MB, Quenzer VK, Ferguson RM, Gilman RH. Rapid, low-technology MIC determination with clinical *Mycobacterium tuberculosis* isolates by using the microplate alamar blue assay. J Clin Microbiol 1998;36(2):362–366.

29. Leite CQF, Beretta ALRZ, Anno IS, Telles MAS. Development of a microdilution method to evaluate *Mycobacterium tuberculosis* drug susceptibility. J Antimicrob Chemother 2003;52:796–800.

Chapter 12
Rapid Screening of Inhibitors of *Mycobacterium tuberculosis* Growth Using Tetrazolium Salts

Anita G. Amin, Shiva K. Angala, Delphi Chatterjee and Dean C. Crick

Abstract With the increased need for novel antimicrobials to improve the existing treatment for tuberculosis, to combat multidrug-resistant tuberculosis, and to address the presence of latent bacilli in a large population throughout the world, which can reactivate and cause active disease, there is a need for rapid, low-cost, high-throughput assays for screening new drug candidates. A microplate-based Alamar blue assay meets these requirements. In addition to the identification of the antimicrobial activities of compounds, determination of their toxicities is important. The high costs involved in testing compounds in whole animal models has led to the development of *in vitro* cytotoxicity assays using human and animal cell lines. Microplate-based Alamar blue and cytotoxicity assays have been applied to search for novel antimicrobials to treat tuberculosis. These methods are described in detail herein.

Keywords Alamar blue · cytotoxicity · MIC · *Mycobacterium tuberculosis* · tetrazolium

12.1 Introduction

Tuberculosis (TB) is the major cause of death due to an infectious agent worldwide, and treatment of this ancient disease now faces new challenges because of the increase in multidrug-resistant tuberculosis (MDR-TB) [1]. Most of the TB treatment drugs, as we know them today, were developed some decades ago. The current treatment for the disease spans a period of 6 to 8 months with a combination four-drug regimen of isoniazid (INH), rifampin (RIF), ethambutol (EMB), and pyrazinamide (PZA), which when followed as recommended is highly effective. However, because of unpleasant side effects of the treatment, many patients find it difficult to continue treatment for such a prolonged period

D.C. Crick

Department of Microbiology, Immunology & Pathology, 1682 Campus Delivery, Colorado State University, Fort Collins, Colorado 80523-1682, USA
e-mail: Dean.Crick@colostate.edu

T. Parish, A.C. Brown (eds.), *Mycobacteria Protocols*,
doi: 10.1007/978-1-59745-207-6_12, © Humana Press, Totowa, NJ 2008

of time. This results in treatment failure and development of drug resistance. The second-line drugs used to treat patients who do not respond to treatment with first-line drugs are less effective and very expensive [2]. In addition, treatment of the disease is complicated by the presence of latent, nonreplicating bacilli harbored by a large portion of the population throughout the world. These bacilli have the potential to reactivate and cause active disease [3, 4], but current TB therapy is mainly effective against replicating and metabolically active bacteria [5, 6]. Therefore, to improve the existing treatment for TB, there is a need to discover novel drugs that would attack the disease-causing bacilli in new ways, overcoming the increasing problems of MDR strains, duration of treatment, and latent infections. According to the guidelines set by Global Alliance for TB drug development [www.tballiance.org (Scientific Blueprint)], selection of a lead compound generally starts with an initial evaluation of its preliminary characteristics. These include *in vitro* determination of the minimum inhibitory concentration (MIC) against *Mycobacterium tuberculosis* and assessment of toxicity using a eukaryotic cell line followed by evaluation of bioavailability and efficacy in animal models. Substantial progress has been made in the past decade in understanding the molecular basis of drug resistance in *M. tuberculosis* [7], but understanding the mechanisms of action of antimicrobial agents is important in designing novel antibiotics that are active against the resistant strains. Many factors are involved in understanding the mode of action of antimicrobial agents, but the first and foremost is the susceptibility of microorganisms to these agents [8].

12.1.1 Minimum Inhibitory Concentration

MICs are considered the gold standard for determining the susceptibility of organisms to antimicrobials and are usually defined as the lowest concentrations of antimicrobial agents that inhibit more than 99% of bacterial growth. MICs are often used to confirm susceptibility or resistance to drugs but can also be used as a research tool to determine the *in vitro* activity of new antimicrobials. For the purpose of this chapter, MIC is defined as the lowest concentration of the antimicrobial that inhibits the visible growth of a microorganism using a microplate-based Alamar blue assay (MABA) system.

Drug-susceptibility testing or MIC determination using solid culture systems such as Lowenstein-Jensen (LJ) medium or Middlebrook agar take about 3 weeks to get results [9, 10]. Various alternative methods have been developed that have dramatically reduced the time required for susceptibility testing from weeks to days [11, 12, 13]. However, two methods of choice for determination of MIC using *M. tuberculosis* have emerged: (1) the radiometric BACTEC 460 TB method using BACTEC 12B vials [14, 15] and (2) the colorimetric microtiter plate–based method using Alamar blue [16, 17]. Both methods can be used to evaluate new compounds reducing the time to complete a test. The BACTEC 460 system is a radiometric assay used to determine susceptibility to an antimicrobial very rapidly and has long been the system of choice, but it is less useful for high-throughput screening because of the high cost and the generation of radioactive waste. The MABA system is a simple, rapid, low-cost, high-throughput system that does not require expensive instrumentation as the growth of bacteria can be measured by a visual color change [16, 17], and this is the method described in this protocol.

12.1.2 Bacterial Growth Inhibition Assay

MABA is used for measuring cell proliferation and viability by monitoring the oxidation-reduction state of the environment of cellular growth. As with the tetrazolium salts, Alamar blue is a soluble redox dye that is stable in culture medium and nontoxic [18]. The oxidized dye is blue and nonfluorescent; upon reduction it turns pink and fluorescent, therefore, growth can be determined by a visual color change or by using a fluorometer. Alamar blue has been successfully used to assess the susceptibility of *M. tuberculosis* to various antimicrobials in several laboratories [16, 17, 19].

12.1.3 Toxicology

The design of new therapeutic regimens relies on preclinical data to choose promising drugs and dosage schedules to be evaluated further in clinical trials. The identification of compounds with potential toxic activity is an important aspect in the testing of new antimicrobials. The relatively high costs, low throughput, and animal distress involved in testing compounds in whole animal models has led to the development of *in vitro* assays making use of human and animal cell lines that can be used for the cytotoxicity determination assays.

12.1.4 Eukaryotic Cytotoxicity Assay

Cytotoxicity may be defined simply as the cell-killing property of a chemical compound and is independent of the mechanism of death. Most cytotoxicity assays measure the amount of cell death that occurs in culture. When cell membranes are compromised, they become porous and allow macromolecules to leak out; these molecules can then be quantitated and used to estimate viability. For example, a typical assay might measure the presence of intracellular enzymes, such as lactate dehydrogenase in the culture supernatant [20]. The reduction of tetrazolium salts is now widely used as a reliable way to determine cytotoxicity. The yellow tetrazolium compound ([3-(4,5-dimethylthiazol-2-yl)-5-(3-carboxymethoxyphenyl)-2-(4-sulfophenyl)-2 H-tetrazolium, inner salt; MTS) and an electron coupling reagent (phenazine ethosulfate; PES), with an enhanced chemical stability, combined with tetrazolium salt makes a stable solution. The tetrazolium compound is reduced by metabolically active cells to a purple formazan product, in part by the action of dehydrogenase enzymes (which generate reducing equivalents such as NADH and NADPH) [21, 22]. The quantity of formazan is measured by a 96-well plate reader at 490 nm absorbance and is directly proportional to the number of viable cells in the culture medium [23].

Two commonly used cell lines for testing cytotoxicity are African green monkey kidney cells (Vero) and human hepatoma (HepG2) cells. Vero cells do not generally modify compounds that are added to them [24], whereas HepG2 cells are capable of modifying compounds [25, 26]. Both cell lines are grown as monolayers, and in order to keep the cell cultures healthy and actively growing, it is necessary to subculture them at regular intervals. The most common method of subcultivation is the use of proteolytic enzymes such as trypsin or collagenase to break the cell to substrate and intercellular connections. After the proteolytic disassociation of cells into a single cell suspension, the cells are diluted with fresh media and transferred into new culture flasks. The cells attach to the surface of the flask and begin to grow and divide and reach near confluency when they are again ready to be subcultured or used for testing purposes.

In this chapter, we describe a rapid *in vitro* method using the MABA to determine MIC values for novel mycobacterial growth inhibitors. In addition, we describe the use of a spectrophotometric method for determining the cytotoxic effects of these growth inhibitors. We also describe how to maintain the eukaryotic cell line and determine the concentration of the test compound causing 50% loss of cell viability (also referred to as IC_{50}).

12.2 Materials

12.2.1 Inhibition of Mycobacterial Growth

12.2.1.1 Preparation of *Mycobacterium tuberculosis* Working Stocks

1. *M. tuberculosis* $H_{37}Rv$ reference strain ATCC 25618 (American Type Culture Collection, USA).
2. Middlebrook OADC (oleic acid, albumin, dextrose, catalase) enrichment: 8.5 g/L sodium chloride, 50.0 g/L bovine albumin (fraction V), 0.04 g/L catalase, 0.3 mL/L oleic acid, and 20.0 g/L dextrose (Sigma, USA) in deionized water. Filter sterilize (*see* **Note 1**).
3. Tween-80 (Sigma): Prepare a 20% (w/v) stock. Filter sterilize through 0.2-μm membrane and store at 4°C.
4. Middlebrook 7 H9 broth (Difco, USA): Dissolve in deionized water at 4.7 g per 900 mL, add 0.2% (v/v) glycerol and autoclave. Supplement with 10% (v/v) OADC and 0.05% (v/v) Tween 80 (*see* **Note 2**).

5. Middlebrook 7 H11 agar (Difco): Dissolve 21.0 g in 900 mL deionized water, add 0.2% (v/v) glycerol, and autoclave. Supplement with 10% (v/v) OADC. Pour 25 mL in each of 100 × 15 mm Petri dishes.
6. Tryptic soy agar (Difco) (*see* **Note 3**).
7. 16 × 150 mm glass culture tubes with rubber-lined screw caps (Kimble, USA).
8. 25 × 8 mm disposable stir bars (VWR International, USA).
9. 1 μL sterile disposable inoculating loops (Nunc, USA).
10. Lowenstein – Jensen medium (LJ) slants (BD Diagnostic).
11. 250 mL and 1000 mL polycarbonate Erlenmeyer flasks with polypropylene screw caps (Nalgene, USA).
12. Polypropylene 2.0 mL cryogenic vials (Corning, USA).

12.2.1.2 Alamar Blue Susceptibility Assay

1. 96-well round-bottom tissue culture plates with low evaporation lid (Becton Dickinson, USA).
2. Sterile deionized water.
3. Middlebrook 7 H9 broth (Difco): Dissolve in deionized water at 4.7 g per 900 mL, add 0.2% (v/v) glycerol, and autoclave. Supplement with 10% (v/v) OADC and 0.05% (v/v) Tween 80 (*see* **Note 2**).
4. Middlebrook OADC (oleic acid, albumin, dextrose, catalase) enrichment: 8.5 g/L sodium chloride, 50.0 g/L bovine albumin (fraction V), 0.04 g/L catalase, 0.3 mL/L oleic acid, and 20.0 g/L dextrose (Sigma) in deionized water. Filter sterilize (*see* **Note 1**).
5. Stock cultures containing 2×10^7 CFU/mL of *M. tuberculosis* $H_{37}Rv$ strain. Store frozen at –80°C.
6. Dimethyl sulfoxide (DMSO) (Sigma).
7. 100% ethanol.
8. Test compounds: Prepare as 20 mg/mL stocks in appropriate solvents such as deionized water, DMSO or ethanol, depending on the solubility of the compound.
9. Isoniazid (INH) (Sigma): 10 mg/mL stock solution in deionized sterile water. Dilute to 25 μg/mL for the working stock.
10. Parafilm.
11. 10X Alamar blue dye solution (Biosource International, USA). Use at 1X final concentration.

12.2.2 Growth of Eukaryotic Cells and Toxicity Test

12.2.2.1 Maintenance of Eukaryotic Cell Lines

1. African green monkey kidney cell line (Vero) ATCC CCL-81 (American Type Culture Collection).

2. Human hepatocellular liver carcinoma cell line (HepG2) ATCC HB-8065 (American Type Culture Collection).
3. RPMI 1640 medium with L-glutamine (Invitrogen, USA).
4. RPMI complete medium. Supplement RPMI 1640 with 1.5 g/L sodium bicarbonate (Sigma), 10 mL/L 100 mM sodium pyruvate (Mediatech, USA), 140 mL/L 100X nonessential amino acids (Mediatech), 100 mL/L 100X penicillin/streptomycin solution (10,000 I.U/10,000 μg/mL) (Mediatech), and 10% (v/v) bovine calf serum (BCS) (Hyclone, USA) (*see* **Note 4, 5**).
5. 1 M KOH in deionized water, filter sterilize and store at 4°C (*see* **Note 4**).
6. Hank's balanced salt solution: without sodium bicarbonate, calcium, and magnesium (Mediatech).
7. Trypsin EDTA 1X solution: 0.05% trypsin, 0.53 mM EDTA in Hank's balanced salt solution.
8. Phosphate-buffered saline (PBS) solution: 0.210 g/L monobasic potassium phosphate, 9 g/L NaCl, 0.726 g/L dibasic potassium phosphate in deionized water, pH 7.2 (Invitrogen).
9. 250 mL 75 cm^2 sterile tissue culture flasks with standard cap.
10. Bright-Line hemocytometer (Hausser Scientific, USA).
11. Compound research microscope (Olympus, Japan).
12. Incubator at 37°C with 5% CO_2 and 75% humidity.
13. Basic inverted microscope (Zeiss, Germany).

12.2.2.2 Cytotoxicity Assay

1. African green monkey kidney cell line (Vero) ATCC CCL-81 (American Type Culture Collection).
2. Human hepatocellular liver carcinoma cell line (HepG2) ATCC HB-8065 (American Type Culture Collection).
3. RPMI 1640 medium with L-glutamine (Invitrogen).
4. RPMI complete medium. Supplement RPMI 1640 with 1.5 g/L sodium bicarbonate (Sigma), 10 mL/L 100 mM sodium pyruvate (Mediatech), 140 mL/L 100X nonessential amino acids (Mediatech), 100 mL/L 100X penicillin/streptomycin solution (10,000 I.U/10,000 μg/mL) (Mediatech), and 10% (v/v) bovine calf serum (BCS) (Hyclone) (*see* **Note 4, 5**).
5. 1 M KOH in deionized water, filter sterilize and store at 4°C (*see* **Note 4**).
6. Hank's balanced salt solution: without sodium bicarbonate, calcium, and magnesium (Mediatech).
7. Trypsin EDTA 1X solution: 0.05% trypsin, 0.53 mM EDTA in Hank's balanced salt solution.
8. PBS solution: 0.210 g/L monobasic potassium phosphate, 9 g/L NaCl, 0.726 g/L dibasic potassium phosphate in deionized water, pH 7.2 (Invitrogen).
9. Sterile water.
10. 250 mL 75 cm^2 sterile tissue culture flasks with standard cap.
11. Bright-Line hemocytometer (Hausser Scientific).

12. Compound research microscope (Olympus).
13. Basic inverted microscope (Zeiss).
14. Incubator at 37°C with 5% CO_2 and 75% humidity.
15. DMSO.
16. Test compounds: Prepare as 20 mg/mL stocks in DMSO. Based on the MIC of the compound to be tested for cytotoxicity, make further dilutions. Make five serial dilutions of each test compound in 1640 RPMI complete medium (*see* **Note 6**), resulting in the final concentration ranging from 1X MIC to 25X MIC.
17. 96-well flat-bottom tissue culture plates with low evaporation lid (Becton-Dickinson).
18. Cell Titer 96 Aqueous One Solution Cell Proliferation Kit (Promega Corporation, USA).
19. Microplate reader with 490-nm wavelength filter.

12.3 Methods

12.3.1 Inhibition of Mycobacterial Growth

12.3.1.1 Preparation of *Mycobacterium tuberculosis* Working Stocks

1. Starting from the ATCC stock of *M. tuberculosis* H37Rv, streak a loopful using a 1-μL inoculating loop onto 7 H11 agar plate.
2. Incubate at 37°C for 3 to 4 weeks to obtain isolated colonies.
3. Pick a single colony from the plate and inoculate into 10 mL of 7 H9 broth in a sterile glass culture tube with a stir bar (*see* **Note 7**).
4. Incubate at 37°C with slow stirring.
5. Spread 0.5 mL from the newly made tube culture onto an LJ slant or 7 H11 agar plate as a low-passage backup.
6. Take periodic readings of the culture at OD_{600} until mid log phase (OD_{600} = 0.6 to 0.9).
7. Transfer 1 mL culture into 10 mL 7 H9 medium.
8. Incubate with slow stirring at 37°C for future use (*see* **Note 8**). Simultaneously, take 10 μL from the 10 mL culture and spread onto a Tryptic soy agar plate (*see* **Note 3**).
9. Divide the rest of the inoculum from the 10 mL 7 H9 mid log-phase culture between two 250-mL Erlenmeyer flasks containing 50 mL 7 H9 medium each. This is passage 1 (*see* **Note 9**). Incubate the flask cultures on a rotary shaker at 125 rpm at 37°C until they reach an OD_{600} of 0.6 to 0.9.
10. For a second scale-up, divide the 50 mL culture between two 1-L Erlenmeyer flasks each containing 500 mL 7 H9 broth. This is passage 2. Spread 10 μL of this culture on a Tryptic soy agar plate to check for contamination (*see* **Note 3**).
11. The 500 mL culture is the working stock and is ready to be frozen in aliquots when it reaches an OD_{600} of 0.6 to 0.9.

12. Add 1.5 mL aliquots of the 500 mL culture to 2.0-mL cryovials, swirling the culture flask frequently to keep the bacteria in suspension.

13. Place the vials in appropriately labeled freezer boxes in the order in which they were filled. Store the culture stocks frozen at –80°C.

14. Take one cryovial from each freezer box and thaw on ice.

15. Make eight 10-fold serial dilutions of each stock culture vial taken from the freezer box.

16. Plate 0.1 mL on 7 H11 agar plates and incubate at 37°C for 2 to 3 weeks.

17. Identify a plate with about 100 to 200 colonies, count the colonies, and calculate CFU/mL, which should be approximately 2×10^7 CFU/mL (*see* **Note 10**).

12.3.1.2 Alamar Blue Susceptibility Assay

1. Take a sterile 96-well round-bottom microplate and add 200 µL sterile deionized water to all of the wells on the outer perimeter (Fig. 12.1) (*see* **Note 11**).

2. Add 98 µL 7 H9 broth in the wells in column 2, rows B to G. Add 50 µL 7 H9 broth to the rest of the wells from B3 to G11 (Fig. 12.1).

3. Add 2 µL of the test compound from the 20 mg/mL stock concentration to the wells in column 2, rows B to F, and 2 µL of INH from the 25 µg/mL working stock concentration to well G2 (Fig. 12.1).

4. Take 50 µL from column wells B2 to G2 and transfer to column wells B3 to G3. Make identical serial dilutions (1:2) up to B10 to G10. Discard the excess

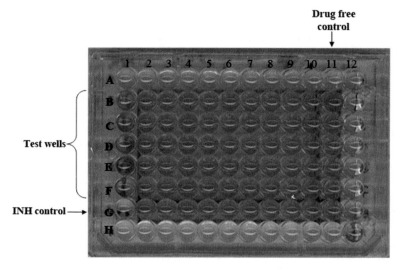

Fig. 12.1 Representative 96-well microplate format for screening compounds to determine their MICs on *M. tuberculosis* H_{37}Rv using MABA

50 μL medium from the wells B10 to G10 so that each well has only 50 μL volume remaining. The wells B11 to G11 serve as a compound free control.

5. Thaw the frozen stock cryovial and adjust to 2×10^7 CFU/mL of *M. tuberculosis* $H_{37}Rv$ on ice.

6. Add 100 μL thawed stock to 10 mL fresh 7 H9 medium and mix well. Add 50 μL of this mixture to wells B2 to G11 so the final concentration is 1×10^5 CFU/mL in each well. The final concentration will range from 200 μg/mL to 0.78 μg/mL for the test compounds (*see* **Note 12**) and 0.25 μg/mL to 0.0009 μg/mL for INH. Negative control wells (B11 to G11) have only the solvent (Section 12.2.1.2, step 8) and bacterial culture added.

7. Seal the plate with Parafilm and incubate at 37°C for 5 days (*see* **Note 13**).

8. Add 10 μL Alamar blue reagent to well B11. Reincubate the plate at 37°C for 24 h. If well B11 turns pink, add the Alamar blue reagent mixture to all the remaining wells in the microplate (*see* **Note 14**).

9. Reseal the plate and incubate for an additional 24 h at 37°C. Record the color change in all of the wells at the end of incubation. Blue indicates no growth, whereas pink indicates growth (*see* **Note 15**).

12.3.2 Growth of Eukaryotic Cells and Toxicity Test

12.3.2.1 Maintenance of Eukaryotic Cell Lines

1. Grow cells in 75 cm^2 tissue culture flasks in 30 mL RPMI 1640 complete media (*see* **Note 16**) at 37°C in a CO_2 incubator.

2. To harvest a cell monolayer, remove and discard the culture medium. Rinse the cell layer with 5 mL PBS. (*see* Notes 16 and 17).

3. Add 2 to 3 mL of trypsin solution to the cell layer and check the progress of the enzyme treatment every 5 min for a maximum of 15 min. Monolayers that are particularly difficult to detach can be placed at 37°C (*see* **Note 18**).

4. Add 8 to 10 mL of RPMI 1640 complete medium to the cell suspension and wash any remaining cells from the bottom of the culture flask. Pipette the cell suspension up and down several times to ensure a single cell dispersion.

5. Count the cells using a hemocytometer as follows: add 10 to 20 μL of medium containing cells to the hemocytometer, count the cells in four large corner squares (these are 1 mm × 1 mm in area) using a microscope at low magnification (10×) (*see* **Note 19**). Take the average of the four counts and multiply by 10^4 to obtain the number of cells per milliliter.

6. Seed a new 75 cm^2 tissue culture flask containing 30 mL RPMI 1640 complete media with 5×10^4 cells.

7. Repeat steps 1 to 4 as the cells near confluence (2 to 3 days).

12.3.2.2 Cytotoxicity Assay

1. Using a near confluent (75% to 95%) 75 cm² tissue culture flask, remove the media and rinse the cells with 5 mL PBS buffer without disturbing the monolayer (*see* **Note 17**).
2. To detach cells from the surface of the flask, add 2 to 3 mL trypsin solution to the cell layer and incubate at 37°C. Check the progress of the enzyme treatment every 5 min for a maximum of 15 min (*see* **Note 18**).
3. Add 5 mL RPMI 1640 complete media and pipette up and down to form a single cell suspension.
4. Count 10 µL aliquot of the cells on a hemocytometer as described in Section 12.3.2.1, step 5.
5. Calculate and adjust cell suspension to 5×10^4 cells/mL by adding fresh RPMI 1640 complete medium.
6. Dispense 100 µL/well in a sterile 96-well flat-bottom plate in wells B2 to G11 (Fig. 12.2).
7. Add 200 µL sterile water to wells A1 to A12, H1 to H12, and B1 to G1 on the outer perimeter of the 96-well plate to limit evaporation (*see* **Note 11**). Add 200 µL RPMI 1640 complete medium in wells from B12 to G12 as a media only (negative) control.

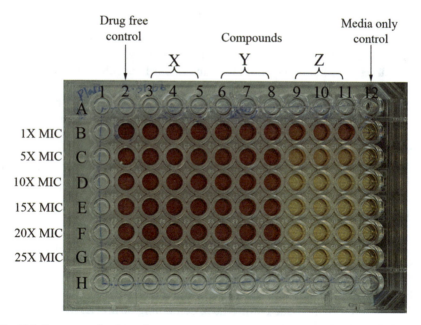

Fig. 12.2 Representative 96-well plate format for screening compounds to determine their effect on the viability of a eukaryotic cell line

8. Incubate plate for 2 h in the CO_2 incubator to allow the cells to attach to the surface of the plate.

9. Add the test compounds to the wells B3 to G11 starting at 1X MIC and increasing to 25X MIC (Fig. 12.2). Each concentration of the test compound should be done in triplicate. Three compounds can be tested in each 96-well plate. Bring up the total volume of wells B2 to G11 to 200 μL by adding an appropriate amount of fresh RPMI 1640 complete medium. The wells B2 to G2 serve as the cells only control.

10. Incubate the plates for 72 h at 37°C in the CO_2 incubator. Visually inspect under the inverted microscope every 24 h to check for contamination or cell lysis.

11. After 72 h remove the media from all wells and wash the cells with 200 μL PBS buffer.

12. Add 100 μL RPMI 1640 complete medium to each well from B2 to G12 (*see* **Note 16**).

13. Add 10 μL thawed Cell Titer 96 Aqueous One Solution Reagent to wells B2 to G12 and incubate at 37°C in CO_2 incubator for 4 h.

14. Read the absorbance at 490-nm wavelength using a microplate reader (*see* **Note 20**).

15. Calculate the IC_{50} value for each test compound. Background absorbance, which is due solely to the reaction of the reagents, should be deducted from the absorbance values of the treated and untreated cells. The mean absorbance of the medium-only control is IC_0. The mean absorbance obtained from the cells-only control is IC_{100} and for the test compounds is IC_{100T}. To obtain the corrected absorbance, IC_0 is subtracted from the IC_{100} and IC_{100T} absorbance values. To determine IC_{50} for each compound, plot the corrected absorbance at 490 nm (y axis) versus concentration of the

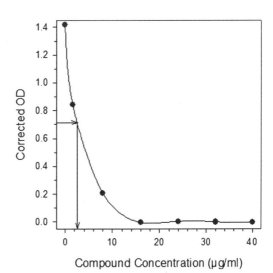

Fig. 12.3 Representative graph showing cytotoxic effect of a compound and determination of its IC_{50} value (~ 2.5 μg/mL)

compound tested (x axis), and calculate the IC_{50} value by determining the x-axis value corresponding with half the difference between the corrected IC_{100} and corrected IC_{100T} absorbance values [27] (Figs. 12.2 and 12.3).

12.4 Notes

1. Gently warm the oleic acid solution until it melts, do not microwave or heat above 55°C. Albumin should be added gradually to avoid clumping. Store OADC at 4°C, protected from light. The OADC supplement is extremely heat-labile and should not be added to hot media.
2. Mycobacteria, particularly *M. tuberculosis*, have a tendency to clump in culture because of the presence of a thick waxy outer cell wall. The addition of a non-ionic detergent to the media reduces the amount of clumping and provides a more homogenous suspension of cells.
3. Tryptic soy agar is used as an initial growth medium to observe colony morphology and develop a pure culture. If the culture has contaminating bacteria in it, they will become apparent in 24 to 48 h.
4. The medium supplements are added to a number of different incomplete media along with 10% (v/v) bovine calf serum to make complete medium. The pH of the complete media should be 7.0. The addition of nonessential amino acids may reduce the pH, and this can be adjusted by adding potassium hydroxide (KOH) solution to the media.
5. The serum should be stored frozen. Before adding it to the RPMI 1640 media, remove the serum bottle from the freezer and allow it to acclimate to the room temperature for about 10 min and place in a 37°C water bath. Excessive temperature can degrade the heat-labile nutrients in the serum. Thawing of serum is crucial to its performance. It is recommended to periodically swirl the serum container while thawing, otherwise cryoprecipitates are formed, which are often insoluble. Filtering serum to remove cryoprecipitates is not recommended as it could result in loss of nutrients.
6. The concentration of the solvent carrier (DMSO) for the test compound in the medium with cells should not exceed 1%; beyond this, the cells start to die because of the toxic effect of the solvent. If a higher concentration of the compound is required, make dilutions using the RPMI 1640 complete medium instead of DMSO. It is recommended to always set up a separate plate to determine that the solvent (DMSO) is at an acceptable (nontoxic) concentration.
7. It is always recommended to start cultures on a solid medium by taking a loopful of the thawed stock and streaking on an appropriate agar plate. After growth, pick a colony from the plate and add to liquid medium as it is sometimes difficult to start the bacterial growth in a liquid medium from the frozen state.
8. The purpose of having a staggered series of backup cultures lagging behind the primary cultures is if the primary culture is contaminated, there is no need to start a 10 mL culture again from the seed stock.
9. Passage number is important, the working stocks should be no greater than passage 6, the lower the better. Every time a culture is subcultured or plated onto a solid media, it is considered a passage.
10. To determine the CFU/mL, take one vial from each freezer box to plate out serial dilutions, keeping track of the boxes the vials come from. This is done in order to avoid any variations in the CFU/mL from cryovials filled at different times.
11. The water is added to the wells on the outer perimeter of the microplate in order to minimize the evaporation of the medium in the test wells during incubation.
12. The concentration of the test compounds used here are representative values and can be changed according to individual requirements and the nature of the compound being tested.

13. Alternately, the plates can be sealed in a Ziplock bag and a damp paper towel placed on top of the plate inside the bag in order to prevent evaporation during incubation.
14. If the B11 well remains blue, the Alamar blue reagent should be added to another control well C11, reincubated for additional 24 h, and checked for color change. This is done to make sure the cells are growing normally in the control wells. If there is no color change even after 48 h, the plates should be discarded and the assay repeated.
15. Wells may appear violet after 24 h incubation but invariably change to pink after extended incubation at 37°C indicating growth. The MIC is defined as the lowest concentration of compound in which the dye remains blue. A standard scanner can be used to generate a permanent record of the plate (Fig. 12.1).
16. PBS and RPMI 1640 complete media are stored at 4°C and should be brought to room temperature before use.
17. Trypsin is the most common disassociating solution, and its action is inhibited by serum. Residual amounts of serum are often responsible for the failure of the trypsin solution to detach the cells from the substrate. To avoid this, the cells are washed with PBS before treatment with trypsin.
18. Cells vary greatly in how fast they come off the flask. After adding trypsin to the cells in the tissue culture flask and incubating for 10 to 15 min, a quick screening of the cell suspension under an inverted microscope is recommended in order to ensure complete detachment of cells from the surface of the tissue culture flask and also to get a suspension of at least 95% single cells. Cells should be trypsinized until they come off the flask, but not longer.
19. It is easy to double-count cells that lie on the lines between individual small squares. For cells that touch the lines defining the squares, only count those touching the top or left side of an individual square.
20. Absorbance values lower than the control cells indicate a reduction in the rate of cell growth. Conversely, a higher absorbance indicates an increase in cell proliferation

Acknowledgments Research in the authors' laboratories was supported by grants AI-057836, AI-37139, AI-18357, AI-46393, AI-49151, and U54 AI-06357 from the National Institute of Allergy and Infectious Diseases, National Institutes of Health.

References

1. Kritski, A.L., Marques, M.J., Rabahi, M.F., Vieira, M.A., Weneck Barroso, E., Carvalho, C.E., Andrade, G de N., Bravo de Souza K., Andrade, L.M., Gontijo, P.P., Riley, L.W. (1996). Transmission of tuberculosis to close contacts of patients with muti-drug-resistant tuberculosis. Am. J. Resp. Crit. Care. Med. 153, 331–335.
2. Espinal, M.A., Kim, S.J., Suarez, P.G., Kam, K.M., Khomenko, A.G., Migliori, G.B., Baez, J., Kochi, A., Dye, C., and Raviglione, M.C. (2000). Standard short-course chemotherapy for drug-resistant tuberculosis: treatment outcomes in 6 countries. JAMA 283, 2537–2545.
3. Stead, W.W. (1997). Pathogenesis of a first episode of chronic pulmonary tuberculosis in man: recrudescence of residuals of the primary infection or exogenous reinfection? Am. Rev. Respir. Dis. 95, 729–745.
4. Tead, W.W., Kerby, G.R., Schlueter, D.P., and Jordahl, C.W. (1968). The clinical spectrum of primary tuberculosis in adults. Confusion with reinfection in the pathogenesis of chronic tuberculosis. Am. Intern. Med. 68, 731–745.
5. Dick, T. (2001). Dormant tubercle bacilli: the key to more effective TB chemotherapy. Antimicrob. Agents Chemother. 47, 117–118.
6. Wayne, L.G and Hayes, L.G. (1996). An *in vitro* model for sequential study of down shift of *Mycobacterium tuberculosis* through two stages of non–replicating persistence. Infect. Immun. 64, 2062–2069.

7. Musser, J (1995). Antimicrobial agent resistance in mycobacteria: molecular genetic insights. Clin. Microbiol. Rev. 8, 496–514.

8. Piatek, A., Telenti, A., Murray, M., et.al (2000). Genotypic analysis of *Mycobacterium tuberculosis* in two distinct populations using molecular beacons: implications for rapid susceptibility testing. Antimicrob. Agents Chemother. 44, 103–110.

9. Canetti, G., Froman. F., Grosset, J., Hauduroy, P., Langerova, M., Mahler, H.T., Meissner, G., Mitchison, D.A., and Sula, L. (1963). Mycobacteria: laboratory methods for testing drug sensitivity and resistance. Bull. W. H. O. 29, 565–578.

10. Canetti, G., Fox, W., Khomenko, A., Mahle, H.T., Menon, N.K., Mitchison, D.A., Rist N., and Smelev, N.A. (1969). Advances in techniques of testing mycobacterial drug sensitivity, and the use of sensitivity tests in tuberculosis control programmes. Bull. W. H. O. 41, 21–43.

11. Norden, M.A., Kurzynski, T.A., Bownds, S.E., Callister, S.M., and Schell, R.F. (1995). Rapid susceptibility testing of *Mycobacterium tuberculosis* (H_{37}Ra) by flow cytometry. J. Clin. Microbiol. 33, 1231–1237.

12. Ranger, A., and Mills, K. (1996). Testing of susceptibility to ethambutol, isoniazid, rifampin and streptomycin by using E test. J. Clin. Microbiol. 34, 1672–1676.

13. Wilson, S.M., Al-Suwaidi, Z., McNerney, R., et al. (1997). Evaluation of a new rapid bacteriophage-based method for the drug susceptibility testing of *Mycobacterium tuberculosis*. Nat. Med. 3, 465–468.

14. Heifets, L.B. (1991). Drug susceptibility tests in the management of chemotherapy of tuberculosis, 89–122. In L.B. Heifets (ed.), Drug susceptibility in the chemotherapy of mycobacterial infections. CRC Press, Inc., Boca Raton, FL.

15. Inderleid, C.B., and Salfinger, M. (1995). Antimicrobial agents and susceptibility tests: mycobactera, 1385–1404. In P.R. Murray, E.J. Baron, M.A. Pfaller, F.C. Tenover, and R. H. Yolken (eds.), Manual of clinical microbiology, 6th ed. ASM Press, Washington, DC.

16. Collins, L.S., and Franzblau, S.G. (1997). Microplate Alamar Blue assay versus BACTEC 460 system for high-throughput screening of compounds against *Mycobacterium tuberculosis* and *Mycobacterium avium*. Antimicrob. Agents Chemother. 41, 1004–1009.

17. Franzblau, S.G., Witzig, R.S., McLaughlin, J.C., Torres, P., Madico, G., Hernandez, A., Degnan, M.T., Cook, M.B., Quenzer, V.K., Ferguson, R.M., and Gilman, R.H. (1998). Rapid, low-technology MIC determination with clinical *Mycobacterium tuberculosis* isolates by using the microplate Alamar Blue assay. J. Clin. Microbiol. 36, 362–366.

18. Ahmed, S.A., Gogal, R.M., and Walsh, J.E. (1994). A new rapid and non-radioactive assay to monitor and determine the proliferation of lymphocytes: an alternative to [^{3}H]thymidine incorporation assay. J. Immunol. Methods 170, 211–224.

19. Yajko, D.M., Medej, J.J., Lancaster, M.V., Sanders, C.A., Cawthon, V.L., Gee, B., Babst, A., and Hadley, W.K. (1995). Colorimetric method for detecting MICs of anti-microbial agents for *Mycobacterium tuberculosis*. J. Clin. Microbiol. 33, 2324–2327.

20. Nachlas, M.M., et al. (1960). The determination of lactic dehydrogenase with a tetra-zolium salt. Anal. Biochem. 1, 317–326.

21. Barltrop, J.A., Owen, T.C., Cory, A.H., and Cory, J.G. (1991). 5-(3-carboxymethoxy-phenyl) -2-(4, 5- menthylthiazoly)-3-(4-sulfophenyl) tetrazolium, inner salt (MTS) and related analogs of 3-(4,5-dimethylthiazolyl)-2.5-diphenyltetrazolium bromide (MTT) reducing to purple water-soluble formazans as cell-viability indicators. Bioorg. Med. Chem. Lett. 1, 611–614.

22. Cory, A.H., Owen, T.C., Barltrop, J.A., and Cory, J.G. (1991). Use of aqueous soluble tetrazolium/formazan assay for cell growth assays in culture. Cancer Commun. 3, 207–212.

23. Mosmann, T. (1993). Rapid colorimetric assay for cellular growth and survival. Application to proliferation and cytotoxicity assays. J. Immunol. Methods 65, 55–63.

24. Yasumura, Y., and Kawakita, Y. (1963). Vero cell line derived from the kidney of a normal, adult, African green monkey (Cercopithecus) Nippon Rinsho. 21, 1209.

25. Ferro, M., Bassi, A.M., and Nanni, G. (1988). Hepatoma cell cultures as *in vitro* models for the hepatotoxicity of xenobiotics. ATLA 16, 32–37.
26. Bassi, A.M., Piana, S., Penco, S., Bosco, O., Brenci, S., and Ferro, M. (1991). Use of an established cell line in the evaluation of the cytotoxic effects of various chemicals. Boll. Soc. It. Biol. Sper. 8, 809–816.
27. Promega Corporation. (1999). Cell Titer 96 Aqueous One Solution Cell Proliferation Assay. Promega Technical Bulletin No. 245, 1–9. Promega Corporation, USA.

Chapter 13
Electroporation of Mycobacteria

Renan Goude and Tanya Parish

Abstract High-efficiency transformation is a major limitation in the study of mycobacteria. The genus *Mycobacterium* can be difficult to transform; this is mainly caused by the thick and waxy cell wall but is compounded by the fact that most molecular techniques have been developed for distantly related species such as *Escherichia coli* and *Bacillus subtilis*. In spite of these obstacles, mycobacterial plasmids have been identified, and DNA transformation of many mycobacterial species has now been described. The most successful method for introducing DNA into mycobacteria is electroporation. Many parameters contribute to successful transformation; these include the species/strain, the nature of the transforming DNA, the selectable marker used, the growth medium, and the conditions for the electroporation pulse. Optimized methods for the transformation of both slow-grower and fast-grower are detailed here. Transformation efficiencies for different mycobacterial species and with various selectable markers are reported.

Keywords electrocompetent cells · selectable marker · transformation efficiency · transforming DNA

13.1 Introduction

A considerable body of knowledge has been accumulated regarding the biology of the mycobacteria, although mycobacterial pathogenic strategies remain poorly understood. The slow growth rate of most species, the impenetrable nature of the cell wall, and the hazardous nature of working with pathogens are largely responsible for our poor understanding of mycobacteria. Genetic techniques are invaluable tools for investigating bacterial biology, but a

R. Goude
Institute of Cell and Molecular Science, Barts and the London, Queen Mary's School
of Medicine and Dentistry, 4 Newark Street, Whitechapel, London, E1 2AT, UK
e-mail: r.goude@qmul.ac.uk

T. Parish, A.C. Brown (eds.), *Mycobacteria Protocols*,
doi: 10.1007/978-1-59745-207-6_13, © Humana Press, Totowa, NJ 2008

prerequisite for all genetic manipulation is the requirement for a facile method of introducing recombinant DNA into cells.

Electroporation involves subjecting cells to a brief high electrical impulse, which allows the entry of DNA, and is the most widely used method for introducing DNA into mycobacterial cells. It produces high efficiencies of transformation [1] and enables the genetic manipulation of both fast-growing and slow-growing mycobacterial species. To date, numerous species of-mycobacteria have been successfully transformed by electroporation including *Mycobacterium tuberculosis* [2, 3, 4, 5, 6, 7, 8], *Mycobacterium bovis* BCG [2, 3, 5, 7, 9, 10, 11, 12, 13, 14, 15, 16], *Mycobacterium vaccae* [9, 10], *Mycobacterium phlei* [13], *Mycobacterium w* [10], *Mycobacterium fortuitum* [13], *Mycobacterium aurum* [17, 18], *Mycobacterium intracellulare* [16, 19], *Mycobacterium parafortuitum* [20], *Mycobacterium marinum* [21], *Mycobacterium avium* [22, 23, 24], and *Mycobacterium smegmatis* [25].

The efficiency of electroporation depends on the mycobacterial species used (Table 13.1). Some species or strains are notoriously difficult to transform, for example some clinical isolates of the *M. avium* complex [24]. A second important factor in electroporation is the resistance marker used on the incoming DNA, as this will determine the maximum efficiency attainable (Table 13.1).

13.1.1 Vectors

Most extrachromosomal vectors used in mycobacteria are based on the pAL5000 replicon from *M. fortuitum*, which has an estimated copy number of five [26]. Shuttle plasmids with an additional origin of replication from *E. coli* are used widely to allow for ease of DNA manipulation in an amenable cloning host [15]. The pAL5000 minimal origin has been defined and useful cloning sites introduced in several version (e.g., pMV261) [5, 27]. pAL5000-based vectors replicate in many mycobacterial species, except members of the *M. avium* complex [19].

Alternative mycobacterial plasmid replicons have been derived from the *Mycobacterium scrofulaceum* plasmid pMSC262 [11]; the *M. avium* plasmid pLR7 [22]; *M. fortuitum* plasmids pJAZ38 and pMF1 [28, 29]; and the linear plasmid, pCLP from *Mycobacterium celatum*. In addition, origins of replication from heterologous species that are capable of replication in both mycobacteria and *E. coli* have been used. These include plasmid pNG2 from *Corynebacterium* [14] and the broad host-range Gram-negative cosmid vector pJRD215 [17].

Integrative vectors are also available for mycobacteria. The major advantage of these is improved stability. The most widely used integrative-plasmid is derived from the mycobacteriophage L5 [15, 30]. The minimal system only requires the phage attachment site (*attP*); the integrase function can be provided either on the same vector or on a second vector. L5-based vectors are stably maintained in *M. smegmatis*, *M. tuberculosis*, and *M. bovis* BCG [30].

Table 13.1 Selectable Markers and Reported Electroporation Efficiencies for Mycobacteria Species

Species	Selection	Efficiency[a]	References
Fast-growers			
M. aurum	Kanamycin	100	[17, 18]
	Streptomycin		[17, 18]
M. fortuitum	Kanamycin		[13]
M. parafortuitum	Kanamycin	300	[20]
	Streptomycin	30	[20]
M. phlei	Kanamycin		[13]
M. smegmatis	Kanamycin	10	[12, 15]
M. smegmatis mc^2 155	Kanamycin	10^5 to 10^6	[2, 9, 16, 25, 41]
	Hygromycin	5×10^3	[10]
	Apramycin	2×10^4	[3, 7]
	Streptomycin		[17]
	Tetracycline		[35]
	Gentamicin		[42]
	Sulfonamide		[42]
	Chloramphenicol[b]		[25]
M. vaccae	Hygromycin	10^3 to 10^5	[9, 10]
Slow-growers			
M. avium	Kanamycin	10^2 to 10^4	[22, 23, 24]
	Hygromycin	10^4	[24]
M. avium subsp.	Kanamycin	100	[23]
M. bovis BCG	Kanamycin	10^3 to 10^5	[2, 5, 9, 11, 12, 15, 16]
	Apramycin	10^3	[3, 7]
	Hygromycin		[10, 14]
	Chloramphenicol[c]		[13]
M. bovis	Kanamycin	10^4	[16]
M. intracellulare	Kanamycin		[16, 19]
	Gentamicin		[19]
M. marinum	Kanamycin	100	[21]
M. tuberculosis	Kanamycin	10^4 to 10^6	[2, 5, 8]
	Apramycin	10^2	[3, 7, 8]
	Hygromycin	10^7	[4, 6]
M. w	Hygromycin	10^3 to 10^5	[10]

[a]Number of transformants per μg of DNA.
[b]Used for screening. Not for direct selection.
[c]In conjunction with kanamycin.

Other integrative vectors include a derivative of mycobacteriophage Ms6 (pEA4), which is stably maintained in *M. smegmatis* [31], and plasmid pSAM2 from S*treptomyces ambofaciens* [32]. An artificial transposon derived from the *M. avium* subsp. *paratuberculosis* insertion sequence IS*900* has also been used to integrate into the chromosome of *M. bovis* BCG and *M. smegmatis*, but the copy number per cell varied from one to five [9, 33].

13.1.2 Selection Markers

Vectors for use in mycobacteria must carry appropriate selectable markers (Table 13.1). The natural resistance of mycobacteria to many antibiotics and the requirement to use stable drugs with a low frequency of spontaneous resistance restrict the alternatives. Genes conferring kanamycin resistance (*aph*) were the first selectable markers used in mycobacteria [15]. However, most slow-growers possess only one ribosomal RNA operon; this unusual situation means that resistance to agents such as kanamycin easily arises by spontaneous mutation in the *rrn* operon itself [34]. This does not occur in fast-growers, which contain two *rrn* operons. Electroporation with shuttle plasmids using *aph* was not achievable in *M. w* and *M. vaccae* [10], although the hygromycin resistance gene from *Streptomyces hygroscopicus* was transformable and is suitable for use in both fast-growers and slow-growers [14]. Apramycin resistance has also been used in *M. tuberculosis* but with low transformation frequencies [3, 7].

Other selectable markers such as gentamicin [19], streptomycin [20], and sulfonamide resistance genes [35] have been used for selection. It has been reported that tetracycline can be used in *M. smegmatis* [35]; however, this antibiotic is not suitable for slow-growing mycobacteria because tetracycline is unstable over the time required for culture (3 to 6 weeks). Ampicillin resistance is not suitable for use in mycobacteria because they are naturally resistant to β-lactams. Chloramphenicol cannot be used for direct selection owing to the high rate of spontaneous mutations, although it has been used in conjunction with other antibiotic-resistance genes [13, 25].

13.1.3 Electroporation Conditions

During electroporation, an electrical impulse is delivered to the cells that produces a reversible rupture of the membrane and allows the entrance of molecules such as DNA. The critical parameters are the electric field strength and the duration of the pulse. Usually, the voltage necessary to obtain pores in the membrane is inversely proportional to the size of the cells. The electric field strength (V/cm) is the difference of potential between the two electrodes and can be modified by changing either the applied voltage or the distance between the electrodes. The third important parameter is the form of the pulse. There are two types of pulse, the square and the exponential decay wave, which are produced by the partial or complete discharge of the capacitor, respectively. Most electroporation apparatuses deliver an exponential decay wave pulse that gives good results but can kill fragile cells. The square wave pulse is less aggressive. For mycobacteria, a single pulse of 2.5 kV, 25 µF, and a resistance of 1000 Ω with an exponential decay wave pulse is usually used.

In this chapter, we describe protocols for electroporation of a slow-growing species, *M. tuberculosis*, and a fast-growing species, *M. smegmatis*. Protocols can be adapted for use in other species, although some optimization of parameters may be required.

13.2 Materials

13.2.1 Electroporation of M. tuberculosis

1. Tween-80 (Sigma, Poole, Dorset, UK): Prepare as a 20% (v/v) stock, filter sterilize through an 0.2-μm membrane, and store at 4°C (*see* **Note 1**).
2. Middlebrook 7H9 broth (Becton Dickinson, Sparks, USA): Dissolve in deionized water at 4.7 g per 900 mL, add 5 mL 10% w/v Tween-80, and autoclave (*see* **Note 2**).
3. Middlebrook OADC enrichment (Becton Dickinson), containing oleic acid, bovine albumin fraction V, dextrose, catalase, and NaCl: store at 4°C (*see* **Note 3**).
4. 7H9-Tw-OADC liquid medium should be prepared by adding 10% (v/v) OADC supplement to 7H9-Tween broth. Autoclaved 7H9 broth can be kept at room temperature for months. OADC is stable at 4°C for months. The OADC should be added to the 7H9-Tween just before use.
5. Roller bottles: 450 cm^2 (Corning, New York, USA).
6. 2 M glycine (Analar grade; Sigma), autoclave (*see* **Note 4**).
7. 10% (w/v) glycerol; sterilize by autoclaving.
8. Electroporation apparatus with pulse controller (*see* **Note 5**).
9. Electroporation cuvettes; 0.2 cm gap electrodes (*see* **Note 5**).
10. DNA in solution (*see* **Notes 6** and **7**); this should be free from salts, enzymes, and other substances. In order to clean up DNA, it can be ethanol precipitated and thoroughly washed with 70% ethanol (this will also remove excess salts). The concentration of DNA should be about 0.2 to 1 mg/mL.
11. Middlebrook 7H10 agar: 19 g agar base per 900 mL, autoclave.
12. Solid medium should be prepared by adding 10% (v/v) OADC supplement to 7H10 agar. Pour plates and use within 1 week.
13. Kanamycin sulfate (Sigma), 50 mg/mL stock (filter sterilize), store at –20°C.
14. Hygromycin B (Roche diagnostics, Mannheim, Germany), obtained as a 50 mg/mL stock in phosphate-buffered saline; store at 4°C in the dark.
15. Selection plates should contain 10 to 30 μg/mL for kanamycin resistance and 50 to 100 μg/mL for hygromycin resistance (*see* Table 13. 2 for other antibiotics).

13.2.2 Electroporation of M. smegmatis

1. Tween-80 (Sigma) prepare as a 20% (v/v) stock, filter sterilize through an 0.2-μm membrane, and store at 4°C (*see* **Note 1**).
2. Lemco-Tw broth: 5 g/L peptone (Oxoid, Hampshire, UK), 5 g/L Lemco powder (Oxoid), 5 g/L NaCl, 5 mL/L 10% (w/v) Tween-80. Autoclave (*see* **Note 2**).

Table 13.2 Antibiotic Selection for Mycobacteria

Antibiotic	Stock solution	Working concentration
Chloramphenicol	34 mg/mL in ethanol	40 µg/mL
Gentamicin	50 mg/mL in water	5 to 20 µg/mL
Hygromycin	50 mg/mL in PBS	50 to 100 µg/mL
Kanamycin	50 mg/mL in water	10 to 30 µg/mL
Streptomycin	20 mg/mL in water	10 to 50 µg/mL
Apramycin	25 mg/mL in water	30 to 50 µg/mL

PBS, phosphate-buffered saline.

3. 10% (w/v) glycerol; sterilize by autoclaving.
4. Electroporation apparatus with pulse controller (*see* **Note 5**).
5. Electroporation cuvettes: 0.2 cm gap electrodes (*see* **Note 5**).
6. DNA in solution (*see* **Notes 6** and **7**); this should be free from salts, enzymes, and other substances. In order to clean up DNA, it can be ethanol precipitated and thoroughly washed with 70% ethanol (this will also remove excess salts). The concentration of DNA should be about 0.2 to 1 mg/mL.
7. Kanamycin sulfate (Sigma), 50 mg/mL stock (filter sterilize), store at –20°C.
8. Hygromycin B (Roche diagnostic), obtained as a 50 mg/mL stock in phosphate-buffered saline; store at 4°C in the dark.
9. Selection plates: Lemco agar (5 g/L peptone, 5 g/L Lemco powder, 5 g/L NaCl and 15 g/L agar) should contain 10 to 50 µg/mL for kanamycin resistance and 50 to 100 µg/mL for hygromycin resistance (*see* Table 13.2 for other antibiotics).

13.3 Methods

13.3.1 Electroporation of M. tuberculosis

Caution: *M. tuberculosis* is pathogenic to humans, therefore appropriate containment facilities should be used for all procedures (*see* **Note 8**).

1. Inoculate 10 mL 7H9-Tw-OADC broth with a loopful of mycobacteria, vortex to disperse cells, and incubate at 37°C for 10 to 15 days (*see* **Notes 1**, **2**, and **9**).
2. Inoculate 100 mL 7H9-Tw-OADC broth in a roller bottle with 1 to 10 mL starter culture and continue incubation at 37°C with rolling at 100 rpm for 5 to 7 days (*see* **Note 10**).
3. Add 0.1 volumes 2 M glycine (final concentration 1.5% w/v) 16 to 24 h before harvesting the cells (*see* **Note 4**).
4. Harvest the cells from 50 mL only of the culture by centrifugation at $3000 \times g$ for 10 min at room temperature (*see* **Note 11**).
5. Wash the cells with 10 mL prewarmed 10% glycerol.
6. Wash the cells with 5 mL prewarmed 10% glycerol.

7. Resuspend the cells in 1 to 5 mL 10% glycerol (*see* **Note 12**).
8. Add 0.5 to 5 µg salt-free DNA in no more than 5 µL volume (*see* **Notes 6 and 7**) to 0.2 mL mycobacterial suspension.
9. Transfer to an 0.2-cm electrode gap electroporation cuvette (*see* **Note 5**).
10. Place the cuvette in the electroporation chamber and subject to a single pulse of 2.5 kV, 25 µF, with the pulse-controller resistance set at 1000 Ω resistance (*see* **Notes 5, 7, 13,** and **14**).
11. Recover cell suspension immediately into 10 mL 7 H9-Tw-OADC. Wash cuvette once to recover all cells (*see* **Note 15**).
12. Incubate at 37°C for 16 h. This step allows expression of any antibiotic-resistance gene carried on the DNA (*see* **Notes 7** and **15**).
13. Harvest the cells by centrifugation at 3000 × *g* for 10 min and plate out suitable dilutions (to give 30 to 300 colonies per plate) on 7 H10 agar plus OADC enrichment and appropriate antibiotic (*see* **Notes 16** and **17**).
14. Incubate plates at 37°C until colonies become visible; this will take approximately 3 weeks (*see* Table 13.3 for other species requirements).
15. Count transformants to calculate transformation efficiency (*see* **Note 18**).
16. Streak out transformants onto solid medium (7 H10-OADC) plus selection (*see* **Notes 19** and **20**).
17. Analyze transformants as required (*see* **Note 20**).

13.3.2 Electroporation of M. smegmatis

1. *M. smegmatis* should be maintained in the laboratory by regular subculture on solid medium (Lemco agar) (*see* **Note 9**).
2. Inoculate 5 mL Lemco-Tw broth with a loopful of mycobacteria and disperse the cells using a vortex (*see* **Note 1**).
3. Incubate at 37°C with shaking (100 rpm) overnight.

Table 13.3 Growth Conditions for Mycobacterial Transformants

Species	Growth temperature (°C)	Length of incubation
Fast-growers		
M. aurum	37	3 to 5 days
M. phlei	37	3 to 5 days
M. smegmatis	37	3 to 5 days
M. vaccae	30	3 to 7 days
Slow-growers		
M. avium	37	2 to 3 weeks
M. bovis BCG	37	3 to 4 weeks
M. intracellulare	37	10 to 14 days
M. tuberculosis	37	3 to 4 weeks
M. w	37	10 to 14 days

4. Inoculate a large-scale culture (100 to 500 mL Lemco-Tw in 250 to 1000 mL conical flask) with a 1/100 dilution of the overnight culture and continue incubation at 37°C with shaking until $OD_{600} = 0.8$ to 1.0 (usually between 16 and 24 h; *see* **Note 21**).

5. Incubate cells on ice for 1.5 h (*see* **Note 22**).

6. Harvest cells by centrifugation at 3000 × *g* for 10 min.

7. Wash cells 3 times in ice-cold 10% glycerol. Reduce the volume each time; for example, for 100 mL, wash one, 20 mL; wash two, 10 mL; and wash three, 5 mL.

8. Resuspend in 1/10 to 1/100 original volume of ice-cold 10% glycerol (*see* **Note 12**).

9. At this stage, cells may be frozen and stored in aliquots at –70°C for future use. Cells frozen in this way should be thawed on ice and used as required (*see* **Note 23**).

10. Add 0.5 to 5 μg salt-free DNA in no more than 5 μL volume (*see* **Notes 6** and **7**) to 0.2 mL mycobacterial suspension and leave on ice for 10 min (*see* **Note 24**).

11. Transfer to an 0.2-cm electrode gap electroporation cuvette (*see* **Note 24**). The cuvette should be chilled on ice before use (*see* **Note 7**).

12. Place the cuvette in the electroporation chamber and subject to one single pulse of 2.5 kV, 25 μF, with the pulse-controller resistance set a 1000 Ω resistance (*see* **Notes 7** and **13**).

13. Put cuvette back on ice for 10 min, transfer cell suspension to a sterile universal bottle, add 5 mL of Lemco-Tw broth, and incubate at 37°C for 2 to 3 h (*see* **Note 15**).

14. Harvest the cells by centrifugation at 3000 × *g* for 10 min and plate out suitable dilutions (to give 30 to 300 colonies per plate) on Lemco agar and appropriate antibiotic (*see* **Note 16**).

15. Incubate plates at 37°C until colonies become visible; this will take 3 to 5 days (Table 13.3).

16. Count transformants to calculate transformation efficiency (*see* **Note 18**).

17. Streak out transformants onto solid medium (Lemco) plus selection or inoculate 5 mL Lemco-Tw broth and selection antibiotic with transformant colonies. Incubate at 37°C with shaking (100 rpm) for 2 to 3 days.

18. Analyze transformants as required (*see* **Notes 19** and **20**).

13.4 Notes

1. Mycobacterial cells, particularly *M. tuberculosis*, have a tendency to clump together in culture; this is owing to the thick waxy nature of the mycobacterial coat. The addition of Tween-80, a non-ionic detergent, to medium reduces the amount of clumping and provides a more homogenous suspension of cells.

2. The medium used for growth of mycobacteria for electroporation is not important, and a variety of different recipes can be used, the most common being Middlebrook 7 H9 supplemented with Tween and OADC.

3. OADC supplement is extremely heat-labile and should only be added to 7 H10 or 7 H9 media after cooling. *M. bovis* BCG can be grown in medium supplemented with ADC (no oleic acid) rather than OADC, but growth is slower. Growth of mycobacterial cultures is enhanced by the provision of up to 10% CO_2 in the air above the medium.

4. For slow-growing species, the addition of glycine (to a final concentration of 1.5%) improves transformation efficiencies [16, 24, 36, 37, 38]. Among Gram-positive bacteria, glycine replaces alanine during peptidoglycan synthesis [39]. Glycine represents a poor substrate for transpeptidation resulting in decreased cell wall cross-linking [39]. Ideally, glycine should be added 16 to 24 hours prior to harvesting.

5. There are many different electroporation devices available commercially; any apparatus that can deliver high-voltage pulses can be used (i.e., 2.5 kV, 25 µF, 1000 Ω). There are also different makes of cuvettes available; although the gap or pathlength may be the same, the maximum volume of the cell suspension can vary from 50 µL to 400 µL. The volume of cell suspension used does not seem to affect the efficiency [14]. We routinely use 200 µL cells in an 0.2-cm cuvette.

6. DNA concentration: The volume of DNA used is critical; for small volumes of cell suspensions, the addition of a large amount of DNA in water will alter the conductivity of the suspension. Therefore, it is important that not more than 5 µL of DNA solution is added to the cell suspension. For a replicating or integrating plasmid in *M. tuberculosis* or *M. smegmatis*, the efficiency of transformation is not affected by the amount of DNA added [16]; addition of 0.5 to 500 ng DNA produces the same efficiency. However, a recent study with *M. avium* showed that an increase in plasmid DNA from 1.5 to 3 µg resulted in 700-fold increase in transformation frequency, and doubling the DNA amount again to 6 µg yielded a further 6.4-fold increase [24]. For homologous recombination, up to 5 µg can be used.

7. Arcing: The use of the pulse-controller apparatus serves to reduce the probability of arcing when using high voltages applied to high-resistance media, although it may still occur. Factors that cause arcing include the presence of lysed cells in the sample and salts in the DNA solution. These factors can be minimized by ensuring that during preparation of electrocompetent cells, the preincubation on ice for fast-growers is no longer than 1.5 h. Also make sure that the outside of the cuvette is dry before placing in the pulse chamber. Always ensure that the DNA for transformation is free from salts and other contaminants; ethanol precipitation and washing with 70% ethanol can be used to clean up DNA, which should preferably be dissolved in sterile deionized, distilled water. The settings for the pulse are important as well. Increasing the parallel resistance to ∞ Ω increases the possibility of arcing; therefore a setting of 1000 Ω produces more consistent results. In some cases, arcing may be violent enough to blow the lid off the electroporation cuvette, dispersing the cell suspension over the inside of the electroporation chamber (thereby creating aerosols).

8. Pathogenic mycobacteria represent an important biohazard, therefore, all culture and genetic manipulation must be carried out in appropriate containment facilities inside a class I safety cabinet. In most countries, genetic manipulation involving pathogenic mycobacteria or their DNA must be met with approval by the relevant authorities. In any case, risk assessment must form the first part of any experiment with pathogenic mycobacteria. A list of mycobacterial species and the type of containment required should be consulted prior to use.

9. Mycobacteria are relatively slow-growing organisms; the fast-growing species have a generation of 2 to 3 h and the slow-growing species of around 20 h. This often leads to a problem with contamination of cultures because many common contaminants have a much quicker doubling time and will rapidly outgrow mycobacteria. It is extremely important to maintain a good aseptic technique, especially with slow-growers. It is often wise to set up duplicate cultures in case one becomes contaminated. Cultures can be checked for purity using acid-fast staining [40].

10. The volume of the culture in a $450 \, cm^2$ Corning roller should not exceed $100 \, mL$. It is important to pre-roll the bottles with the media 24 h before inoculation. This is to check for contamination and leaks. Cells should grow until late log phase (e.g., for *M. tuberculosis*, 7 days if inoculated 1/100 and 5 days if inoculated 4/100). For *M. avium*, the transformation frequency is maximal at early log phase of growth [24].

11. For slow-growing species, the cells can be kept at room temperature, but electroporation at 37°C increases the transformation efficiency [16, 24]. We routinely prewarm the glycerol washes to 37°C and perform the centrifugation and pulse steps at room temperature.

12. For the transformation of a replicative or integrative plasmid, cells can be resuspended in 1/10 of the original volume (i.e., 5 mL). However, when homologous recombination is required, it is recommended to resuspend the cells in 1/50 of the original volume (i.e., 1 mL).

13. Pulse conditions: The use of a pulse controller in addition to the electroporation apparatus allows control over the parallel resistance and therefore the time constant; higher parallel resistance produces a longer time constant. Observations have shown that the optimum time constant is 15 to 25 ms (1000 Ω resistance) [1]. The use of 0.2-cm electrode gap cuvettes as opposed to 0.4-cm gap cuvettes results in a higher field strength [1]. The electroporation medium also has an effect on the time constant. Use of 10% glycerol provides a high-resistance medium, allowing longer time constants to be achieved. Mycobacteria have chemically resistant cell walls that are difficult to lyse, thus they are able to survive high voltage even when pulses have long time constants.

14. When working with pathogenic organisms, it is imperative that the pulse is delivered with the electroporation chamber placed inside a safety cabinet and that appropriate disinfectants are at hand.

15. The dilution of cells immediately after the pulse is important. Cells should be diluted at least 10-fold and incubated for several hours prior to plating. Omission of this step leads to greatly reduced efficiencies [41]. Presumably, the dilution allows better recovery from the pulse and therefore greater survival of transformants. Slow-growers should be incubated for 16 h and fast-growers for 2 to 3 h.

16. The problem of clumping is important when plating out cells after electroporation. It is important to ensure that resistant colonies have arisen from a single cell, so the cells must be thoroughly resuspended before plating. Appropriate dilutions may also help to alleviate this problem by thinning the cell suspension. If the cells are not diluted, it can be very difficult to visualize truly resistant colonies against a background lawn of sensitive cells. This is owing to aggregation, which protects some of the untransformed cells from the effects of the antibiotic.

17. Because slow-growing organisms take up to 6 weeks to form colonies from single cells, it is important to pour plates thickly and to wrap them securely in Parafilm or place them in sealable bags to prevent drying out during the long incubation period. Cycloheximide can be added to plates (100 μg/mL) to prevent fungal contamination. The long incubation period also means that antibiotic-containing plates should be freshly poured for each experiment to minimize the decay of antibiotic activity.

18. Several factors affect the efficiency of transformation; these include the growth phase of cells when harvested, electroporation media, and the field strength and time constant of delivered pulse. The efficiency of electroporation depends on the choice of DNA for transformation; some vectors have been unable to transform particular mycobacterial species, and the efficiency often depends on the choice of the selectable marker. We routinely obtain efficiencies of 10^7 to 10^8 per g DNA using hygromycin or streptomycin resistance in *M. tuberculosis* and 10^5 to 10^6 in *M. smegmatis*.

19. Transformants may be inoculated directly into liquid medium, but we find that they grow more rapidly if they are first streaked out to get good growth on plates.

20. Spontaneous kanamycin resistance can often be a problem with *M. tuberculosis*, owing to mutations in the *rrn* operon, of which slow-growing species possess only one. A control electroporation with no DNA, to check for the frequency of such mutants, can also be

included. Problems of spontaneous resistance will apply to all antibiotics that act on *rrn* operon (e.g., streptomycin). For most applications, three transformants should be streaked and analyzed. It is important to check the identity of extrachromosomal plasmids after transformation, as deletions and rearrangements are common.

21. In general, mycobacterial cultures should be removed from the incubator when in the late logarithmic phase of growth.

22. For fast-growing species, once cultures have reached the required stage of growth, they should be removed from the incubator and incubated on ice for 1.5 h prior to harvesting. This results in a fourfold increase in transformation efficiency [1]. Longer incubations on ice result in reduced efficiency, probably owing to increased cell lysis. This may also increase the possibility of arcing during the pulse delivery.

23. It is recommended that competent cells that have been thawed from frozen should be harvested and resuspended in fresh 10% glycerol prior to use. The transformation efficiency often increases after freezing the cells.

24. Pulse delivery: It is important to have an even cell suspension for electroporation because any clumping of cells will lead to arcing and reduced transformation efficiency. During the standing time on ice prior to pulse delivery, the cells may settle in the tube, and it is necessary to redistribute them using a pipette or a vortex immediately prior to the high-voltage pulse. This step serves both to resuspend the cells and to ensure thorough mixing of the DNA. Care must be taken to ensure that no bubbles are introduced between the two electrodes of the cuvette.

References

1. Jacobs, W.R. Jr., Kalpana, G.V., Cirillo, J.D., Pascopella, L., Snapper, S.B., Udani, R.A., Jones, W., Barletta, R.G. and Bloom, B.R. (1991) Genetic systems for mycobacteria. Methods Enzymol, **204**, 537–555.
2. Kalpana, G.V., Bloom, B.R. and Jacobs, W.R., Jr. (1991) Insertional mutagenesis and illegitimate recombination in mycobacteria. Proc Natl Acad Sci *U S A*, **88**, 5433–5437.
3. Paget, E. and Davies, J. (1996) Apramycin resistance as a selective marker for gene transfer in mycobacteria. *J Bacteriol*, **178**, 6357–6360.
4. Parish, T., Gordhan, B.G., McAdam, R.A., Duncan, K., Mizrahi, V. and Stoker, N.G. (1999) Production of mutants in amino acid biosynthesis genes of *Mycobacterium tuberculosis* by homologous recombination. *Microbiology*, **145**, 3497–3503.
5. Ranes, M.G., Rauzier, J., Lagranderie, M., Gheorghiu, M. and Gicquel, B. (1990) Functional analysis of pAL5000, a plasmid from *Mycobacterium fortuitum*: construction of a "mini" *mycobacterium-Escherichia coli* shuttle vector. *J Bacteriol*, **172**, 2793–2797.
6. Yuan, Y., Crane, D.D., Simpson, R.M., Zhu, Y.Q., Hickey, M.J., Sherman, D.R. and Barry, C.E. III. (1998) The 16-kDa alpha-crystallin (Acr) protein of *Mycobacterium tuberculosis* is required for growth in macrophages. *Proc Natl Acad Sci U S A*, **95**, 9578–9583.
7. Consaul, S.A. and Pavelka, M.S. Jr. (2004) Use of a novel allele of the *Escherichia coli* aacC4 aminoglycoside resistance gene as a genetic marker in mycobacteria. *FEMS Microbiol Lett*, **234**, 297–301.
8. Pashley, C.A. and Parish, T. (2003) Efficient switching of mycobacteriophage L5-based integrating plasmids in *Mycobacterium tuberculosis*. *FEMS Microbiol Lett*, **229**, 211–215.
9. Dellagostin, O.A., Wall, S., Norman, E., O'Shaughnessy, T., Dale, J.W. and McFadden, J. (1993) Construction and use of integrative vectors to express foreign genes in mycobacteria. *Mol Microbiol*, **10**, 983–993.

10. Garbe, T.R., Barathi, J., Barnini, S., Zhang, Y., Abou-Zeid, C., Tang, D., Mukherjee, R. and Young, D.B. (1994) Transformation of mycobacterial species using hygromycin resistance as selectable marker. *Microbiology*, **140**, 133–138.

11. Goto, Y., Taniguchi, H., Udou, T., Mizuguchi, Y. and Tokunaga, T. (1991) Development of a new host vector system in mycobacteria. *FEMS Microbiol Lett*, **67**, 277–282.

12. Matsuo, K., Yamaguchi, R., Yamazaki, A., Tasaka, H., Terasaka, K., Totsuka, M., Kobayashi, K., Yukitake, H. and Yamada, T. (1990) Establishment of a foreign antigen secretion system in mycobacteria. *Infect Immun*, **58**, 4049–4054.

13. Qin, M., Taniguchi, H. and Mizuguchi, Y. (1994) Analysis of the replication region of a mycobacterial plasmid, pMSC262. *J Bacteriol*, **176**, 419–425.

14. Radford, A.J. and Hodgson, A.L. (1991) Construction and characterization of a *Mycobacterium-Escherichia coli* shuttle vector. *Plasmid*, **25**, 149–153.

15. Snapper, S.B., Lugosi, L., Jekkel, A., Melton, R.E., Kieser, T., Bloom, B.R. and Jacobs, W.R., Jr. (1988) Lysogeny and transformation in mycobacteria: stable expression of foreign genes. *Proc Natl Acad Sci U S A*, **85**, 6987–6991.

16. Wards, B.J. and Collins, D.M. (1996) Electroporation at elevated temperatures substantially improves transformation efficiency of slow-growing mycobacteria. *FEMS Microbiol Lett*, **145**, 101–105.

17. Hermans, J., Martin, C., Huijberts, G.N., Goosen, T. and de Bont, J.A. (1991) Transformation of *Mycobacterium aurum* and *Mycobacterium smegmatis* with the broad host-range gram-negative cosmid vector pJRD215. *Mol Microbiol*, **5**, 1561–1566.

18. Houssaini-Iraqui, M., Lazraq, M.H., Clavel-Seres, S., Rastogi, N. and David, H.L. (1992) Cloning and expression of *Mycobacterium aurum* carotenogenesis genes in *Mycobacterium smegmatis*. *FEMS Microbiol Lett*, **69**, 239–244.

19. Marklund, B.I., Speert, D.P. and Stokes, R.W. (1995) Gene replacement through homologous recombination in *Mycobacterium intracellulare*. *J Bacteriol*, **177**, 6100–6105.

20. Hermans, J., Suy, I.M.L. and De Bont, J.A.M. (1993) Transformation of Gram-positive microorganisms with the Gram-negative broad-host-range cosmid vector pJRD215. *FEMS Microbiol Lett*, **108**, 201–204.

21. Talaat, A.M. and Trucksis, M. (2000) Transformation and transposition of the genome of *Mycobacterium marinum*. *Am J Vet Res*, **61**, 125–128.

22. Beggs, M.L., Crawford, J.T. and Eisenach, K.D. (1995) Isolation and sequencing of the replication region of *Mycobacterium avium* plasmid pLR7. *J Bacteriol*, **177**, 4836–4840.

23. Foley-Thomas, E.M., Whipple, D.L., Bermudez, L.E. and Barletta, R.G. (1995) Phage infection, transfection and transformation of *Mycobacterium avium* complex and *Mycobacterium paratuberculosis*. *Microbiology*, **141**, 1173–1181.

24. Lee, S.H., Cheung, M., Irani, V., Carroll, J.D., Inamine, J.M., Howe, W.R. and Maslow, J.N. (2002) Optimization of electroporation conditions for *Mycobacterium avium*. *Tuberculosis (Edinb)*, **82**, 167–174.

25. Snapper, S.B., Melton, R.E., Mustafa, S., Kieser, T. and Jacobs, W.R., Jr. (1990) Isolation and characterization of efficient plasmid transformation mutants of *Mycobacterium smegmatis*. *Mol Microbiol*, **4**, 1911–1919.

26. Labidi, A., Dauguet, C., Goh, K.S. and David, H.L. (1984) Plasmid profiles of *Mycobacterium fortuitum* complex isolates. *Curr Microbiol*, **11**, 235–240.

27. Stover, C.K., de la Cruz, V.F., Fuerst, T.R., Burlein, J.E., Benson, L.A., Bennett, L.T., Bansal, G.P., Young, J.F., Lee, M.H., Hatfull, G.F. et al. (1991) New use of BCG for recombinant vaccines. *Nature*, **351**, 456–460.

28. Bachrach, G., Colston, M.J., Bercovier, H., Bar-Nir, D., Anderson, C. and Papavinasasundaram, K.G. (2000) A new single-copy mycobacterial plasmid, pMF1, from *Mycobacterium fortuitum* which is compatible with the pAL5000 replicon. *Microbiology*, **146**(Pt 2), 297–303.

29. Gavigan, J.A., Ainsa, J.A., Perez, E., Otal, I. and Martin, C. (1997) Isolation by genetic labeling of a new mycobacterial plasmid, pJAZ38, from *Mycobacterium fortuitum*. *J Bacteriol*, **179**, 4115–4122.

30. Lee, M.H., Pascopella, L., Jacobs, W.R. Jr. and Hatfull, G.F. (1991) Site-specific integration of mycobacteriophage L5: integration-proficient vectors for *Mycobacterium smegmatis*, *Mycobacterium tuberculosis*, and bacille Calmette-Guerin. *Proc Natl Acad Sci U S A*, **88**, 3111–3115.

31. Anes, E., Portugal, I. and Moniz-Pereira, J. (1992) Insertion into the *Mycobacterium smegmatis* genome of the*aph* gene through lysogenization with the temperate mycobacteriophage Ms6. *FEMS Microbiol Lett*, **74**, 21–25.

32. Martin, C., Mazodier, P., Mediola, M.V., Gicquel, B., Smokvina, T., Thompson, C.J. and Davies, J. (1991) Site-specific integration of the *Streptomyces* plasmid pSAM2 in *Mycobacterium smegmatis*. *Mol Microbiol*, **5**, 2499–2502.

33. England, P.M., Mazodier, P., Mediola, M.V., Gicquel, B., Smokvina, T., Thompson, C.J. and Davies, J. (1991) Site-specific integration of the *Streptomyces* plasmid pSAM2 in *Mycobacterium smegmatis*. *Mol Microbiol*, **5**, 2499–2502.

34. Bottger, E.C. (1994) Resistance to drugs targeting protein synthesis in mycobacteria. *Trends Microbiol*, **2**, 416–421.

35. Hatfull, G.F. (1993) Genetic transformation of mycobacteria. *Trends Microbiol*, **1**,310–314.

36. Aldovini, A., Husson, R.N. and Young, R.A. (1993) The uraA locus and homologous recombination in *Mycobacterium bovis* BCG. *J Bacteriol*, **175**, 7282–7289.

37. Hermans, J., Boschloo, J.G. and de Bont, J.A.M. (1990) Transformation of *Mycobacterium aurum* by electroporation: the use of glycine, lysozyme and isonicotinic acid hydrazide in enhancing transformation efficiency. *FEMS Microbiol Lett*, **72**, 221–224.

38. Husson, R.N., James, B.E. and Young, R.A. (1990) Gene replacement and expression of foreign DNA in mycobacteria. *J Bacteriol*, **172**, 519–524.

39. Hammes, W., Schleifer, K.H. and Kandler, O. (1973) Mode of action of glycine on the biosynthesis of peptidoglycan. *J Bacteriol*, **116**, 1029–1053.

40. Cruickshank, R. (1965) *Medical Microbiology: A Guide to the Laboratory Diagnosis and Control of Infection,* 11th ed., E & S Livingstone Limitated, London.

41. David, M., Lubinsky-Mink, S., Ben-Zvi, A., Ulitzur, S., Kuhn, J. and Suissa, M. (1992) A stable *Escherichia coli-Mycobacterium smegmatis* plasmid shuttle vector containing the mycobacteriophage D29 origin. *Plasmid*, **28**, 267–271.

42. Gormley, E.P. and Davies, J. (1991) Transfer of plasmid RSF1010 by conjugation from *Escherichia coli* to *Streptomyces lividans* and *Mycobacterium smegmatis*. *J Bacteriol*, **173**, 6705–6708.

Chapter 14
Ins and Outs of Mycobacterial Plasmids

Farahnaz Movahedzadeh and Wilbert Bitter

Abstract The importance of plasmids for molecular research cannot be under-estimated. These double-stranded DNA units that replicate independently of the chromosomal DNA are as valuable to bacterial geneticists as a carpenter's hammer. Fortunately, today the mycobacterial research community has a number of these genetic tools at its disposal, and the development of these tools has greatly accelerated the study of mycobacterial pathogens. However, working with mycobacterial cloning plasmids is still not always as straightforward as working with *Escherichia coli* plasmids, and therefore a number of precautions and potential pitfalls will be discussed in this chapter.

Keywords antibiotic resistance marker · integrating vector · Mycobacterium · natural plasmid · shuttle vector

14.1 Introduction

14.1.1 Natural Plasmids of Mycobacterium

Natural plasmids are found in many bacteria, and mycobacteria are no exception. Most plasmids are found in environmental mycobacterial species, such as *Mycobacterium fortuitum* and *Mycobacterium avium* [1]. It has been reported that approximately 50% of *M. avium* strains contain plasmids [1], sometimes carrying multiple plasmids [2]. However, thus far there is no evidence that species belonging to the *M. tuberculosis* complex contain natural plasmids [3]. The identification of natural plasmids is problematic because they generally have a low copy number, and it is difficult to isolate sufficient amounts of plasmid DNA from mycobacteria. Solutions to this problem include the use of special transposons, for example, Tn*552 kan* •*ori*; this transposon carries a resistance marker and an *Escherichia coli* plasmid origin of replication, so

F. Movahedzadeh
Institute for Tuberculosis Research, College of Pharmacy, Rm 412, University of Illinois at Chicago, 833 S. Wood St., Chicago, Illinois, USA 60612-7231
e-mail: movahed@uic.edu

T. Parish, A.C. Brown (eds.), *Mycobacteria Protocols*,
doi: 10.1007/978-1-59745-207-6_14, © Humana Press, Totowa, NJ 2008

that DNA isolated from mycobacteria can be used in an *in vitro* transposition and subsequently transformed into *E. coli*. Another complicating factor is that mycobacterial plasmids can be linear [4], with terminal inverted repeats that are covalently bound with protein (invertons) [4].

Mycobacterial plasmids in use can be grouped in two families based on their origin of replication: the pMSC262 family and the pAL5000 family. The pMSC262 family, named after a plasmid isolated from *Mycobacterium scrofulaceum,* is widespread and can be found in many different mycobacteria. For instance, several *M. avium* plasmids [2] and the *Mycobacterium ulcerans* virulence-associated plasmid pMUM001 belong to this family [5]. In addition, the only linear mycobacterial plasmid that has been identified (pCLP) belongs to this group [4]. Although these plasmids belong to the same family, they are not members of the same incompatibility group. For example, plasmids pCK5 and pBP10, both belonging to the pMSC262 family, can stably coexist [2]. This means that coexistence of different pMSC262 plasmids needs to be determined experimentally.

pAL5000 is a small cryptic plasmid isolated from *M. fortuitum* [6]. Although it has no homology with other mycobacterial plasmids, it forms the archetype of a plasmid family that includes natural plasmids of *Rhodococcus* sp. (pFAJ2600) and *Bifidobacterium* sp. (pMB1) [7]. This group is distantly related to the ColE2 replicon [7]. The pAL5000 backbone is most frequently used in mycobacterial shuttle plasmids and is compatible with all pMSC262-type plasmids [7].

The function of most natural plasmids in mycobacteria is unknown. Plasmids of the *Mycobacterium avium* complex have been suggested to be involved in virulence, because (i) most *M. avium* isolates from AIDS patients carry plasmids [8] and (ii) the *M. avium* strain LR25 cured of its three natural plasmids showed reduced virulence in mice when compared with the wild-type strain [9]. However, thus far no clear virulence-related function can be ascribed to these plasmids. In contrast, *M. ulcerans*, an important human pathogen, contains a 174-kb plasmid (pMUM001), which is crucial for virulence [10]. Six genes of pMUM001, accounting for 60% of the total plasmid sequence, are involved in the biosynthesis of mycolactone, an unusual cytotoxic macrolide. This secreted macrolide is responsible for the formation of the characteristic lesions formed by *M. ulcerans* [11]. The closely related species *Mycobacterium pseudoshottsii, Mycobacterium liflandii,* and some strains of *Mycobacterium marinum* produce a related mycolactone toxin, and it has been suggested that the genes responsible are also located on virulence plasmids [12]. However, these plasmids have not been studied in detail.

An interesting property of some mycobacterial plasmids is that they have been shown to be transferred horizontally between mycobacterial species. For instance, pGB9.2, which belongs to the pMSC262 family, was shown to be self-transmissible between different mycobacteria [13]. However, there is no clear genetic information on this plasmid that would predict its conjugative ability. It is possible that the chromosomally encoded conjugation system of mycobacteria could play a role in this process [14].

14.1.2 Mycobacterial Recombinant Plasmids

For the genetic analysis of mycobacteria, it is pivotal to have the availability of good cloning vectors. These cloning vectors should (i) replicate in *E. coli* and mycobacteria or be designed to integrate into the mycobacterial genome, (ii) contain selectable marker(s), and (iii) contain suitable unique restriction sites used for cloning.

A number of plasmids based on different replicons are available. For instance, the broad-host range plasmid RSF1010, with an IncQ replicon, is able to replicate in both *E. coli* and a number of different mycobacteria [15]. However, plasmids derived from RSF1010 are not used extensively because of their low copy number and relatively large size (approximately 9 kb). Therefore, most mycobacterial cloning vectors are shuttle plasmids, having two replicative origins, one active in *E. coli* and one in mycobacteria. These plasmids are generally called shuttle plasmids. An advantage of using these plasmids is that, although they have a low copy number in mycobacteria, they have a high copy number in *E. coli*. Therefore, plasmid isolation and manipulation is relatively efficient in the latter species.

The origin of replication from pAL5000 has been used extensively in mycobacterial shuttle plasmids. The minimal replicon of pAL5000 is a 1.5-kb fragment containing two genes, *repA* and *repB,* and the origin of replication (Fig. 14.1) [16]. Most available shuttle plasmids have a larger fragment of pAL5000 that also includes orf5 (the *EcoRV-HpaI* fragment of pAL5000; Fig. 14.1). Shuttle plasmids derived from pAL5000 have a copy number of between two and five in *Mycobacterium smegmatis* [7].

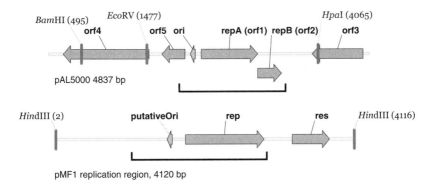

Fig. 14.1 Schematic representation of **(A)** the pAL5000 plasmid and **(B)** the replication element of plasmid pMF1. The minimal replication element of each plasmid is indicated. Most pAL5000-based shuttle plasmids contain the *HpaI-EcoRV* fragment of pAL5000. The minimal replication element of pMF1 does not give the same stability as the entire *Hind*III fragment depicted in this figure. Abbreviations: bp, base pairs; orf, open reading frame; ori, origin of replication

Vectors with a replicative origin of the pMCS262 family have not been developed extensively, but Bachrach et al. [17] have generated a useful cloning vector using the minimal replicative origin of pMF1. This general purpose cloning vector, pBP10, contains a 2.1-kb fragment of pMF1 (Fig. 14.1) and is stably inherited in the absence of selection pressure. This plasmid is compatible with the pAL5000 replicon. pMF1 derivatives have a copy number of one in *M. smegmatis*.

14.2 Types of Mycobacterial Recombinant Plasmids

14.2.1 Traditional Cloning Vectors

A number of cloning systems have been developed based on *M. fortuitum* plasmid pAL5000. pMV261 is a widely used *E. coli*–mycobacterial shuttle vector for the expression of foreign genes [18]. It contains the Tn903-derived kanamycin resistance gene, the *M. bovis* BCG *hsp60* promoter (P_{hsp60}), an *E. coli* origin of replication, a mycobacterial origin of replication, and a transcription terminator. pYUB12 is another popular *E. coli*–mycobacterial shuttle plasmid and also a derivative of pAL5000 [19]; it carries the chloramphenicol resistance gene of pACYC184 and the kanamycin resistance gene from transposon Tn5.

Although the hsp60 promoter has been used successfully for gene expression in mycobacteria, there have been problems with stability [20]. There are two related problems with this promoter: first, that expression levels could be too high, and second, that the promoter DNA has the tendency to become rearranged or deleted. Alternative expression systems use the Ag85a promoter, as used in pEM37 and pFM209 (unpublished plasmids constructed by E. Machowksi and F. Movahedzadeh, respectively) and in pAPA3 [21], or inducible promoters (*see* Chapters 16 and 17). One such inducible expression systems is the acetamidase promoter from *M. smegmatis* [22], which is inducible in the presence of exogenous acetamide (*see* Chapter 16). Another useful inducible promoter is the tetracycline-inducible system, which functions very well in mycobacteria and can also be used to generate conditional expression [23, 24, 25].

For promoter studies, plasmids have been generated to allow promoter fusions to a promoterless chloramphenicol acetyltransferase (CAT) reporter gene [26], *lacZ*gene [27], *gfp* [28], or *xylE* [29] (*see* Chapter 18).

14.2.2 Conditionally Replicating Vectors for Allelic Exchange and Transposon Delivery

Insertional mutagenesis is a fundamental technique for understanding gene function. Several strategies have been developed to create transposon mutant libraries in both *M. smegmatis* and *M. tuberculosis*. One approach uses conditional vectors that are unable to replicate in the host organism under specific conditions. For example, temperature-sensitive (ts) versions of the pAL5000 replicon isolated by chemical mutagenesis are able to replicate in *E. coli* at all temperatures but only in *M. smegmatis* at the permissive temperature of 30°C; no replication is seen at 41°C [30]. This ts replicon was used to generate a large number of insertional mutations in *M. smegmatis* [31]. Ts plasmids have also been used for site-specific integration and allelic exchange [32] (*see* Chapter 15).

Although ts vectors work efficiently in *M. smegmatis*, they are weakly thermosensitive in *M. tuberculosis*, because of the restricted temperature range of this bacterium. Therefore, a two-step strategy employing the ts vector in combination with the *sacB* gene has been developed for allelic exchange and transposon mutagenesis in *M. tuberculosis* [32, 33]. The *sacB* gene, originally from *Bacillus subtilis*, encodes the enzyme levansucrase, which is toxic for bacteria in sucrose-containing medium. Therefore, *sacB* is often used as a counter-selective marker to allow selection against both single-crossover and illegitimate recombination events. Although sucrose selection works very efficiently in mycobacteria, point mutations and deletions in the *sacB* gene are easily isolated [34, 35, 36, 37]. The *rpsL+* gene has also been used as a negative selection marker [38]. A wild-type copy of this gene confers sensitivity to the antibiotic streptomycin in a dominant fashion so that, if a streptomycin resistant host strain is used, complementation with the wild-type *rpsL+* gene will result in streptomycin sensitivity.

In addition to conditional plasmids, a novel and highly efficient method of DNA delivery, such as for transposon mutagenesis, can also be obtained using conditionally replicating mycobacteriophages [39], such as the mariner transposon containing Mycomar phage [40] (*see* Chapter 21).

The generation of directed mutations can also be achieved using nonreplicating ColE1-based vectors. Parish and Stoker [37] developed an efficient method for generating mutants in mycobacteria by allelic replacement based on a two-step strategy using the pNIL and pGOAL vectors (*see* Chapter 20).

14.2.3 Integrating Vectors

A valuable addition to the spectrum of mycobacterial cloning vectors is the integration vectors, which generally include a phage integrase encoding gene and a phage attachment site (*attP*). The most widely used of these is derived from mycobacteriophage L5 [41]. The phage integrase works in combination with the mycobacterial integration host factor (mIHF) to effect recombination between the *attP* site and the *attB* site on the bacterial chromosome, resulting in the chromosomal integration of the entire plasmid (Fig. 14.2). The advantages of integration systems are (i) the inserted gene is present in a single copy; (ii) the inserted gene is stably inherited in the absence of antibiotic selection; and

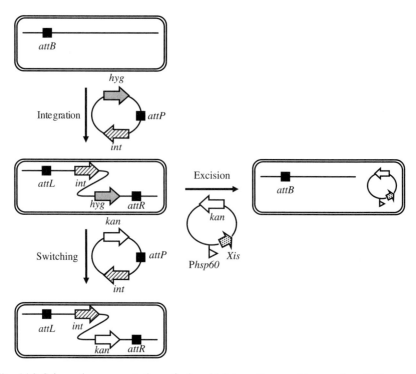

Fig. 14.2 Schematic representation of plasmid integration, excision, and switching using mycobacteriophage L5-based vectors. Integration [42] is dependent on the presence of a chromosomal *attB* site, a plasmid *attP* site, plasmid encoded integrase (int), and chromosomal encoded mycobacterial integration host factor. Excision of this plasmid is based on plasmid encoded excisionase (*xis*), under control of the strong hsp60 promoter (Phsp60) and the proteins needed for integration [48]. Excised plasmid is circularized but is not maintained in *mycobacterium*. Switching of integrated vectors is most efficient if an integrating plasmid carrying the integrase encoding gene, the *attP* site, and a different antibiotic resistance marker is introduced. hyg, hygromycin resistance gene; kan, kanamycin resistance gene

(iii) chromosomal integration is highly efficient. Integration vectors developed for mycobacteria include those using the *attP* and *int* gene of the temperate mycobacteriophage L5 [42]. Reported disadvantages of L5-derived vectors are (i) the integration locus is unfavorable for transcription of integrated genes using native promoters [43] and (ii) the integrated vector can be lost if the insert contains the integrase gene function [44]. To avoid the latter problem, dual integration vectors are available that separate the *attP* site from the integrase encoding gene [44]. Similar vectors have been constructed that use integration systems from other mycobacteriophages such as Ms6 [45] and other bacteriophages, such as the *Streptomyces coelicolor* phage *ΦC31* [43]. These vectors may be used for complementing *M. tuberculosis* mutants, constructing merodiploid strains, integrating recombinant antigens, and reporter constructs in mycobacteria.

An integration vector results in a stably inherited genotype. However, there are still some possibilities for further manipulation. For instance, the L5 *attP* integration site is longer than the *attB* site, therefore integration of this vector will result in two incomplete *att* sites that are neither functional as a donor nor as an acceptor sequence for integration (Fig. 14.2) [31]. To be able to insert two independent plasmids, Saviola and Bishai [46] developed a plasmid with an additional bacterial attachment site (*attB*). When this plasmid is integrated in the genome, it can accept the integration of an additional plasmid by having a plasmid borne *attB* site. Alternatively, the original integrated vector can also be efficiently replaced by another integrating vector with the same *attP* site (switching; Fig. 14.2) [47]. Finally, the integrated vector can also be removed from the chromosome, and the removed chromosomal vector can even be rescued using the mycobacteriophage L5 excisionase (Fig. 14.2) [48]. In this case, plasmid rescue is performed *in vitro*, using isolated chromosomal DNA, purified excisionase, integrase, and mIHF. After incubation, the DNA is isolated and used to transform *E. coli* cells. This application is especially useful if integrated gene libraries are used.

14.3 Using *Mycobacterium* Plasmids

Plasmids for use in mycobacteria are generally constructed in *E. coli*, the workhorse of molecular genetics, and have to be transported into the appropriate host strain. Introduction of plasmid DNA can be achieved by different methods. Generally, electroporation is the method of choice to introduce plasmid DNA into mycobacteria [49] (*see* Chapter 13). In addition, plasmids can also be introduced by conjugation. For example, the conjugation of the IncQ plasmid RSF1010 from *E. coli* to *M. smegmatis* has been described [15]. However, conjugation of plasmids directly from *E. coli* to species of the *M. tuberculosis* complex has not been reported. Finally, DNA can also be introduced by using mycobacteriophages [39].

14.3.1 Antibiotic Markers for Selection

As described above, each vector needs a selection marker. Although the cell wall of mycobacteria is generally impermeable for hydrophilic compounds, there are still a number of antibiotics that can be used successfully in order to select for plasmid-containing cells. A list of the antibiotic selection markers that are used in cloning vectors is shown in Table 14.1. The most important group of antibiotics for mycobacteria are the aminoglycosides, such as hygromycin, apramycin, gentamicin, and kanamycin. One problem of using aminoglycoside selection markers, other than hygromycin, is the problem of cross-resistance [50]. Therefore, these antibiotic markers usually cannot be used on two

Table 14.1 Antibiotic Selection Markers in Mycobacteria

Antibiotic	Gene	Origin	Concentration
Apramycin	*aacC3*	*Salmonella* spp.	20 to 40 µg/mL
	aacC4	*Escherichia coli*	
Chloramphenicol	*cat*	Tn9	25 to 50 µg/mL
Gentamicin	*aacC1*	Tn*1696*	5 to 40 µg/mL
Hygromycin B	*Hyg*	*Streptomyces hygroscopicus*	50 to 100 µg/mL
Kanamycin	*aph*	Tn5 or Tn*903*	20 to 50 µg/mL
Viomycin	*Vph*	Tn*4560*	30 µg/mL
Zeocin	*Ble*	*Streptoalloteichus hindustanus*	25 to 100 µg/mL

compatible plasmids/transposons. Another problem of these markers is that they also result in (partial) resistance to amikacin, the most popular antibiotic used to selectively kill extracellular mycobacteria in tissue culture experiments.

Generally, the antibiotics most used for vector selection are hygromycin and kanamycin. The former antibiotic is generally more expensive but has fewer problems. For example, plasmids carrying the hygromycin resistance gene are more stable than those carrying kanamycin resistance genes [7]. Probably, the *aph* gene (both originating from Tn5 and Tn*903*), conferring kanamycin resistance, results in a larger metabolic burden for the host bacterium. In addition, the hygromycin resistance gene does not affect amikacin sensitivity.

To obtain optimal expression of antibiotic resistance, resistance genes may have to be expressed from mycobacterial promoters. This strategy generally results in increased antibiotic resistance, which also results in increased numbers of transformants and faster growth. However, using such promoters may also result in a higher metabolic burden.

14.3.2 Plasmid Stability

Plasmids containing the pAL5000 replicon are extremely stable in most mycobacterial species, especially if these plasmids contain the hygromycin resistance marker [7]. In fact, if you regularly would like to cure your strain from such a plasmid, it is advisable to use a pAL5000 derivate with the counter-selection gene *sacB*, which allows positive selection of cured strains on sucrose-containing media.

Plasmids containing pMCS262 replicons show more variation in stability, which is dependent on the host species. The stability of mycobacterial plasmids can also depend on the insert. Based on our observations (unpublished data), some plasmid inserts can be extremely unstable in mycobacteria. Electroporation of these constructs will result in antibiotic-resistant colonies that contain the plasmid, but often the plasmids isolated from these colonies show spontaneous deletions in various sizes. Therefore, it is advised to check the plasmid for its integrity after introduction.

14.3.3 Plasmids in Different Mycobacterial Species

The pAL5000 replicon is active in a broad range of mycobacteria, including both fast-growing and slow-growing species. Thus far, the only species reported not to support this replicon is *Mycobacterium intracellulare* [7]. The results with plasmids containing a pMSC262 replicon are more variable. Derivatives of pMF1 can be maintained in *M. smegmatis*, *M. marinum*, *M. bovis*, and *M. tuberculosis*, whereas derivatives of pLR7 and pMSC262 seem to be unable to replicate in *M. smegmatis* [51, 52]. Because *M. smegmatis* (strain mc^{2}155) is the workhorse of mycobacteria, this severely diminishes the usability of these plasmids. Kirby et al. [2] also reported difficulties in introducing the pVT2 plasmid of the pMSC262 family in *M. smegmatis*, with colonies appearing only after 14 days of growth instead of 3 days. However, when these transformants were cured of their plasmids and re-transformed, colonies appeared after 4 days. This result shows that *M. smegmatis* apparently acquires a mutation during the first transformation that allows replication of pVT2 and perhaps also other pMSC262 family plasmids. These data show that mycobacterial (shuttle) plasmids based on pMSC262 replicons cannot be used automatically for all mycobacterial species.

References

1. Jucker, M. T. & Falkinham, J. O. III. (1990). Epidemiology of infection by nontuberculous mycobacteria IX. Evidence for two DNA homology groups among small plasmids in *Mycobacterium avium*, *Mycobacterium intracellulare*, and *Mycobacterium scrofulaceum*. *Am. Rev. Respir. Dis.* **142**, 858–62.
2. Kirby, C., Waring, A., Griffin, T. J., Falkinham, J. O. III, Grindley, N. D. & Derbyshire, K. M. (2002). Cryptic plasmids of *Mycobacterium avium*: Tn552 to the rescue. *Mol. Microbiol.* **43**, 173–86.
3. Zainuddin, Z. F. & Dale, J. W. (1990). Does *Mycobacterium tuberculosis* have plasmids? *Tubercle* **71**, 43–9.
4. Le Dantec, C., Winter, N., Gicquel, B., Vincent, V. & Picardeau, M. (2001). Genomic sequence and transcriptional analysis of a 23-kilobase mycobacterial linear plasmid: evidence for horizontal transfer and identification of plasmid maintenance systems. *J. Bacteriol.* **183**, 2157–64.
5. Stinear, T. P., Pryor, M. J., Porter, J. L. & Cole, S. T. (2005). Functional analysis and annotation of the virulence plasmid pMUM001 from *Mycobacterium ulcerans*. *Microbiology* **151**, 683–92.
6. Labidi, A., David, H. L. & Roulland-Dussoix, D. (1985). Restriction endonuclease mapping and cloning of *Mycobacterium fortuitum* var. fortuitum plasmid pAL5000. *Ann. Inst. Pasteur. Microbiol.* **136B**, 209–15.
7. Pashley, C. & Stoker, N. G. Plasmids in mycobacteria. *In*: Hatfull, G. F. and Jacobs, W.R. (eds.), Molecular Genetics of Mycobacteria, ASM Press, 2000: 55–68.
8. Crawford, J. T. & Bates, J. H. (1986). Analysis of plasmids in *Mycobacterium avium-intracellulare* isolates from persons with acquired immunodeficiency syndrome. *Am. Rev. Respir. Dis.* **134**, 659–61.
9. Gangadharam, P. R., Perumal, V. K., Crawford, J. T. & Bates, J. H. (1988). Association of plasmids and virulence of *Mycobacterium avium* complex. *Am. Rev. Respir. Dis.* **137**, 212–4.

10. Stinear, T. P., Mve-Obiang, A., Small, P. L., Frigui, W., Pryor, M. J., Brosch, R., Jenkin, G. A., Johnson, P. D., Davies, J. K., Lee, R. E., Adusumilli, S., Garnier, T., Haydock, S. F., Leadlay, P. F. & Cole, S. T. (2004). Giant plasmid-encoded polyketide synthases produce the macrolide toxin of *Mycobacterium ulcerans*. *Proc. Natl. Acad. Sci. U. S. A.* **101**, 1345–9.

11. George, K. M., Chatterjee, D., Gunawardana, G., Welty, D., Hayman, J., Lee, R. & Small, P. L. (1999). Mycolactone: a polyketide toxin from *Mycobacterium ulcerans* required for virulence. *Science* **283**, 854–7.

12. Ranger, B. S., Mahrous, E. A., Mosi, L., Adusumilli, S., Lee, R. E., Colorni, A., Rhodes, M. & Small, P. L. (2006). Globally distributed mycobacterial fish pathogens produce a novel plasmid-encoded toxic macrolide, mycolactone F. *Infect. Immun.* **74**(11), 6037–45.

13. Harth, G., Maslesa-Galic, S. & Horwitz, M. A. (2004). A two-plasmid system for stable, selective-pressure-independent expression of multiple extracellular proteins in mycobacteria. *Microbiology* **150**, 2143–51.

14. Wang, J., Parsons, L. M. & Derbyshire, K. M. (2003). Unconventional conjugal DNA transfer in mycobacteria. *Nat. Genet.* **34**, 80–4.

15. Gormley, E. P. & Davies, J. (1991). Transfer of plasmid RSF1010 by conjugation from *Escherichia coli* to *Streptomyces lividans* and *Mycobacterium smegmatis*. *J. Bacteriol.* **173**, 6705–8.

16. Stolt, P. & Stoker, N. G. (1996). Functional definition of regions necessary for replication and incompatibility in the *Mycobacterium fortuitum* plasmid pAL5000. *Microbiology* **142**(Pt 10), 2795–802.

17. Bachrach, G., Colston, M. J., Bercovier, H., Bar-Nir, D., Anderson, C. & Papavinasa-sundaram, K. G. (2000). A new single-copy mycobacterial plasmid, pMF1, from *Mycobacterium fortuitum* which is compatible with the pAL5000 replicon. *Microbiology* **146**(Pt 2), 297–303.

18. Stover, C. K., de la Cruz, V. F., Fuerst, T. R., Burlein, J. E., Benson, L. A., Bennett, L. T., Bansal, G. P., Young, J. F., Lee, M. H., Hatfull, G. F. & et al. (1991). New use of BCG for recombinant vaccines. *Nature* **351**, 456–60.

19. Snapper, S. B., Melton, R. E., Mustafa, S., Kieser, T. & Jacobs, W. R., Jr. (1990). Isolation and characterization of efficient plasmid transformation mutants of *Mycobacterium smegmatis*. *Mol. Microbiol.* **4**, 1911–9.

20. Al-Zarouni, M. & Dale, J. W. (2002). Expression of foreign genes in *Mycobacterium bovis* BCG strains using different promoters reveals instability of the *hsp60* promoter for expression of foreign genes in *Mycobacterium bovis* BCG strains. *Tuberculosis* (Edinb) **82**, 283–91.

21. Parish, T., Roberts, G., Laval, F., Schaeffer, M., Daffe, M. & Duncan, K. (2007). Functional Complementation of the Essential Gene *fabG1* of *Mycobacterium tuberculosis* by *Mycobacterium smegmatis fabG* but Not *Escherichia coli fabG*. *J. Bacteriol.* **189**, 3721–8.

22. Parish, T., Liu, J., Nikaido, H. & Stoker, N. G. (1997). A *Mycobacterium smegmatis* mutant with a defective inositol monophosphate phosphatase gene homolog has altered cell envelope permeability. *J. Bacteriol.* **179**, 7827–33.

23. Carroll, P., Muttucumaru, D. G. & Parish, T. (2005). Use of a tetracycline-inducible system for conditional expression in *Mycobacterium tuberculosis* and *Mycobacterium smegmatis*. *Appl. Environ. Microbiol.* **71**, 3077–84.

24. Ehrt, S., Guo, X. V., Hickey, C. M., Ryou, M., Monteleone, M., Riley, L. W. & Schnappinger, D. (2005). Controlling gene expression in mycobacteria with anhydrotetracycline and Tet repressor. *Nucleic Acids Res.* **33**, e21.

25. Blokpoel, M. C., Murphy, H. N., O'Toole, R., Wiles, S., Runn, E. S., Stewart, G. R., Young, D. B. & Robertson, B. D. (2005). Tetracycline-inducible gene regulation in mycobacteria. *Nucleic Acids Res.* **33**, e22.

26. Das Gupta, S. K., Bashyam, M. D. & Tyagi, A. K. (1993). Cloning and assessment of mycobacterial promoters by using a plasmid shuttle vector. *J. Bacteriol.* **175**, 5186–92.

27. Timm, J., Lim, E. M. & Gicquel, B. (1994). *Escherichia coli*-mycobacteria shuttle vectors for operon and gene fusions to *lacZ*: the pJEM series. *J. Bacteriol.* **176**, 6749–53.
28. Valdivia, R. H., Hromockyj, A. E., Monack, D., Ramakrishnan, L. & Falkow, S. (1996). Applications for green fluorescent protein (GFP) in the study of host-pathogen interactions. *Gene* **173**, 47–52.
29. Kenney, T. J. & Churchward, G. (1996). Genetic analysis of the *Mycobacterium smegmatis rpsL* promoter. *J. Bacteriol.* **178**, 3564–71.
30. Guilhot, C., Gicquel, B. & Martin, C. (1992). Temperature-sensitive mutants of the Mycobacterium plasmid pAL5000. *FEMS Microbiol. Lett.* **77**, 181–6.
31. Guilhot, C., Otal, I., Van Rompaey, I., Martin, C. & Gicquel, B. (1994). Efficient transposition in mycobacteria: construction of *Mycobacterium smegmatis* insertional mutant libraries. *J. Bacteriol.* **176**, 535–9.
32. Pelicic, V., Reyrat, J. M. & Gicquel, B. (1996). Generation of unmarked directed mutations in mycobacteria, using sucrose counter-selectable suicide vectors. *Mol. Microbiol.* **20**, 919–25.
33. Pelicic, V., Jackson, M., Reyrat, J. M., Jacobs, W. R. Jr., Gicquel, B. & Guilhot, C. (1997). Efficient allelic exchange and transposon mutagenesis in *Mycobacterium tuberculosis*. *Proc. Natl. Acad. Sci. U. S. A.* **94**, 10955–60.
34. Papavinasasundaram, K. G., Colston, M. J. & Davis, E. O. (1998). Construction and complementation of a *recA* deletion mutant of *Mycobacterium smegmatis* reveals that the intein in *Mycobacterium tuberculosis recA* does not affect RecA function. *Mol. Microbiol.* **30**, 525–34.
35. Pedulla, M. L. & Hatfull, G. F. (1998). Characterization of the *mIHF* gene of *Mycobacterium smegmatis*. *J. Bacteriol.* **180**, 5473–7.
36. Pavelka, M. S. Jr. & Jacobs, W. R. Jr. (1999). Comparison of the construction of unmarked deletion mutations in *Mycobacterium smegmatis*, *Mycobacterium bovis* bacillus Calmette-Guerin, and *Mycobacterium tuberculosis* H37Rv by allelic exchange. *J. Bacteriol.* **181**, 4780–9.
37. Parish, T. & Stoker, N. G. (2000). Use of a flexible cassette method to generate a double unmarked *Mycobacterium tuberculosis tlyA plcABC* mutant by gene replacement. *Microbiology* **146**(Pt 8), 1969–75.
38. Sander, P., Meier, A. & Bottger, E. C. (1995). *rpsL+*: a dominant selectable marker for gene replacement in mycobacteria. *Mol. Microbiol.* **16**, 991–1000.
39. Bardarov, S., Kriakov, J., Carriere, C., Yu, S., Vaamonde, C., McAdam, R. A., Bloom, B. R., Hatfull, G. F. & Jacobs, W. R. Jr. (1997). Conditionally replicating mycobacteriophages: a system for transposon delivery to *Mycobacterium tuberculosis*. *Proc. Natl. Acad. Sci. U. S. A.* **94**, 10961–6.
40. Rubin, E. J., Akerley, B. J., Novik, V. N., Lampe, D. J., Husson, R. N. & Mekalanos, J. J. (1999). *In vivo* transposition of mariner-based elements in enteric bacteria and mycobacteria. *Proc. Natl. Acad. Sci. U. S. A.* **96**, 1645–50.
41. Snapper, S. B., Lugosi, L., Jekkel, A., Melton, R. E., Kieser, T., Bloom, B. R. & Jacobs, W. R. Jr. (1988). Lysogeny and transformation in mycobacteria: stable expression of foreign genes. *Proc. Natl. Acad. Sci. U. S. A.* **85**, 6987–91.
42. Mahenthiralingam, E., Marklund, B. I., Brooks, L. A., Smith, D. A., Bancroft, G. J. & Stokes, R. W. (1998). Site-directed mutagenesis of the 19-kilodalton lipoprotein antigen reveals no essential role for the protein in the growth and virulence of *Mycobacterium intracellulare*. *Infect. Immun.* **66**, 3626–34.
43. Murry, J., Sassetti, C. M., Moreira, J., Lane, J. & Rubin, E. J. (2005). A new site-specific integration system for mycobacteria. *Tuberculosis (Edinb)* **85**, 317–23.
44. Springer, B., Sander, P., Sedlacek, L., Ellrott, K. & Bottger, E. C. (2001). Instability and site-specific excision of integration-proficient mycobacteriophage L5 plasmids: development of stably maintained integrative vectors. *Int. J. Med. Microbiol.* **290**, 669–75.
45. Vultos, T. D., Mederle, I., Abadie, V., Pimentel, M., Moniz-Pereira, J., Gicquel, B., Reyrat, J. M. & Winter, N. (2006). Modification of the mycobacteriophage Ms6 attP core

allows the integration of multiple vectors into different tRNAala T-loops in slow- and fast-growing mycobacteria. *BMC Mol. Biol.* **7**, 47.

46. Saviola, B. & Bishai, W. R. (2004). Method to integrate multiple plasmids into the mycobacterial chromosome. *Nucleic Acids Res.* **32**, e11.
47. Pashley, C. A. & Parish, T. (2003). Efficient switching of mycobacteriophage L5-based integrating plasmids in *Mycobacterium tuberculosis*. *FEMS Microbiol. Lett.* **229**, 211–5.
48. Lewis, J. A. & Hatfull, G. F. (2000). Identification and characterization of mycobacteriophage L5 excisionase. *Mol. Microbiol.* **35**, 350–60.
49. Parish, T. & Stoker, N. G. Electroporation of mycobacteria. *In* Nickoloff, J. A. (ed.), Methods in Molecular Biology: Electroporation Protocols for Microorganisms, Vol. 47. Humana Press, 1995: 237–52.
50. Consaul, S. A. & Pavelka, M. S. Jr. (2004). Use of a novel allele of the *Escherichia coli aacC4* aminoglycoside resistance gene as a genetic marker in mycobacteria. *FEMS Microbiol. Lett.* **234**, 297–301.
51. Goto, Y., Taniguchi, H., Udou, T., Mizuguchi, Y. & Tokunaga, T. (1991). Development of a new host vector system in mycobacteria. *FEMS Microbiol. Lett.* **67**, 277–82.
52. Beggs, M. L., Crawford, J. T. & Eisenach, K. D. (1995). Isolation and sequencing of the replication region of *Mycobacterium avium* plasmid pLR7. *J. Bacteriol.* **177**, 4836–40.

Chapter 15
The Use of Temperature-Sensitive Plasmids in Mycobacteria

Damien Portevin, Wladimir Malaga and Christophe Guilhot

Abstract The construction of allelic exchange mutants is one of the major strategies to decipher the function of a defined gene. In this chapter, protocols are described to perform allelic exchange in the mycobacterial species *Mycobacterium smegmatis* and *Mycobacterium tuberculosis* using thermosensitive counterselectable vectors. The antibiotic resistance cassette used makes it possible to efficiently rescue the marker, leaving an unmarked mutation on the chromosome. In addition, a method is provided to construct conditional *M. smegmatis* mutants and to test whether or not a gene is essential.

Keywords allelic exchange · conditional mutants · essentiality testing · mycobacteria · thermosensitive plasmids

15.1 Introduction

Allelic exchange mutagenesis is of particular interest for deciphering the biology of mycobacteria, such as the model strain *Mycobacterium smegmatis* or the human pathogen *Mycobacterium tuberculosis*. However, homologous recombination occurs at low frequencies, below 10^{-3} events/cell/generation [1]. Therefore, the successful isolation of allelic exchange mutants, in which the wild-type allele on the chromosome has been replaced by a mutated allele, is dependent on the ability of the genetic tools and protocols used to circumvent the low transformation rate of mycobacteria (10^2 to 10^5 transformants per microgram of DNA) [2] and to enable the efficient detection or selection of allelic exchange mutants among the total population of transformants.

C. Guilhot
Institut de Pharmacologie et de Biologie Structurale, CNRS, 205 route de Narbonne, 31077 Toulouse Cedex, France
e-mail: Christophe.Guilhot@ipbs.fr

T. Parish, A.C. Brown (eds.), *Mycobacteria Protocols*,
doi: 10.1007/978-1-59745-207-6_15, © Humana Press, Totowa, NJ 2008

15.1.1 Thermosensitive Plasmids for Allelic Exchange Mutagenesis

A thermosensitive (*ts*) plasmid was obtained after chemical mutagenesis of the mycobacterial plasmid pAL5000 [3]. This plasmid is stably maintained in *M. smegmatis* or *M. tuberculosis* at 32°C or below but is lost when the temperature of growth is above 39°C [3]. Derivatives of this plasmid were used in conjunction with the *sacB* counterselectable marker to successfully achieve allelic exchange mutagenesis in various mycobacterial strains [4, 5, 6].

The insertional mutagenesis methods described here rely on the transfer of an antibiotic resistance marker cassette onto the mycobacterial chromosome. Insertion of a marker facilitates selection of the mutation and allows the tracking of a specific recombinant strain in experiments that have several strains containing different markers. However, insertion of an antibiotic cassette is often associated with a strong polar effect on the expression of downstream genes, and it excludes the use of this antibiotic marker for subsequent genetic manipulation. We therefore developed a genetic tool for simple rescue of the antibiotic marker and production of unmarked mutation [7]. This tool relies on the transposon γδ site-specific recombination system, which allows the very efficient resolution of the cointegrate intermediate formed during the transposition of γδ [8]. This resolution is catalyzed by the resolvase TnpR through site-specific recombination between two *res* sites in direct orientation [8]. We engineered an antibiotic resistance cassette that contains two *res* sites flanking a kanamycin resistance gene (the *res-Ωkm-res* cassette) and a thermosensitive plasmid, pWM19, allowing expression of the *tnpR* gene in mycobacteria [7]. The *res-Ωkm-res* cassette can be used to generate insertional mutation by allelic exchange both in *M. smegmatis* and in *M. tuberculosis*. Upon expression of the resolvase TnpR from plasmid pWM19 in the mutant strain, *res-Ωkm-res* is excised efficiently leaving behind a single *res* sequence (approximately 100 bp) at the mutated locus. This system can be used for simple construction of unmarked mutant both in *M. smegmatis* and in *M. tuberculosis* or repeated use of the same antibiotic cassette to generate multiple mutants.

15.1.2 Construction of Thermosensitive Mutants of M. smegmatis

One problem with insertional mutagenesis is that gene disruption is not possible when the product of the target gene is essential for growth in laboratory conditions. In this case, it is important to establish that failure to obtain an insertion within the gene of interest is not due to technical problems but is because the gene is essential. Genetic disruption of a putative essential gene requires the previous transfer in that strain of a functional copy of the gene. In the method proposed here, the functional copy is provided *in trans* on a *ts* plasmid. This construct enables one to establish that loss of the gene or its product is associated with a growth arrest and therefore to demonstrate that the target gene is essential [5, 9, 10, 11]. This latter protocol is of special interest in the validation of new drug targets. There are four steps in this method: (i) construction of a merodiploid strain in which the targeted gene is duplicated on the chromosome, with one copy functional and the other one disrupted; (ii) transfer of a third copy of the targeted gene on a *ts* plasmid; (iii) removal of the chromosomal functional copy; (iv) testing if the gene is essential. The method proposed uses a suicide vector to perform the first step. As recombination may take place in a locus putatively essential for mycobacterial survival, the allelic exchange substrate (AES) has to be carefully designed. For instance, the choice of the antibiotic cassette used to disrupt the target gene appears important. After insertion, the genetic structure of this chromosomal region will be modified. Therefore, polar effects of the insertion on the expression of downstream gene may occur.

15.2 Materials

15.2.1 Allelic Exchange Using ts Vector

Plasmids pPR23 and pPR27 are published and available upon request from B. Gicquel (Unité de Génétique Mycobactérienne, Institut Pasteur, 25 rue du Dr. Roux, 75015 Paris cedex 15, France).

Plasmids pCG122 and pWM19 can be obtained from C. Guilhot (Institut de Pharmacologie et de Biologie Structurale, CNRS, 205 route de Narbonne, 31077 Toulouse cedex 4, France).

1. *E. coli* strains suitable for recombinant genetic methods (for instance, DH5α). If plasmids or cassettes containing *res* sites are used, we use *E. coli* strains devoid of transposon γδ (for instance, C600).
2. *M. tuberculosis* strains; for example H37Rv (ATCC) or *M. smegmatis*, mc^{2}155 [12].
3. PCR cloning vector, such as pGEM-T cloning kit (Promega, Lyon, France).
4. Taq DNA polymerase, deoxynucleotides, and polymerase chain reaction apparatus.
5. DNA agarose gel electrophoresis equipment.
6. Primers A and B (*see* **Notes 1** and **2**).
7. Restriction enzymes X and Y (*see* **Notes 1** and **2**).
8. *Pme*I.
9. Suicide vector carrying the counterselectable *sacB*; for example, pJQ200 [13] (Fig. 15.1).
10. Thermosensitive vectors carrying the counterselectable marker *sacB*; for example, pPR23 (Fig. 15.1) or pPR27 [4].
11. Kanamycin antibiotic resistance cassette obtained from plasmid pCG122 [8] or hygromycin resistance gene [14].
12. Thermosensitive plasmid, pWM19, allowing expression of transposon γδ resolvase, TnpR [8].

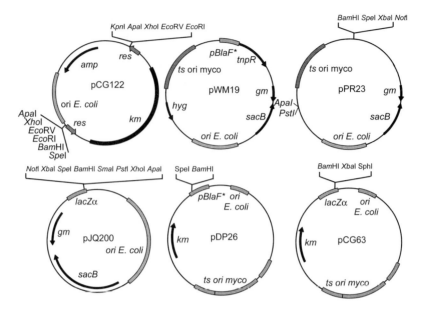

Fig. 15.1 Plasmids used to perform allelic exchange mutagenesis in mycobacteria. Vectors pPR23 and pJQ200 are used to select for allelic exchange. Plasmid pWM19 allows production of TnpR from transposon γδ and excision of an antibiotic cassette flanked by transposon γδ *res* sites. Plasmid pCG122 provides an excisable antibiotic resistance cassette. Vectors pCG63 and pDP26 are used in the construction of conditional *M. smegmatis* mutants. Abbreviations: amp, ampicillin resistance gene; *res*, target sequence from transposon γδ resolvase; *km*, kanamycin resistance gene; *hyg*, hygromycin resistance gene; *gm*, gentamicin resistance gene; *sacB*, gene conferring sensitivity to sucrose to mycobacteria; *pBlaF**, mycobacterial promoter [17]

13. 10% (v/v) Tween 80: mix 10 mL Tween 80 with 90 mL distilled water, filter sterilize.

14. 7H11 plates: Dissolve 21 g Middlebrook 7H11 medium (Fisher Scientific Bioblock, Illkirch, France) in 900 mL deionized water. Add 5 mL glycerol. Autoclave. When media is cooled to ~50°C, add 100 mL oleic albumin dextrose catalase (OADC) enrichment. Mix well and pour in standard plastic Petri dishes.

15. 7H9 medium: Dissolve 4.7 g Middlebrook 7H9 broth base (Fisher Scientific Bioblock) in 900 mL deionized water and add 2 mL glycerol, mix well, and autoclave to sterilize. Add 100 mL albumin dextrose catalase (ADC) immediately before use.

16. *E. coli* competent cells.

17. 1.5-mL sterile Microfuge tubes.

18. Benchtop Microfuge.

19. Luria-Bertani (LB) media (Fisher Scientific Bioblock): To distilled water add 10 g tryptone (pancreatic digest of casein), 5 g yeast extract, 5 g NaCl, and make up to 1 L [supplement with 5 mL 10% (v/v) Tween 80 when used

for *M. smegmatis*]. Autoclave to sterilize. For solid medium (LB-agar) add 15 g agar before autoclaving. Allow to cool and pour onto Petri dishes.

20. Antibiotic stock solutions: ampicillin (100 mg/mL), gentamicin (15 mg/mL), kanamycin (40 mg/mL), hygromycin (50 mg/mL). All antibiotics are diluted at 1/1000 v/v in the culture media.

21. Sucrose: Add to the culture medium to final concentration of 5% or 2% w/v.

22. Incubators set at 30°C, 32°C, 37°C, 39°C, and 42°C.

23. Electrotransformation cuvettes and an electroporation system.

24. 10% (v/v) glycerol: dissolve 10 mL glycerol in 90 mL distilled water and autoclave to sterilize.

15.2.2 Construction of Thermosensitive Mutants of M. smegmatis

Plasmids pCG63 are published and available upon request from B. Gicquel (Unité de Génétique Mycobactérienne, Institut Pasteur, 25 rue du Dr. Roux, 75015 Paris cedex 15, France).

Plasmid pDP26 can be obtained from C. Guilhot (Institut de Pharmacologie et de Biologie Structurale, CNRS, 205 route de Narbonne, 31077 Toulouse cedex 4, France).

1. *E. coli* strains suitable for recombinant genetic methods (for instance, DH5α). If plasmids or cassettes containing *res* sites are used, we use *E. coli* strains devoid of transposon γδ (for instance C600).

2. *M. smegmatis* strain, mc^{2}155 [12].

3. Suicide vector carrying the counterselectable *sacB* e.g. pJQ200 [13] (Fig. 15.1).

4. Kanamycin resistance cassette obtained from plasmid pCG122 [8] or hygromycin resistance gene [14].

5. Thermosensitive complementation vectors; for example, pCG63 [15, 16] or pDP26 [9], a plasmid derived from pCG63 and containing a mycobacterial expression cassette derived from pMIP12 [17] (Fig.15.1).

15.3 Methods

15.3.1 Allelic Exchange Using ts Vector

15.3.1.1 Construction of the Allelic Exchange Substrate (AES) for Recombination

An allelic exchange substrate (AES) is composed of an antibiotic resistance gene flanked by DNA fragments homologous to the targeted gene (Fig. 15.2) (*see* **Notes 1** and **2**).

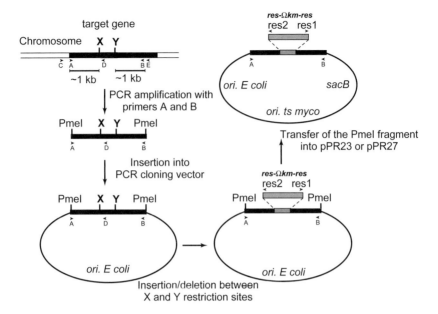

Fig. 15.2 General strategy for the construction of a *ts/sacB* plasmid for allelic exchange mutagenesis in mycobacteria. Both primers A and B contain *Pme*I restriction site at their 5' end. Primers C, D, and E are used for the PCR screening of allelic exchange mutants and are specific to the target gene. Primers res1 (5'-GCTCTAGAGCAACCGTCCGAAATAT-TATAAA-3') and res2 (5'-GCTCTAGATCTCATAAAAATGTATCCTAAATCAAA-TATC-3') are located within the *res-Ωkm-res* cassette and are used for the PCR screening of both allelic exchange mutants carrying this cassette and unmarked mutant containing the *res* site (method described in Section 15.3.1 and Fig. 15.3)

1. Using primers A and B, amplify by PCR a DNA fragment larger than 2 kb from the mycobacterial strain to be mutated (*see* **Notes 1** and **2**) [18].
2. Insert the amplified fragment into a PCR cloning vector (such as the pGEM-T vector) [18].
3. Digest the plasmid containing the amplified fragment using restriction enzymes X and Y [18].
4. Insert the *res-Ωkm-res* cassette obtained from plasmid pCG122 (Fig. 15.1) (*see* **Note 3**). The choice of restriction enzyme will vary for each construct.
5. Recover the AES from the cloning vector using *Pme*I and insert it into one of the *ts/sacB* vectors pPR23 (Fig. 15.1) or pPR27.

15.3.1.2 Allelic Exchange in *M. smegmatis* (*see* Note 4)

1. Inoculate 125 μL of a saturated culture of *M. smegmatis* in 50 mL fresh LB medium containing 0.05% Tween 80.
2. Incubate at 37°C for approximately 15 h to reach an absorbance between 0.4 and 0.6.

3. Place the culture on ice for 30 min.
4. Centrifuge the culture for 10 min. at $3000 \times g$.
5. Wash the pelleted bacteria twice with 50 mL dicionized water containing 0.05% Tween 80.
6. Resuspend bacteria in 1 mL ice-cold water (*see* **Note 5**).
7. Add approximately 1 µg salt-free plasmid (pPR23 or pPR27 derivative containing the AES) to 150 µL of this suspension.
8. Transfer to an 0.2-cm electrode-gap electroporation cuvette.
9. Place cuvette in electroporation chamber and subject to one single pulse of 2.5 kV, 25 µF, and 200 Ω.
10. Transfer into 5 mL LB medium (without antibiotic) supplemented with 0.05% Tween 80 and incubate overnight at 30°C.
11. Plate serial dilutions of the transformation mixture on LB agar plates containing kanamycin and incubate at 30°C for 3 to 4 days.
12. Pick a few resistant colonies (2 to 3 transformants) and grow them to saturation in 5 mL LB medium with kanamycin and 0.05% Tween 80 for 2 to 3 days at 30°C.
13. Plate serial dilutions of these cultures on LB agar plates containing kanamycin and 5% sucrose (w/v). Incubate at 42°C for 2 to 4 days.
14. Pick 10 kanamycin- and sucrose-resistant clones and culture in 5 mL liquid medium at 37°C (*see* **Note 6**).
15. Screen for allelic exchange event by PCR (*see* **Note 7**).

15.3.1.3 Allelic Exchange in *M. tuberculosis* (*see* Note 8)

1. Inoculate 500 µL of a saturated culture of *M. tuberculosis* in 50 mL 7H9 ADC medium containing 0.05% Tween 80 and incubate at 37°C for approximately 10 days.
2. Centrifuge the culture for 10 min at $3000 \times g$ at room temperature.
3. Wash the pelleted bacteria twice with 50 mL dieionized water containing 0.05% Tween 80 and once with 10% glycerol, 0.05% Tween 80. All these steps are performed at room temperature.
4. Resuspend bacteria in 1 mL 10% glycerol (*see* **Note 9**).
5. Add approximately 1 µg (in 1 to 5 µL) of salt-free plasmid (pPR23 or pPR27 derivative containing the AES) to 100 µL competent cells.
6. Transfer to an 0.2-cm electrode-gap electroporation cuvette.
7. Place cuvette in electroporation chamber and subject to one single pulse of 2.5 kV, 25 µF, and 200 Ω.
8. Transfer into 5 mL 7H9 ADC medium (without antibiotic) containing 0.05% Tween 80 and incubate 48 h at 32°C.
9. Plate serial dilutions of the transformation mixture on 7H11 OADC plates containing kanamycin and incubate at 32°C for 5 to 7 weeks (*see* **Note 10**). Approximately 10^4 transformants/µg of DNA are usually obtained.
10. Pick a few resistant colonies (2 to 3 transformants) and grow them in 5 mL 7H9 ADC medium with kanamycin and 0.05% Tween 80 for 3 weeks at 32°C.

11. Plate serial dilutions of these cultures onto 7H11 OADC solid medium containing kanamycin and 2% sucrose (w/v). Incubate at 39°C for 5 to 7 weeks (*see* **Notes 10** and **11**).
12. Pick several sucrose- and kanamycin-resistant clones and screen for allelic exchange events by PCR (*see* Section 15.3.1.4 and **Note 12**).

15.3.1.4 PCR Screening to Find Allelic Exchange Mutants

1. Perform PCR screening to find allelic exchange mutants (Fig. 15.3).
2. Mutants should be screened using PCR with various primers pairs; for example, pairs corresponding with the targeted gene and others located within the cassette used to disrupt the gene.
3. Primers should be designed upstream and downstream of the fragment amplified for AES construction (Fig. 15.3).

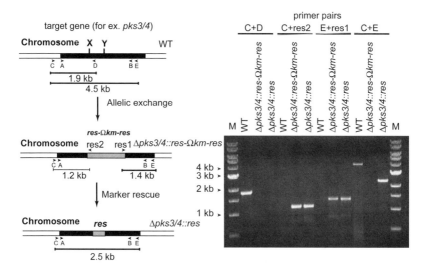

Fig. 15.3 General screening procedure to identify allelic exchange mutant. In the example shown, gene *pks3/4* was disrupted in *M. tuberculosis* H37Rv by internal deletion of a 2-kb fragment and insertion of the *res-Ωkm-res* cassette. The marker was then rescued by transient expression of TnpR. PCR fragments were obtained with four pairs of primer allowing discrimination of WT, marked and unmarked mutants. In this example, strain WT corresponds with H37Rv, *Δpks3/4::res-Ωkm-res* with the *pks3/4* mutant containing an internal deletion and an insertion of the *res-Ωkm-res* cassette, and *Δpks3/4::res* with the unmarked *pks3/4* mutant. The AES was obtained by PCR amplification using primer A (5'-GCTCTA-GAGTTTAAACCTCGGCTGAGCAACTACAC-3') and primer B (5'-GCTCTAGAGTT-TAAACCACCGGTGACGAGATAGG-3') followed by insertion of the *res-Ωkm-res* cassette between two internal XhoI sites (in this case, X and Y were identical and corresponded with XhoI). Allelic exchange mutant and unmarked mutant were screened by PCR using primers C (5'-CCTGAGCTCACTCCCGAAGC-3'), D (5'-TGTTAAA-CATTTTTCGCAGTTCCT-3'), E (5'-TCACCGCATTCCACCTGGATA-3'), res1, and res2 (Fig. 15.2). M corresponds with the 1-kb ladder (Biolabs)

4. Extract DNA from the clones to be tested as described in Chapter 1 (*see* **Note 3**).
5. Amplify using standard PCR protocols.

15.3.1.5 Rescue of the *res-Ωkm-res* Cassette and Production of Unmarked Mutation

1. Electroporate the *M. tuberculosis* or *M. smegmatis* mutant (from Sections 15.3.1.2 and 15.3.1.3) with approximately 1 µg pWM19.
2. Transfer the transformation mixture into 5 mL 7H9 (with ADC) supplemented with 0.05% Tween 80 and incubate for 48 h at 32°C for *M. tuberculosis*; or LB supplemented with 0.05% Tween 80 and incubate overnight at 30°C for *M. smegmatis*.
3. Plate serial dilutions of the transformation mixture on solid medium plates containing hygromycin and incubate at 32°C for 4 to 6 weeks for *M. tuberculosis* or at 30°C for 3 to 4 days for *M. smegmatis* (*see* **Note 14**).
4. Pick one resistant colony and grow to saturation in 5 mL liquid medium without antibiotic at 32°C for 2 weeks for *M. tuberculosis* or for 2 to 3 days at 30°C for *M. smegmatis*.
5. Plate serial dilutions of this culture on solid medium without antibiotic and containing 2% sucrose (w/v) for *M. tuberculosis* or 5% sucrose (w/v) for *M. smegmatis*. Incubate at 39°C for 4 weeks for *M. tuberculosis* or at 42°C for 2 to 4 days for *M. smegmatis*.
6. Pick 10 sucrose-resistant clones and test them for kanamycin and hygromycin sensitivity.
7. Check for excision of the *res-Ωkm-res* by PCR (*see* Section 15.3.1.4 and Fig. 15.3).

15.3.2 Testing if a Gene Is Essential in M. smegmatis *(Fig. 15.4)*

15.3.2.1 Construction of a Conditional Mutant in *M. smegmatis* (*see* Note 15)

1. Clone the gene of interest in a *ts* mycobacterial plasmid, pCG63 or pDP26.
2. Clone the AES containing the antibiotic resistance gene (for instance, the hygromycin resistance gene) into a suicide vector carrying the counterselectable marker *sacB* (e.g., pJQ200) (*see* **Note 15**).
3. Electroporate approximately 1 µg of the AES containing plasmid into *M. smegmatis* mc^{2}155 (*see* Section 15.3.1.2, steps 1 to 9).
4. Transfer into 5 mL LB liquid medium (without antibiotic) and incubate for 4 h at 37°C.
5. Plate serial dilutions of the transformation mixture on LB agar plates containing the appropriate antibiotic (hygromycin) and incubate at 37°C for 3 to 4 days (*see* **Note 7**).
6. Pick a few antibiotic resistant clones and grow them in 5 mL liquid medium for 2 to 4 days at 37°C.

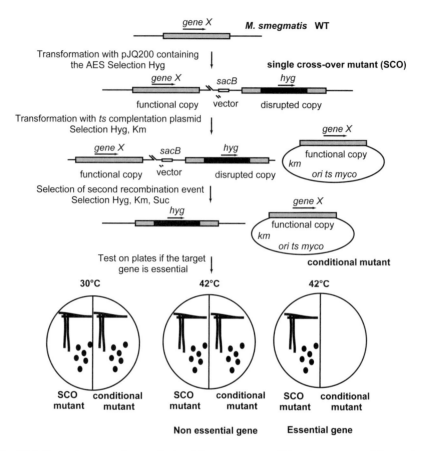

Fig. 15.4 General strategy for producing *M. smegmatis* conditional mutants and for testing the target gene is essential. The principle of this strategy is to generate a conditional mutant in which the chromosomal copy of the target gene is disrupted and a functional copy is provided on a *ts* plasmid. First a recombinant strain is generated after a single recombination event between the target gene and a mutated copy of this gene carried by a nonreplicating plasmid. Second, a third copy of the gene of interest is transferred on a *ts* plasmid in this recombinant strain. Third, the second recombination event between the two chromosomal copies of the target gene is selected leaving only the disrupted allele on the chromosome. Fourth, the essentiality of the target gene is tested by streaking the conditional mutant on plates and by incubating them at various growth temperatures

7. Extract DNA (*see* Chapter 1) (*see* **Note 13**) and analyze these clones by PCR. Select one that has undergone integration of the plasmid by single event of homologous recombination (*see* **Note 8**).

8. Electroporate this recombinant strain with approximately 1 µg of *ts* complementation plasmid.

9. Plate serial dilutions of the transformation mixture on LB agar plates containing the appropriate antibiotics (corresponding with the complementation plasmid and AES) and incubate at 30°C for 3 to 5 days.

10. Pick one or two transformants and grow them in 5 mL liquid medium with antibiotic for 2 to 4 days at 30°C.
11. Plate serial dilutions of the transformation mixture on LB agar plates containing the appropriate antibiotics (corresponding with the complementation plasmid and AES) and 5% sucrose. Incubate at 30°C for 3 to 5 days.
12. Pick 10 antibiotic- and sucrose-resistant clones and grow them in 5 mL liquid medium with antibiotics for 2 to 4 days at 30°C (*see* **Note 2**).
13. Analyze them by PCR to test whether the second recombination occurred on the chromosome leaving only the disrupted copy of the target gene (*see* **Note 13**).

15.3.2.2 Testing if a Gene Is Essential (Fig. 15.4)

At high temperature, such as at 42°C, pCG63 derivatives are not able to replicate in *M. smegmatis* [3]. Therefore, the complementation plasmid is progressively lost during cell division, leaving bacteria with only the chromosomal mutated allele of the target gene. Consequently, these conditional mutants are unable to form colonies (*see* **Note 19**).

1. Pick a colony from the conditional mutant strain and streak it onto two LB plates containing the appropriate antibiotic for the resistance marker inserted within the target gene (*see* **Note 20**).
2. Incubate one plate at 30°C and the second one at 42°C for 2 to 4 days. Check for colonies. If the gene is essential, colonies form at 30°C but not at 42°C. As a control, single crossover strains can be included (*see* **Note 19**).

15.4 Notes

1. The relevant length of homologous DNA to be cloned for gene replacement may depend on the gene that is targeted. We have had reasonable success with about 1 kb of homologous DNA on each side of the selectable marker.
2. Usually, DNA primers A and B are designed in such a way that the amplified fragment contains two restriction sites (X and Y) located approximately in the middle of the fragment (Fig. 15.2). These sites will be used to clone the antibiotic resistance cassette, *res-Ωkm-res*. Alternatively, two independent fragments may be cloned. We designed the oligonucleotides used for the PCR amplification by systematically adding a PmeI restriction site (GTTTAAAC) to the 5′ end. This restriction site is absent from most mycobacterial genomes and allows easy excision of AES from the cloning vector.
3. Other antibiotic resistance genes can be used, such as the hygromycin resistance gene, but if they are not flanked by *res* sites from transposon γδ, excision using pWM19 is not possible.
4. In hyper-transformable strains of *M. smegmatis,* such as mc^{2}155, a suicide vector can be an easier alternative to get allelic exchange mutants. However, the *ts/sacB* vector is useful in less-transformable strains of *M. smegmatis* (unpublished data).
5. At this stage, cells may be stored in aliquots at −80°C. For this purpose, add glycerol at a final concentration of 10% and freeze the aliquots in liquid nitrogen before transferring them to −80°C.

6. Because a mutation in *sacB* may lead to spontaneous sucrose-resistant clones, it can be useful to add a reporter gene such as *xylE* or *gfp* to the *ts/sacB* vector. These reporter genes will provide an additional visual selection, allowing the preselection of a sub-population of colonies in which both the *sacB* gene and the second reporter gene have been lost. Thus *sacB* mutants will be distinguishable from allelic exchange mutants.

7. Allelic exchange mutants are screened by PCR analysis using various primers pairs (some corresponding with the targeted gene and others located within the cassette used to disrupt the gene). Primers are designed upstream and downstream of the fragment amplified for AES construction (Fig. 15.3). DNA is extracted from the clones to be tested as described in Chapter 1 (*see* **Note 12**). PCR amplification is performed according to the manufacturer's recommendations using standard conditions.

8. Despite numerous attempts, we have never obtained allelic exchange in *M. bovis* BCG using this method, although the same vector was used successfully in *M. tuberculosis*. The reason for the failure of this protocol in *M. bovis* BCG is unclear because others have reported successful use of this system in *M. bovis* BCG [19].

9. Aliquots of the remaining electro-competent cells can be stored at −80°C.

10. To avoid desiccation of the plates, wrap them in aluminum foil, place in sealable plastic bags or wrap Parafilm around the edges.

11. Because *M. tuberculosis* exhibits a more restricted temperature growth range than *M. smegmatis,* plates are incubated at 39°C, which is the highest temperature giving no significant growth defect.

12. From our experience, the number of clones to analyze before getting the desired allelic exchange mutant is highly variable. We generally start with 10 clones and then continue by tens until we obtain a mutant clone. The use of a second reporter gene (*see* **Note 6**) allows the preselection of clones, thus limiting the number of clones to analyze.

13. For speed screening, crude DNA extracts are easily obtained by three successive freezing/boiling cycles of a small sample. Pipette 100 µL of a saturated liquid culture in a 1.5-mL tube and centrifuge at 4000 rpm in a Microfuge. Resuspend the pelleted bacteria in the same volume of water and immerse tubes alternately in liquid nitrogen and boiling water. Repeat this step two times and pellet bacterial residue by a centrifugation pulse. Use a few microliters (1 to 5 µL) of the supernatant for the PCR reaction.

14. Steps 3 to 4 can be avoided by adding hygromycin directly to the transformation mixture after step 2 and by incubating for 3 days at 30°C for *M. smegmatis* or 12 days at 32°C for *M. tuberculosis.* Serial dilutions of this mixture are then plated onto solid medium containing sucrose but without antibiotic as described in step 5. This alternative method is faster, but successful achievement of each step cannot be evaluated.

15. As the target locus is putatively essential for growth, the AES has to be carefully designed. After the first homologous recombination event, one copy of the gene to be disrupted has to remain functional (i.e., the gene should be downstream of a functional promoter and should not be truncated). One way to achieve this is to use a DNA fragment that spans the entire target gene for the AES (Fig. 15.4). In addition, the choice of the antibiotic cassette might be determinant for the success of this experiment if the target gene is part of an operon (*see* **Note 6**).

16. If the targeted essential gene is suspected to be part of an operon, strong polar effects may result from the insertion event. In this case, the use of a resistance marker without transcription terminator when constructing the AES and insertion of this cassette in the same orientation as the operon may allow expression of downstream genes. However, the transcriptional regulation will be lost in this construct.

17. In three independent experiments with essential genes, we obtained between 50 and 1000 tranformants per microgram.

18. At this step, it is useful to test if target gene disruption can be obtained at 30°C without a complementation plasmid. For this purpose, plate serial dilutions of the first recombination event strain on 5% sucrose at 30°C and screen around 100 sucrose resistant colonies

(*see* **Note 3**). The absence of allelic exchange mutant is a good indication that the target gene may be essential.

19. Derivatives of the *ts* plasmid pCG63 are unable to replicate at 42°C. When bacteria are dividing, one of the daughter cells will keep a copy of the plasmid, whereas the other one will be plasmid-free. The former will go to another round of cell division in contrast with the plasmid-free daughter, which will stop to grow and to divide. This phenomenon can be easily observed after inoculation of a liquid culture (containing the antibiotic corresponding with the AES) at low absorbance (below 0.01) and incubation at 42°C. If the target gene is essential, culture absorbance over time will give a linear growth curve, in contrast with the usual exponential curve. In addition, the number of colony-forming units in the culture does not increase. This may be an advantage because quite a large amount of depleted cells can be obtained. However, it is noteworthy that the bacteria in these cultures are not synchronized.

20. Alternatively, grow a 5-mL starter-culture until saturation at 30°C in liquid medium containing the antibiotics corresponding with the AES and complementation plasmids. Streak a loop of this saturated liquid culture onto two LB plates containing the antibiotic corresponding with the resistance cassette inserted within the target gene.

References

1. McFadden, J. (1996) Recombination in mycobacteria. *Mol. Microbiol.* **21**, 205–211.
2. Parish, T., and Stoker, N.G. (1998) Electroporation of mycobacteria, in *Mycobacteria protocols* (Parish, T. and Stoker, N.G., eds.) Humana Press, Inc., Totowa, NJ, pp. 129–144.
3. Guilhot, C., Gicquel, B. and Martin, C. (1992) Temperature-sensitive mutants of the *Mycobacterium* plasmid pAL5000. *FEMS Microbiol. Lett.* **98**, 181–186.
4. Pelicic, V., Jackson, M., Reyrat, J.M., Jacobs Jr, W.R., Gicquel, B. and Guilhot; C. (1997) Efficient allelic exchange and transposon mutagenesis in *Mycobacterium tuberculosis*. *Proc. Natl. Acad. Sci. U.S.A.* **94**, 10955–10960.
5. Jackson, M., Crick, D.C. and Brennan, P.J. (2000) Phosphatidylinositol is an essential phospholipid of mycobacteria. *J. Biol. Chem.* **275**, 30092–30099.
6. Irani, V.R., Lee, S.H., Eckstein, T.M., Inamine, J.M., Belisle, J.T. and Maslow, J.N. (2004) Utilization of a *ts-sacB* selection system for the generation of a *Mycobacterium avium* serovar-8 specific glycopeptidolipid allelic exchange mutant. *Ann. Clin. Microbiol. Antimicrob.* **30**, 3–18.
7. Malaga, W., Perez, E. and Guilhot, C. (2003) Production of unmarked mutations in mycobacteria using site-specific recombination. *FEMS Microbiol. Lett.* **219**, 261–268.
8. Reed, R.R. (1981) Transposon-mediated site-specific recombination: a defined in vitro system. *Cell* **25**, 713–719.
9. Portevin, D., de Sousa-D'Auria, C., Houssin, C., Grimaldi, C., Chami, M., Daffé, M. and Guilhot, C. (2004) A polyketide synthase catalyzes the last condensation step of mycolic acid biosynthesis in mycobacteria and related organisms. *Proc. Natl. Acad. Sci. U.S.A.* **101**, 314–319.
10. Portevin, D., de Sousa-D'Auria, C., Montrozier, H., Houssin, C., Stella, A., Lanéelle, M.A., Bardou, F., Guilhot, C., and Daffé, M. (2005) The acyl-AMP ligase FadD32 and AccD4-containing acyl-CoA carboxylase are required for the synthesis of mycolic acids and essential for mycobacterial growth: identification of the carboxylation product and determination of the acyl-CoA carboxylase components. *J. Biol. Chem.* **280**, 8862–8874.

11. Chalut, C., Botella, L., de Sousa-D'Auria, C. Houssin, C. and Guilhot, C. (2006) The nonredundant roles of two 4'-phosphopantetheinyl transferases in vital processes of Mycobacteria. *Proc. Natl. Acad. Sci. U.S.A.* **103**, 8511–8516.

12. Snapper, S.B., Melton, R.E., Mustapha, S., Kieser, T. and Jacobs Jr, W.R. (1990) Isolation and characterization of efficient plasmid transformation mutants of *Mycobacterium smegmatis*. *Molec. Microbiol.* **4**, 1911–1919.

13. Quandt, J., and Hynes, M.F. (1993) Versatile suicide vectors which allow direct selection for gene replacement in Gram-negative bacteria. *Gene* **127**, 15–21.

14. Zalacain, M., Gonzalez, A., Guerrero, M.C., Maltaliano, R.J., Malpartida, F. and Jimenez, A. (1986) Nucleotide sequence of the hygromycin B phosphotransferase gene from *Streptomyces hygroscopicus*. *Nucleic Acids Res.* **14**, 1565–1581.

15. Guilhot, C., Gicquel, B., Davies, J. and Martin, C. (1992) Isolation and analysis of *IS6120*, a new insertion sequence from *Mycobacterium smegmatis*. *Mol. Microbiol.* **6**, 107–113.

16. Guilhot, C., Otal, I., van Rompaey, I., Martín, C. and Gicquel, B. (1994) Efficient transposition in mycobacteria: construction of *Mycobacterium smegmatis* insertional mutant libraries. *J. Bacteriol.* **176**, 535–539.

17. Le Dantec, C., Winter, N., Gicquel, B., Vincent, V. and Picardeau, M. (2001) Genomic sequence and transcriptional analysis of a 23-kb mycobacterial linear plasmid: evidence for horizontal transfer and identification of plasmid maintenance systems. *J. Bacteriol.* **183**, 2157–2164.

18. Sambrook, J., and Russell, D.W. (2001) Molecular Cloning. A laboratory manual. Cold Spring Harbor Laboratory Press, Cold Spring Harbor, NY.

19. Pethe, K., Alonso, S., Biet, F., Delogu, G., Brennan, M.J., Locht, C. and Menozzi, F.D. (2001) The heparin-binding haemagglutinin of *M. tuberculosis* is required for extrapulmonary dissemination. *Nature* **412**, 190–194.

Chapter 16
Heterologous Expression of Genes in Mycobacteria

James A. Triccas and Anthony A. Ryan

Abstract Elucidating the function of mycobacterial proteins, determining their contribution to virulence, and developing new vaccine candidates has been facilitated by systems permitting the heterologous expression of genes in myco-bacteria. *Mycobacterium bovis* bacille Calmette Guérin (BCG) and *Mycobacterium smegmatis* have commonly been employed as host systems for the heterologous expression of mycobacterial genes as well as genes from other bacteria, viruses, and mammalian cells. Vectors that permit strong, constitutive expression of genes have been developed, and more recently systems that allow tightly regulated induction of gene expression have become available. In this chapter, we describe two complementary techniques relevant to the field of gene expression in mycobacteria. We first outline the methodology used for the expression and specific detection of recombinant products expressed in BCG. The expression vectors described use an epitope tag fused to the C-terminal end of the foreign protein, ablating the need for additional reagents to detect the recombinant product. Second, we describe the inducible expression of genes in recombinant *M. smegmatis* and the subsequent purification of gene products using affinity chromatography.

Keywords c-myc · gene expression · Mycobacteria · *Mycobacterium bovis* BCG · *Mycobacterium smegmatis* · Ni^{2+}-affinity chromatography · protein purification · recombinant protein · Western blotting

16.1 Introduction

Expression of genes in recombinant mycobacteria is a powerful tool to aid elucidation of gene function and permit the development of novel vaccine candidates to combat important infectious diseases. *Mycobacterium bovis*

J.A. Triccas
Discipline of Infectious Diseases and Immunology, University of Sydney,
Camperdown, NSW 2006, Australia
e-mail: jamiet@infdis.usyd.edu.au

T. Parish, A.C. Brown (eds.), *Mycobacteria Protocols*,
doi: 10.1007/978-1-59745-207-6_16, © Humana Press, Totowa, NJ 2008

bacille Calmette Guérin (BCG) can be used as a host vector for major protective antigens from a variety of viruses, bacteria, and parasites, and a number of these recombinant BCG (rBCG) strains can generate protective immune responses [1, 2]. The majority of systems employed to express recombinant genes in mycobacteria are based on early expression vectors that permit strong, constitutive expression of recombinant products in mycobacterial host strains [3, 4]. More recently, inducible expression systems have became available and have been used for applications ranging from gene silencing [5] to analysis of global gene regulation [6]. Vectors are now available that permit the expression and purification of recombinant proteins from mycobacterial hosts [7], and new-generation expression systems allow the specific detection of any protein expressed in mycobacteria [8]. In this chapter, we describe the procedure for the expression and specific detection of recombinant products expressed in *Mycobacterium bovis* BCG. The expression vectors described allow the detection of proteins expressed in mycobacteria independent of specific reagents to recognize the protein of interest. The second part of the chapter details one procedure for the purification of recombinant proteins from mycobacterial hosts.

16.1.1 Expression of Heterologous Genes in Recombinant Mycobacterium bovis *BCG*

The standard cloning vector used in this protocol is the pJEX55 expression vector, which is depicted in Figure 16.1A. The vector contains a number of important elements that are required for expression of recombinant genes in mycobacteria. The vector is based on the pMV261 plasmid [3] and contains origins of replication for expression in mycobacteria (oriM) and *E. coli*, (oriE) together with the *aph* gene conferring resistance to kanamycin. Expression is driven by the strong *Mycobacterium bovis hsp60* promoter. Detection of recombinant products is facilitated by an 11-residue C-terminal portion of the human c-myc protooncogene, which is specifically recognized by the 9E10 mAb. Incorporation of this tag into the pJEX55 vector series allows the detection of all recombinant proteins successfully expressed in this system and alleviates the requirement for polyclonal/monoclonal antibodies specifically recognizing the target protein.

Figure 16.1B shows the expression of the *M. tuberculosis* Ag85B protein from pJEX55 in BCG Pasteur. Three of the five clones selected contained a single band of the expected size after detection with the 9E10 mAb. The identity of the expressed protein was confirmed by immunoblotting with an anti-Ag85B polyclonal antibody (Fig. 16.1B). Lack of expression in two of the five clones may be due to plasmid instability associated with use of the *hsp60* promoter to drive gene expression in mycobacteria [9]. We routinely detect expression of "secreted" products in the lysates of recombinant clones, as we find the method

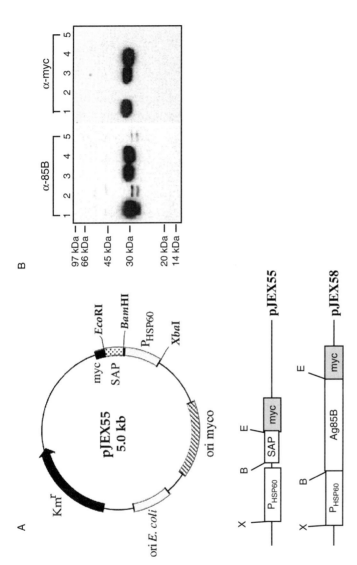

Fig. 16.1 Detection of recombinant proteins expressed in BCG using the pJEX55 expression vector. (**A**) Schematic representation of the pJEX55 vector for gene expression in mycobacteria. The vector permits fusion of genes to residues 409–419 of the c-myc protein (myc) under control of the *M. bovis* Hsp60 promoter (P_{hsp60}). X, *XbaI*; B, *BamHI*; E, *EcoRI*. The human SAP protein used to introduce the c-myc tag into the vectors is also shown. (**B**) Immunoblotting of recombinant BCG:Ag85B clones with antibodies recognizing the *M. tuberculosis* Ag85B protein (α-85B) or the c-myc epitope tag (α-myc). Lanes 1 to 5 represent cell lysates (20 μg per lane) of five independent recombinant BCG:Ag85B clones (Reprinted from Spratt, J.M., Ryan, A.A., Britton, W.J., and Triccas, J.A. [2005] Epitope-tagging vectors for the expression and detection of recombinant proteins in mycobacteria. *Plasmid* **53**, 269–273, with permission from Elsevier.)

described the most efficient for screening a large number of recombinant bacterial clones. We normally confirm secretion of proteins by Western blotting using the procedures described, however culture supernatants prepared are used in the place of cell lysates.

16.1.2 Expression of Heterologous Genes in Recombinant Mycobacterium smegmatis

Expression of mycobacterial proteins in mycobacteria represents the ideal way to produce proteins that most closely resemble the native counterpart [10, 11], and an important procedure is the purification of recombinant proteins from mycobacterial hosts. The protocol described in this chapter outlines the purification of the *Mycobacterium leprae* 35-kDa protein from recombinant *M. smegmatis* by one-step Ni^{2+}-affinity chromatography. The expression system employs the pJAM2 vector, an *E. coli*–mycobacterial shuttle vector that incorporates the inducible promoter of the *M. smegmatis* acetamidase gene (Fig. 16.2A) [7]. Gene expression is induced by the addition of acetamide to cultures of recombinant *M. smegmatis* containing pJAM2-derived expression plasmids. Using this system, we have overexpressed the *M. leprae* 35-kDa protein in recombinant *M. smegmatis* (Fig. 16.2B). Ni^{2+}-affinity chromatography was performed by virtue of a 6-histidine tag, which is present in the vector and is fused to the C-terminal end of expressed proteins. Sodium dodecyl sulfate–polyacrylamide gel electrophoresis (SDS-PAGE) of the final product revealed predominately a single species to be purified (Fig. 16.2B, lane 3). The purified product was reactive with the anti-*M. leprae* 35-kDa protein mAb CS38 (Fig. 16.2C, lane 3). The latter part of this chapter describes in detail the methodology for expression and purification of recombinant plasmids using the pJAM2 system.

16.2 Materials

16.2.1 Cloning of Genes into Expression Vectors

1. pJAM2 [7] or pJEX55 [8] expression vectors.
2. *M. tuberculosis* genomic DNA. (Obtained through NIH NIAID Contract No. HHSN266200400091C, entitled "Tuberculosis Vaccine Testing and Research Materials," awarded to Colorado State University.)
3. PCR reagents; oligonucleotide primers, dNTPs, Taq DNA polymerase (Roche Diagnostics, Castle Hill, Australia).
4. Restriction enzymes and T4 DNA ligase (Roche Diagnostics).
5. *Escherichia coli* strain DH5α.

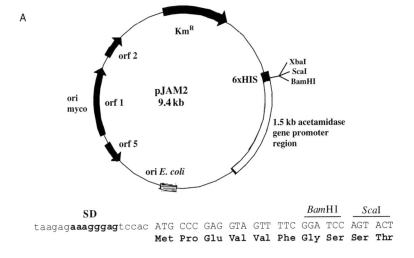

SD   *Bam*HI   *Sca*I

```
taagagaaagggagtccac ATG CCC GAG GTA GTT TTC GGA TCC AGT ACT
                     Met Pro Glu Val Val Phe Gly Ser Ser Thr
```

*Xba*I

```
TCT AGA CAC CAC CAC CAC CAC CAC TGA
Ser Arg His His His His His His  *
```

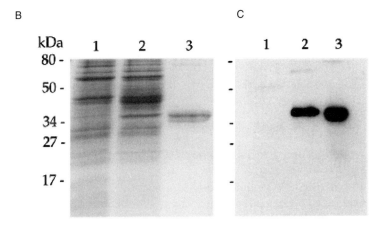

Fig. 16.2 Heterologous expression of the *M. leprae* 35-kDa protein from the acetamidase promoter in *M. smegmatis*. **(A)** Genetic organization of the pJAM2 expression vector. The vector map and nucleotide sequence of the multicloning site and surrounding regions are shown. The Shine-Dalgarno sequence (SD) is shown in bold type. **(B)** SDS-PAGE of bacterial sonicates and purified protein and **(C)** immunoblotting of a corresponding gel with the anti–*M. leprae* 35-kDa mAb CS38. Lane 1, *M. smegmatis* harboring pJAM4 grown in the absence of acetamide to induce expression of the 35-kDa protein; lane 2, *M. smegmatis* harboring pJAM4 grown in the presence of acetamide; lane 3, purified *M. leprae* 35-kDa protein. (Reprinted from Triccas, J.A., Parish, T., Britton, W.J., and Gicquel, B. [1998] An inducible expression system permitting the efficient purification of a recombinant antigen from *Mycobacterium smegmatis. FEMS Microbiol. Lett.* **167**, 151–156, with permission from Blackwell Publishing.)

6. Accuprep Plasmid Extraction Kit (Bioneer Life Sciences Corporation, Rockville, MD).
7. Luria broth: tryptone 10 g/L (Sigma, St. Louis, MO), yeast extract 5 g/L (Sigma), and NaCl 10 g/L (Sigma).
8. Kanamycin sulfate (Km) (Sigma) made as a stock solution of 25 mg/mL in triple-distilled water (TDW) and used at a final concentration of 25 µg/mL.
9. Agarose and DNA gel electrophoresis equipment.

16.2.2 Introduction of Plasmids into Mycobacteria and Sample Preparation

1. 4-mm electroporation cuvette (Bio-Rad Laboratories, Hercules, CA).
2. 10% (v/v) glycerol (Sigma).
3. Gene Pulsar (Bio-Rad) or similar.
4. Middlebrook 7H9 broth medium (Difco Laboratories, Detroit, MI) supplemented with 0.5% glycerol, 0.05% Tween 80, and 10% (v/v) albumin-dextrose-catalase (ADC) (Difco).
5. Middlebrook 7H11 medium (Difco) supplemented with 10% (v/v) oleic acid plus ADC (Difco).
6. 24-well tissue culture plate (BD Biosciences, Bedford, MA).
7. Phosphate-buffered saline (PBS) (pH 7.2) prepared by dissolving 8 g NaCl, 0.2 g KCL, 1.44 g Na_2HPO_4, and 0.24 g KH_2PO_4 in 1 L TDW.
8. Branson 250 Sonifier with 3-mm Microtip probe (Branson, Danbury, CT).
9. BCA Protein Assay kit (Pierce).

16.2.3 Detection of Recombinant Proteins by Western Blotting

1. Standard SDS-PAGE buffers and equipment.
2. Broad range prestained molecular markers, 7 kDa to 209 kDa (Bio-Rad).
3. Biotrace PVDF (Polyvinylidene fluoride) membrane (Pall Corporation, Pensacola, FL).
4. Super Signal West Pico Chemiluminescent Substrate (Pierce Biotechnology, Rockford, IL).
5. Anti-c-myc 9E10 monoclonal antibody (mAb) (Santa Cruz Biotechnology, Santa Cruz, CA) or Penta-HIS antibody (Qiagen Australia, Doncaster, Victoria, Australia).
6. Sheep anti-mouse IgG-HRP (Amersham Biosciences, Buckinghamshire, England).
7. Blocking solution: 3% w/v skim milk powder (Devondale, Brunswick, Australia) in PBS.
8. Antibody binding solution: 1% w/v skim milk powder (Devondale) in PBS.

9. 5x Reducing buffer: 0.625 M Tris buffer (pH 6.8), 1.25% w/v SDS, 12.5% v/v glycerol, 1 mM DTT (dithiothreitol), 1 mg/mL bromophenol blue.
10. Supersignal West Pico Chemiluminescent Substrate: peroxide buffer and lumino/enhancer solution (Pierce Biotechnology, Rockford, IL).

16.2.4 Purification of Proteins from M. smegmatis

1. PBS pH 7.2 (*see* Section 16.2.2, step 7).
2. Ni-NTA resin (Qiagen).
3. Sonication buffer: 5% v/v glycerol, 0.5 M NaCl, 5 mM $MgCl_2$.
4. Middlebrook 7H9 broth medium (Difco Laboratories) supplemented with 0.5% glycerol, 0.05% Tween 80, and 10% (v/v) albumin-dextrose-catalase (ADC) (Difco).
5. Branson 250 Sonifier with 13-mm disruptor (Branson).
6. Wash buffers: 5 mM, 20 mM, and 40 mM imidazole in PBS.
7. Elution buffer: 200 mM imidazole in PBS.
8. 12-14 K Cellusep Dialysis membrane (Membrane Filtration Products, Seguim, TX).
9. FD1 freeze dryer (Dynavac Engineering, NSW, Australia).

16.3 Methods

16.3.1 Cloning of Genes into Expression Vectors

1. Amplify gene of interest gene from mycobacterial genomic DNA (10 ng per reaction) using primers specific for the target sequence (20 pg of primer per reaction). Incorporate into the primer sequence the required restriction enzyme sites for cloning into the desired expression vector (*see* **Note 1**).
2. Digest the amplified product with appropriate restriction enzymes. Use 5 units of restriction enzyme per microgram of amplified DNA and incubate for 1 h at 37°C.
3. Digest pJEX55 or pJAM2 with the appropriate restriction enzymes (*see* **Notes 1** and **2**).
4. Ligate digested PCR product into digested vector (pJEX55 or pJAM2). Use 1 unit of T4 DNA ligase per reaction and ligate at 14°C for 16 h.
5. Transform plasmid into *E. coli* DH5α, plate onto LB plates containing Km and incubate overnight at 37°C.
6. Select single colonies, grow overnight in LB with Km and prepare plasmids using Accuprep Plasmid Extraction Kit or similar.
7. Visualize insert by restriction enzyme digestion and confirm by sequencing.

16.3.2 Introduction of Plasmids into Mycobacteria and Sample Preparation

1. Grow *M. bovis* BCG (Pasteur strain) or *M. smegmatis* mc^{2}155 to an OD_{600} of 0.5 in 200 mL of 7H9 broth.
2. Centrifuge at 4000 × *g* for 5 min (room temperature for BCG, 4°C for *M. smegmatis*) and resuspend cells in 100 mL of 10% glycerol.
3. Centrifuge cells twice, resuspending the pellet initially in 50 mL 10% glycerol and a second time in 25 mL 10% glycerol.
4. Centrifuge at 4000 × *g* for 5 min and resuspended cells in 2 mL 10% glycerol.
5. Add 2 μg plasmid DNA to 200 μL competent cells in a 4-mm electroporation cuvette.
6. For BCG transformation, deliver a pulse of 25 mF, 2.5 kV, and 600 Ohm resistance. For *M. smegmatis*, the settings are 25 mF, 2.5 kV, and 200 Ohm resistance.
7. Transfer cells immediately to 4 mL 7H9 broth + ADC and incubate at 37°C for 18 h (BCG) or 3 h (*M. smegmatis*). Plate onto 7H11 + OADC plates containing Km.
8. Select recombinant colonies after 3 weeks growth for BCG or 3 days for *M. smegmatis*.
9. Grow approximately 20 transformants in 24-well tissue culture plates at 37°C containing 2 mL 7H9 + ADC + Km for 2 weeks for BCG or 3 to 4 days for *M. smegmatis* (*see* **Note 3**).
10. Take approximately 1.5 mL of each transformant from 16.2.9 and centrifuge in a 1.5-mL tube (6000 × *g*, 5 min). Wash 2 times with 1.5 mL PBS in order to remove BSA from the cells. The remaining 0.5 mL culture can be used for propagation of positive clones selected by Western blotting (*see* Section 16.3.3).
11. Resuspend the final cell pellet in 200 μL PBS, and sonicate on ice 3 times for 1 min (100 W).
12. Determine protein concentrations using the BCA assay kit.

16.3.3 Detection of Recombinant Proteins by Western Blotting

1. Take 20 μg cell lysate (prepared in Section 16.3.2) and boil for 5 min at 95°C in 5x reducing buffer. Load reduced lysates onto a 12% SDS-PAGE gel (*see* **Note 4**).
2. Perform SDS-PAGE on samples and transfer to a PVDF membrane [12].
3. Block membranes with blocking solution (*see* Section 16.2.3, step 7) for 1 h at room temperature (RT). Wash for 5 min with 20 mL PBS. Repeat wash 2 times.
4. For detection of proteins expressed in pJEX55, incubate with the 9E10 antibody (1 μg/mL) in 10 mL of antibody binding solution (*see* **Note 5**). For plasmids expressed in pJAM2. expression is detected with the Penta-HIS

antibody (1 µg/mL) in 10 mL antibody binding solution. Incubate membrane for 1 h at RT.

5. Wash membrane 3 times with PBS. Incubate with sheep anti-mouse IgG-HRP at 1 µg/mL in antibody binding solution for 1 h at RT.

6. Prepare chemiluminescent substrate by adding 2 mL super signal stable peroxide buffer to 2 mL lumino/enhancer solution. Expose membrane to the substrate for 5 min.

7. Wrap the membrane in cling wrap and expose to X-ray film for 2 to 30 min.

8. Develop the film.

16.3.4 Purification of Proteins from Mycobacterium smegmatis

1. Grow 1 liter of *M. smegmatis* carrying the expression plasmid in 7H9 + ADC + Km + 0.2% acetamide (*see* **Note 6**).

2. Harvest cells by centrifugation at 10,000 × g for 20 min. Resuspend in 25 mL sonication buffer.

3. Sonicate cells 3 times for 1 min (100 W). Centrifuge (10,000 × g, 20 min) and collect the soluble fraction.

4. Apply lysate to 2 mL Ni-NTA resin and wash consecutively with 20 mL 5 mM, 20 mM, and 40 mM imidazole in PBS.

5. Elute proteins with elution buffer. Collect 10 × 2-mL fractions.

6. Identify fractions containing eluted protein by SDS-PAGE analysis of 20 µL sample from each fraction.

7. Pool samples containing recombinant product and dialyze overnight against 0.1x PBS. Concentrate samples by freeze-drying.

8. Resuspend freeze-dried samples in TDW at 1/10 the original dialysis volume. The integrity of the purified protein is determined by SDS-PAGE electrophoresis and Western blotting (*see* Section 16.3.3). Protein concentration is determined by the BCA assay kit.

16.4 Notes

1. Primers for cloning into both pJEX55 and pJAM2 must be designed with no stop codon to allow in-frame fusion to the c-terminal c-myc epitope tag or 6-HIS tag, respectively. For cloning, we typically digest pJEX55 with *Bam*HI and *Eco*RI. This digestion removes the SAP gene, which was used in the original construction of pJEX55 to introduce the c-myc epitope tag. All restriction sites in the pJAM2 multicloning region of pJAM2 are suitable for cloning purposes.

2. The plasmids described in this chapter are maintained episomally, however stable expression of recombinant genes in mycobacteria may be facilitated by the use of vectors that integrate into the mycobacterial chromosome (reviewed in Ref. 13). Typically, these systems result in reduced level of expression compared with multicopy plasmids. However, integrative vectors based on the *M. fortuitum pBlaF** promoter have demonstrated high-level expression of HIV genes in rBCG [14].

3. When using plasmids such as pJEX55 that use the *hsp60* promoter, we suggest the screening of a number of recombinant mycobacterial colonies (more than 10), as we have noted variable expression for certain proteins (Fig. 16.1). In some cases, around 80% to 90% of recombinant colonies express the protein of interest, whereas for other expression plasmids the number of positive recombinant clones is as low as 10%. This may be due to plasmid instability associated with use of the *hsp60* promoter to drive gene expression in mycobacteria [9]. We have not observed this problem with other promoters such as the *Mycobacterium fortuitum pBlaF** promoter [15]. However, once a positive clone is selected, expression appears to be stable, and bacteria can be continually passaged and stored for long periods of time without loss of gene expression.

4. Culture supernatants can be used in the place of cell lysates to detect expressed proteins. To prepare culture supernatants for Western blotting, we grow recombinant BCG clones in 7H9 medium as described (*see* Section 16.3.2), however BSA is omitted from the ADC enrichment. Twenty milliliter cultures are grown rolling for 10 to 14 days at 37°C, after which cultures are centrifuged (4000 × *g*, 10 min) and supernatant collected. Culture supernatants are concentrated approximately 10-fold using a Nanosep 3 K Omega centrifugal column (Pall Corporation, Ann Arbor, MI) according to the manufacturer's instructions.

5. We favor the use of the c-myc tag/9E10 mAb combination over other systems for detection of recombinant products in mycobacteria. Immunoblotting of BCG lysates with the anti-HIS mAb or the anti-hemagglutinin mAb 12CA5 resulted in the appearance of numerous cross-reactive protein bands (data not shown). We have attempted to use the 9E10 mAb to purify myc-tagged proteins by affinity purification by using the mAb coupled to cyanogen bromide-activated Sepharose 4B. We are unable to purify myc-tagged Ag85B using this column, however this may be a protein-specific effect and does not necessarily indicate that the 9E10 mAb is unsuitable for this purpose. We have previously used mAb affinity chromatography to successfully purify a number of proteins from mycobacterial hosts [10, 11, 16].

6. The *M. smegmatis* acetamidase promoter has been used to facilitate inducible expression of genes in *M. smegmatis* [7, 17] and can allow sufficient level of gene induction to permit purification of the recombinant product (*see* Section 16.3.4 and Ref. 7). The expression system adapted by Daugelat et al. (2003) contains all the genetic elements responsible for the regulation of acetamidase expression and may provide increased levels of gene induction compared with the system described in this chapter. The *M. smegmatis* acetamidase promoter does appear unstable in *M. tuberculosis* and BCG [7, 18], however the promoter has been successfully used for identification of genes controlled by *M. tuberculosis* SigF [6]. Tetracycline-inducible expression systems are now available for use in mycobacteria and have proved useful to construct conditional knockouts of mycobacterial genes [19, 20]. Use of these tetracycline-inducible systems to drive high-level expression of proteins for subsequent purification is yet to be tested.

Acknowledgments Work described in his chapter was supported by the National Health and Medical Research Council of Australia.

References

1. Dietrich, G., Viret, J.F., and Hess, J (2003) Novel vaccination strategies based on recombinant *Mycobacterium bovis* BCG. *Int. J. Med. Microbiol.* **292**, 441–451.
2. Ohara, N. and Yamada, T. (2001) Recombinant BCG vaccines. *Vaccine* **19**, 4089–4098.
3. Stover, C.K., de la Cruz, V.F., Fuerst, T.R., Burlein, J.E., Benson, L.A., Bennett, L.T., Bansal, G.P., Young, J.F., Lee, M.H., and Hatfull, G.F. (1991) New use of BCG for recombinant vaccines. *Nature* **351**, 456–460.

4. Winter, N., Lagranderie, M., Rauzier, J., Timm, J., Leclerc, C., Guy, B., Kieny, M.P., Gheorghiu, M., and Gicquel, B. (1991) Expression of heterologous genes in *Mycobacterium bovis* BCG: induction of a cellular response against HIV-1 Nef protein. *Gene* **109**, 47–54.
5. Parish, T. and Stoker, N.G. (1997) Development and use of a conditional antisense mutagenesis system in mycobacteria. *FEMS Microbiol. Lett.* **154**, 151–157.
6. Manabe, Y.C., Chen, J.M., Ko, C.G., Chen, P., and Bishai, W.R. (1999) Conditional sigma factor expression, using the inducible acetamidase promoter, reveals that the *Mycobacterium tuberculosis sigF* gene modulates expression of the 16-kilodalton alpha-crystallin homologue. *J. Bacteriol.* **181**, 7629–7633.
7. Triccas, J.A., Parish, T., Britton, W.J., and Gicquel, B. (1998) An inducible expression system permitting the efficient purification of a recombinant antigen from *Mycobacterium smegmatis*. *FEMS Microbiol. Lett.* **167**, 151–156.
8. Spratt, J.M., Ryan, A.A., Britton, W.J., and Triccas, J.A. (2005) Epitope-tagging vectors for the expression and detection of recombinant proteins in mycobacteria. *Plasmid* **53**, 269–273.
9. Al-Zarouni, M. and Dale, J.W. (2002) Expression of foreign genes in *Mycobacterium bovis* BCG strains using different promoters reveals instability of the *hsp60* promoter for expression of foreign genes in *Mycobacterium bovis* BCG strains. *Tuberculosis* **82**, 283–291.
10. Roche, P.W., Winter, N., Triccas, J.A., Feng, C.G., and Britton, W.J. (1996) Expression of *Mycobacterium tuberculosis* MPT64 in recombinant *Myco. smegmatis: purification, immunogenicity and application to skin tests for tuberculosis*. *Clin. Exp. Immunol.* **103**, 226–232.
11. Triccas, J.A., Roche, P.W., Winter, N., Feng, C.G., Butlin, C.R., and Britton, W.J. (1996) A 35-kilodalton protein is a major target of the human immune response to *Mycobacterium leprae*. *Infect. Immun.* **64**, 5171–5177.
12. Sambrook, J., Fritsch, E.F., and Maniatis, T. (1989) *Molecular Cloning: A Laboratory Manual*. Cold Spring Harbor, NY: Cold Spring Harbor Laboratory Press.
13. Dennehy, M. and Williamson, A.L. (2005) Factors influencing the immune response to foreign antigen expressed in recombinant BCG vaccines. *Vaccine* **23**, 1209–1224.
14. Mederle, I., Bourguin, I., Ensergueix, D., Badell, E., Moniz-Peireira, J., Gicquel, B., and Winter, N. (2002) Plasmidic versus insertional cloning of heterologous genes in *Mycobacterium bovis* BCG: impact on in vivo antigen persistence and immune responses. *Infect. Immun.* **70**, 303–314.
15. Timm, J.M., Perilli, G., Duez, C., Trias, J., Orefici, G., Fattorini, L., Amicosante, G., Oratore, A., Joris, B., Frere, J.M., Pugsley, A.P., and Gicquel, B. (1994) Transcription and expression analysis, using *lacZ* and *phoA* gene fusions, of *Mycobacterium fortuitum* β-lactamase genes cloned from a natural isolate and a high-level, β-lactamase producer. *Mol. Microbiol.* **12**, 491–504.
16. Triccas, J.A., Winter, N., Roche P.W., Gilpin, A., and Britton, W.J. (1998) Molecular and immunological analyses of the *Mycobacterium avium* homologue of the highly antigenic *Mycobacterium leprae* 35 kDa protein. *Infect. Immun.* **66**, 2684–2690.
17. Daugelat, S., Kowall, J., Mattow, J., Bumann, D., Winter, R., Hurwitz, R., and Kaufmann, S.H. (2003) The RD1 proteins of *Mycobacterium tuberculosis*: expression in *Mycobacterium smegmatis* and biochemical characterization. *Microbes Infect.* **5**, 1082–1095.
18. Brown, A.C. and Parish, T. (2006) Instability of the acetamide-inducible expression vector pJAM2 in *Mycobacterium tuberculosis*. *Plasmid* **55**, 81–86.
19. Carroll, P., Muttucumaru, D.G., and Parish, T. (2005) Use of a tetracycline-inducible system for conditional expression in *Mycobacterium tuberculosis* and *Mycobacterium smegmatis*. *Appl. Environ. Microbiol.* **71**, 3077–3084.
20. Ehrt, S., Guo, X.V., Hickey, C.M., Ryou, M., Monteleone, M., Riley, L.W., and Schnappinger, D. (2005) Controlling gene expression in mycobacteria with anhydrotetracycline and Tet repressor. *Nucleic Acids Res.* **33**, e21.

Chapter 17
Inducible Expression Systems for Mycobacteria

Christopher M. Sassetti

Abstract A wide variety of inducible expression systems have been designed for Gram-negative bacteria, but adapting these systems to phylogenetically distinct species, such as mycobacteria, has proved notoriously difficult. Mycobacteria belong to a class of high G+C Gram-positive bacteria known as actinomycetes. Although comparatively few genetic tools are available for these organisms, those that do exist are more likely to be adaptable for use in mycobacteria. A compelling example of this rationale is the recent description of a tetracycline-responsive element from corynebacteria that functions in mycobacteria. Here we describe the use of two additional mycobacterial expression systems that are derived from endogenous regulons of *Streptomyces* and *Rhodococcus* spp. Each of the currently available systems has specific advantages and limitations, and the conditions that recommend the use of each will be discussed.

Keywords gene regulation · inducible expression · overexpression

17.1 Introduction

17.1.1 Inducible Gene Expression in Mycobacteria

The ability to experimentally perturb gene expression is an indispensable genetic tool. Inducible expression systems are commonly used for the ectopic expression of genes of interest, for the inhibition of endogenous gene expression via antisense methodologies, and for the overexpression of proteins to be purified for biochemical or structural studies [1].

Although many tools for the genetic manipulation of mycobacteria have been developed [2], only recently have useful regulated expression systems been

C.M. Sassetti
Department of Microbiology and Molecular Genetics, University of Massachusetts
Medical School, 55 Lake Avenue North Worcester, MA 01655, USA
e-mail: Christopher.sassetti@umassmed.edu

T. Parish, A.C. Brown (eds.), *Mycobacteria Protocols*,
doi: 10.1007/978-1-59745-207-6_17, © Humana Press, Totowa, NJ 2008

developed that can be used in both rapidly growing environmental species and slow-growing pathogens. The first system to be described was based on an endogenous acetamide-dependent regulon of *Mycobacterium smegmatis* [3]. Whereas this has proved a useful tool in *M. smegmatis*, it has rarely been used in *Mycobacterium tuberculosis* due to its relatively high basal transcription rate and instability [3, 4].

More recently, three tetracycline-responsive systems have been developed [5, 6, 7]. All depend on a constitutively expressed tetracycline repressor (TetR) protein, although the origin of the *tetR* genes differ. Repression is relieved by the addition of the comparatively nontoxic inducer, anhydrotetracycline (aTET). Expression of exogenous genes can be induced ~100-fold, and these systems are useful in *M. tuberculosis*. Below we describe two additional regulatable promoters that, in conjunction with these aTET- promoters, form a complementary set of tools for regulating gene expression in mycobacteria.

17.1.2 Nitrile-Responsive Gene Expression

Rhodococcus rhodochrous is a saprophytic organism that is used extensively in the industrial production of acrylamide and nicotinamide. Upon encountering nitrile-containing chemicals, this organism produces large amounts of a nitrilase, encoded by the *nitA* gene, which detoxifies a broad range of these compounds [8, 9]. Under inducing conditions, this single enzyme can account for 35% of total cellular protein, and this powerful inducible system has been used to overexpress industrially important enzymes in *Streptomyces* spp. [10]. We have adapted the NitA expression system for use in both fast-growing and slow-growing mycobacterial spp.

In *R. rhodochrous*, *nitA* is divergently transcribed from a gene encoding a regulatory protein, NitR [8]. The relevant elements in the *nitA* promoter have been mapped, and NitR is sufficient to confer inducible transcription from the *nitA* promoter to a variety of *Streptomyces* spp. NitR is homologous to the well-characterized AraC protein of *Escherichia coli*, which acts as both a positive and negative regulator of the arabinose catabolic regulon and is the basis of the pBAD series *Escherichia coli* expression plasmids [11]. For AraC, the shift from repression to activation is mediated by allosteric interactions with hexose sugars [12]. It has been hypothesized that NitR is similarly regulated by inducer binding [8], but no direct experimental data is available.

17.1.3 Thiostrepton-Responsive Gene Expression

The first inducible expression system to be used broadly in actinomycetes is based on the thiostrepton (TSR) regulon of *Streptomyces lividans* [13]. Vectors encoding the cis-acting factors necessary for regulation have been generated previously, allowing ectopic expression of cloned genes in *Streptomyces* species that contain the trans-acting regulatory proteins [14]. We have constructed shuttle vectors containing all necessary regulatory elements that can be used in mycobacteria (pTSR). Though not as robust as the pNIT system described above, TSR regulation remains useful when multiple compatible systems are required.

The TSR regulon of *S. lividans* is controlled by the *tipA* gene, which encodes two proteins [15]. The larger TipAL protein is a member of the MerR/SoxR family of transcriptional regulators. Proteins of this family share a similar domain architecture consisting of an N-terminal DNA-binding domain and a C-terminal ligand interaction domain. Most family members regulate transcription by altering DNA structure upon ligand binding. The TipAL protein shares this activity and in addition may directly increase the affinity of RNA polymerase for the promoter sequence [16].

The smaller peptide produced from the *tipA* transcript, TipAS, consists of only the C-terminal ligand-binding domain of TipAL. Because TipAL auto-induces its own expression upon ligand binding, the TipAS protein is thought to function by sequestering inducer and dampening this amplified response.

17.1.4 Description of Shuttle Vectors

The vectors depicted in Figure 17.1 contain the elements necessary for replication in *E. coli* and mycobacteria. Both episomal (based on the pAL5000 origin of replication that is maintained at ~10 copies/cell [17]) and integrating (based on the phage L5 or C31 integrase systems) plasmids have been generated (*see* **Note 1**).

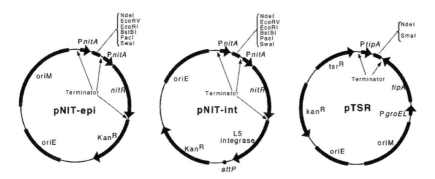

Fig. 17.1 pNIT and pTSR expression plasmids. pNITepi and pNITint are episomal and integrating versions of the nitrile-inducible expression plasmids. Abbreviations: P*nitA*, inducible *nitA* promoter; P*tipA*, inducible *tipA* promoter; P*groEL*, constitutive mycobacterial promoter; ori, episomal origin of replication kan[R], kanamycin resistance gene; tsr[R], thiostrepton resistance gene; attP, phage L5 attachment site

As depicted, the nitrile-responsive pNIT vectors place the *nitR* gene under the control of the inducible P*nitA* promoter. Thus, under uninduced conditions, the NitR is expressed at low levels, decreasing the basal transcription of the gene to be ectopically expressed. In addition, this organization should result in positive feedback upon induction that is likely to be responsible for the bistable nature of this system (see below). In both pNIT vectors, the expression cassette is isolated between strong transcriptional terminators to minimize the effects of promoters located elsewhere in the plasmids or in regions flanking the chromosomal insertion site.

The thiostrepton-responsive promoter used in the construction of pTSR was obtained from the *Streptomyces* expression plasmid pIJ8600 [14]. The *tipA* gene was amplified from the chromosome of *Streptomyces coelicolor* and cloned downstream of a constitutive mycobacterial promoter. The TSR resistance gene from *Streptomyces azureus* is also required to protect mycobacteria from TSR toxicity (*see* **Note 2**).

17.1.5 Induction Conditions for pNIT

A large family of nitrile-containing compounds can be used to activate NitR in *R. rhodochrous*. Rather surprisingly, unrelated lactam compounds are also effective, safe, and inexpensive [18]. We have found that only a subset of the compounds that are active in *R. rhodochrous* are useful in mycobacteria, presumably due to permeability differences. The relative usefulness of these inducers in mycobacteria is discussed below.

Thus far, the most active compound that we have identified for use in mycobacteria is isovaleronitrile (IVN), which results in maximal induction of the pNIT system at very low concentrations (Table 17.1). This compound is equally effective in *Mycobacterium smegmatis, Mycobacterium bovis* BCG, and *Mycobacterium tuberculosis* IVN can also be used to induce reporter gene expression during intracellular growth in cultured macrophages. Although nitriles such as IVN are toxic to both bacteria and humans, the concentration used for induction is orders of magnitude lower than that which causes inhibition of bacterial growth, and safety concerns are mitigated by the use of dilute stock solutions.

Similar induction can be achieved in *M. smegmatis* using IVN or 0.2% (18 mM) ε-caprolactam. The latter compound is relatively nontoxic and non-volatile. Furthermore, ε-caprolactam is used for the industrial production of nylon polymers and is therefore very inexpensive. However, because none of the lactam inducers that we have tested (*see* **Note 3**) are useful in *M. bovis* BCG or *M. tuberculosis*, we have found it most convenient to use IVN for all

Table 17.1 Optimal Inducer Concentrations for the pNIT System

Inducer	*M. smegmatis*	*M. bovis* BCG and *M. tuberculosis*
Isovaleronitrile (IVN)	10 μM	30 μM
ε-Caprolactam	18 mM	not active

mycobacterial species. ε-Caprolactam may be useful for certain applications where volatile compounds, such as IVN, are inconvenient. For example, ε-caprolactam can be mixed into agar plates and stored for long periods at 4°C.

17.1.6 Induction Conditions for pTSR

Optimal induction of a GFP reporter in *M. smegmatis* can be achieved using 1 μg/mL TSR. Under these conditions ~25-fold induction of reporter gene expression is achieved (*see* **Note 4**).

17.1.7 Titration of Expression

Single-cell analysis has only recently been used to investigate the behavior of inducible expression systems. In general, under suboptimal induction conditions, one of two types of behavior is apparent. In some cases, titratable expression at the single-cell level is apparent, as the entire population homogenously expresses low levels of reporter [19]. In other cases, such as in the *lac* regulon of *E. coli*, a bistable expression pattern is observed [20]. In this case, two distinct populations are present: one that maximally expresses the reporter and another where no expression is apparent. The latter systems behave essentially as binary switches and are common in both endogenous regulatory circuits and artificial expression systems. Thus, the proper characterization of expression systems at the single-cell level is important for understanding the effects of ectopically expressed genes.

We have used fluorescence-activated cell sorting, FACS, to investigate the behavior of the pNIT system under half-maximal induction, and this system clearly behaves as a binary switch. Under these conditions, 50% of the population is "on" and 50% is "off." This is in contrast with the tetracycline-inducible system described by Ehrt et al. [7], which we have found to produce titratable expression at the single-cell level (authors unpublished observations).

17.1.8 Comparison of Inducible Systems

Clearly there is no "one size fits all" expression system for every application. Table 17.2 lists the relevant attributes of the pNIT and pTSR systems described above in comparison with the tetracycline-inducible system (pTET) described by Ehrt and colleagues [7]. Other TET systems [5, 6] were not included in our experimental comparison, but based on published literature, all TET regulons are expected to behave similarly.

Table 17.2 Comparison of Inducible Expression Systems

	pNIT	pTET	pTSR
Regulator	NitR	TetR	TipAL
Regulatory mechanism	Allosteric activator	Allosteric repressor	Allosteric activator
Inducer	ε-Caprolactam/IVN	Anhydrotetracycline	Thiostrepton
Inducibility	>100-fold	>100-fold	25-fold
Maximal expression	++++	+++	++
Titratable in single cells?	Bistable	Titratable	Unknown
Inducible during macrophage infection?	+	+	Unknown

17.2 Materials

17.2.1 Mycobacterial Culture

1. 20% (v/v) Tween-80. Dilute 20 mL Tween-80 in 80 mL deionized water. Mix thoroughly on a magnetic stirrer (heat to 50°C if necessary), and filter sterilize. Store at room temperature.
2. AD enrichment. 10x stock solution: Dissolve 50 g of bovine serum albumin (Fraction V; Serologicals Corp. Norcross, GA), 20 g dextrose, and 85 g NaCl in 100 mL deionized water. Filter sterilize and store at 4°C.
3. Middlebrook OADC enrichment. 10x stock solution (Becton Dickinson. Sparks, MD).
4. Kanamycin sulfate (Sigma, St. Louis. MO). 1000x stock solution (50 mg/mL): dissolve 500 mg in 10 mL deionized water and filter sterilize. Store at 4°C.
5. Hygromycin sulfate (A.G. Scientific, San Diego, CA). Supplied as 2000X stock solution (100 mg/mL)
6. 7H9 broth. Dissolve 4.7 g 7H9 powder (Difco, Detroit, MI) in 900 mL deionized water. Add 2 mL glycerol and autoclave or filter sterilize. Store at room temperature.
7. 7H9 broth + AD for growth of *M. smegmatis* and *M. bovis* BCG. To 900 mL 7H9 broth add 100 mL 10x AD enrichment and 2.5 mL 20% Tween-80. Store at 4°C.
8. 7H9 broth + OADC for growth of *M. tuberculosis*. To 900 mL 7H9 broth add 100 mL 10x OADC enrichment and 2.5 mL 20% Tween-80. Store at 4°C.
9. 7H10 agar plates. Add 19 g 7H10 agar and 5 mL glycerol to 900 mL deionized water and autoclave. Cool melted agar to 50°C and add either AD or OADC enrichment (AD for *M. smegmatis* or *M bovis* BCG, and OADC for *M. tuberculosis*). Immediately pour into Petri dishes.

17.2.2 Stock Solutions of Inducers

1. Isovaleronitrile (Sigma). 1000x stock solution: dilute the concentrated reagent 10,000-fold in dimethyl sulfoxide (DMSO). Store at 4°C (*see* **Note 5**).
2. ε-Caprolactam (Sigma). 100x stock solution (20% w/v): Dissolve 200 g in 1 L deionized water and filter sterilize. Store at room temperature.
3. Thiostrepton (Sigma). 1000x stock solution: dissolve at 1 mg/mL in DMSO. Store at –20°C.

17.2.3 Plasmids

pNIT and pTSR plasmds are available from the author (Fig. 17.1).

17.3 Methods

17.3.1 Construction and Transformation of Expression Plasmids

1. Ligate the DNA fragment to be expressed in the multiple cloning site of pNIT or pTSR plasmids (*see* Fig. 17. 1 and **Notes 6** and **7**).
2. Electroporate appropriate plasmid into a suspension of mycobacterial cells (*see* **Note 8** and Chapter 13).
3. Dilute transformed bacteria in ~2 mL 7H9 broth and culture at 37°C with agitation to allow for the expression of antibiotic resistance gene. *M. smegmatis* should be allowed 3 h, and *M. bovis* or *M. tuberculosis* should be allowed 16 h.
4. Concentrate the bacteria by centrifugation at $4000 \times g$ for 10 min, and spread the entire population on appropriate selective plates.
5. Incubate plates until colonies are visible (3 days for *M. smegmatis*, 14 days for BCG, or 21 days for *M. tuberculosis*).
6. Culture single colonies into 7H9 broth and appropriate antibiotic. If desired, bacterial stocks can be stored at –80°C after the addition of 10% glycerol.

17.3.2 Gene Induction in Broth Culture (see Note 9)

1. Culture bacteria in 7H9 broth (*see* **Note 10**) with 20 μg/mL kanamycin at 37°C with agitation until the desired growth phase is reached (*see* **Note 11**).
2. Add 1/1000 volume of appropriate inducer stock (*see* Sections 17.1.5 and 17.1.6).
3. Return cultures to 37°C with agitation to assay for expression.
4. At appropriate time points (*see* **Note 12**), aliquots of cells should be removed and assayed for induction. Reporter genes, such as green fluorescent protein (GFP) or luciferase, can be easily assayed using ~100 μL of culture.

Alternatively, induction can be assessed by sodium dodecyl sulfate-polyacrylamide gel electrophoresis (SDS-PAGE) after lysing 5-ml aliquots of cells in a "Fastprep" instrument.

17.4 Notes

1. An integrating version of the pTSR has also been constructed, which uses the streptomyces phage C31 integration system. This construct is derived from a relatively uncharacterized plasmid [14] and therefore a map is not available.
2. Thiostrepton (TSR) is toxic to mycobacteria unless the *tsr* resistance gene of *Streptomyces azureus* is expressed. Although the *tsr* gene is clearly active in mycobacteria, TSR cannot be used to select for plasmid bearing transformants, due to a high rate of spontaneous resistance. Thus, two antibiotic resistance genes (TSR and kanamycin resistance) must be included on the pTSR plasmid.
3. Thus far, we have tested ε-caprolactam, δ-valerolactam, and γ-butyrolactam. ε-Caprolactam is the most effective for induction in *M. smegmatis*, but none are useful in *M. tuberculosis* or *M. bovis* BCG. Only isovaleronitrile potently induces the pNIT system in these slow-growing organisms.
4. The usefulness of the pTSR system in *M. tuberculosis* has not yet been determined.
5. Concentrated isovaleronitrile is highly toxic and should be handled in a fume hood. The fumes from dilute stock solutions pose little risk, but gloves should be worn when handling.
6. We and others have noted that it is difficult to achieve high-level gene expression after integrating plasmids into the phage L5 attB site. Although functional complementation is often achievable, overexpression generally requires the use of episomal plasmids.
7. If ectopic protein expression is desired, a gene of interest can be cloned in frame with the ATG codon encoded in the *Nde*I site of the pNIT plasmids. This codon corresponds with the start site of the native nitA transcript and is preceded by its ribosome binding site (RBS). This option is not available for pTSR vectors, and the inserted DNA fragment should include a both a start codon and RBS.
8. Many electroporation conditions have been successfully used to transform mycobacteria. This procedure should be optimized by each user depending on the available equipment. Using standard conditions, we routinely obtain >1000 transformants from each electroporation. For higher efficiencies, *see* Chapter 13.
9. It is also possible to induce transcription from pNIT during growth on solid media by adding 0.2% ε-caprolactam to the agar immediately before pouring the plates. It should also be possible to use nitrile inducers or TSR, but this has not yet been optimized.
10. We generally use Middlebrook 7H9 broth and 7H10 agar for cultivation of mycobacteria and pNIT induction. We have noted high basal expression levels when using rich media, such as LB, presumably due to a stimulatory compound in the media. Thus, care should be taken when complex media is substituted. Although the plasmids described in Fig. 17.1 replicate stably in both mycobacteria and *E. coli*, appropriate antibiotics should always be included during the passage of plasmid-bearing strains.
11. As with virtually any expression system, the inducibility of pNIT depends on the growth phase of the bacteria. We have found, however, that detectable induction is apparent upon IVN addition even in late-stationary-phase *M. smegmatis* and in *M. bovis* BCG during intracellular growth in macrophages. Thus, pNIT may be particularly useful when inducible gene expression is required in slowly replicating cells.
12. The time points used will vary depending on the goal of the experiment. Using green fluorescent protein as a reporter, both pNIT and pTSR expression is maximal within 9 h in *M. smegmatis* and 24 h in *M. tuberculosis*. Using these conditions, we routinely observe >100-fold induction of reporter gene expression in all mycobacterial species tested.

References

1. Balabas, P. and Lorence, A. (2004) *Recombinant Gene Expression*, 2nd cd. Humana Press. Totowa, NJ.
2. Hatfull, G. H. and Jacobs, W. R. Jr. (2000) *Molecular Genetics of Mycobacteria*. ASM Press. Washington, DC.
3. Parish, T., Mahenthiralingam, E. Draper, P., Davis, E. O., and Colston, M. J. (1997) Regulation of the inducible acetamidase gene of *Mycobacterium smegmatis*. *Microbiology*. **143**, 2267–76.
4. Brown, A. C., and Parish, T. (2006) Instability of pJAM2-based expression vectors in *Mycobacterium tuberculosis*. *Plasmid*. **55**, 81–6.
5. Blokpoel, M. C., Murphy, H. N., O'Toole, R., Wiles, S., Runn, E. S., Stewart, G. R., Young, D. B., and Robertson, B. D. (2005) Tetracycline-inducible gene regulation in mycobacteria. *Nucleic Acids Res*. **33**, e22.
6. Carroll, P., Muttucumaru, D. G., and Parish, T. (2005) Use of a tetracycline-inducible system for conditional expression in *Mycobacterium tuberculosis* and *Mycobacterium smegmatis*. *Appl Environ Microbiol*. **71**, 3077–84.
7. Ehrt, S., Guo, X. V., Hickey, C. M., Ryou, M., Monteleone, M., Riley, L. W., and Schnappinger, D. (2005) Controlling gene expression in mycobacteria with anhydrotetracycline and Tet repressor. *Nucleic Acids Res*. **33**, e21.
8. Komeda, H., Hori, Y., Kobayashi, M., and Shimizu, S. (1996) Transcriptional regulation of the *Rhodococcus rhodochrous* J1 nitA gene encoding a nitrilase. *Proc Natl Acad Sci U S A*. **93**, 10572–7.
9. Nagasawa, T., Wieser, M., Nakamura, T., Iwahara, H., Yoshida, T., and Gekko, K. (2000) Nitrilase of *Rhodococcus rhodochrous* J1. Conversion into the active form by subunit association. *Eur J Biochem*. **267**, 138–44.
10. Herai, S., Hashimoto, Y., Higashibata, H., Maseda, H., Ikeda, H., Omura, S., and Kobayashi, M. (2004) Hyper-inducible expression system for streptomycetes. *Proc Natl Acad Sci U S A*. **101**, 14031–5.
11. Guzman, L. M., Belin, D., Carson, M. J., and Beckwith, J. (1995) Tight regulation, modulation, and high-level expression by vectors containing the arabinose PBAD promoter. *J Bacteriol*. **177**, 4121–30.
12. Lee, N. L., Gielow, W. O., and Wallace, R. G. (1981) Mechanism of araC auto-regulation and the domains of two overlapping promoters, Pc and PBAD, in the L-arabinose regulatory region of *Escherichia coli*. *Proc Natl Acad Sci U S A*. **78**, 752–6.
13. Murakami, T., Holt, T. G., and Thompson, C. J. (1989) Thiostrepton-induced gene expression in *Streptomyces lividans*. *J Bacteriol*. **171**, 1459–66.
14. Takano, E., White, J., Thompson, C. J., and Bibb, M. J. (1995) Construction of thiostrepton-inducible, high-copy-number expression vectors for use in Streptomyces spp. *Gene*. **166**, 133–7.
15. Chiu, M. L., Folcher, M., Katoh, T., Puglia, A. M., Vohradsky, J., Yun, B. S., Seto, H., and Thompson, C. J. (1999) Broad spectrum thiopeptide recognition specificity of the *Streptomyces lividans* TipAL protein and its role in regulating gene expression. *J Biol Chem*. **274**, 20578–86.
16. Chiu, M. L., Viollier, P. H., Katoh, T., Ramsden, J. J., and Thompson, C. J. (2001) Ligand-induced changes in the *Streptomyces lividans* TipAL protein imply an alternative mechanism of transcriptional activation for MerR-like proteins. *Biochemistry*. **40**, 12950–8.
17. Ranes, M. G., Rauzier, J., Lagranderie, M., Gheorghui, M., and Gicquel, B. (1990) Functional analysis of pAL5000, a plasmid from *Mycobacterium fortuitum*: construction of a "mini" mycobacterium-*Escherichia coli* shuttle vector. *Journal of Bacteriology*. **172**, 2793–7.

18. Nagasawa, T., Nakamura, T., and Yamada, N. (1990) ε-caprolactam, a new powerful inducer for the formation of *Rhodococcus rhodochrous* J1 nitrilase. *Arch Microbiol.* **155**, 13–17.

19. Chan, P. F., O'Dwyer, K. M., Palmer, L. M., Ambrad, J. D., Ingraham, K. A., So, C., Lonetto, M. A., Biswas, S., Rosenberg, M., Holmes, D. J., and Zalacain, M. (2003) Characterization of a novel fucose-regulated promoter (PfcsK) suitable for gene essentiality and antibacterial mode-of-action studies in *Streptococcus pneumoniae. J Bacteriol.* **185**, 2051–8.

20. Ozbudak, E. M., Thattai, M., Lim, H. N., Shraiman, B. I., and Van Oudenaarden, A. (2004) Multistability in the lactose utilization network of *Escherichia coli. Nature.* **427**, 737–40.

Chapter 18
Assaying Promoter Activity Using LacZ and GFP as Reporters

Paul Carroll and Jade James

Abstract The ability of bacteria to survive in a variety of different niches is due, in part, to their ability to respond and adapt to the environment. Extracellular signals are recognized by bacilli, and their responses are generally conducted at the transcript level. RNA polymerases recognize specific promoter regions on the genome and initiate transcription. Therefore, the analysis of gene expression is paramount to understanding the biology of an organism. In the case of pathogens, gene expression can alter during the course of the infection, and, therefore, specific targets can be identified for drug development. Promoter activity can be determined by cloning a promoter sequence upstream of a reporter gene and assaying the reporter activity, either from whole cells or from cell lysates. This chapter describes two reporter systems (GFP and LacZ) used for determining promoter activity that have been widely used in mycobacteria.

Keywords β-galactosidase · cell-free extract · fluorescence · GFP · LacZ · live bacilli · mycobacteria · reporter · vector

18.1 Introduction

Bacteria are able to respond to environmental stimuli and react accordingly. Changes to temperature, pH, oxygen, and nutrient availability elicit cellular responses in both the proteome and transcriptome, in osmotic pressure, and, in the case of pathogenic bacteria, in host interactions [1, 2, 3, 4]. Reporter systems are powerful tools for investigating transcriptional regulation in response to environmental signals. Such systems are often plasmid based with a promoterless reporter gene, encoding an assayable enzyme that can be detected directly or indirectly. When a functional promoter is present, the protein will be expressed, and activity can be detected by either colorimetric or fluorometric assays.

P. Carroll
Institute of Cell and Molecular Science, Barts and the London, Queen Mary's School of Medicine and Dentistry, 4 Newark Street, Whitechapel, London E1 2AT, UK
e-mail: p.carroll@qmul.ac.uk

T. Parish, A.C. Brown (eds.), *Mycobacteria Protocols*,
doi: 10.1007/978-1-59745-207-6_18, © Humana Press, Totowa, NJ 2008

The identification and characterization of the replicating plasmid pAL5000, from *Mycobacterium fortuitum*, allowed genetic tools to be developed [5, 6, 7, 8, 9]. This has allowed the development of many important genetic systems for the study of mycobacterial biology (*see* Chapter 14).

18.1.1 Fluorescent Proteins

Fluorescence has been a valuable tool in biological research since the 1950s when antibodies were tagged to aid detection [10]. Fluorescent reporters have been extensively used to determine transcriptional activity, protein localization, screening microbial population dynamics *in situ*, comparative genomics, transcriptomics, and proteomics. Green fluorescent protein (GFP) was first isolated in 1960 from *Aequorea victoria*, but it was not until 1994 that GFP was first used to report gene expression in prokaryotes and eukaryotes [11].

To date, GFP has been used in bacteria [11], yeast [12], fungi [13], plants, *Drosophila* [14], *Caenorhabditis elegans* [11], zebrafish [15], and in mammalian cell lines [16]. The expression of fluorescent proteins causes no disruption to cell metabolism or toxicity [17]. Newly synthesized GFP matures by an autocatalytic reaction to form a light-sensitive domain located in the center of the protein, termed the *fluorophore*. The fluorophore responds to light, with a major excitement peak at a wavelength of 395 nm and an emission peak at 510 nm. GFP is very stable and can be detected in real-time; these properties make GFP an ideal reporter, as it can be used to both qualify and quantify gene expression. However, it must be noted that oxygen is required for mature fluorophore emission, although it has been demonstrated that fluorescence can be visualized when the fluorescent protein is recovered from anaerobic conditions [18]. Therefore, in limiting oxygen conditions, fluorescence can be qualified, but whether it can be quantified is open to question.

Other fluorescent proteins are available, and the recognition that amino acid substitutions in the fluorophore can generate color variations, which emit at different fluorescent wavelengths, has led to an increase in the number of variants available. These include enhanced green (EGFP) [19], yellow (YFP) [20], and red (RFP) [21]. To determine temporal expression, fluorescent proteins have also been modified to include the *ssrA* tag at the C-terminal [22, 23]. SsrA, also referred as 10Sa RNA or tmRNA, tags are recognized by SmpB (small protein B) and are targeted for proteolytic degradation [24]. Because these fluorescent proteins are rapidly degraded, temporal gene expression can be monitored. In addition, different bacterial recognition peptides can be used to modulate the half-life of the protein [22, 23].

18.1.2 β-Galactosidase

The LacZ reporter systems encode β-galactosidase, which is responsible for the degradation of β-galactosyl linkages. LacZ activity can be monitored by cleavage of the colorless substrate 5-bromo-4-chloro-3-indolyl-β-D-galactopyranoside (X-gal) by β-galactosidase to galactose and a blue insoluble product (used to identify promoter activity in living cells on solid media), or *o*-nitrophenyl-β-D-galactopyranoside (ONPG) to *o*-nitrophenol (ONP), a soluble yellow product that can be quantitated by reading at 410 nm (used to identify promoter activity in cell-free extracts). The advantage of this system is that β-galactosidase is a stable protein that can be assayed simply by measuring color changes. Mycobacteria do not naturally produce β-galactosidase, and therefore no background enzyme activity is present. The LacZ reporter system has been widely used in mycobacterial system to assess promoter activity [25, 26, 27]; it has also been used to conduct rapid antimycobacterial drug screening [28] and to screen for potential vaccine strains [29].

18.1.3 Reporters Used in Mycobacteria

Fluorescent proteins have been used extensively in mycobacterial species. GFP was first used to quantify the expression levels of genes in laboratory media or during infection of macrophages [30]. In addition, the fluorescent marker allowed the visualization of intracellular mycobacteria, and counting by epifluorescence microscopy, confocal scanning microscopy, fluorescent microscopy, and flow cytometry [30]. To date, GFP has been used to study mechanisms such as signal transduction [31, 32, 33, 34], host-pathogen responses [35], and secreted protein localization and quantification [36]. Unstable GFP variants have also been generated for use in mycobacteria for temporal gene expression [22, 23].

LacZ was the first reporter successfully applied to mycobacteria [37] and has been used extensively to report expression levels of many genes in a variety of mycobacterial species [38, 39, 40, 41, 42]. Other reporter systems, such as luciferase from *Vibrio harveyi* (*luxAB*), chloromphenicol acetyltransferase (CAT), and the product of *xylE*, catechol 2,3-dioxygenase (CDO), have also been described for use in mycobacteria [43, 44, 45, 46].

The effects of various environmental conditions on promoter can be assayed in mycobacteria by using a reporter probe. In a defined environment, the effect of different carbon and nitrogen sources on the promoter can be assayed. In addition, the promoter's response to oxygen availability can be determined as well as the activity in comparison with the age of the culture. Many different defined environments can be generated to study the promoter's response to stimuli. In this chapter, we will describe the methodologies used to assay promoter activity using either GFP or LacZ reporters.

We have used two vectors extensively, namely pFLAME, carrying a GFP reporter and kanamycin resistance on a replicating plasmid (Fig. 18.1A), and pSM128, a LacZ reporter plasmid with streptomycin resistance that integrates into the *Mycobacterium* genome at a specific locus (Fig. 18.1B). We have used both these systems to determine the activity of both native and foreign

Fig. 18.1 Diagrammatic representation of the GFP and LacZ reporter plasmids pFLAME and pSM128, respectively. Cloning sites for promoter insertion displayed. *oriE*, origin of replication for *E. coli*; *oriM*, origin of replication for mycobacteria; strep[R], streptomycin resistance cassette; kan[R], kanamycin resistance cassette; trpA, terminator of trpA; T4ter, terminator of T4

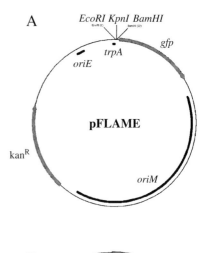

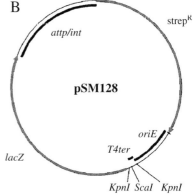

promoter responses to different environmental stimuli. We describe how to assay reporter activity from both cell-free extracts and from live mycobacteria.

18.2 Materials

18.2.1 Construction of Reporter Vectors

1. *Escherichia coli* strains suitable for cloning procedures.
2. Mycobacterial species amenable to genetic manipulation; for example, *Mycobacterium tuberculosis* H37Rv (ATCC 25618) or *Mycobacterium smegmatis* mc^2-155.
3. *E. coli* cloning system; for example, pSC-A (Stratagene, La Jolla, USA).
4. Mycobacterial reporter vector; for example, the replicating pFLAME [23] (*gfp*) replicating vector or the integrating pSM128 [47] (*lacZ*) plasmid.
5. PCR primers specific for the promoter region and restriction sites for cloning.
6. PCR reagents; that is, *Pfu*, dNTPs, DMSO, MgCl$_2$, and double-distilled water.

7. Thermocycler.
8. Gel electrophoresis equipment.
9. Plasmid DNA purification kit.
10. Electroporation materials and equipment (*see* Chapter 13).
11. LB media: To 800 mL distilled water add 10 g tryptone (pancreatic digest of casein), 5 g yeast extract, 5 g NaCl, and make up to 1 L. Autoclave to sterilize. For solid medium (LB-agar, Sigma, Dorset, UK), add 15 g agar before autoclaving. Allow to cool and pour onto Petri dishes.
12. Appropriate antibiotics. Streptomycin: dissolve 1 g in 20 mL distilled water, filter sterilize to give a stock solution of 50 µg/mL. Store in 500-µL aliquots at –20°C, use in media to give a final concentration of 20 µg/mL for pSM128 for *E. coli* and mycobacteria. Kanamycin: Dissolve 1 g in 20 mL distilled water, filter sterilize to give a stock solution of 50 µg/mL. Store in 500-µL aliquots at –20°C, use in media to give a final concentration of 20 µg/mL for mycobacteria harboring pFLAME3, 50 µg/mL for *E. coli*.
13. 20% (v/v) Tween 80: mix 20 mL Tween 80 with 80 mL distilled water, filter sterilize.
14. 7H10 plates: Dissolve 19 g Middlebrook 7H10 medium (Difco, Becton, Dickinson and Company, Sparks, USA) in 900 mL deionized water. Add 8.4 mL 60% glycerol (v/v). Autoclave. When media is cooled to ~50°C, add 100 mL oleic albumin dextrose catalase (OADC, Becton, Dickinson and Company, Sparks, USA) enrichment. Mix well and pour in standard plastic Petri dishes.
15. 7H9-Tw medium: Dissolve 4.7 g Middlebrook 7H9 broth base (Difco) in 900 mL deionized water and add 2 mL glycerol, mix well, and filter sterilize. Add 5 mL 20% (v/v) Tween 80 (final concentration of 0.05% v/v). Add 100 mL OADC immediately before use.
16. Lemco media: To 800 mL distilled water add 10 g Bacto-peptone (Becton, Dickinson and Company), 5 g NaCl (Sigma), and 5 g Lemco powder (Oxoid, Hampshire, UK). Add 5 mL 20% (v/v) Tween 80 (Sigma) (final concentration of 0.05% v/v) and make up to 1 L. Autoclave to sterilize. For solid medium (Lemco-agar), add 15 g agar in place of Tween 80 before autoclaving. Allow to cool and pour onto Petri dishes.
17. X-gal (Sigma) (5-bromo-4-chloro-3-indolyl-β-D-galactopyranoside) at 50 µg/mL.

18.2.2 Recovery of Transformants Harboring the Reporter Constructs

1. Cell disruption system; for example, FastPrep FP120 (Q-BIOGENE, MP Biomedicals, Illkirch, France).
2. Lysing matrix B tubes (Q-BIOGENE).
3. Sterile 1.5-mL tubes.
4. Tris-HCl (Sigma), 10 mM, pH 8.

18.2.3 *Assaying Reporter Expression*

18.2.3.1 Assaying GFP Expression in Live Cells

1. Tris-HCl (Sigma), 10 mM, pH 8.
2. Glass screw-top tubes (VWR, Leicestershire, UK) that allow fluorescent readings (e.g., 12 × 100-mm tubes, 1-mm wall thickness).
3. Spectrofluorometer; for example, Shimadzu Duisburg, Germany RF-1501 (Shimadzu).

18.2.3.2 Assaying GFP Fluorescence in Cell-Free Extracts

1. Tris-HCl, 10 mM, pH 8.
2. 96-well optical bottomed plates for fluorescence (e.g., Microplate Nunc, Rochester, NY, USA FluoroNunc 96).
3. Fluorometric plate reader; for example, FLUOstar OPTIMA (BMG Labtech, Germany), with excitation at 490 nm and emission at 510 nm

18.2.3.3 β-Galactosidase Activity Assays

1. Water bath set at 37°C.
2. 1-mL cuvettes, which can allow readings at 410 nm.
3. Spectrophotometer with a 410-nm filter.
4. Polystyrene tubes with polyethylene caps, 110 × 16-mm tubes (e.g., Falcon 2054 tubes; Becton, Dickinson and Company).
5. Z-buffer: 60 mM Na_2HPO_4, 40 mM NaH_2PO_4, 10 mM KCl, and 1 mM $MgSO_4$. Filter sterilize. Store at 4°C.
6. 4 mg/mL Ortho-nitrophenyl-β-galactoside (ONPG, Sigma). Filter sterilize. Store at –20°C
7. 1 M sodium bicarbonate ($NaHCO_3$).

18.3 Methods

The following protocols have been used in *M. tuberculosis* and *M. smegmatis*. However, they should be applicable to all mycobacterial species. The correct containment procedures should be applied when manipulating the microorganism.

18.3.1 *Construction of Reporter Vectors*

1. Isolate the target promoter sequence and amplify by PCR (*see* **Note 1**).
2. Clone the target promoter into an *E. coli* cloning vector and transform into *E. coli* (*see* **Note 2**).
3. Excise fragment with suitable restriction enzyme(s) for the vector selected for use and clone into the correspondingly digested reporter vector.

4. Determine orientation of promoter region by diagnostic restriction digests or other methods (*see* **Note 3**).
5. Prepare competent mycobacteria (*see* Chapter 13).
6. Electroporate both the promoter/reporter construct and also the empty reporter vector (as a control) into mycobacteria (*see* Chapter 13 and **Note 4**).
7. Plate recovered transformants onto Lemco agar (for *M. smegmatis*) or 7H10/OADC (for *M. tuberculosis*) supplemented with 20 µg/mL kanamycin (for pFLAME-based constructs) or 20 µg/mL streptomycin (for pSM128-based constructs).
8. Isolate at least three individual transformants for each plasmid transformed (*see* **Note 5**). Streak onto solid media containing 20 µg/mL kanamycin (for pFLAME-based constructs) or 20 µg/mL streptomycin (for pSM128-based constructs).
9. Confirm transformants harbor the correct vector by isolating the plasmid and confirming its integrity by restriction digests, PCR, and/or sequencing (*see* **Note 6**).
10. Inoculate 10 mL starter cultures of Lemco broth (for *M. smegmatis*) or 7H9-Tw (for *M. tuberculosis*) containing 20 µg/mL kanamycin (for pFLAME-based constructs) or 20 µg/mL streptomycin (for pSM128-based constructs) with a loop of cells and incubate at 37°C with shaking at ~100 rpm until confluent culture has grown (typically overnight for *M. smegmatis*, 1 week for *M. tuberculosis*).
11. Use 1 mL of the starter cultures from step 10 to inoculate new 10 mL cultures. Promoter activity can be induced or repressed by different growth conditions, such as supplementing the media with different nitrogen and carbon sources or other substances. In addition, the effects of oxygen concentration can be assayed (*see* **Note 7**).

18.3.2 Recovery of Transformants Harboring Reporter Constructs

18.3.2.1 Preparation of Cell-Free Extracts from Colonies Growing on Plates

1. Add 1 mL 10 mM Tris, pH 8, to a Lysing Matrix B tube (Q-BIOGENE).
2. Scrape mycobacterial growth from the plate using a loop and transfer into the Lysing Matrix B tube.
3. Tightly fasten the top and transfer to a FastPrep (Q-BIOGENE) (*see* **Note 8**).
4. Lyse cells at speed 6.0 for 30 s (*see* **Note 9**).
5. Transfer tubes to a microcentrifuge and spin at 16,000 × *g* for 5 min.
6. Remove the cell-free extract to a sterile 1.5-mL tube (*see* **Notes 9** and **10**).

18.3.2.2 Preparation of Cell-Free Extracts from Liquid Cultures

1. Centrifuge liquid cultures at 2500 × *g* to 3000 × *g* for 10 min at room temperature to pellet bacteria.
2. Wash pellet with the original culture volume of 10 mM Tris, pH 8.

3. Pellet bacteria as in step 1.
4. Resuspend cell pellet in 1 mL 10 mM Tris, pH 8.
5. Transfer the cells to a lysing Matrix B tube (Q-BIOGENE).
6. Tightly fasten the top and transfer to a FastPrep (Q-BIOGENE).
7. Lyse cells at speed 6.0 for 30 s.
8. Transfer tubes to a microcentrifuge and spin at $16,060 \times g$ for 5 min.
9. Remove the cell-free extract to a sterile 1.5-mL tube (see **Notes 9** and **10**).

18.3.3 Assaying Reporter Expression

18.3.3.1 Assaying GFP Expression in Live Cells (see Note 11)

1. Centrifuge liquid cultures at $2500 \times g$ to $3000 \times g$ for 10 min to pellet bacteria.
2. Wash pellet with the original culture volume of 10 mM Tris, pH 8.
3. Resuspend cells in 3 mL 10 mM Tris (pH 8) and transfer to a 12-mm screw-cap glass tube.
4. Read optical density at 580 nm.
5. Read fluorescence using spectrofluorometer, excitation at 490 nm and emission at 510 nm.
6. Calculate relative activity by dividing fluorescence by the OD_{580}.

18.3.3.2 Assaying GFP Fluorescence in Cell-Free Extracts

1. Determine the protein concentration of the cell-free extracts from Section 18.3.2 (see **Note 12**).
2. Add 100 µL cell-free extract to a glass-bottom polystyrene microtiter plate (see **Note 13**).
3. Measure fluorescence using a fluorometric plate reader; for example, FLUOstar OPTIMA with excitation at 490 nm and emission at 510 nm.
4. Raw data can be normalized by deleting the background reading (e.g., 10 mM Tris) from the sample data.
5. Divide the fluorescent readings by the protein concentration to give the relative fluorescence per milligram of total protein.

18.3.3.3 β-Galactosidase Activity Assays and Analysis

1. Determine the protein concentration in the cell-free extracts from Section 18.3.2 (see **Note 12**).
2. Add 900 µL Z-buffer to a Falcon 2054 polystyrene tube. In addition, a negative control containing 900 µL Z-buffer and 100 µL sterile distilled water should be included. This should be subject to the same reaction conditions as the other samples.

3. Add 100 μL cell-free extracts from Section 18.3.2 (*see* **Note 14**).
4. Incubate tubes for 5 min at 37°C (*see* **Note 15**).
5. Add 200 μL ONPG solution to start the reaction.
6. Incubate for up to 90 min at 37°C, check for color development periodically (*see* **Note 16**).
7. Add 500 μL 1 M NaHCO₃ to stop the reaction (*see* **Note 17**).
8. Transfer reaction mixture to a cuvette.
9. Measure the optical density at 420 nm. The spectrometer should be blanked using Z-buffer. The negative control should then be measured and this reading subtracted from the sample readings to give an accurate measurement
10. To determine the β-galactosidase activity, the following formula is used:

$$(OD_{420} \times 1.7)/(t \times v \times p \times 0.0045)$$

where t = reaction time (min), v = volume of cell-free extract (mL), and p = total protein concentration (mg/mL) [48, 49]. Activity is expressed in Miller units.

18.4 Notes

1. The size of the fragment, and the location of the promoter region within the fragment, will have an effect on reporter expression. It is advised that the promoter should be located as close as possible to the 3′ end of the fragment. Fragments between 300 and 400 bp have been shown to work adequately. In addition, when amplifying the target region, it is recommended to use a proofreading polymerase and to sequence the amplified region.
2. We routinely use pSC-A (Stratagene), but many other cloning systems are available.
3. Plasmids such as pFLAME [23], a replicating vector, and pSM128 [47], an integrating vector, are sequenced and have unique restriction enzyme sites for cloning. We routinely clone fragments directionally into the pFlame vectors in the *Eco*RI-*Bam*HI or *Eco*RI-*Kpn*I sites. Fragments can be cloned into pSM128 via using the blunt-ended *Sca*I site. Orientation should be determined by either sequencing or restriction enzyme digestion.
4. The reporter vector without the insert should also be electroporated into the mycobacteria to act as a negative control.
5. It is recommended to assay at least three independent transformants. If the transformants are assayed in duplicate, then statistical analysis, such as the Student's *t*-test, can be performed.
6. Mycobacteria, especially *M. tuberculosis*, have been shown to rearrange transformed DNA [50, 51, 52, 53, 54]. Therefore, plasmids should be recovered and tested by using restriction enzyme digests or sequencing. Low yields of plasmid DNA are recovered from mycobacterial plasmid preparations, therefore recovered plasmids can be transformed into *E. coli*. Integrated vectors, such as pSM128, can be removed from the chromosome by electroporating with a plasmid that expresses excisonase or with an empty integrating vector, recovering the cells as per normal and preparing plasmid DNA from the transformation mix. Integration can also be assessed by Southern hybridization.
7. If the strain is going to be grown under different conditions than from the initial culture, a washing step with 10 mM Tris-HCl, pH 8, is advised to reduce the carryover from the culture. If LacZ is used as the reporter, solid media can be supplemented with X-gal and expression can be visualized.

8. Other lysing methodologies can be used, such as using the Mini Beadbeater (Biospec Products). Samples should be processed twice and stored on ice for at least 1 min. The Mini Beadbeater machine allows only one sample to be processed per run, in comparison with the 12 that can be processed by the FastPrep FP120 (Q-BIOGENE) simultaneously.

9. If using pathogenic mycobacteria, such as *M. tuberculosis,* then filter the cell-free extract through an 0.2-µm filter unit before removing from the containment laboratory.

10. To reduce protease activity, samples should be transferred to sterile containers and stored at –20°C until use. The extract can be further stabilized by using protease inhibitors.

11. An advantage of measuring GFP expression from live bacteria is that activity can be followed over a period of times without lysing the cells.

12. Protein concentration can be assayed using a variety of different methodologies. We typically use the BCA kit from Pierce, Rockford, USA.

13. The solution used to generate the cell lysates (i.e., 10 mM Tris) should be added to a separate well to act as a blank.

14. Smaller volumes or diluted extracts can be assayed, although 100 µL is the optimal volume of cell-free extract.

15. The reactions should be incubated in a water bath at 37°C for efficient heat transfer.

16. The color change is from colorless to yellow; decrease reaction time if strong yellow color is achieved too quickly, as the optical reading from the spectrometer will be saturated. If promoter activity is very high, the sample should be diluted 1/10 or 1/100 and the assay repeated.

17. The reaction mixture, once stopped, is stable for up to 2 h.

References

1. Sherman, D. R., Voskuil, M., Schnappinger, D., Liao, R., Harrell, M. I. & Schoolnik, G. K. (2001). Regulation of the *Mycobacterium tuberculosis* hypoxic response gene encoding alpha -crystallin. *Proc Natl Acad Sci U S A.* 98, 7534–9.

2. Betts, J. C., Lukey, P. T., Robb, L. C., McAdam, R. A. & Duncan, K. (2002). Evaluation of a nutrient starvation model of *Mycobacterium tuberculosis* persistence by gene and protein expression profiling. *Mol Microbiol.* 43, 717–31.

3. Fisher, M. A., Plikaytis, B. B. & Shinnick, T. M. (2002). Microarray analysis of the *Mycobacterium tuberculosis* transcriptional response to the acidic conditions found in phagosomes. *J Bacteriol.* 184, 4025–32.

4. Talaat, A. M., Lyons, R., Howard, S. T. & Johnston, S. A. (2004). The temporal expression profile of *Mycobacterium tuberculosis* infection in mice. *Proc Natl Acad Sci U S A.* 101, 4602–7.

5. Labidi, A., David, H. L. & Roulland-Dussoix, D. (1985). Restriction endonuclease mapping and cloning of *Mycobacterium fortuitum* var. fortuitum plasmid pAL5000. *Ann Inst Pasteur Microbiol.* 136B, 209–15.

6. Snapper, S. B., Melton, R. E., Mustafa, S., Kieser, T. & Jacobs, W. R. Jr. (1990). Isolation and characterization of efficient plasmid transformation mutants of *Mycobacterium smegmatis. Mol Microbiol.* 4, 1911–9.

7. Hinshelwood, S. & Stoker, N. G. (1992). An *Escherichia coli-Mycobacterium* shuttle cosmid vector, pMSC1. *Gene.* 110, 115–8.

8. Guilhot, C., Gicquel, B. & Martin, C. (1992). Temperature-sensitive mutants of the Mycobacterium plasmid pAL5000. *FEMS Microbiol Lett.* 77, 181–6.

9. Gicquel-Sanzey, B., Moniz-Pereira, J., Gheorghiu, M. & Rauzier, J. (1989). Structure of pAL5000, a plasmid from *M. fortuitum* and its utilization in transformation of mycobacteria. *Acta Leprol.* 7, 208–11.

10. Coons, A. H. & Kaplan, M. H. (1950). Localization of antigen in tissue cells; improvements in a method for the detection of antigen by means of fluorescent antibody. *J Exp Med.* 91, 1–13.

11. Chalfie, M., Tu, Y., Euskirchen, G., Ward, W. W. & Prasher, D. C. (1994). Green fluorescent protein as a marker for gene expression. *Science.* 263, 802–5.

12. Atkins, D. & Izant, J. G. (1995). Expression and analysis of the green fluorescent protein gene in the fission yeast *Schizosaccharomyces pombe. Curr Genet.* 28, 585–8.

13. Lim, C. R., Kimata, Y., Oka, M., Nomaguchi, K. & Kohno, K. (1995). Thermosensitivity of green fluorescent protein fluorescence utilized to reveal novel nuclear-like compartments in a mutant nucleoporin NSP1. *J Biochem (Tokyo).* 118, 13–7.

14. Yeh, E., Gustafson, K. & Boulianne, G. L. (1995). Green fluorescent protein as a vital marker and reporter of gene expression in Drosophila. *Proc Natl Acad Sci U S A.* 92, 7036–40.

15. Amsterdam, A., Lin, S. & Hopkins, N. (1995). The *Aequorea victoria* green fluorescent protein can be used as a reporter in live zebrafish embryos. *Dev Biol.* 171, 123–9.

16. Rizzuto, R., Brini, M., Pizzo, P., Murgia, M. & Pozzan, T. (1995). Chimeric green fluorescent protein as a tool for visualizing subcellular organelles in living cells. *Curr Biol.* 5, 635–42.

17. Ikawa, M., Kominami, K., Yoshimura, Y., Tanaka, K., Nishimune, Y. & Okabe, M. (1995). A rapid and non-invasive selection of transgenic embryos before implantation using green fluorescent protein (GFP). *FEBS Lett.* 375, 125–8.

18. Hansen, M. C., Palmer, R. J. Jr., Udsen, C., White, D. C. & Molin, S. (2001). Assessment of GFP fluorescence in cells of *Streptococcus gordonii* under conditions of low pH and low oxygen concentration. *Microbiology.* 147, 1383–91.

19. Reichel, C., Mathur, J., Eckes, P., Langenkemper, K., Koncz, C., Schell, J., Reiss, B. & Maas, C. (1996). Enhanced green fluorescence by the expression of an *Aequorea victoria* green fluorescent protein mutant in mono- and dicotyledonous plant cells. *Proc Natl Acad Sci U S A.* 93, 5888–93.

20. Daabrowski, S., Brillowska, A. & Kur, J. (1999). Use of the green fluorescent protein variant (YFP) to monitor MetArg human proinsulin production in *Escherichia coli. Protein Expr Purif.* 16, 315–23.

21. Matz, M. V., Fradkov, A. F., Labas, Y. A., Savitsky, A. P., Zaraisky, A. G., Markelov, M. L. & Lukyanov, S. A. (1999). Fluorescent proteins from nonbioluminescent *Anthozoa* species. *Nat Biotechnol.* 17, 969–73.

22. Triccas, J. A., Pinto, R. & Britton, W. J. (2002). Destabilized green fluorescent protein for monitoring transient changes in mycobacterial gene expression. *Res Microbiol.* 153, 379–83.

23. Blokpoel, M. C., O'Toole, R., Smeulders, M. J. & Williams, H. D. (2003). Development and application of unstable GFP variants to kinetic studies of mycobacterial gene expression. *J Microbiol Methods.* 54, 203–11.

24. Dulebohn, D. P., Cho, H. J. & Karzai, A. W. (2006). Role of conserved surface amino acids in binding of SmpB protein to SsrA RNA. *J Biol Chem.* 281, 28536–45.

25. Jain, V., Sujatha, S., Ojha, A. K. & Chatterji, D. (2005). Identification and characterization of *rel* promoter element of *Mycobacterium tuberculosis. Gene* 351, 149–57.

26. Haydel, S. E., Benjamin, W. H., Jr., Dunlap, N. E. & Clark-Curtiss, J. E. (2002). Expression, autoregulation, and DNA binding properties of the *Mycobacterium tuberculosis* TrcR response regulator. *J Bacteriol.* 184, 2192–203.

27. Parish, T., Turner, J. & Stoker, N. G. (2001). amiA is a negative regulator of acetamidase expression in *Mycobacterium smegmatis. BMC Microbiol.* 1, 19.

28. Kumar, D., Srivastava, B. S. & Srivastava, R. (1998). Genetic rearrangements leading to disruption of heterologous gene expression in mycobacteria: an observation with *Escherichia coli* beta-galactosidase in *Mycobacterium smegmatis* and its implication in vaccine development. *Vaccine* 16, 1212–5.

29. Srivastava, R., Kumar, D., Subramaniam, P. & Srivastava, B. S. (1997). beta-Galactosidase reporter system in mycobacteria and its application in rapid antimycobacterial drug screening. *Biochem Biophys Res Commun.* 235, 602–5.

30. Dhandayuthapani, S., Via, L. E., Thomas, C. A., Horowitz, P. M., Deretic, D. & Deretic, V. (1995). Green fluorescent protein as a marker for gene expression and cell biology of mycobacterial interactions with macrophages. *Mol Microbiol.* 17, 901–12.

31. Via, L. E., Curcic, R., Mudd, M. H., Dhandayuthapani, S., Ulmer, R. J. & Deretic, V. (1996). Elements of signal transduction in *Mycobacterium tuberculosis*: in vitro phosphorylation and *in vivo* expression of the response regulator MtrA. *J Bacteriol.* 178, 3314–21.

32. Zahrt, T. C. & Deretic, V. (2000). An essential two-component signal transduction system in *Mycobacterium tuberculosis*. *J Bacteriol.* 182, 3832–8.

33. O'Toole, R., Smeulders, M. J., Blokpoel, M. C., Kay, E. J., Lougheed, K. & Williams, H. D. (2003). A two-component regulator of universal stress protein expression and adaptation to oxygen starvation in *Mycobacterium smegmatis*. *J Bacteriol.* 185, 1543–54.

34. Roberts, E. A., Clark, A., McBeth, S. & Friedman, R. L. (2004). Molecular characterization of the *eis* promoter of *Mycobacterium tuberculosis*. *J Bacteriol.* 186, 5410–7.

35. Danelishvili, L., Poort, M. J. & Bermudez, L. E. (2004). Identification of *Mycobacterium avium* genes up-regulated in cultured macrophages and in mice. *FEMS Microbiol Lett.* 239, 41–9.

36. Cowley, S. C. & Av-Gay, Y. (2001). Monitoring promoter activity and protein localization in *Mycobacterium* spp. using green fluorescent protein. *Gene.* 264, 225–31.

37. Barletta, R. G., Snapper, B., Cirillo, J. D., Connell, N. D., Kim, D. D., Jacobs, W. R. & Bloom, B. R. (1990). Recombinant BCG as a candidate oral vaccine vector. *Res Microbiol.* 141, 931–9.

38. Timm, J., Lim, E. M. & Gicquel, B. (1994). *Escherichia coli-mycobacteria* shuttle vectors for operon and gene fusions to *lacZ*: the pJEM series. *J Bacteriol.* 176, 6749–53.

39. Timm, J., Perilli, M. G., Duez, C., Trias, J., Orefici, G., Fattorini, L., Amicosante, G., Oratore, A., Joris, B., Frere, J. M., et al. (1994). Transcription and expression analysis, using *lacZ* and *phoA* gene fusions, of *Mycobacterium fortuitum* beta-lactamase genes cloned from a natural isolate and a high-level beta-lactamase producer. *Mol Microbiol.* 12, 491–504.

40. Dellagostin, O. A., Esposito, G., Eales, L. J., Dale, J. W. & McFadden, J. (1995). Activity of mycobacterial promoters during intracellular and extracellular growth. *Microbiology.* 141, 1785–92.

41. Ainsa, J. A., Martin, C., Cabeza, M., De la Cruz, F. & Mendiola, M. V. (1996). Construction of a family of *Mycobacterium/Escherichia coli* shuttle vectors derived from pAL5000 and pACYC184: their use for cloning an antibiotic-resistance gene from *Mycobacterium fortuitum*. *Gene.* 176, 23–6.

42. Hotter, G. S., Wilson, T. & Collins, D. M. (2001). Identification of a cadmium-induced gene in Mycobacterium bovis and *Mycobacterium tuberculosis*. *FEMS Microbiol Lett.* 200, 151–5.

43. Curcic, R., Dhandayuthapani, S. & Deretic, V. (1994). Gene expression in mycobacteria: transcriptional fusions based on *xylE* and analysis of the promoter region of the response regulator *mtrA* from *Mycobacterium tuberculosis*. *Mol Microbiol.* 13, 1057–64.

44. Das Gupta, S. K., Bashyam, M. D. & Tyagi, A. K. (1993). Cloning and assessment of mycobacterial promoters by using a plasmid shuttle vector. *J Bacteriol.* 175, 5186–92.

45. Sarkis, G. J., Jacobs, W. R., Jr. & Hatfull, G. F. (1995). L5 luciferase reporter mycobacteriophages: a sensitive tool for the detection and assay of live mycobacteria. *Mol Microbiol.* 15, 1055–67.

46. Blokpoel, M. C., Murphy, H. N., O'Toole, R., Wiles, S., Runn, E. S., Stewart, G. R., Young, D. B. & Robertson, B. D. (2005). Tetracycline-inducible gene regulation in mycobacteria. *Nucleic Acids Res.* 33, e22.

47. Dussurget, O., Timm, J., Gomez, M., Gold, B., Yu, S. W., Sabol, S. Z., Holmes, R. K., Jacobs, W. R. & Smith, I. (1999). Transcriptional control of the iron-responsive *fxbA* gene by the mycobacterial regulator IdeR. *J Bacteriol.* 181, 3402–3408.

48. Miller, J. H. (1972). *Experiments in Molecular Genetics*. Cold Spring Harbor Laboratory Press, Cold Spring Harbor, NY.

49. Miller, J. H. (1972). Assay of beta-galactosidase activity. In *Experiments in Molecular Genetics*. Cold Spring Harbor Laboratory Press, Cold Spring Harbor, NY.

50. Brown, A. C. & Parish, T. (2006). Instability of the acetamide-inducible expression vector pJAM2 in *Mycobacterium tuberculosis*. *Plasmid*. 55, 81–6.

51. Chawla, M. & Das Gupta, S. K. (1999). Transposition-induced structural instability of *Escherichia coli-mycobacteria* shuttle vectors. *Plasmid*. 41, 135–40.

52. Medeiros, M. A., Dellagostin, O. A., Armoa, G. R., Degrave, W. M., De Mendonca-Lima, L., Lopes, M. Q., Costa, J. F., McFadden, J. & McIntosh, D. (2002). Comparative evaluation of *Mycobacterium vaccae* as a surrogate cloning host for use in the study of mycobacterial genetics. *Microbiology*. 148, 1999–2009.

53. Springer, B., Sander, P., Sedlacek, L., Ellrott, K. & Bottger, E. C. (2001). Instability and site-specific excision of integration-proficient mycobacteriophage L5 plasmids: development of stably maintained integrative vectors. *Int J Med Microbiol*. 290, 669–75.

54. Chan Kwo Chion, C. K., Askew, S. E. & Leak, D. J. (2005). Cloning, expression, and site-directed mutagenesis of the propene monooxygenase genes from *Mycobacterium* sp. strain M156. *Appl Environ Microbiol*. 71, 1909–14.

Chapter 19
Construction of Unmarked Deletion Mutants in Mycobacteria

Houhui Song, Frank Wolschendorf and Michael Niederweis

Abstract Site-specific recombinases such as the *Saccharomyces cerevisiae* Flp and the P1 phage Cre proteins have been increasingly used for the construction of unmarked deletions in bacteria. Both systems consist of an antibiotic resistance gene flanked by recognition sites in direct orientation and a curable plasmid for temporary expression of the respective recombinase gene. In this chapter, we describe strategies and methods of how to use sequence-specific recombination mediated by Flp and Cre to construct mutants of *Mycobacterium smegmatis*, *Mycobacterium bovis* BCG, and *Mycobacterium tuberculosis*.

Keywords allelic exchange · Cre/loxP · Flp/FRT · homologous recombination · sequence-specific recombination

19.1 Introduction

19.1.1 Gene Deletions in Mycobacteria

To understand mycobacterial pathogenesis at the molecular level, efficient and specific genetic systems for recombination, mutagenesis, and complementation are required. In particular, the ability to construct mutants by allelic exchange is imperative to characterize the function of a particular gene. Because homologous recombination is a rare event in mycobacteria compared with that in other bacteria, efficient systems for delivery of template DNA are necessary. Considerable progress in constructing allelic exchange mutants in mycobacteria has been achieved using conditionally replicating temperature sensitive plasmids [1] or specialized transducing mycobacteriophages [2]. These initial systems were improved by the use of multiple markers to increase the counterselection efficiency and by reporter genes to indicate the presence or absence of the allelic exchange cassette [3].

M. Niederweis
Department of Microbiology, University of Alabama at Birmingham, 609 Bevill
Biomedical Research Building, 845 19th Street South, Birmingham, AL 35294, U.S.A.
e-mail: mnieder@uab.edu

T. Parish, A.C. Brown (eds.), *Mycobacteria Protocols*,
doi: 10.1007/978-1-59745-207-6_19, © Humana Press, Totowa, NJ 2008

Another problem arises when the aim is to analyze the functions of redundant genes because only a few resistance genes are functional in mycobacteria [4]. Because of their superior efficiency, the *hyg* gene from *Streptomyces hygroscopicus* and the *aph* gene are used for almost all knock-out experiments in mycobacteria [5]. Two strategies are available to re-use these resistance markers. One is based on two consecutive allelic exchange reactions. This is a tedious work both for construction and analysis of the mutants. Sequence-specific recombination provides a faster and more efficient strategy. Several site-specific recombination systems are used in *Escherichia coli*. The most frequently used system is the Flp/*FRT* system from the 2-μm plasmid of *Saccharomyces cerevisiae* [6]. In addition, the Cre/*loxP* system of the bacteriophage P1 [7], the TnpR/*res* system of the γδ transposon [8], and the ParA/*res* system of the broad-host-range plasmid RP4 [9] are known. Thus far, the γδ resolvase and the Flp recombinase have been shown to be functional in mycobacteria [4, 10]. In this chapter, we focus on the use of the sequence-specific recombinases Flp and Cre to construct unmarked mutants in *Mycobacterium smegmatis*, *Mycobacterium bovis* BCG, and *Mycobacterium tuberculosis*.

19.1.2 General Strategy

In order to construct a deletion mutant with as little site effects as possible, the gene of interest and its chromosomal location need to be analyzed using bioinformatic methods. In particular, the following considerations are important: First, the influence of the deletion on the expression of flanking genes should be minimized. To this end, putative promoters, transcription terminators, and inverted repeats as potential regulatory regions should be left intact. Second, the length of the deletion is important. Obviously, the shorter the deleted region, the less likely are polar effects on expression of downstream genes. However, short deletions increase the likelihood of expression of partially functional proteins by the mutant strain or of unwanted recombination events with similar genes in the same strain either on the chromosome or on plasmids in case of complementation experiments.

19.1.3 Construction of Unmarked Deletions in M. smegmatis

The general strategy to create unmarked deletions in mycobacteria is depicted in Fig. 19.1. For *M. smegmatis*, we used this strategy to delete the four *msp* porin genes [11]. To this end, we established the Flp/*FRT* system consisting of the basic allelic exchange vector pMN252, which carries the hygromycin resistant cassette flanked by two *FRT* sites in direct orientation and the Flp expression vector pMN234 as shown in Fig. 19.2. The individual steps of this procedure are described using the deletion of the *mspC* gene as an example. Other genes can be deleted in a similar manner. Here, we used the *rpsL* gene for

Fig. 19.1 Strategy to create unmarked deletions in mycobacteria. SCO, single crossover; DCO, double crossove

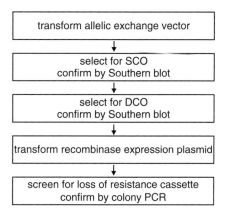

| transform allelic exchange vector |
| select for SCO confirm by Southern blot |
| select for DCO confirm by Southern blot |
| transform recombinase expression plasmid |
| screen for loss of resistance cassette confirm by colony PCR |

counterselection against the plasmids carrying the allelic exchange substrate and the *flp* expression cassette. Therefore, we first chose the *rpsL* mutant strain *M. smegmatis* SMR5, a streptomycin-resistant derivative of *M. smegmatis* mc^{2}155, as the parent strain for all deletion experiments. Then, we constructed the plasmid pMN249, which carries a *FRT-hyg-FRT* cassette with the two *FRT* sites (5′**GAAGTTCCTATAC**TTTCTAGA**GAATAGGAACTTC**3′) as direct repeats, about 1000-bp fragments of upstream and downstream regions of the *mspC* gene, the thermosensitive origin of replication (oriM$_{ts}$), and the *rpsL* gene as a counterselectable marker. The plasmid pMN249 was transformed into *M. smegmatis* SMR5. Single crossover (SCO) clones were selected after growing

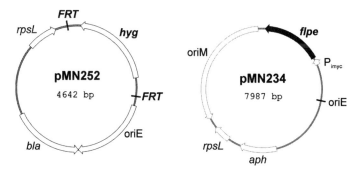

Fig. 19.2 Plasmids used for application of the Flp recombinase in *M. smegmatis* [4]. The plasmid pMN252 is a basic suicide vector for allelic exchange and contains the hygromycin resistance gene *hyg* flanked by *FRT* sites for excision by the Flp recombinase. Upstream and downstream fragments of the gene of interest should be inserted to flank *hyg* gene. Insertion of a temperature-sensitive pAL5000 origin [13] for replication in mycobacteria will yield a replicating version of this allelic exchange vector. The plasmid pMN234 carries the expression cassette for the Flp recombinase gene (*flpe*). The *rpsL* gene is a counterselectable marker encoding the streptomycin sensitivity. oriE, origin for replication in *E. coli*; oriM, the pAL5000 origin for replication in mycobacteria; *aph*, kanamycin resistance gene

of *M. smegmatis* at 39°C to 41°C to stop the replication of the plasmid and identified by Southern blot hybridization of chromosomal DNA. Then, double crossover (DCO) clones were selected on streptomycin plates and identified by Southern blot hybridization of chromosomal DNA. Competent cells of the DCO clone were transformed with the Flp expression plasmid pMN234 (Fig. 19.2), and the *FRT-hyg-FRT* gene was excised from the DCO clone. The plasmid pMN234 was removed by selection on streptomycin plates. The resulting clones were streaked on plates with and without hygromycin to confirm the loss of the *hyg* gene. A final Southern blot was performed to confirm the SCO, DCO, and the marker-free mutant.

19.1.4 *Construction of an Unmarked* ompATb *Deletion Mutant of* M. tuberculosis

The Flp system does not function in *M. bovis* BCG and *M. tuberculosis*, probably because *flp* exhibits a very low G+C content of only 38% compared with the average G+C content of >65% in mycobacteria [4]. The Cre recombinase can be used as an alternative sequence-specific recombinase in order to construct unmarked gene deletion mutants in *M. bovis* BCG and *M. tuberculosis*. The Cre/ *loxP* recombination system consists of the basic allelic exchange vector pML339, which carries the hygromycin-resistant cassette flanked by two *loxP* sites in direct orientation and the Cre expression vector pCreSacB1 as depicted in Fig. 19.3. To

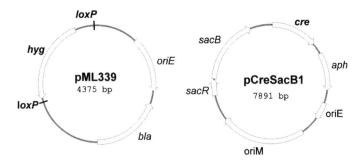

Fig. 19.3 Plasmids used for application of the Cre recombinase in mycobacteria. Plasmid pML339 is a basic suicide vector for allelic exchange and contains the hygromycin resistance gene (*hyg*) flanked by *loxP* sites for excision by the Cre recombinase. Upstream and downstream fragments of the gene of interest should be inserted to flank *hyg* gene. Insertion of a temperature-sensitive origin for replication in mycobacteria will yield a replicating version of this allelic exchange vector. The plasmid pCreSacB1 carries the expression cassette for the Cre recombinase gene (*flpe*). The other markers are *sacB* gene as a counterselectable marker for sucrose sensitivity; oriE, origin for replication in *E. coli*; oriM, the pAL5000 origin for replication in mycobacteria; *aph*, kanamycin resistance gene

demonstrate the usefulness of the Cre recombinase in slow-growing mycobacteria, the *ompATb* (*Rv0899*) genes of *M. tuberculosis* and *M. bovis* BCG were deleted. The *sacB* gene was chosen for counterselection against the plasmids carrying the allelic exchange substrate and the Cre expression cassette. This allowed the use of the wild-type *M. bovis* BCG and *M. tuberculosis* strains for the deletion experiments. Then, we constructed a plasmid that carried a *loxP-hyg-loxP* cassette with the two *loxP* sites (5′ ATAACTTCGTATAATGTATGCTATACGAAGTTAT 3′) as direct repeats, about 1000-bp fragments of upstream and downstream regions of the *ompATb* gene, the thermosensitive origin of replication (oriM$_{ts}$), and the *sacB* gene as a counterselectable marker. The plasmid pML563 was transformed into *M. bovis* BCG and *M. tuberculosis*. SCO clones were selected after growth of the bacteria at 39°C for *M. bovis* BCG or 41°C for *M. tuberculosis* (*see* **Note 1**) to stop the replication of the plasmid and identified by Southern blot hybridization of chromosomal DNA. Then, DCO clones were selected on sucrose plates and identified by Southern blot hybridization of chromosomal DNA. Competent cells of the DCO clones were transformed with the Cre expression plasmid pCreSacB1 (Fig. 19.3), and the *loxP-hyg-loxP* gene was excised from the DCO clone. The plasmid pCreSacB1 was removed by counterselection on sucrose plates. The resulting clones were streaked on plates with and without hygromycin to confirm the loss of the *hyg* gene. Final Southern blots were performed to confirm the SCOs, DCOs, and the marker-free mutants.

19.2 Materials

19.2.1 Growth and Preparation of M. smegmatis Competent Cells

1. *M. smegmatis* SMR5 strain (a streptomycin-resistant derivative of *M. smegmatis* mc²155).
2. 7H10 plates: Dissolve 19 g Middlebrook 7H10 medium (Difco, USA) in 1 L deionized water. Add 8.4 mL 60% glycerol (v/v). Autoclave and pour in standard plastic Petri dishes.
3. 20% (v/v) Tween 80: mix with distilled water, filter sterilize.
4. 7H9-Tw medium: Dissolve 4.7 g broth base in 1 L deionized water. Add 3.3 mL 60% glycerol (v/v), mix well, and autoclave for 10 min. Add 5 mL 20% (v/v) Tween 80 before use (final concentration of 0.05% v/v) (*see* **Note 2**).
5. Test tubes for preculture and 1-L flasks (Corning Incorporated, USA) for large culture.
6. Wash buffer: 10% (v/v) glycerol, 0.05% (v/v) Tween 80, autoclave for 20 min.
7. Beckman Coulter (Allegra X-12R, USA) Centrifuge (3000 × g at 4°C).
8. 500-mL centrifuge tubes (Beckman) and 50-mL centrifuge tubes (Greiner, USA).
9. Prechilled 1.5-mL tubes (Denville, USA).

19.2.2 Growth and Preparation of M. bovis BCG and M. tuberculosis Competent Cells

1. *M. bovis* BCG (Pasteur strain, ATCC 35739) and *M. tuberculosis* H37Rv.
2. 7H10-OADC plates: Dissolve 19 g Middlebrook 7H10 medium (Difco) in 900 mL deionized water. Add 8.4 mL 60% glycerol (v/v). Autoclave. When media is cooled to ~50°C, add 100 mL oleic albumin dextrose catalase (OADC) enrichment prewarmed to 50°C and wait until medium becomes hand-warm. Mix well, add antibiotics if needed, and pour in standard plastic Petri dishes.
3. 20% (v/v) Tween 80: mix with distilled water, filter sterilize.
4. 7H9-OADC-Tw medium: Dissolve 4.7 g Middlebrook 7H9 broth base (Difco) in 900 mL deionized water and add 3.3 mL 60% glycerol, mix well, and autoclave. Add 5 mL 20% (v/v) Tween 80 before use (final concentration of 0.05% v/v) (*see* **Note 2**). Add 100 mL OADC immediately before use.
5. 50-mL (25 cm^2) tissue culture flasks (Greiner).
6. 490-cm^2 roller bottle (Corning Incorporated).
7. Wash buffer: 10% (v/v) glycerol, 0.05% (v/v) Tween 80, autoclave for 20 min.
8. Beckman Coulter (Allegra X-12R) Centrifuge (3000 × g at 4°C).
9. 50-mL centrifuge tubes (Greiner).
10. Prechilled 1.5-mL tubes (Denville).

19.2.3 Electroporation of Mycobacteria

1. Prechilled 1.5-mL tubes (Denville).
2. Plasmid DNA (100 to 500 ng in a volume not exceeding 10 μL).
3. Electroporation cuvettes with a 4-mm gap.
4. Electroporation equipment capable of producing a pulse of 2.5 kV, 1000 Ω, 25 μF.
5. 7H9-Hyg-Kan-Tw for *M. smegmatis:* Dissolve 4.7 g Middlebrook 7H9 broth base (Difco) in 1 L deionized water. Add 3.3 mL 60% glycerol (v/v), mix well, and autoclave for 10 min. Add 5 mL 20% (v/v) Tween 80 before use (final concentration of 0.05% v/v) (*see* **Note 2**). Add 50 μg/mL hygromycin and 30 μg/mL kanamycin.
6. 7H9-OADC-Hyg-Kan-Tw for *M. bovis* BCG or *M. tuberculosis:* Dissolve 4.7 g Middlebrook 7H9 broth base (Difco) in 900 mL deionized water and add 3.3 mL 60% glycerol, mix well, and autoclave. Add 5 mL 20% (v/v) Tween 80 before use (final concentration of 0.05% v/v) (*see* **Note 2**). Add 100 mL OADC and 50 μg/mL hygromycin and 30 μg/mL kanamycin immediately before use.

7. 7H10-Hyg-Kan plates for *M. smegmatis:* Dissolve 19 g Middlebrook 7H10 medium (Difco) in 1 L deionized water. Add 8.4 mL 60% glycerol (v/v). Autoclave. When cooled, add 50 μg/mL hygromycin and 30 μg/mL kanamycin and pour in standard plastic Petri dishes.

8. 7H10-OADC-Hyg-Kan plates for *M. bovis* BCG or *M. tuberculosis*: Dissolve 19 g Middlebrook 7H10 medium (Difco) in 900 mL deionized water. Add 8.4 mL 60% glycerol (v/v). Autoclave. When media is cooled to ~50°C, add 100 mL of oleic albumin dextrose catalase (OADC) enrichment prewarmed to 50°C and wait until medium becomes hand-warm. Mix well, add 50 μg/mL hygromycin and 30 μg/mL kanamycin immediately before use and pour in standard plastic Petri dishes.

19.2.4 Selection of SCO Clones in M. smegmatis

7H10-Hyg-Kan plates: Dissolve 19 g Middlebrook 7H10 medium (Difco) in 1 L deionized water. Add 8.4 mL 60% glycerol (v/v). Autoclave. When cooled, add 50 μg/mL hygromycin and 30 μg/mL kanamycin and pour in standard plastic Petri dishes.

1. 7H9-Hyg-Kan-Tw: Dissolve 4.7 g Middlebrook 7H9 broth base (Difco) in 1 L deionized water. Add 3.3 mL 60% glycerol (v/v), mix well, and autoclave for 10 min. Add 5 mL 20% (v/v) Tween 80 before use (final concentration of 0.05% v/v) (*see* **Note 2**). Add 50 μg/mL hygromycin and 30 μg/mL kanamycin.
2. Incubator (set at 32°C or 39°C).

19.2.5 Selection of DCO Clones in M. smegmatis

7H10-Hyg-Strep plates: Dissolve 19 g Middlebrook 7H10 medium (Difco) in 1 L deionized water. Add 8.4 mL 60% glycerol (v/v). Autoclave. When cooled, add 50 μg/mL hygromycin, 400 μg/mL streptomycin, and pour in standard plastic Petri dishes.

1. 7H9-Hyg-Kan-Tw: Dissolve 4.7 g Middlebrook 7H9 broth base (Difco) in 1 L deionized water. Add 3.3 mL 60% glycerol (v/v), mix well, and autoclave for 10 min. Add 5 mL 20% (v/v) Tween 80 before use (final concentration of 0.05% v/v) (*see* **Note 2**) and 50 μg/mL hygromycin and 30 μg/mL kanamycin.
2. 7H9-Hyg-Strep-Tw: Dissolve 4.7 g Middlebrook 7H9 broth base (Difco) in 1 L deionized water. Add 3.3 mL 60% glycerol (v/v), mix well, and autoclave for 10 min. Add 5 mL 20% (v/v) Tween 80 before use (final concentration of 0.05% v/v) (*see* **Note 2**). Add 50 μg/mL hygromycin and 400 μg/mL streptomycin.
3. Incubator (set at 39°C).

19.2.6 Excision of FRT-hyg-FRT Cassette in M. smegmatis Using the Flp Recombinase

1. Plasmid pMN234, available from our laboratory on request.
2. 7H10-Kan plates: Dissolve 19 g Middlebrook 7H10 medium (Difco) in 1 L deionized water. Add 8.4 mL 60% glycerol (v/v). Autoclave. When cooled, add 30 µg/mL kanamycin and pour in standard plastic Petri dishes.
3. 7H9-Kan-Tw: Dissolve 4.7 g Middlebrook 7H9 broth base (Difco) in 1 L deionized water. Add 3.3 mL 60% glycerol (v/v), mix well, and autoclave for 10 min. Add 5 mL 20% (v/v) Tween 80 before use (final concentration of 0.05% v/v) (*see* **Note 2**). Add 30 µg/mL kanamycin.
4. 250-mL Pyrex flasks (Corning Incorporated).
5. Prechilled 1.5-mL tubes.
6. Electroporation cuvettes with a 4-mm gap.
7. Electroporation equipment capable of producing a pulse of 2.5 kV, 1000 Ω, 25 µF.
8. Wash buffer: 10% (v/v) glycerol, 0.05% (v/v) Tween 80, autoclave for 20 min.
9. Beckman Coulter (Allegra X-12R) Centrifuge (3000 × *g* at 4°C).
10. 50-mL centrifuge tubes (Greiner).
11. Prechilled 1.5-mL tubes.
12. Incubator (set at 37°C).

19.2.7 Removal of the Plasmid pMN234 from M. smegmatis

1. 7H9-Kan-Tw: Dissolve 4.7 g Middlebrook 7H9 broth base (Difco) in 1 L deionized water. Add 3.3 mL 60% glycerol (v/v), mix well, and autoclave for 10 min. Add 5 mL 20% (v/v) Tween 80 before use (final concentration of 0.05% v/v) (*see* **Note 2**). Add 30 µg/mL kanamycin.
2. 7H10-Strep plates: Dissolve 19 g Middlebrook 7H10 medium (Difco) in 1 L deionized water. Add 8.4 mL 60% glycerol (v/v). Autoclave. When cooled, add 40 µg/mL streptomycin and pour in standard plastic Petri dishes.
3. 7H10-Hyg plates: Dissolve 19 g Middlebrook 7H10 medium (Difco) in 1 L deionized water. Add 8.4 mL 60% glycerol (v/v). Autoclave. When cooled, add 50 µg/mL hygromycin and pour in standard plastic Petri dishes.
4. 250-mL Pyrex flasks (Corning Incorporated).

19.2.8 Selection of SCO Clones in M. bovis BCG or M. tuberculosis

1. 7H9-OADC-Hyg-Kan-Tw: Dissolve 4.7 g Middlebrook 7H9 broth base (Difco) in 900 mL deionized water and add 3.3 mL 60% glycerol, mix well, and autoclave. Add 5 mL 20% (v/v) Tween 80 before use (final concentration of 0.05% v/v) (*see* **Note 2**). Add 100 mL OADC and 50 µg/mL hygromycin and 30 µg/mL kanamycin immediately before use.

2. 7H10-OADC-Hyg plates: Dissolve 19 g Middlebrook 7H10 medium (Difco) in 900 mL deionized water. Add 8.4 mL 60% glycerol (v/v). Autoclave. When media is cooled to ~50°C, add 100 mL oleic albumin dextrose catalase (OADC) enrichment prewarmed to 50°C and wait until medium becomes hand-warm. Mix well, add 50 µg/mL hygromycin immediately before use and pour in standard plastic Petri dishes.
3. Replicating plasmid (e.g., pML563) 100 to 500 ng in a volume not exceeding 10 µL.
4. Prechilled 1.5-mL tubes.
5. Electroporation cuvettes with a 4-mm gap.
6. Electroporation equipment capable of producing a pulse of 2.5 kV, 1000 Ω, 25 µF.
7. Wash buffer: 10% (v/v) glycerol, 0.05% (v/v) Tween 80, autoclave for 20 min.
8. Beckman Coulter (Allegra X-12R) Centrifuge (3000 × g at 4°C).
9. 50-mL centrifuge tubes (Greiner).
10. Prechilled 1.5-mL tubes.
11. Incubator (set at 34°C or 39°C for *M. bovis* BCG or 41°C for *M. tuberculosis*).

19.2.9 Selection of DCO Clones in M. bovis BCG or M. tuberculosis

1. 7H9-OADC-Hyg-Tw: Dissolve 4.7 g Middlebrook 7H9 broth base (Difco) in 900 mL deionized water and add 3.3 mL 60% glycerol, mix well, and autoclave. Add 5 mL 20% (v/v) Tween 80 before use (final concentration of 0.05% v/v) (*see* **Note 2**). Add 100 mL OADC and 50 µg/mL hygromycin and 30 µg/mL kanamycin immediately before use.
2. 7H10-OADC-Hyg-Suc plates: Dissolve 19 g Middlebrook 7H10 medium (Difco) in 900 mL deionized water. Add 8.4 mL 60% glycerol (v/v). Autoclave. When media is cooled to ~50°C, add 100 mL oleic albumin dextrose catalase (OADC) enrichment prewarmed to 50°C and wait until medium becomes hand-warm. Mix well, add 50 µg/mL hygromycin and 25 mL 50% sucrose (w/v) to give a final concentration of 2% immediately before use and pour in standard plastic Petri dishes.
3. Incubator (set at 39°C for *M. bovis* BCG or 41°C for *M. tuberculosis*).

19.2.10 Excision loxP-hyg-loxP Cassette in M. bovis BCG or M. tuberculosis Using the Cre Recombinase

1. 7H9-OADC-Kan-Tw: Dissolve 4.7 g Middlebrook 7H9 broth base (Difco) in 900 mL deionized water and add 3.3 mL 60% glycerol, mix well, and autoclave. Add 5 mL 20% (v/v) Tween 80 before use (final concentration of

0.05% v/v (*see* **Note 2**). Add 100 mL OADC and 30 μg/mL kanamycin immediately before use.

2. 7H10-OADC-Kan plates: Dissolve 19 g Middlebrook 7H10 medium (Difco) in 900 mL deionized water. Add 8.4 mL 60% glycerol (v/v). Autoclave. When media is cooled to ~50°C, add 100 mL oleic albumin dextrose catalase (OADC) enrichment prewarmed to 50°C and wait until medium becomes hand-warm. Mix well, add 50 μg/mL hygromycin immediately before use and pour in standard plastic Petri dishes.

3. pCreSacB1 (100 to 500 ng in a volume not exceeding 10 μL).

4. Prechilled 1.5-mL tubes.

5. Electroporation cuvettes with a 4-mm gap.

6. Electroporation equipment capable of producing a pulse of 2.5 kV, 1000 Ω, 25 μF.

7. Wash buffer: 10% (v/v) glycerol, 0.05% (v/v) Tween 80, autoclave for 20 min.

8. Beckman Coulter (Allegra X-12R) Centrifuge (3000 × g at 4°C).

9. 50-mL centrifuge tubes (Greiner).

10. Prechilled 1.5-mL tubes.

11. Incubator (set at 37°C).

19.2.11 Removal of Plasmid pCreSacB1 from M. bovis BCG or M. tuberculosis

1. 7H9-OADC-Tw: Dissolve 4.7 g Middlebrook 7H9 broth base (Difco) in 900 mL deionized water and add 3.3 mL 60% glycerol, mix well, and autoclave. Add 5 mL 20% (v/v) Tween 80 before use (final concentration of 0.05% v/v) (*see* **Note 2**). Add 100 mL OADC immediately before use.

2. 7H10-OADC-Suc plates: Dissolve 19 g Middlebrook 7H10 medium (Difco) in 900 mL deionized water. Add 8.4 mL 60% glycerol (v/v). Autoclave. When media is cooled to ~50°C, add 100 mL oleic albumin dextrose catalase (OADC) enrichment prewarmed to 50°C and wait until medium becomes hand-warm. Mix well and add 25 mL 50% sucrose (w/v) to give a final concentration of 2% immediately before use and pour in standard plastic Petri dishes.

3. 7H10-OADC-Hyg plates: Dissolve 19 g Middlebrook 7H10 medium (Difco) in 900 mL deionized water. Add 8.4 mL 60% glycerol (v/v). Autoclave. When media is cooled to ~50°C, add 100 mL oleic albumin dextrose catalase (OADC) enrichment prewarmed to 50°C and wait until medium becomes hand-warm. Mix well and add 50 μg/mL hygromycin immediately before use and pour in standard plastic Petri dishes.

4. 7H10-OADC plates: Dissolve 19 g Middlebrook 7H10 medium (Difco) in 900 mL deionized water. Add 8.4 mL 60% glycerol (v/v). Autoclave. When media is cooled to ~50°C, add 100 mL oleic albumin dextrose catalase (OADC) enrichment prewarmed to 50°C and wait until medium becomes hand-warm. Mix well and pour in standard plastic Petri dishes.

5. Incubator (set at 37°C).

19.2.12 Preparation of Chromosomal DNA from Mycobacteria

1. Chloroform/methanol solution (3:1, v:v).
2. Phenol/chloroform/isoamyl alcohol (25:24:1, v:v), pH 8.
3. GTC buffer: 4 M guanidium thiocyanate, 100 mM Tris pH 7.5, 0.5% sarcosyl (w/v), 1% β-mercaptoethanol.

19.2.13 Colony PCR

1. PuReTaq Ready-To-Go PCR beads (GE Healthcare, USA).
2. 100% DMSO (Sigma).

19.3 Methods

19.3.1 Growth and Preparation of M. smegmatis Competent Cells

1. Plate *M. smegmatis* from a frozen stock on 7H10 plates.
2. Incubate plate in a sealed bag at 37°C for 3 days.
3. Inoculate a preculture of 4 mL 7H9-Tw in a test tube with cells from the plate, vortex well, and grow overnight on a shaker at 37°C.
4. Inoculate 200 to 400 mL 7H9-Tw with 1 mL preculture from step 3.
5. Determine the initial OD_{600} of the culture in order to estimate the time when an OD_{600} of 0.5 to 0.8 will be reached (the double time of *M. smegmatis* is about 3 h).
6. Allow culture to grow at 37°C, 200 rpm.
7. When culture reaches an OD_{600} of 0.5 to 0.8, harvest cells by centrifugation at ~3000 × g for 10 min at 4°C.
8. Wash pellet with wash buffer at 1/10 of original culture volume.
9. Centrifuge suspension for 10 min at 3000 × g at 4°C.
10. Wash pellet with wash buffer at 1/20 of original volume.
11. Centrifuge suspension for 10 min at 3000 × g at 4°C.
12. Resuspend pellet in wash buffer at 1/100 of original culture volume.
13. Make 100 µL aliquots in prechilled 1.5-mL tubes.
14. Freeze at –80°C immediately.

19.3.2 Growth and Preparation of Competent Cells of M. bovis BCG and M. tuberculosis

1. Streak *M. bovis* BCG or *M. tuberculosis* from a frozen stock on 7H10-OADC plates.
2. Seal plates with Parafilm, foil, or put in a Ziplock bag and incubate at 37°C for up to 2 to 3 weeks (*see* **Note 3**).

3. Inoculate a preculture of 10 mL 7H9-OADC-Tw in a 30-mL cell culture flask with a disposable loop.
4. Incubate at 37°C for 1 week or until culture is dense, shaking flask once a day.
5. Inoculate 100 to 200 mL 7H9-OADC-Tw in a 490-cm^3 roller bottle with 1 to 2 mL preculture of step 4. Incubate at 37°C with rolling until an OD_{600} of 0.5 to 0.8 is reached.
6. Centrifuge culture for 10 min at 3000 × g at (room temperature) RT.
7. Wash pellet with wash buffer at 1/10 original culture volume.
8. Centrifuge suspension for 10 min at 3000 × g at 4°C.
9. Wash pellet with wash buffer at 1/20 volume.
10. Centrifuge suspension for 10 min at 3000 × g at 4°C.
11. Resuspend pellet in wash buffer at 1/100 of original culture volume.
12. Make 100-μL aliquots in prechilled 1.5-mL tubes.
13. Freeze at –80°C immediately.

19.3.3 Electroporation of Mycobacteria

1. Take a frozen aliquot of competent cells and allow to defrost. Keep the melted aliquot on ice.
2. Add 100 to 500 ng (not exceed 10 μL) of plasmid DNA.
3. Transfer cells and DNA in a prechilled electroporation cuvette (–20°C). Allow cuvette to incubate on ice for 10 to 20 min (*see* **Note 4**).
4. Use the following settings to perform the electroporation: 2.5 kV, 1000 Ω, 25 μF.
5. Add 100 μL 7H9 (for *M. smegmatis*) or 7H9-OADC (for *M. bovis* BCG or *M. tuberculosis*) medium into the electroporation cuvette and transfer content into a sterile 1.5-mL centrifuge vial.
6. Incubate for 3 h at 37°C for *M. smegmatis* or for 24 h for *M. bovis* BCG.
7. Plate serial dilutions on selective 7H10 (for *M. smegmatis*) or 7H10-OADC (for *M. bovis* BCG or *M. tuberculosis*) plates.

19.3.4 Selection of SCO Clones in M. smegmatis

1. Transform a temperature-sensitive replicating plasmid (e.g., pMN249; *see* **Note 5**) into *M. smegmatis* competent cells according to Section 19.3.3.
2. Plate on 7H10-Hyg-Kan plates and incubate at 32°C for 4 to 5 days.
3. Pick up one colony from the plates and inoculate into liquid culture (7H9-Hyg-Kan-Tw).
4. Incubate at 32°C to an OD_{600} of 1.0.
5. Plate dilutions onto 7H10-Hyg-Kan plates and incubate at 39°C for 4 to 5 days.

6. Pick up 10 to 20 colonies and make liquid culture at 39°C in 7H9-Hyg-Kan-Tw.
7. Prepare chromosomal DNA, perform Southern blot or PCR to confirm SCO.

19.3.5 Selection of DCO Clones in M. smegmatis

1. Inoculate one SCO in 7H9-Hyg-Kan-Tw.
2. Incubate at 39°C to an OD_{600} of 1.0 is reached.
3. Plate serial dilutions onto 7H10-Hyg-Strep plates.
4. Incubate at 39°C for 4 to 5 days.
5. Pick up 10 to 20 single colonies and use to inoculate 7H9-Hyg-Strep-Tw.
6. Incubate at 39°C to an OD_{600} of 1.0 is reached.
7. Prepare chromosomal DNA extractions (see Chapter 1) and perform Southern blot to confirm the allelic exchange.

19.3.6 Excision of the FRT-hyg-FRT Cassette in M. smegmatis Using the Flp Recombinase

1. Transform the Flp expression plasmid pMN234 into DCO cells containing the *FRT-hyg-FRT* resistance cassette (*see* Section 19.3.5).
2. Plate electroporation as serial dilutions onto 7H10-Kan plates.
3. Incubate at 37°C for 4 to 5 days.
4. Pick 10 to 20 single colonies and use to inoculate 7H9-Kan-Tw
5. Prepare chromosomal DNA (see Chapter 1).
6. Perform Southern blot or colony PCR to confirm the removal of *FRT-hyg-FRT* cassette (*see* **Note 6**).

19.3.7 Removal of the Plasmid pMN234 from M. smegmatis

1. Inoculate 7H9-Kan-Tw with one colony from Section 19.3.6 that has been identified to have lost the *FRT-hyg-FRT* cassette.
2. Incubate at 37°C until an OD600 of 1.0 is reached.
3. Plate serial dilutions onto 7H10-Strep plates to counter select against pMN234. Incubate at 37°C for 4 to 5 days.
4. Streak 40 to 50 colonies onto 7H10-Hyg, 7H10-Kan, and 7H10-Strep plates.
5. Incubate at 37°C.
6. Colonies that grow on 7H10-Strep but not on 7H10-Hyg or 7H10-Kan are correct (*see* **Note 7**).

19.3.8 Selection of SCO Clones in **M. bovis** *BCG or* **M. tuberculosis**

1. Transform a replicating plasmid (e.g., pML563) into *M. bovis* BCG or *M. tuberculosis* competent cells according to Section 19.3.3 (*see* **Note 8**).
2. Plate onto 7H10-OADC-Hyg plates, incubate at 34°C for 3 to 4 weeks (see **Notes 9** and **10**).
3. Pick one colony from the plates. Inoculate into 7H9-OADC-Hyg-Tw.
4. Incubate at 34°C until an OD_{600} of 1.0 is reached.
5. Plate serial dilutions onto 7H10-OADC-Hyg plates.
6. Incubate at 41°C for 3 to 4 weeks (see **Note 1**).
7. Pick up 10 to 20 colonies, make liquid culture at 41°C in 7H9-OADC-Hyg-Tw to prepare chromosomal DNA, perform Southern blot or colony PCR to confirm SCO.

19.3.9 Selection of DCO Clones in **M. bovis** *BCG* or **M. tuberculosis**

1. Inoculate one SCO into 7H9-OADC-Hyg-Tw.
2. Incubate at 41°C until an OD_{600} of 1.0 is reached.
3. Plate serial dilutions onto 7H10-OADC-Hyg-Suc plates.
4. Incubate at 41°C for 3 to 4 weeks.
5. Pick 10 to 20 single colonies.
6. Inoculate 7H9-OADC-Hyg-Tw.
7. Incubate at 41°C for 3 to 4 weeks.
8. Prepare chromosomal DNA and perform Southern blot or colony PCR to confirm the allelic exchange.

19.3.10 Excision of the **loxP-hyg-loxP** *Cassette in* **M. bovis** *BCG or* **M. tuberculosis** *Using the Cre Recombinase*

1. Transform pCreSacB1 into competent cells of one DCO strain (Section 19.3.9).
2. Plate onto 7H10-OADC-Kan plates, incubate at 37°C for 3 to 4 weeks.
3. Pick up 10 to 20 single colonies, use to inoculate 7H9-OADC-Kan-Tw.
4. Incubate at 37°C for 3 to 4 weeks.
5. Prepare chromosomal DNA, perform Southern blot or Colony PCR to confirm the removal of *loxP-hyg-loxP* cassette.

19.3.11 Removal of Plasmid pCreSacB1 from **M. bovis** *BCG* or **M. tuberculosis**

1. Inoculate one colony from Section 19.3.10 that has been confirmed to have lost the *loxP-hyg-loxP* cassette into 7H9-OADC-Tw media.

2. Incubate at 37°C until an OD_{600} of 1.0 is reached.
3. Plate serial dilutions onto on 7H10-OADC-Suc plates to counterselect for pCreSacB1.
4. Incubate at 37°C for 3 to 4 weeks.
5. Pick up 40 to 50 colonies, streak on both 7H10-OADC-Hyg, and 7H10-OADC plates. The colonies that grow on 7H10-OADC but not on 7H10-OADC-Hyg will be correct.

19.3.12 Colony PCR

The PuReTaq Ready-To-Go PCR beads kit is recommended for colony PCR, and the manufacturer's protocol should be followed.

1. To each PCR bead add 20 μL water, 1 μL of each primer (10 pmol/μL), 1 μL 100% DMSO (v/v), and 2 μL of a dense bacterial culture (*see* **Note 11**). The final reaction volume is 25 μL.
2. Use a typical PCR program; for example, pre-denature for 5 min at 94°C, 30 cycles of denaturation at 94°C for 30 s, annealing 55°C for 30 s, and extension 72°C for 2 min, final extension for 10 min at 72°C.

19.4 Notes

1. To increase the efficiency of counterselection against plasmids with the temperature-sensitive origin of replication, a temperature of 41°C for H37Rv instead of 39°C should be used. The growth rates of H37Rv at 41°C and 37°C are similar, whereas *M. bovis* BCG grows much slower at 41°C compared with 37°C.
2. Tween 80 is added to reduce clumping of slowly growing mycobacteria in subsequent liquid cultures.
3. Because of the slow growth of mycobacteria, the agar plates quickly dry if kept in an incubator for more than 3 days. Therefore, put the plates into a Ziplock bag or wrap them between two layers of aluminum foil if a longer incubation time is needed.
4. Incubate the cells with the DNA for at least 10 min at room temperature before electroporation to achieve optimal transformation efficiency.
5. The plasmid pMN249 is a derivate of pMN252 (Fig. 19.2), and it contains a temperature-sensitive origin of replication (PAL5000$_{ts}$) for mycobacteria, a hygromycin resistance gene (*hyg*) flanked by *FRT* sites for excision by the Flp recombinase, about 1000-bp fragments of upstream and downstream regions of the *mspC* gene, and the *rpsL* gene as a counterselectable marker.
6. The excision of FRT-flanked genes by the Flp recombinase in *M. smegmatis* can be as low as 1%. Therefore, it is useful to streak at least 100 clones. Because of this, we find that colony PCR provides the best rapid screen to determine which transformants have lost the *FRT*-hyg-*FRT* cassette. Work is in progress to improve the efficiency of the Flp recombinase in mycobacteria.
7. Counterselection against the wild-type *rpsL* gene on streptomycin plates can lead to the selection of spontaneously streptomycin-resistant mutants. Therefore, strains should always be tested for kanamycin resistance after counterselection to ensure that the plasmid has been lost.

8. Generation of mutants of *M. bovis* BCG or *M. tuberculosis* by homologous recombination using suicide vectors requires transformation efficiencies of greater than 10^5 transformants/µg DNA. For some strains, this many be difficult to achieve. Therefore, we recommend the use of replicating vectors in these organisms.

9. In contrast with *M. smegmatis*, direct selection for DCO candidates is not efficient in slowly growing mycobacteria. Using the method described in this chapter, it is essential to construct first a SCO clone.

10. It was shown that the thermosensitive pAL5000 origin of replication gives optimal results if used at 32°C in *M. smegmatis*. However, slowly growing mycobacteria such *M. bovis* BCG or *M. tuberculosis* grow very poorly at 32°C. Therefore, the growth temperature should be increased to at least 34°C.

11. Colony PCR can be done either with cells from plate or from liquid culture. DMSO should be added to the reaction mixture to increase the amplification efficiency of the G+C-rich mycobacterial DNA.

Acknowledgments This work was supported by grant AI063432 from the National Institutes of Health.

References

1. Pelicic, V., Jackson, M., Reyrat, J. M., Jacobs, W. R. Jr., Gicquel, B., and Guilhot, C. (1997) Efficient allelic exchange and transposon mutagenesis in *Mycobacterium tuberculosis. Proc Natl Acad Sci U S A* **94**, 10955–10960.

2. Bardarov, S., Bardarov S. Jr., Pavelka M. S. Jr., Sambandamurthy, V., Larsen, M., Tufariello, J., Chan, J., Hatfull, G., and Jacobs W. R. Jr. (2002) Specialized transduction: an efficient method for generating marked and unmarked targeted gene disruptions in *Mycobacterium tuberculosis, M. bovis* BCG and *M. smegmatis. Microbiology* **148**, 3007–3017.

3. Machowski, E. E., Dawes, S., and Mizrahi, V. (2005) TB tools to tell the tale—molecular genetic methods for mycobacterial research. *Int J Biochem Cell Biol* **37**, 54–68.

4. Stephan, J., Stemmer, V., and Niederweis, M. (2004) Consecutive gene deletions in *Mycobacterium smegmatis* using the yeast FLP recombinase. *Gene* **343**, 181–190.

5. Kana, B. D., and Mizrahi, V. (2004) Molecular genetics of *Mycobacterium tuberculosis* in relation to the discovery of novel drugs and vaccines. *Tuberculosis* **84**, 63–75.

6. Merlin, C., McAteer, S., and Masters, M. (2002) Tools for characterization of *Escherichia coli* genes of unknown function. *J Bacteriol* **184**, 4573–4581.

7. Hasan, N., Koob, M., and Szybalski, W. (1994) *Escherichia coli* genome targeting, I. Cre-lox-mediated in vitro generation of ori- plasmids and their in vivo chromosomal integration and retrieval. *Gene* **150**, 51–56.

8. Tsuda, M. (1998) Use of a transposon-encoded site-specific resolution system for construction of large and defined deletion mutations in bacterial chromosome. *Gene* **207**, 33–41.

9. Denome, S. A., Elf, P. K., Henderson, T. A., Nelson, D. E., and Young, K. D. (1999) *Escherichia coli* mutants lacking all possible combinations of eight penicillin binding proteins: viability, characteristics, and implications for peptidoglycan synthesis. *J Bacteriol* **181**, 3981–3993.

10. Malaga, W., Perez, E., and Guilhot, C. (2003) Production of unmarked mutations in mycobacteria using site-specific recombination. *FEMS Microbiol Lett* **219**, 261–268.

11. Stephan, J., Bender, J., Wolschendorf, F., Hoffmann, C., Roth, E., Mailander, C., Engelhardt, H., and Niederweis, M. (2005) The growth rate of *Mycobacterium smegmatis* depends on sufficient porin-mediated influx of nutrients. *Mol Microbiol* **58**, 714–730.

12. Sander, P., Meier, A., and Boettger, E. C. (1995) *rpsL+*: a dominant selectable marker for gene replacement in mycobacteria. *Mol Microbiol* **16,** 991–1000.
13. Guilhot, C., Gicquel, B. and Martin, C. (1992) Temperature-sensitive mutants of the Mycobacterium plasmid pAL5000. *FEMS Microbiol Lett* **98**, 181–186.

Chapter 20
Construction of Targeted Mycobacterial Mutants by Homologous Recombination

Sharon L. Kendall and Rosangela Frita

Abstract The ability to select genes to knock out of mycobacterial genomes has greatly improved our understanding of mycobacteria. This chapter describes a method for doing this. The gene (including a 1-kb flanking region) is cloned into a pNIL series vector and disrupted by deletion or insertion of a cassette. A selection of marker genes obtained from the pGOAL series of vectors are inserted into the pNIL vector to create a suicide delivery system. This delivery vector is introduced into mycobacteria where the disrupted version of the gene replaces the wild-type version by a two-step homologous recombination process. The method involves selecting for a single crossover event followed by selection of double crossovers. Single crossovers have incorporated plasmid marker genes and are sucrose^S, kanamycin^R and blue on media containing X-gal. Double crossovers have lost plasmid markers and are sucrose^R, kanamycin^S and white on media containing X-gal.

Keywords allelic replacement · gene knock-out · homologous recombination · mycobacteria · mutant

20.1 Introduction

20.1.1 Mutant Making in Mycobacteria: A Historical Perspective

Reverse genetics, an approach that aims to find the function of a gene by disruption and phenotypic characterization, is a very powerful way of studying bacterial gene function. The application of reverse genetics to the study of mycobacterial gene function has greatly contributed to the understanding of genes involved in virulence in pathogenic mycobacteria.

S.L. Kendall
Department of Pathology and Infectious Diseases, The Royal Veterinary College,
Royal College Street, London, NW1 0TU, United Kingdom
e-mail: skendall@rvc.ac.uk

T. Parish, A.C. Brown (eds.), *Mycobacteria Protocols*,
doi: 10.1007/978-1-59745-207-6_20, © Humana Press, Totowa, NJ 2008

Both targeted and random mutagenesis strategies have been used to generate mycobacterial mutant strains (see later). Random mutagenesis strategies have been used to identify genes involved in virulence in the slow-growing mycobacteria including *Mycobacterium tuberculosis, Mycobacterium marinum, Mycobacterium avium*, and *Mycobacterium avium* subsp. *paratuberculosis* [1, 2, 3, 4, 5, 6, 7]. Additionally, similar approaches have been used in the fast-growing *Mycobacterium smegmatis* [8]. Random mutation by the use of insertion sequences has contributed greatly to our understanding of mycobacteria, but it requires large-scale screening of phenotypes. Targeted gene disruption offers the ability to probe specific pathways and has been very useful in the demonstration of gene essentiality [9, 10, 11], for rational attenuation in the creation of vaccine strains [12, 13, 14], and it also has potential for the mutation of specific amino acid residues by site-directed mutagenesis [15].

20.1.2 Approaches to Making Mutant Strains

Chemical mutagenesis systems using physical and chemical DNA-damaging agents have been used in both slow-growing and fast-growing mycobacteria with *N*-methyl-*N'*-Nitro-*N*-nitrosoguanidine (NTG) being the most widely used [16]. However, whereas this approach is quick, cheap, and efficient, it does not allow rational selection of the gene to be mutated and may introduce multiple lesions.

The use of random transposon mutagenesis in mycobacteria was first described more than a decade ago in the fast-growing *M. smegmatis* [17, 18] and later in the slow-growing species *Mycobacterium bovis* BCG and *M. tuberculosis* [6, 19]. This approach has been very effective in the generation of marked mutant libraries for large-scale phenotypic screens [5, 6, 20] but again does not allow rational selection of the gene(s) to be disrupted.

Allelic exchange by homologous recombination allows specific genes to be targeted for mutagenesis. A schematic of gene replacement by homologous recombination using both a one-step and two-step strategy is given in Figure 20.1. The first report of mutagenesis by homologous recombination in mycobacteria was in *M. smegmatis* [21, 22]. In this study, the authors used a one-step method (Fig. 20.1a) to replace the *pyrF* gene with *pyrF::aph*. Gene replacement by homologous recombination in the slower-growing mycobacteria was found to be more difficult and resulted in high frequencies of illegitimate recombination, single crossover (SCO) events, or unstable double crossover (DCO) events [23]. However, the technique was eventually successful in the slow-growing *M. bovis* BCG where the *ureC* gene was replaced with *ureC::aph* using a one-step method [24]. The first reported *M. tuberculosis* allelic exchange mutant used a one-step method to replace *leuD3* gene with *leuD3::aph* [25].

This chapter focuses on gene disruption by homologous recombination in both slow-growing (*M. tuberculosis*) and fast-growing (*M. smegmatis*) species

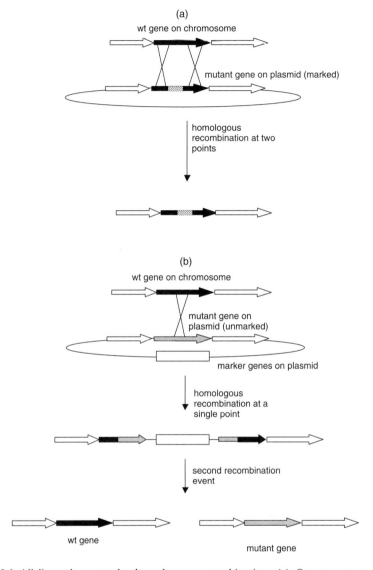

Fig. 20.1 Allelic replacement by homologous recombination. **(a)** One-step strategy for the generation of marked mutations. A selectable marker (hashed lines) is used to disrupt the gene to be replaced and is cloned on a delivery vector. A DCO event leads to the wild-type gene being replaced by the mutant gene in a single step. The technique relies on the ability to select for the recombination event by selecting for the inserted marker. **(b)** Two-step strategy for the generation of unmarked mutations. A deleted/mutated version of the gene to be replaced is cloned onto a delivery vector containing suitable marker genes. A single recombination event gives rise to SCOs containing two copies of the gene (both mutant and wild type) and the suicide vector integrated into the genome. A second recombination event gives rise to DCOs that are either wild type or mutant. This second event is selected for by the use of counterselectable markers on the suicide plasmid. (Adapted from Muttucumaru, D. G., and Parish, T. [2004] The molecular biology of recombination in mycobacteria: what do we know and how can we use it? *Curr Issues Mol Biol* 6, 145–57.)

Table 20.1 The Uses of the pNIL/pGOAL Suicide Delivery System Described in the Literature

Use	Gene	Reference
Auxotrophic mutants	*metB, proC, trpD*	[14]
Transcription factor mutants	*trcS, kdpDE, narL, dosR, tcrXY, Rv3220c; senX3-regX3; amiA*	[33, 34, 35]
Demonstration of essentiality	*hemZ; glnE*	[10, 11]
Cell division and chromosome replication	*ftsZ; nrdF2, nrdZ*	[36, 37]
Respiratory network	*qcrCAB, ctaC, ctaD1, ctaDII*	[38]
Transport systems	*Rv1747*	[39]
Lipid synthesis	*papA5*	[40]
Macrophage entry	*yrbE1B*	[41]
DNA repair	*dnaE2*	[42]
Stringent response	*rel*	[43]
Major antigens	*hspx*	[44]

using a two-step method (Fig. 20.1b). Two-step methods are more efficient than one-step methods and work better in the slow-growing species where the frequency of homologous recombination is low. Additionally, two-step methods allow for unmarked mutations to be made. We use a suicide delivery system (pNIL/pGOAL) that has been previously described [26]. This system was originally used to generate unmarked mutations in phospholipases and a hemolysin but has since been used successfully to generate mutants in a number of important processes (Table 20.1).

20.1.3 Homologous Recombination in Mycobacteria

Homologous recombination occurs between two copies of identical or highly homologous sequences and is mediated by RecA [27]. The DNA recombination repair pathway has been extensively studied in mycobacteria, and a recent review of homologous recombination in mycobacteria can provide the reader with current knowledge on the process in mycobacteria [28]. As stated above, initial attempts to use homologous recombination in the slow-growing myco-bacteria were problematic, and the reasons for this remain largely unknown. However, a number of strategies have been used to overcome this. These include the use of long flanking regions [19, 25], the use of damaged DNA substrates such as UV-treated or alkali-denatured [29], the use of counterselectable markers [30], and the development of improved delivery strategies such as temperature-sensitive phages and plasmids, suicide (nonreplicating) plasmids, and plasmid incompatibility systems [23, 31, 32].

20.1.4 Selectable Markers

The selection steps used in generating a mutant by allelic exchange requires the use of selectable markers present on the plasmid. Selectable markers can be used for positive or negative selection. In mycobacteria, positive selection markers include genes conferring resistance to kanamycin, hygromycin B, chloramphenicol, and gentamicin. Negative selection markers are genes that have deleterious effects to the bacteria when certain substances are present in the medium. In mycobacteria, negative markers that have been used include *rpsL* (streptomycin sensitivity), *katG* (isoniazid sensitivity), and *sacB* (sucrose sensitivity) [23]. *pyrF* has been used as both a positive marker and a negative marker with positive selection in the absence of uracil and negative selection in the presence of 5-fluoro orotic acid (5-FOA) [21]. *sacB* is the most widely used negative marker. The advantage of using *sacB* over the antibiotic sensitivity markers is that the strain does not have to be antibiotic resistant to begin with. However, a disadvantage is that spontaneous *sacB* mutants arise at a high frequency [28].

Reporter genes and suicide vectors have been used in combination with the selection markers described above as a means of rapidly identifying clones that have incorporated the plasmid. Both *xylE* and *lacZ* have been successfully used in mycobacteria [23]. Colonies expressing *xylE* turn yellow when sprayed with catechol. However, the indicator color of *xylE* may be easily confused with the color of mycobacterial colonies if expressed from a weak promoter. Colonies expressing *lacZ* turn blue when growing in the presence of 5-bromo-4-chloro-3-indolyl-β-D-galactopyranoside (X-gal), and therefore this reporter is normally a better choice than *xylE*. The system described here uses a suicide (nonreplicating) delivery plasmid using the selectable markers *hyg*, *kan*, *lacZ*, and *sacB* [26].

Selective markers that are commonly used in *Escherichia coli* often need to be expressed from mycobacterial promoters to be functional. Promoters used for the expression of the selectable markers used in the pNIL/pGOAL system described here are *hsp60* (used to drive the expression of *sacB*) and *Ag85a* (used to drive the expression of *lacZ*). A schematic of the selection process for the pNIL/pGOAL system is given in Figure 20.2.

20.1.5 Suicide Plasmids

Vectors that cannot replicate in the host species are called suicide vectors. The vectors presented here can replicate in *E. coli* but cannot replicate in mycobacteria. The target gene is first cloned into a cloning vector (pNIL series) where the mutation is generated, and then the markers necessary for selection are introduced by cloning a cassette from a second set of plasmids (pGOAL series). The advantage of this system over others is that fewer cloning steps are needed

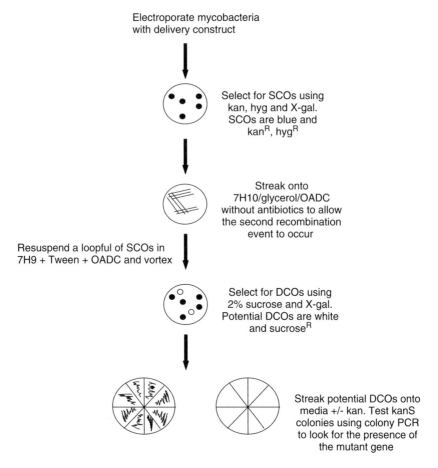

Fig. 20.2 Steps in the selection strategy for gene replacement using the pNIL/pGOAL system. The suicide delivery vector is electroporated into competent mycobacteria. The plasmid cannot survive autonomously, so selecting for the presence of the plasmid marker genes (*kan, hyg, lacZ*) selects for those colonies that have undergone a single recombination event. SCOs (blue, kan^R, hyg^R) are plated onto media without selection while the second recombination event occurs. DCOs are selected for by plating onto media containing X-gal and sucrose. DCOs have lost the vector and are white and sucrose^R. Finally, colonies are patch tested for sensitivity to kanamycin, and kan^S colonies are screened for the wild type (wt) or mutant version of the gene by colony PCR

making it easier to find appropriate restriction sites. This system also has the advantage of easily exchanging the marker gene cassette for other marker genes previously incorporated into the pGOAL series [26]. A schematic of the cloning steps along with the maps of the pNIL/pGOAL vectors are given in Figure 20.3.

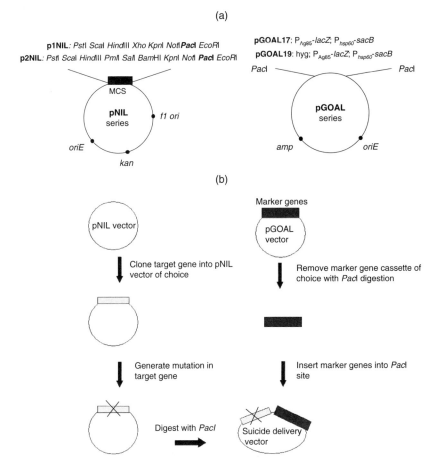

Fig. 20.3 **(a)** Features of the pNIL and pGOAL series of vectors. The pNIL vectors can be selected for using kanamycin in both mycobacteria and *E. coli*. The vectors can only replicate in *E. coli*. The p1NIL and p2NIL vectors differ only in their multiple cloning sites. Both multiple cloning sites contain *PacI* for insertion of the *PacI* marker cassette from one of the vectors in the pGOAL series. The pGOAL series of vectors differ in the selectable markers present on the *PacI* cassette (*see* Note 1). **(b)** Generation of the suicide delivery construct. The target gene is cloned in a pNIL series vector and a deletion is made. The *PacI* cassette is introduced from one of the pGOAL series of vectors

20.1.6 Protocols Presented Here

The protocols will describe the construction of the suicide delivery vector, UV treatment of DNA prior to electroporation, and the selection process. As electroporation is being considered in detail elsewhere in this book (*see* Chapter 13), we do not describe this in detail here.

20.2 Materials

20.2.1 Construction of Suicide Delivery Vector

1. Mycobacterial genomic DNA.
2. Primers designed to amplify the gene to be deleted.
3. p1NIL or p2NIL cloning vector (Fig. 20.3a).
4. pGOAL17 or pGOAL19, for marked or unmarked mutations, respectively (*see* Fig. 20.3a and **Note 1**).

Plasmids can be obtained on request from Prof. N. Stoker (Department of Pathology and Infectious Diseases, The Royal Veterinary College, Royal College Street, London, NW1 OUT, UK) or from Prof. T. Parish (Centre for Infectious Disease, Institute of Cell & Molecular Science, Queen Mary University of London, 4 Newark St., London, E1 2 AT UK).

20.2.2 UV Irradiation of DNA

1. UV Stratalinker 1800 (Stratagene, La Jolla, CA, USA) (*see* **Note 2**).
2. Plasmid DNA to be treated.

20.2.3 Electroporation of Mycobacteria

1. Electroporation equipment (*see* **Note 3**).
2. Competent mycobacteria (*see* Chapter 13).

20.2.4 Two-Step Selection Process

1. Kanamycin B (Sigma-Aldrich, Poole, Dorset, UK). Make stock at 50 mg/mL in distilled water, filter sterilize, and store in aliquots at –20°C. Use at 50 µg/mL for *E. coli* and at 20 µg/mL for mycobacteria.
2. 5-Bromo-4-chloro-3-indolyl-β-D-galactopyranoside (X-gal) (Sigma-Aldrich). Make stock at 50 mg/mL in dimethylformamide or DMSO. Store at –20°C and protect from light (*see* **Note 4**). Use at 50 µg/mL for *E. coli* and mycobacteria.
3. Hygromycin B (Roche, West Sussex, UK). Comes as stock at 50 mg/mL in PBS. Store at 4°C and protect from light (*see* **Note 5**). Use at 200 µg/mL in *E. coli* and 50 to 100 µg/mL in mycobacteria.
4. Sucrose. Make stock at 50% (w/v) and filter sterilize. Use at a final concentration of 2% (w/v) for *M. tuberculosis* and 10% (w/v) for *M. smegmatis*.
5. Tween-80. Make stock at 10% (w/v) and filter sterilize. Use at a final concentration of 0.05% (w/v).
6. Oleic albumin dextrose catalase (OADC) supplement (Becton Dickinson, Plymouth, Devon, UK). Comes sterile. Store at 4°C and use at a final concentration of 10% v/v.

7. Glycerol. Make a 50% v/v stock in water and filter sterilize. Store at 4°C and use at a final concentration of 0.5% in 7 H9-Tw.

8. 7 H9-Tw medium: Dissolve 4.7 g Middlebrook 7 H9 broth base (Difco, West Molsey, Surrey, UK) in 900 mL deionized water and add 5 mL 50% glycerol, mix well, and filter sterilize. Add 5 mL 20% (v/v) Tween 80 (final concentration of 0.05% v/v). Add 100 mL OADC immediately before use (*see* **Note 6**).

9. 7 H10 plates: Dissolve 19 g Middlebrook 7 H10 medium (Difco) in 900 mL deionized water. Add 8.4 mL 60% glycerol (v/v). Autoclave. When media is cooled to ~50°C, add 100 mL OADC enrichment (*see* **Note 6**). Mix well and pour in standard plastic Petri dishes.

10. Lemco media (*see* **Note 7**): To 800 mL distilled water add 10 g Bacto peptone, 5 g NaCl, 5 g Lemco powder. Add 5 mL 20% (v/v) Tween 80 (final concentration of 0.05% v/v) and make up to 1 L. Autoclave to sterilize. For solid medium (Lemco-agar), add 15 g agar in place of Tween 80 before autoclaving. Allow to cool and pour onto Petri dishes.

11. Recovery media: 10 mL 7 H9-Tw in 50-mL sterile conical tube for each electroporation.

12. 1-mm sterile glass beads (Biospec, Bartlesville, OK, USA).

20.3 Methods

20.3.1 Construction of Delivery Constructs

1. Amplify the gene to be mutated from genomic DNA and clone into pNIL vector. The choice of pNIL vectors depends on the restriction site required (*see* **Note 8**).

2. Generate mutation in the gene (*see* **Note 9**). Check the construct by restriction mapping or by sequencing.

3. Choose appropriate pGOAL cassette and clone into *Pac*I site of pNIL containing the mutant version of the target gene (*see* **Note 1**).

4. Make a pure preparation of suicide delivery construct (*see* **Note 10**).

20.3.2 UV irradiation of DNA

1. Aliquot 1 to 5 µg plasmid DNA inside microcentrifuge tubes or 96-well plates. Subject to 100 mJ/cm^2 of UV energy (*see* **Note 2**)

20.3.3 Electroporation of Mycobacteria

1. Place 1 to 5 µg of DNA in an electroporation cuvette and add 200 µL of electrocompetent mycobacteria. Two electroporations should be done, one with UV-treated DNA and one with untreated DNA (*see* **Note 2**).

2. Deliver a pulse using electroporation equipment (*see* **Note 3**). Use the following settings to perform the electroporation: 2.5 kV, 1000 Ω, 25 μF.
3. Transfer electroporation cuvette contents into recovery media.
4. Add 200 μL 7 H9-Tw medium into the electroporation cuvette to wash. Transfer to the recovery media.
5. Incubate for 2 h at 37°C for *M. smegmatis* or for 24 h for *M. tuberculosis*.

20.3.4 Two-Step Selection Process

1. Plate out the entire transformation mix onto 7 H10 plates containing kanamycin, hygromycin, and X-gal (*see* **Note 11**).
2. Seal plates with Parafilm or autoclave tape and put them into plastic bags. For *M. tuberculosis*, incubate at 37°C for 3 to 5 weeks until colonies appear. For *M. smegmatis* incubate at 37°C for 3 to 5 days until colonies appear. Blue kan^R hyg^R colonies are the SCOs (*see* **Note 12**).
3. Pick two SCOs and restreak onto fresh 7 H10 without any antibiotics. Seal the plates as previously and incubate at 37°C for 1 to 2 weeks for *M. tuberculosis* and 2 to 3 days for *M. smegmatis* (*see* **Note 13**).
4. Take a loopful from the above plates and resuspend in a universal containing 1 mL of 1-mm glass beads and 3 mL of 7 H9/OADC/Tween. Vortex vigorously until the cells are resuspended. Leave for aerosols to settle.
5. Plate out serial dilutions (10^0 to 10^{-2}) into plates containing X-gal with and without sucrose (*see* **Note 14**). Seal and incubate for 3 to 5 weeks until colonies appear. The white sucrose^R colonies are potential DCOs (*see* **Note 15**).
6. Patch test the potential DCOs for kanamycin sensitivity. Streak 40 colonies onto 7 H10 and 7 H10 containing kanamycin (*see* **Note 16**).
7. Screen the kanamycin-sensitive colonies using colony PCR (*see* **Note 17**) followed by Southern blotting.

20.4 Notes

1. The vector routinely used in our laboratory is pGOAL19 for the generation of unmarked mutations. The *PacI* excisable cassette contained in this vector contains a *hyg* resistance gene. This allows the use of two antibiotics (kanamycin and hygromycin) for the selection of SCOs. For mutations marked with the *hyg* resistance gene, pGOAL17 will be the most appropriate.
2. UV treatment may lead to the incorporation of point mutations on the treated DNA. Although the frequency of such an event is probably low (we do not see any loss of LacZ function as a result of UV treatment), it could be argued that the mutant phenotype is due to an extra mutation that occurred on the DNA during the UV treatment. If this is a concern, then the relevant region of the SCO can be sequenced. We routinely electroporate 1 μg of UV-treated and 1 μg of untreated plasmid. If we get SCO with both treated and nontreated, we take both forward to the DCO selection and screening stage. In the case of a DCO being obtained from both treated and untreated plasmid, we proceed with the DCO obtained from the untreated plasmid. Alternative cross-linkers are supplied by

GE Healthcare, Hoefer, MIDSCI, Spectronics Corporation, Ultra-Lum, and UVP. Alkali treatment of DNA has also been used to encourage homologous recombination. DNA is incubated with 0.2 M NaOH and 0.2 mM EDTA at 37°C for 30 min. The DNA is then ethanol precipitated prior to electroporation. This method has been used successfully, but the frequency of homologous recombination is lower than that by UV treatment.

3. We use the Bio-Rad Gene Pulser (Bio-Rad, Hemel Hempstead, Herts, UK) and the Flowgen Easyject system (Lichfield, Staffordshire, UK). We recommend that electroporation protocols be optimized by the user depending on which particular piece of equipment is used. Equivalent electroporators can be supplied by Amaxa, Ambion, and BTX.

4. X-gal is light sensitive and will degrade on exposure to light and multiple freeze-thaw cycles. A yellow coloration that darkens occurs as X-gal degrades.

5. Hygromycin B is also light sensitive and will degrade on exposure to light.

6. OADC should only be added to media that has been cooled to 50°C to 55°C. OADC added to hot agar will produce formaldehyde fumes. Poured plates should be kept in the dark.

7. Middlebrook media can also be used for the growth of *M. smegmatis*, but this requires the use of expensive OADC supplements.

8. The target gene cloned in the pNIL vector should include enough flanking region on either side of the gene so that homologous recombination can occur. The minimum length of homologous region has not been determined, but we routinely use 1 kb on either side of the target gene.

9. To create the desired mutation on the target gene, one can first clone the target gene and flanking regions as one fragment into the pNIL vector and then generate the required deletion. This can be done by inverse PCR, or, if there are suitable restriction enzyme sites, by restriction digestion. Alternatively, two fragments containing the flanking region of the target gene can be brought together to generate a deletion.

10. A highly pure and concentrated preparation of the suicide delivery construct is needed for UV treatment and electroporation. We routinely use commercial plasmid purification kits in our laboratory that give a high quality and concentrated plasmid DNA yield. These kits use the same principle of the standard alkaline lysis method but use silica membranes as a means of purifying the DNA instead of the classic phenol extractions. We routinely use up to 5 μg in no more than 5 μL for UV treatment followed by electroporation.

11. The use of two antibiotics (kanamycin and hygromycin) prevents the problem of spontaneous antibiotic-resistant colonies. This can be problematic particularly in *M. tuberculosis* when such long incubation times (3 to 4 weeks) are needed for colony formation and the species has such a high spontaneous resistance frequency.

12. Colonies may take time to become blue or only show a light-blue color. These colonies may be replated on fresh plates containing X-gal. A more intense blue color can be seen if the plates are put at 4°C overnight.

13. If the mutant has a growth defect, then the wild-type colonies will outgrow the mutant. Therefore, the plates should not be left to grow for more than 2 weeks at 37°C and should be used immediately in the next step.

14. Spontaneous *sacB* mutants arise with a frequency similar to secondary homologous recombination events, therefore plating with X-gal is necessary. The blue colonies forming on the sucrose and X-gal plates are SCOs that have become spontaneously resistant to sucrose rather than true DCOs. The X-gal–only plates are there to check that the SCOs are sucrose sensitive. A reduction of 10^{-4} to 10^{-5} CFU should be seen on the sucrose plates.

15. These colonies can be very small pin-prick colonies. Plates should be incubated until colonies are large enough to patch test.

16. We use 50-mm Petri dishes. The plate can be comfortably divided into eight sectors and eight colonies streaked onto each plate.

17. Colony PCR is a quick method to check for DCOs. A loopful of cells is resuspended in 400 µL sterile distilled water and vortexed vigorously for 3 min. The cells are killed by heating for 6 min at 100°C and then passed through an 0.2-µm syringe filter for removal from the category 3 laboratory (in the case of *M. tuberculosis*). The maximum amount of the cell filtrate should be used in a PCR reaction to check for the deletion.

References

1. Shin, S. J., Wu, C. W., Steinberg, H., and Talaat, A. M. (2006) Identification of novel virulence determinants in *Mycobacterium paratuberculosis* by screening a library of insertional mutants. *Infect Immun* 74, 3825–33.
2. Mehta, P. K., Pandey, A. K., Subbian, S., El-Etr, S. H., Cirillo, S. L., Samrakandi, M. M., and Cirillo, J. D. (2006) Identification of *Mycobacterium marinum* macrophage infection mutants. *Microb Pathog* 40, 139–51.
3. Ruley, K. M., Ansede, J. H., Pritchett, C. L., Talaat, A. M., Reimschuessel, R., and Trucksis, M. (2004) Identification of *Mycobacterium marinum* virulence genes using signature-tagged mutagenesis and the goldfish model of mycobacterial pathogenesis. *FEMS Microbiol Lett* 232, 75–81.
4. Philalay, J. S., Palermo, C. O., Hauge, K. A., Rustad, T. R., and Cangelosi, G. A. (2004) Genes required for intrinsic multidrug resistance in *Mycobacterium avium*. *Antimicrob Agents Chemother* 48, 3412–8.
5. Sassetti, C. M., and Rubin, E. J. (2003) Genetic requirements for mycobacterial survival during infection. *Proc Natl Acad Sci U S A* 100, 12989–94.
6. McAdam, R. A., Quan, S., Smith, D. A., Bardarov, S., Betts, J. C., Cook, F. C., Hooker, E. U., Lewis, A. P., Woollard, P., Everett, M. J., Lukey, P. T., Bancroft, G. J., Jacobs W. R. Jr., and Duncan, K. (2002) Characterization of a *Mycobacterium tuberculosis* H37Rv transposon library reveals insertions in 351 ORFs and mutants with altered virulence. *Microbiology* 148, 2975–86.
7. Li, Y., Miltner, E., Wu, M., Petrofsky, M., and Bermudez, L. E. (2005) A *Mycobacterium avium* PPE gene is associated with the ability of the bacterium to grow in macrophages and virulence in mice. *Cell Microbiol* 7, 539–48.
8. Chen, J. M., German, G. J., Alexander, D. C., Ren, H., Tan, T., and Liu, J. (2006) Roles of Lsr2 in colony morphology and biofilm formation of *Mycobacterium smegmatis*. *J Bacteriol* 188, 633–41.
9. Movahedzadeh, F., Smith, D. A., Norman, R. A., Dinadayala, P., Murray-Rust, J., Russell, D. G., Kendall, S. L., Rison, S. C., McAlister, M. S., Bancroft, G. J., McDonald, N. Q., Daffe, M., Av-Gay, Y., and Stoker, N. G. (2004) The *Mycobacterium tuberculosis ino1* gene is essential for growth and virulence. *Mol Microbiol* 51, 1003–14.
10. Parish, T., Schaeffer, M., Roberts, G., and Duncan, K. (2005) HemZ is essential for heme biosynthesis in *Mycobacterium tuberculosis*. *Tuberculosis (Edinb)* 85, 197–204.
11. Parish, T., and Stoker, N. G. (2000) *glnE* is an essential gene in *Mycobacterium tuberculosis*. *J Bacteriol* 182, 5715–20.
12. Hernandez Pando, R., Aguilar, L. D., Infante, E., Cataldi, A., Bigi, F., Martin, C., and Gicquel, B. (2006) The use of mutant mycobacteria as new vaccines to prevent tuberculosis. *Tuberculosis* 86, 203–10.
13. Perez, E., Samper, S., Bordas, Y., Guilhot, C., Gicquel, B., and Martin, C. (2001) An essential role for *phoP* in *Mycobacterium tuberculosis* virulence. *Mol Microbiol* 41, 179–87.
14. Smith, D. A., Parish, T., Stoker, N. G., and Bancroft, G. J. (2001) Characterization of auxotrophic mutants of *Mycobacterium tuberculosis* and their potential as vaccine candidates. *Infect Immun* 69, 1142–50.

15. Vilcheze, C., Wang, F., Arai, M., Hazbon, M. H., Colangeli, R., Kremer, L., Weisbrod, T. R., Alland, D., Sacchettini, J. C., and Jacobs, W. R. Jr. (2006) Transfer of a point mutation in *Mycobacterium tuberculosis inhA* resolves the target of isoniazid. *Nat Med* 12, 1027–9.

16. Brooks, L. A. (1998) Chemical mutagenesis of mycobacteria. *Methods Mol Biol* 101, 175–86.

17. Guilhot, C., Otal, I., Van Rompaey, I., Martin, C., and Gicquel, B. (1994) Efficient transposition in mycobacteria: construction of *Mycobacterium smegmatis* insertional mutant libraries. *J Bacteriol* 176, 535–9.

18. Martin, C., Timm, J., Rauzier, J., Gomez-Lus, R., Davies, J., and Gicquel, B. (1990) Transposition of an antibiotic resistance element in mycobacteria. *Nature* 345, 739–43.

19. McAdam, R. A., Weisbrod, T. R., Martin, J., Scuderi, J. D., Brown, A. M., Cirillo, J. D., Bloom, B. R., and Jacobs, W. R., Jr. (1995) In vivo growth characteristics of leucine and methionine auxotrophic mutants of *Mycobacterium bovis* BCG generated by transposon mutagenesis. *Infect Immun* 63, 1004–12.

20. Rengarajan, J., Bloom, B. R., and Rubin, E. J. (2005) Genome-wide requirements for *Mycobacterium tuberculosis* adaptation and survival in macrophages. *Proc Natl Acad Sci U S A* 102, 8327–32.

21. Husson, R. N., James, B. E., and Young, R. A. (1990) Gene replacement and expression of foreign DNA in mycobacteria. *J Bacteriol* 172, 519–24.

22. Kalpana, G. V., Bloom, B. R., and Jacobs, W. R., Jr. (1991) Insertional mutagenesis and illegitimate recombination in mycobacteria. *Proc Natl Acad Sci U S A* 88, 5433–7.

23. Machowski, E. E., Dawes, S., and Mizrahi, V. (2005) TB tools to tell the tale-molecular genetic methods for mycobacterial research. *Int J Biochem Cell Biol* 37, 54–68.

24. Reyrat, J. M., Berthet, F. X., and Gicquel, B. (1995) The urease locus of *Mycobacterium tuberculosis* and its utilization for the demonstration of allelic exchange in *Mycobacterium bovis* bacillus Calmette-Guerin. *Proc Natl Acad Sci U S A* 92, 8768–72.

25. Balasubramanian, V., Pavelka, M. S., Jr., Bardarov, S. S., Martin, J., Weisbrod, T. R., McAdam, R. A., Bloom, B. R., and Jacobs, W. R., Jr. (1996) Allelic exchange in *Mycobacterium tuberculosis* with long linear recombination substrates. *J Bacteriol* 178, 273–9.

26. Parish, T., and Stoker, N. G. (2000) Use of a flexible cassette method to generate a double unmarked *Mycobacterium tuberculosis tlyA plcABC* mutant by gene replacement. *Microbiology* 146 (Pt 8), 1969–75.

27. Cox, M. M. (1999) Recombinational DNA repair in bacteria and the RecA protein. *Prog Nucleic Acid Res Mol Biol* 63, 311–66.

28. Muttucumaru, D. G., and Parish, T. (2004) The molecular biology of recombination in mycobacteria: what do we know and how can we use it? *Curr Issues Mol Biol* 6, 145–57.

29. Hinds, J., Mahenthiralingam, E., Kempsell, K. E., Duncan, K., Stokes, R. W., Parish, T., and Stoker, N. G. (1999) Enhanced gene replacement in mycobacteria. *Microbiology* 145, 519–27.

30. Pelicic, V., Reyrat, J. M., and Gicquel, B. (1996) Generation of unmarked directed mutations in mycobacteria, using sucrose counter-selectable suicide vectors. *Mol Microbiol* 20, 919–25.

31. Pashley, C. A., Parish, T., McAdam, R. A., Duncan, K., and Stoker, N. G. (2003) Gene replacement in mycobacteria by using incompatible plasmids. *Appl Environ Microbiol* 69, 517–23.

32. Pelicic, V., Reyrat, J. M., and Gicquel, B. (1998) Genetic advances for studying *Mycobacterium tuberculosis* pathogenicity. *Mol Microbiol* 28, 413–20.

33. Parish, T., Smith, D. A., Kendall, S., Casali, N., Bancroft, G. J., and Stoker, N. G. (2003) Deletion of two-component regulatory systems increases the virulence of *Mycobacterium tuberculosis*. *Infect Immun* 71, 1134–40.

34. Parish, T., Smith, D. A., Roberts, G., Betts, J., and Stoker, N. G. (2003) The senX3-regX3 two-component regulatory system of *Mycobacterium tuberculosis* is required for virulence. *Microbiology* 149, 1423–35.

35. Parish, T., Turner, J., and Stoker, N. G. (2001) *amiA* is a negative regulator of acetamidase expression in *Mycobacterium smegmatis*. *BMC Microbiol* 1, 19.

36. Chauhan, A., Madiraju, M. V., Fol, M., Lofton, H., Maloney, E., Reynolds, R., and Rajagopalan, M. (2006) *Mycobacterium tuberculosis* cells growing in macrophages are filamentous and deficient in FtsZ rings. *J Bacteriol* 188, 1856–65.

37. Dawes, S. S., Warner, D. F., Tsenova, L., Timm, J., McKinney, J. D., Kaplan, G., Rubin, H., and Mizrahi, V. (2003) Ribonucleotide reduction in *Mycobacterium tuberculosis*: function and expression of genes encoding class Ib and class II ribonucleotide reductases. *Infect Immun* 71, 6124–31.

38. Matsoso, L. G., Kana, B. D., Crellin, P. K., Lea-Smith, D. J., Pelosi, A., Powell, D., Dawes, S. S., Rubin, H., Coppel, R. L., and Mizrahi, V. (2005) Function of the cytochrome bc1-aa3 branch of the respiratory network in mycobacteria and network adaptation occurring in response to its disruption. *J Bacteriol* 187, 6300–8.

39. Curry, J. M., Whalan, R., Hunt, D. M., Gohil, K., Strom, M., Rickman, L., Colston, M. J., Smerdon, S. J., and Buxton, R. S. (2005) An ABC transporter containing a forkhead-associated domain interacts with a serine-threonine protein kinase and is required for growth of *Mycobacterium tuberculosis* in mice. *Infect Immun* 73, 4471–7.

40. Onwueme, K. C., Ferreras, J. A., Buglino, J., Lima, C. D., and Quadri, L. E. (2004) Mycobacterial polyketide-associated proteins are acyltransferases: proof of principle with *Mycobacterium tuberculosis* PapA5. *Proc Natl Acad Sci USA* 101, 4608–13.

41. Shimono, N., Morici, L., Casali, N., Cantrell, S., Sidders, B., Ehrt, S., and Riley, L. W. (2003) Hypervirulent mutant of *Mycobacterium tuberculosis* resulting from disruption of the *mce1* operon. *Proc Natl Acad Sci U S A* 100, 15918–23.

42. Boshoff, H. I., Reed, M. B., Barry, C. E., 3rd, and Mizrahi, V. (2003) DnaE2 polymerase contributes to in vivo survival and the emergence of drug resistance in Mycobacterium tuberculosis. *Cell* 113, 183–93.

43. Primm, T. P., Andersen, S. J., Mizrahi, V., Avarbock, D., Rubin, H., and Barry, C. E. III (2000) The stringent response of Mycobacterium tuberculosis is required for long-term survival. *J Bacteriol* 182, 4889–98.

44. Hu, Y., Movahedzadeh, F., Stoker, N. G., and Coates, A. R. (2006) Deletion of the Mycobacterium tuberculosis alpha-crystallin-like hspX gene causes increased bacterial growth in vivo. *Infect Immun* 74, 861–8.

Chapter 21
Phage Transposon Mutagenesis

M. Sloan Siegrist and Eric J. Rubin

Abstract Phage transduction is an attractive method of genetic manipulation in mycobacteria. φMycoMarT7 is well suited for transposon mutagenesis as it is temperature sensitive for replication and contains T7 promoters that promote transcription, a highly active transposase gene, and an *Escherichia coli* oriR6 K origin of replication. Mycobacterial transposon mutant libraries produced by φMycoMarT7 transduction are amenable to both forward and reverse genetic studies. In this protocol, we detail the preparation of φMycoMarT7, including a description of the phage, reconstitution of the phage, purification of plaques, preparation of phage stock, and titering of phage stock. We then describe the transduction procedure and finally outline the isolation of individual transposon mutants.

Keywords library · mutagenesis · mycobacteria · phage · transduction · transposon

21.1 Introduction

Transposons are highly useful tools for bacterial genetic studies. They have several advantages over chemical mutagenesis. First, mutants can be easily separated from wild type cells by use of an antibiotic marker. Second, whereas the locations of the small changes produced by chemical mutagenesis can be difficult to identify, transposons mark their sites of insertion and can be easily isolated. Third, transposons can be constructed so that recipient strains contain only a single mutation. Finally, although these elements generally disrupt genes into which they insert, they can be engineered to have useful properties such as the ability to form transcriptional or translational fusions.

E.J. Rubin
HSPH, Immunology & Infectious Disease, 200 Longwood Ave, Armenise 439, Boston, MA 02115, USA
e-mail: erubin@hsph.harvard.edu

T. Parish, A.C. Brown (eds.), *Mycobacteria Protocols*,
doi: 10.1007/978-1-59745-207-6_21, © Humana Press, Totowa, NJ 2008

To create a transposon mutagenesis system for mycobacteria, we searched for a transposon that would produce random mutations at high frequency. We additionally sought an element that would produce stable mutants so that, once integrated, the transposon would not be lost. To this end, we engineered a eukaryotic transposon first isolated from the blowfly *Haematobia irritans* [1]. This element, *Himar1*, is a member of the mariner family of transposable elements. Like all mariner transposons, *Himar1* uses a cut-and-paste mechanism of transposition, resulting in a simple structure at the point of insertion. In addition, it has little sequence specificity beyond a TA dinucleotide at the insertion site that is duplicated during transposition. Together with David Lampe, we found that this element transposes efficiently in a variety of bacteria, and we isolated a hyperactive transposase that produces high-frequency transposition [1].

Although transposon mutagenesis is a powerful tool for making marked, highly complex mutant libraries, until recently, this and other genetic manipulations were difficult in slow-growing, clumpy mycobacteria. Delivery of transposons via suicide vectors has been moderately successful but is dependent on electroporation efficiency and is therefore challenging in many slow-growing mycobacterial species. Phage transduction is an attractive alternative as infection of every cell generates a large number of mutants. This system is especially potent when using a conditionally replicating mutant bacteriophage, a valuable reagent first developed in William Jacobs' lab [2].

Starting with the engineered *Himar1* transposon, we constructed ϕMycoMarT7 with the following properties (Fig 21.1):

- A kanamycin resistance cassette located within the transposon allows selection of resistant mutants.
- The temperature sensitivity of the bacteriophage promotes very efficient delivery of DNA. Because this defective phage cannot replicate or integrate into the genome, the phage is always lost. Thus, the only recipient cells that have stable antibiotic resistance have an integrated copy of the transposon.
- The transposase gene is located outside of the transposon boundaries. Once transposition occurs, the transposase gene is lost along with the phage vector. Thus, the element is incapable of transposing to a new site in the chromosome once integrated.
- Short inverted repeats flanking the element allow recognition by the transposase while producing the smallest possible insertion.
- The *E. coli* oriR6K origin of replication allows replication of DNA as a plasmid in appropriate pir+ *E. coli* strains and facilitates cloning insertions out from chromosomal DNA [3].
- Outward-facing T7 promoters allow *in vitro* transcription into the adjacent chromosomal DNA [4]. These can be used to map the location of transposon insertions in individual mutant strains or in large pools. They serve as the basis for the transposon site hybridization (TraSH) procedure [4].

This chapter describes the creation of a mycobacterial transposon library by ϕMycoMarT7 transduction as well as a method to isolate individual mutants.

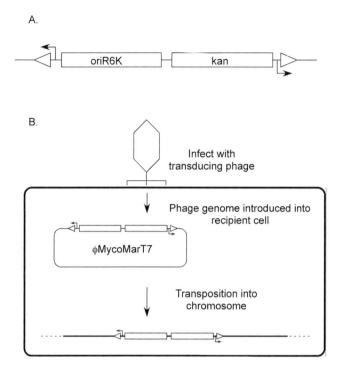

Fig. 21.1 A, representation of φMycoMarT7 showing the kanamycin gene, inverted repeats, T7 promoters, and *E. coli* oriR6 K origin of replication. B, representation of transduction

The methods below outline (1) the preparation of the phage stock, (2) the transduction of *M. tuberculosis* or *M. bovis* BCG, and (3) the isolation of individual transposon mutants. As outlined above, φMycoMarT7 is temperature sensitive for replication and carries a transposon that encodes a kanamycin resistance gene flanked by 29-bp inverted repeats, T7 promoters that promote transcription into adjacent chromosomal DNA [4], the highly active C9 *Himar1* transposase gene [1], and *E. coli* oriR6 K origin of replication [3]. The transposon sequence has been deposited in GenBank under accession number AF411123 (Fig. 21.1). φMycoMarT7 is available as a DNA stock from the Rubin laboratory.

21.2 Materials

21.2.1 Phage Stock Preparation

1. φMycoMarT7 is available as a DNA stock from the Rubin laboratory. (Please send requests by mail to 200 Longwood Ave, Armenise 439, Boston, MA 02115 USA; or by e-mail to erubin@hsph.harvard.edu.)

2. QIAGEN Plasmid Purification kit (Qiagen Valencia, CA) or similar.
3. Top Agar: Dissolve 0.6 g agar in 100 mL distilled water and autoclave. After autoclaving, add 20 μL 1 M sterile $CaCl_2$ (to give 2 mM final concentration).
4. 100 × 15 mm Petri dishes.
5. *Mycobacterium smegmatis* mc^{2}155 competent cells.
6. *Mycobacterium smegmatis* mc^{2}155 saturated culture.
7. MP Buffer: 50 mM Tris-HCl (pH 7.5), 150 mM NaCl, 10 mM $MgSO_4$, 2 mM $CaCl_2$. Filter sterilize.
8. 20% (v/v) Tween-80: mix 20 mL Tween-80 with 80 mL distilled water, filter sterilize.
9. 7 H10 plates: Dissolve 19 g Middlebrook 7 H10 medium (Difco Franklin Lakes, NJ) in 900 mL deionized water. Add 5 mL glycerol (v/v). Autoclave. When medium is cooled to ~50°C, add 100 mL albumin dextrose catalase (ADC) or oleic albumin dextrose catalase (OADC) enrichment. Mix well and pour in standard plastic Petri dishes.
10. 7 H9 medium: Dissolve 4.7 g Middlebrook 7 H9 broth base (Difco) in 900 mL deionized water and add 2 mL glycerol, mix well, and filter sterilize. Add 5 mL 20% (v/v) Tween-80 (final concentration of 0.05% v/v). Add 100 mL ADC or OADC immediately before use.
11. *E. coli* competent cells.
12. Sterile toothpicks.
13. 1.5 mL sterile microfuge tubes.
14. Benchtop microfuge.
15. Lysogeny Broth (LB) media (Difco): To distilled water add 10 g tryptone (pancreatic digest of casein), 5 g yeast extract, 5 g NaCl, and make up to 1 L. Autoclave to sterilize. For solid medium, add 15 g agar before autoclaving. Allow to cool and pour onto Petri dishes.
16. 0.2 μm filter and syringe.

21.2.2 Mycobacterial Transduction

1. *Mycobacterium bovis* BCG, *M. tuberculosis, or M. smegmatis.*
2. 1 L roller bottles (Corning Corning, NY).
3. Kanamycin: 20 μg/mL (for mycobacteria) or 50 μg/mL (for *E. coli*). Prepare a stock concentration of 50 mg/mL by dissolving 1 g in 20 mL distilled water, filter sterilize, and store at 4°C.
4. 2 mL cryovials (Nunc Nalge Nunc, Rochester, NY).
5. Prewarmed 7 H9 (no Tween-80) medium: Dissolve 4.7 g Middlebrook 7 H9 broth base (Difco) in 900 mL deionized water and add 2 mL glycerol, mix well, and filter sterilize. Do not add Tween-80. Add 100 mL ADC (*M. smegmatis*) or OADC (*M. bovis* BCG or *M. tuberculosis*) immediately before use. Prewarm to 37°C prior to use.

6. MP Buffer: 50 mM Tris-HCl (pH 7.5), 150 mM NaCl, 10 mM MgSO$_4$, 2 mM CaCl$_2$. Filter sterilize. Warm to 37°C prior to use.
7. 10% (v/v) glycerol: dissolve 10 mL glycerol in 90 mL distilled water and autoclave to sterilize.
8. Large Petri dishes.
9. 7H10 plates: Dissolve 19 g Middlebrook 7H10 medium (Difco) in 900 mL deionized water. Add 8.4 mL 60% glycerol (v/v). Autoclave. When media is cooled to ~50°C, add 100 mL albumin dextrose catalase (ADC) or oleic albumin dextrose catalase (OADC) enrichment. Add kanamycin to a final concentration of 20 µg/mL. Mix well and pour in large plastic Petri dishes.
10. 4 mm glass beads (Walter Stern Inc. Port Washington, NY): autoclave.
11. Aerosol resistant pipette tips.

21.2.3 Transposon Mutant Isolation

1. SacII restriction enzyme (New England Biolabs Ipswitch, MA).
2. ATP (New England Biolabs).
3. T4 DNA ligase (New England Biolabs).
4. Buffer 4 (New England Biolabs).
5. Dialysis membranes (Millipore Billerica, MA).
6. Vacuum microfuge.
7. DH5α λ pir E. coli competent cells [6].
8. 96-deep-well plates (Corning).
9. 50 × 15 mm Petri dishes.
10. 50 mL conical tubes.
11. 96-well PCR plates and adhesive foil sealing covers.
12. Taq polymerase and buffer (TaKaRa).
13. 4 mm glass beads (Walter Stern Inc.).
14. Oligonucleotide primers. 5′-GCCTTCTTGACGAGTTCTTCTGAG-3′ and 5′-GCTCTACGTGGGAGTCGGACAATGTTG -3′ (see Note 1).
15. Kanamycin: 20 µg/mL (for mycobacteria) or 50 µg/mL (for E. coli). Prepare a stock concentration of 50 mg/mL by dissolving 1 g in 20 mL distilled water, filter sterilize, and store at 4°C.
16. LB agar (Difco): To distilled water add 10 g tryptone (pancreatic digest of casein), 5 g yeast extract, 5 g NaCl, and 15 g agar and make up to 1 L. Autoclave to sterilize. Allow to cool, add kanamycin to give a final concentration of 50 µg/mL, and pour onto Petri dishes.
17. 7H9 medium: Dissolve 4.7 g Middlebrook 7H9 broth base (Difco) in 900 mL deionized water and add 2 mL glycerol, mix well, and filter sterilize. Add 5 mL 20% (v/v) Tween-80 (final concentration of 0.05% v/v). Add 100 mL ADC or OADC and kanamycin to a final concentration of 20 µg/mL immediately before use.
18. Inkwells (Nagle Nunc, Rochester, NY).

19. QIAGEN Plasmid Purification kit (Qiagen) or similar.
20. Sterile distilled water.

21.3 Methods

For schematics of the procedures outlined in this section, please refer to
Figures 21.2 and 21.3.

21.3.1 Phage Stock Preparation

21.3.1.1 φMycoMarT7-Phage Reconstitution

1. φMycoMarT7 is available as a DNA stock. Transform plasmid into *E. coli*
   competent cells and recover 1 h at 37°C in 1 mL LB medium.
2. Plate on LB-agar containing 50 μg/mL kanamycin.
3. Incubate at 37°C overnight.
4. Scrape bacteria from the plate and inoculate 5 mL LB medium containing
   Kanamycin.
5. Incubate at 37°C overnight with shaking at ~200 rpm.
6. Prepare plasmid DNA using a commercial mini-prep kit, as per manufacturer's instructions.
7. Prepare top agar and cool to 42°C.
8. Transform plasmid into competent *M. smegmatis* by electroporation.

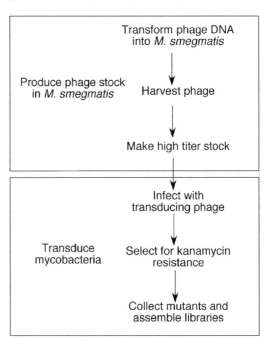

Fig. 21.2 Schematic
of φMycoMarT7 stock
preparation and
transduction into
mycobacteria

Identify transposon insertion sites

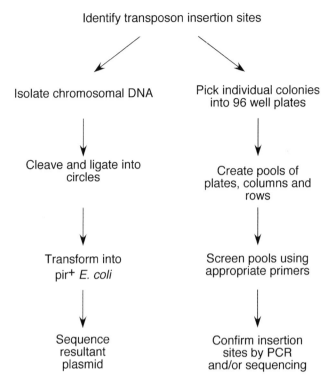

Fig. 21.3 Schematic of transposon mutant isolation. The first column outlines procedures to follow if the investigator has a mutant with an interesting phenotype and wishes to map the transposon insertion site. The second column is guide for the investigator who wants to isolate a transposon mutant in a particular gene of interest for phenotypic characterization

9. Recover cells in 1 mL 7 H9 medium supplemented with glycerol and ADC but *not* Tween-80 for 3 h at 37°C (*see* **Note 2**).
10. Centrifuge the transformed *M. smegmatis* to pellet cells.
11. Resuspend in 150 μL 7 H9 medium supplemented with glycerol and ADC but *not* Tween-80.
12. Add 10 or 100 μL of transformation to 3.5 mL of cooled top agar.
13. Pour onto LB plates and incubate at 30°C. Although a few big plaques may appear earlier, allow 48 h for small plaques.

21.3.1.2 Plaque Purification

1. Dilute a saturated culture of *M. smegmatis* 1:1000 in 7 H9 medium and grow at 37°C with shaking at ∼100 rpm for approximately 24 h to stationary phase.
2. Prepare top agar and cool to 42°C.

3. Add 500 μL fresh stationary phase culture (from step 1) to 7 mL top agar and pour onto two LB plates. Allow plates to dry for a few hours.
4. Patch 5 to 10 plaques (from Section 21.3.1.1) onto both plates with sterile toothpicks.
5. Incubate one plate at 30°C and the other at 37°C for 36 to 48 h. Most or all plaques will appear only at the lower temperature (*see* **Note 3**).
6. Temperature-sensitive (ts) plaques will appear on the plate incubated at 30°C but not on the one incubated at 37°C.
7. Excise the agar containing a ts clone and crush in 250 μL MP buffer in a 1.5 mL Microfuge tube.
8. Pellet the agar by centrifugation at room temperature in a microcentrifuge.
9. Sterilize the supernatant by passing through using an 0.2 μm filter attached to a syringe and store at 4°C.
10. Dilute a saturated culture of *M. smegmatis* 1:1000 in 7 H9 medium and grow at 37°C with shaking at ~100 rpm for approximately 24 h to stationary phase.
11. Prepare top agar and cool to 42°C.
12. Pellet cells from 500 μL of fresh stationary phase *M. smegmatis* in a 1.5 mL Microfuge tube for 1 min at top speed.
13. Wash pellet twice with an equal volume of MP buffer (*see* **Note 2**).
14. Resuspend cells in 500 μL MP buffer (*see* **Note 2**).
15. Dilute 5 μL of the stored temperature-sensitive clone in 50 μL of MP buffer. Store the remainder at 4°C.
16. Make 10-fold dilutions for the temperature-sensitive clone in 50 μL of MP buffer down to 10^{-5}.
17. Mix 100 μL of the washed *M. smegmatis* with 10 μL of each temperature-sensitive clone dilution (10^0 to 10^{-5}).
18. Add 100 μL of phage and bacteria mixture to 3.5 mL of cooled top agar, pour onto separate LB plates, and incubate at 30°C. Determine the dilution that yields near-confluent plaques.

21.3.1.3 Phage Stock Preparation

1. Dilute a saturated culture of *M. smegmatis* 1:1000 in 7 H9 medium and grow at 37°C with shaking at ~100 rpm for approximately 24 h to stationary phase.
2. Prepare top agar and cool to 42°C.
3. Pellet cells from 500 μL of fresh stationary phase *M. smegmatis* in a 1.5-mL microfuge tube for 1 min at top speed.
4. Wash pellet twice with an equal volume of MP buffer (*see* **Note 2**).
5. Resuspend cells in 500 μL MP buffer (*see* **Note 2**).
6. Phage stock should be generated from the TS clone identified in Section 21.3.1.2. Mix the washed *M. smegmatis* with enough phage to create near-confluent plaques (as determined in Section 21.3.1.2). Add 100 μL of phage

and bacteria mixture to 3.5 mL of cooled top agar and pour on a 7 H10 plate (*see* **Note 4**). Prepare five plates.

7. Incubate at 30°C until "lacy" (approximately 48 h).
8. Flood each plate with 3 mL MP buffer (*see* **Note 5**) and gently rock plates at 4°C for several hours.
9. Collect the MP buffer plate stock plate stock, sterilize with an 0.2 μm filter and store at 4°C.

21.3.1.4 Phage Stock Titering

1. Dilute a saturated culture of *M. smegmatis* 1:1000 in 7 H9 medium and grow at 37°C with shaking at ~100 rpm for approximately 24 h to stationary phase (*see* **Note 6**).
2. Prepare top agar and cool to 42°C.
3. To prepare a lawn of *M. smegmatis*, add 250 μL fresh stationary phase culture to 3.5 mL top agar, and pour onto an LB plate. Allow plate to dry for a few hours.
4. Make 10-fold dilutions of phage stock (from 21.3.1.3) in 100 μL MP buffer.
5. Spot 10 μL of each dilution onto the plate.
6. Dry and incubate at 30°C until plaques appear (approximately 48 h). Titer should be greater than 5×10^{10} phage/mL (*see* **Note 6**).

21.3.2 Mycobacterial Transduction

Once the preparation of ϕMycoMarT7 is complete, the next step is transducing the transposon into target mycobacteria. This section describes infection for *M. bovis* BCG or *M. tuberculosis*. Modifications to the protocol for transduction of *M. smegmatis* are detailed in **Note 7**.

1. Grow 100 mL of *M. tuberculosis* or *M. bovis* BCG culture in 7 H9 medium in a roller bottle to OD_{600} 0.8 to 1.0.
2. Spin down 50 mL of culture and wash with an equal volume of MP buffer warmed to 37°C (*see* **Note 2**).
3. Resuspend in 5 mL MP buffer, removing a 200 μL aliquot to serve as a control.
4. Add approximately 10^{11} phage or MP buffer control using an aerosol-resistant pipette tip and incubate for 3 to 4 h at 37°C. The volume of phage added depends on titer obtained in Section 21.3.1.4 and is ideally less than 2 mL.
5. Freeze transduced cells in 200 μL aliquots of 10% glycerol at –80°C.
6. Make 10-fold dilutions of a thawed aliquot in 7 H9 medium and plate on 7 H10 agar. Wrap plates in foil and incubate at 37°C for 2 to 3 weeks for *M. bovis* BCG or 3 to 4 weeks for *M. tuberculosis*.
7. Calculate the titer of the transduced bacteria (*see* **Note 6**).

8. Once the transduction titer is established, plate enough transduced bacteria onto five large 7 H10 plates supplemented with 20 µg/mL kanamycin to yield ≥100,000 kanamycin-resistant colonies. Plating with glass beads gives the most even distribution of colonies across the plate and facilitates picking individual clones. To do this, allow plate to dry such that there is no condensation on the lid. Carefully pour 10 to 20 4 mm autoclaved glass beads on the surface of the agar. Add approximately 300 to 500 µL of culture. Shake firmly in back-and-forth and side-to-side motions for a few minutes until the surface of the plate is evenly covered. Dry plate until there is no more liquid "trail" left behind moving beads. Pour off beads into waste container.

9. Wrap plates in foil and incubate at 37°C for 10 to 12 days for *M. bovis* BCG or 2 to 3 weeks for *M. tuberculosis*. The colonies should be small but clearly visible.

10. To make frozen stocks of the transposon library, scrape plate contents into 50 mL conical tubes containing 7 H9 medium. Rock several hours at room temperature to ensure proper mixing (*see* **Note 8**). Make multiple aliquots in cryovials and store at –80°C. Each stock should be thawed only once to retain library diversity.

21.3.3 Transposon Mutant Isolation

Mapping mutant pools created by φMycoMarT7 transduction is possible via the transposon site hybridization (TraSH) technique but is beyond the scope of the current chapter [4]. This section will instead focus on the isolation of individual transposon mutants (*see* **Note 9**).

21.3.3.1 Transposon Insertion Mapping

1. Prepare genomic DNA from transposon mutant (*see* **Note 10**).
2. Digest 1 µg of the genomic DNA with *Sac*II in a 30 µL volume containing NEB Buffer 4 overnight at 37°C.
3. Inactivate the enzyme for 20 min at 65°C.
4. To ligate the chromosomal pieces into circles for cloning, add 4 µL 10 mM ATP, 1 µL NEB Buffer 4, 1 µL T4 DNA ligase, 4 µL dH$_2$O, and ligate mixture overnight at 16°C.
5. Inactivate the enzyme for 20 min at 65°C.
6. Reduce volume to approximately 15 µL by drying in speed vacuum.
7. Fill a Petri dish with dH$_2$O and float a dialysis membrane on the surface of the water. Gently pipette ligation mix onto the membrane and allow dialysis to occur for 20 to 30 min.
8. Transform dialyzed ligation mix into DH5α λ *pir E. coli* competent cells and plate entire transformation on LB plates with 50 µg/mL kanamycin.
9. Incubate overnight at 37°C.

10. Pick a colony and inoculate 5 mL LB with 50 μg/mL kanamycin.
11. Incubate overnight at 37°C with shaking at ~200 rpm.
12. Prepare plasmid DNA. Sequence the insertion site using a primer that anneals to the transposon sequence (*see* **Note 11**).

21.3.3.2 Specific Transposon Mutant Isolation

1. Incubate the plated *M. bovis BCG* or *M. tuberculosis* library generated in Section 21.3.2 at 37°C for 3 weeks.
2. Fill individual 96-deep-well plates with 400 μL 7 H9 medium supplemented with 20 μg/mL kanamycin. Pick transposon mutants using sterile toothpicks into individual wells. A diverse collection of multiple-sized colonies is desired (*see* **Note 12**).
3. Incubate at 37°C for 4 to 6 weeks without agitation to allow all mutants to grow. To prevent fungal contamination, do not open plates during this time.
4. When the library is cultivated to the desired point, the collection should be replicated and the original plates frozen for long-term storage. Use a multi-channel pipette to transfer 50 μL from the original plate to a second 96-deep-well plate containing 300 μL 7 H9 medium with 20 μg/mL kanamycin. Incubate 1 week at 37°C and store at –80°C.
5. At the same time, an aliquot of the original cultures should be inactivated so that it may be removed from the BL-3 for future transposon insertion sequencing or screening. To do this, transfer an additional 50 μL from the original plate to a 96-well PCR plate. Seal with adhesive foil, heat for 2 h at 80°C to inactivate the bacilli, and store at –80°C.
6. Transfer well contents for PCR screening by creating column, row, and whole plate pools [7]. Combine a small amount of heat-inactivated culture from each well in a given row into a single well on a different plate. Repeat for every row on every plate. Repeat the entire procedure for every column on every plate. Finally, combine a small amount of heat-inactivated culture from each well in an entire plate into a single well on a different plate. Repeat for every plate. In this way, each original mutant colony will be represented in three separate pools. Maintain pools in a 96-well format for further manipulation.
7. In the first round, carry out PCR reaction using whole plate pools. Amplify target of interest by using a primer that hybridizes close to the gene and a second that hybridizes within the inverted repeats flanking the transposon sequence (*see* **Note 13**). Add 1 μL of template in a total volume of 20 μL with Taq polymerase and buffer (TaKaRa). PCR conditions: 95°C for 5 min followed by 30 cycles of 95°C for 30 s, 65°C for 30 s, 70°C for 2 min; and a final extension of 70°C for 10 min.
8. Once the pool containing the gene of interest is identified, repeat the amplification using the appropriate column and row pools.
9. To confirm target isolation, inoculate an inkwell containing 10 mL 7 H9 medium with 20 μg/mL kanamycin with the predicted transposon mutant from the original plate. Repeat the PCR and sequence.

21.4 Notes

1. 5′-GCCTTCTTGACGAGTTCTTCTGAG-3′ binds within the kanamycin resistance gene of φMycoMarT7 and is used for sequencing the junction of the transposon insertion. 5′-GCTCTACGTGGGAGTCGGACAATGTTG -3′ binds within the inverted repeats flanking the transposon sequence of φMycoMarT7 and is used for PCR screening for specific transposon mutants.

2. Alternatively, 7H9 supplemented with glycerol and ADC but *not* Tween-80 can be used in place of MP buffer Although Tween-80 is often used to reduce clumping in mycobacterial cultures, it inhibits phage attachment and therefore must be washed out of the cells for phage stock preparation and transduction. Likewise, infection with phage requires the presence of calcium that is present in MP buffer.

3. Because the mutation that produces temperature sensitivity in the phage can revert, it is important to assay the plaques that will be used to make stocks to ensure they remain mutants. To do so, incubate one at 30°C and the other at 37°C for 36 to 48 h. Most or all plaques will appear only at the lower temperature. As the transposase gene is active in *E. coli* and the only marker encoded in the transposon, this plasmid tends to be unstable upon passage.

4. *M. smegmatis* can typically be grown on LB, but 7H10 agar should be used if the stock is intended for *M. tuberculosis* or *M. bovis* BCG transduction.

5. Flood with a smaller volume of MP buffer if a higher titer is desired. The titer of the phage stock = [(1000 µL)(number of plaques)]/[(amount plated in µL)(dilution)]. For example, the titer of a stock in which there are 12 plaques in a spot containing 10 µL of 10^{-8} dilution of a φMycoMarT7 suspension = [(1000 µL)(12)]/[(10 µL)(10^{-8})] = 1.2×10^{10} plaques/mL.

6. The titer of the transduced bacteria = [(1000 µL)(number of colonies)]/[(amount plated in µL)(dilution)].

7. *M. smegmatis* may be grown in 7H9 containing ADC or no additives at all in lieu of the more expensive OADC. The culture should reach an OD_{600} of 1.0 to 1.2. After pelleting the cells and washing twice in MP buffer, resuspend in 1/10 original volume of MP buffer, removing an aliquot to serve as a control. Warm to 37°C in water bath and add approximately 1 mL of 5×10^{10} phage or MP buffer control. Plate the phage infection on 5 to 10 large LB plates supplemented with 20 µg/mL kanamycin and 0.05% Tween-80. As transduction efficiency in *M. smegmatis* is lower than *M. bovis* BCG or *M. tuberculosis,* this should yield 10,000 to 50,000 kanamycin-resistant colonies.

8. We have also had success vortexing the solution. The goal of either method is to mix the scraped library thoroughly into a homogenous solution.

9. For a reverse genetics approach, it is crucial to be able to map the genetic lesion of interest. φMycoMarT7 contains the oriR6K *E. coli* origin of replication for this purpose. Plasmids that contain this sequence can only replicate in *pir+* strains of *E. coli.* Although we have had success by degenerate PCR, direct cloning of the region surrounding the transposon insertion is our preferred method for mapping transposon insertions. Section 21.3.3.1 reviews this cloning strategy. For a forward genetics approach, it is important to begin with the genetic lesion of interest. Section 21.3.3.2 describes procedures to cultivate, maintain, and store an arrayed library of individual transposon mutants. Such a library can be used for large-scale sequencing [5] or to screen for individual mutants.

10. For optimal results, the concentration of DNA should be above 80 ng/µL and the purity as measured by $OD_{260/280}$ ratio should be above 1. Of the two parameters, DNA purity has more of an effect on the number of colonies obtained from the subsequent ligation.

11. The sequencing primer 5′-GCCTTCTTGACGAGTTCTTCTGAG-3′ listed in Section 21.2.3 binds within the kanamycin gene of φMycoMarT7. We have also had success with the primer 5′-CGCTTCCTCGTGCTTTACGGTATCG-3′.

12. At least two-fold excess coverage of the genome is desirable as contamination is common in slow-growing species in 96-well plates.

13. The second primer allows amplification independent of orientation.

Acknowledgments The authors thank Dr. C. Sassetti, J. Lane, Dr. S. Fortune, and A. Garces for method development. We also thank J. Murry and Dr. M. Unnikrishnan for critical reading of this manuscript.

References

1. Lampe, D.J., Akerley, B.J., Rubin, E.J., Mekalanos, J.J., Robertson, H.M. (1999) Hyperactive transposase mutants of the Himar1 mariner transposon. *Proc Natl Acad Sci USA* **96**, 11428–11433.

2. Bardarov, S., Kriakov, J., Carriere, C., Yu, S., Vaamonde, C., McAdam, R.A., Bloom, B.R., Hatfull, G.F., Jacobs, W.R. Jr. (1997) Conditionally replicating mycobacteriophages: a system for transposon delivery to *Mycobacterium tuberculosis*. *Proc Natl Acad Sci USA* **94**, 10961–10966.

3. Rubin, E.J., Akerley, B.J., Novik, V.N., Lampe, D.J., Husson, R.N., Mekalanos, J.J. (1999) *In vivo* transposition of mariner-based elements in enteric bacteria and mycobacteria. *Proc Natl Acad Sci USA* **96**, 1645–1650.

4. Sassetti, C.M., Boyd, D.H., Rubin, E.J. (2001) Comprehensive identification of conditionally essential genes in mycobacteria. *Proc Natl Acad Sci USA* **98**, 12712–12717.

5. Elliott, S.J. and Kaper, J.B. (1997) Role of type 1 fimbriae in EPEC infections. *Microb Pathog* **23**, 113–118.

6. Lane, J.M. and Rubin, E.J. (2006) Scaling down: a PCR-based method to efficiently screen for desired knockouts in a high density *Mycobacterium tuberculosis* picked mutant library. *Tuberculosis* **86**, 310–313.

7. Lamichhane, G., Zignol, M., Blades, N.J., Geiman, D.E., Dougherty, A., Grosset, J., Broman, K.W., Bishai, W.R. (2003) A postgenomic method for predicting essential genes at subsaturation levels of mutagenesis: application to *Mycobacterium tuberculosis*. *Proc Natl Acad Sci USA* **100**, 7213–7218.

Chapter 22
Gene Essentiality Testing in *Mycobacterium smegmatis* Using Specialized Transduction

Apoorva Bhatt and William R. Jacobs Jr.

Abstract Conditional expression–specialized transduction essentiality test (CESTET) is a genetic tool used to determine essentiality of individual genes in *Mycobacterium smegmatis*. CESTET combines specialized transduction, a highly efficient gene knockout method, with the utility of the inducible acetamidase promoter. In a merodiploid strain containing a second integrated copy of an essential gene under the control of the acetamidase promoter, transductants (gene knockouts of the native chromosomal copy of a gene) are obtained only in the presence of acetamide in the selection medium. Furthermore, effects of loss of essential gene function can then be studied by growing the transductants in medium depleted of acetamide.

Keywords acetamidase · conditional expression · essentiality testing · phages · specialized transduction

22.1 Introduction

Specialized transduction is an efficient and relatively quick method of generating null mutants in mycobacteria [1]. The substrate for allelic exchange (a gene of interest disrupted by a selectable marker) is first cloned into a shuttle phasmid that contains a mycobacteriophage genome with a temperature-sensitive mutation. When introduced into mycobacteria, it can replicate as a phage at 30°C but not at the nonpermissive temperature of 37°C. Thus, the recombination substrate can be delivered to nearly all cells in the recipient mycobacterium population by infection, which is then followed by selecting transductants that have undergone a homologous recombination event on selective, antibiotic-containing media at 37°C. The efficiency of this gene knockout tool has been harnessed in combination with conditional gene expression using the acetamidase promoter

A. Bhatt
School of Biosciences, University of Birmingham, Edgbaston, Birmingham, B15 2TT, United Kingdom
e-mail: a.bhatt@bham.ac.uk

T. Parish, A.C. Brown (eds.), *Mycobacteria Protocols*,
doi: 10.1007/978-1-59745-207-6_22, © Humana Press, Totowa, NJ 2008

for testing gene essentiality in *Mycobacterium smegmatis* [2]. The acetamidase promoter is a chemically inducible promoter that is induced in the presence of acetamide [3, 4]. This gene essentiality testing method is termed CESTET (conditional expression–specialized transduction essentiality test) to distinguish it from other methods that make use of the acetamidase promoter [5, 6].

A preliminary step for any gene essentiality testing tool is to determine whether the gene under study is a *bona fide* candidate for the essentiality test (i.e., there should be some evidence that suggests that the gene is essential). Usually, this may be inferred by the inability to generate knockout mutants of the gene in a wild-type strain. Additionally, genome-wide transposon mutagenesis studies in *Mycobacterium tuberculosis* can also give clues as to which *M. smegmatis* homologues may be essential [7, 8]. The initial step in CESTET is the construction of a merodiploid strain containing a second copy of the putative essential gene (*peg*) under the control of the acetamidase promoter. First, the acetamidase promoter and *peg* are cloned into a single-copy integrating vector. The vector, pMV306, contains a gene encoding the mycobacteriophage L5 integrase (*int*), an attachment site (*attP*), a kanamycin resistance gene, and replicates as a plasmid in *Escherichia coli* [9] (Fig. 22.1). When electroporated into *M. tuberculosis*, *Mycobacterium bovis*, or *M. smegmatis*, pMV306 integrates as a single copy into the corresponding integration site (*attB*) in the bacterial chromosome. A single copy of *peg*, driven by the acetamidase promoter, can thus be integrated into the mycobacterial chromosome in this manner.

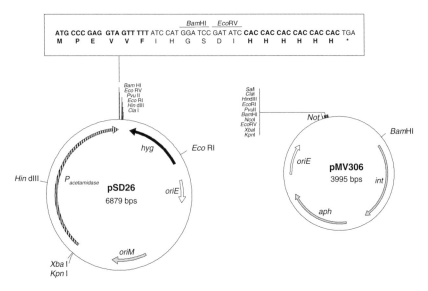

Fig. 22.1 Maps of pSD26 and pMV306. *aph*, kanamycin resistance cassette; *hyg*, hygromycin resistance cassette; $P_{acetamidase}$, inducible acetamidase promoter; *oriE*, ColE1 origin of replication in *E. coli*; *oriM*, mycobacterial replicon

In parallel, a specialized transducing phage (phΔ*peg*) containing an allelic exchange substrate designed to replace the target gene with a *Streptomyces hygroscopicus*–derived hygromycin resistance marker is generated. The merodiploid strain is then infected with the knockout phage, followed by selection at 37°C on plates containing hygromycin and acetamide. If a gene is essential, the ability to obtain transductants in a merodiploid strain is dependent on acetamidase-driven expression of the gene. Thus, if the gene is essential, colonies of hygromycin-resistant transductants (knockouts of the native copy of the gene) can only be obtained on plates containing acetamide (Fig. 22.2).

The protocols described in this chapter are specifically designed to generate merodiploid strains using pMV306 (*see* **Note 1**). In this particular strategy, *peg* is first cloned downstream of the acetamidase promoter in the mycobacterial expression plasmid pSD26 [10] (Fig. 22.1), from which a continuous fragment containing the acetamidase promoter and *peg* is cloned into pMV306. PCR conditions and selection of enzyme sites will depend largely on the gene to be amplified. Thus, PCR amplification, restriction digests, ligation, purification of

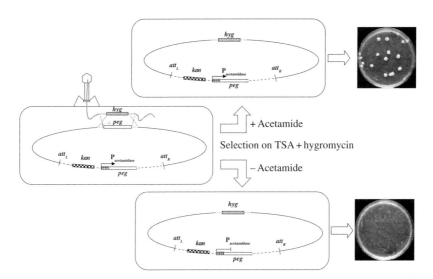

Fig. 22.2 Schematic representation of CESTET. After specialized transduction using a phage containing an allelic recombination substrate to replace *peg*, the transduction mix is split into two equal parts, which are plated on hygromycin-containing medium with or without acetamide. For essential genes, transductants grow only on plates containing acetamide. *attL* and *attR*, left and right junctions of the phage L5 attachment and chromosomal integration site; *kan*, kanamycin resistance cassette; *hyg*, hygromycin resistance cassette; $P_{acetamidase}$, inducible acetamidase promoter. The interrupted arc represents the integrative vector, the solid arc represents the mc^{2}155 chromosome, and the darker gray strand represents phage DNA. (Adapted from Bhatt, A., Kremer, L., Dai, A. Z., Sacchettini, J. C. & Jacobs, W. R. Jr. [2005]. Conditional depletion of KasA, a key enzyme of mycolic acid biosynthesis, leads to mycobacterial cell lysis. *J Bacteriol* **187**, 7596–7606.)

digested DNA fragments, which are standard methods, will not be discussed in detail here.

The open reading frame (ORF) for *peg* can cloned either as a *Bam*HI-*Cla*I fragment, *Bam*HI-*Eco*RV, or *Bam*HI-*Pvu*II into the corresponding sites of pSD26. For this purpose, *peg* is first PCR amplified from genomic DNA using primers that have the appropriate enzyme sites incorporated at the 5′-end. An important consideration while designing primers is that after cloning into pSD26, the start codon of *peg* should be in frame with the N-terminal portion of AmiE (Fig. 22.1). The PCR product is cloned into a standard PCR fragment-cloning and sequencing vector, and the sequence is checked for errors. The sequenced-verified fragment can then be excised using the appropriate restriction enzymes for subsequent cloning in pSD26.

Once a clone of *peg* is obtained in pSD26, the entire fragment containing the acetamidase promoter and downstream *peg* can be excised using the *Xba*I site upstream of the acetamidase promoter and one of the following sites: *Cla*I, *Eco*RV, *Pvu*II, *Hin*dIII, or *Eco*RI. The choice of the latter sites depends on which of these sites are present in the sequence of *peg* (if one or more recognition sites are present in the ORF, that particular enzyme cannot be used). The entire fragment can then be cloned in pMV306 digested with *Xba*I and one of the chosen downstream enzymes to yield the plasmid pMV306P$_{acetamidase}$*peg*. For an alternate cloning strategy, *see* **Note 2**.

22.2 Materials

22.2.1 Generation of High Titers of ph∆peg Mycobacteriophage Particles

1. Allelic Exchange Substrate (AES) plasmid (*see* **Note 3**) designed for replacement of *peg* with a hygromycin resistance gene cassette.
2. Phasmid DNA of phAE159, a temperature-sensitive derivative of mycobacteriophage TM4 [1]. (Available from Jordan Kriakov and W.R. Jacobs Jr., Albert Einstein College of Medicine, Bronx, NY.)
3. *Escherichia coli* HB101.
4. Electrocompetent *M. smegmatis* mc^{2}155 [*ept* mutant that has high electroporation-transformation efficiencies [11]].
5. Luria Bertani (LB) broth (Becton Dickinson and Co., Sparks, MD).
6. LB agar (LB broth containing 1.5% agar).
7. Tryptic Soy Broth (TSB; Becton Dickinson and Co., Sparks, MD).
8. 20% Tween-80 (Fisher Scientific, NJ Somerville; dissolved vol:vol in prewarmed deionized H$_2$O, filter sterilized through an 0.2-μM filter, and stored at room temperature).
9. Sterile 20% Maltose stock solution.
10. Sterile 1 M MgSO$_4$ stock.

11. Hygromycin B (Roche Diagnostics, Indianapolis IN). Supplied as a stock at 50 mg/mL in phosphate buffered saline (PBS). Store at 4°C and protect from light. Use at 150 μg/mL for selection of *E. coli*.
12. T4 DNA Ligase (New England BioLabs, Ipswich, MA, or other supplier).
13. *Pac*I restriction enzyme (New England BioLabs, Ipswich, MA or other supplier).
14. Middlebrook 7H9 basal agar (Middlebrook 7H9 broth, Becton Dickinson and Co., Sparks, MD; without added OADC supplement but containing 1.5% Difco Agar).
15. Soft Agar (Middlebrook 7H9 broth containing 0.6% Difco Agar).
16. Mycobacteriophage (MP) Buffer: 50 mM Tris/HCl containing 150 mM NaCl, 10 mM MgCl$_2$, and 2 mM CaCl$_2$; pH 7·6. The buffer is filter sterilized through an 0.2-μM filter and stored at room temperature (RT).
17. Max Plax lambda phage *in vitro* packaging extract (Epicentre Biotechnologies, Madison WI).
18. QIA-Prep Spin plasmid miniprep kit (QIAGEN Valencia, CA).
19. 0.2-μM filter.

22.2.2 Construction of a Merodiploid Strain

1. *M. smegmatis* mc^{2}155 [11].
2. pSD26 plasmid vector (Fig. 22.1; *see* **Note 4**). Available from Sabine Daugelat (Department of Immunology, Max Planck Institute for Infection Biology, Schumannstrasse 21-22, 10117, Berlin, Germany) [10].
3. PCR reagents.
4. Primers with enzyme sites incorporated into the 5′ ends (*Bam*HI-*Cla*I, *Bam*HI-*Eco*RV, or *Bam*HI-*Pvu*II).
5. Standard PCR fragment cloning and sequencing vector, with accompanying host *E. coli* competent cells, appropriate media, and antibiotics required for culture.
6. Restriction enzymes (*Bam*HI, *Cla*I,*Xba*I, *Eco*RV, *Pvu*II, *Hin*dIII, *Eco*RI).
7. pMV306: single-copy integrative vector in *M. smegmatis* and a multicopy plasmid in *E. coli* [9] (Fig. 22.1). The kanamycin resistance gene is used for selection (20 μg/mL for *M. smegmatis* and 40 μg/mL for *E. coli*). (Available from W.R. Jacobs, Albert Einstein College of Medicine, Bronx, NY.)
8. Tryptic Soy Broth (TSB; Becton Dickinson and Co.).
9. Tryptic Soy Agar (TSA); TSB containing 1.5% agar.
10. Kanamycin sulfate is dissolved in deionized water to make a 20 mg/mL stock, filter sterilize and stored in aliquots at –20°C. A final concentration of 20 μg/mL is used in TSB or TSA for selection of *M. smegmatis*.
11. 2-mL cryovials.
12. Sterile 50% glycerol (v/v).
13. 20% Tween-80 (Fisher Scientific; dissolved vol:vol in prewarmed deionized H$_2$O, filter sterilized through an 0.2-μM filter and stored at RT).

22.2.3 Growth and Transduction of Mycobacterium smegmatis

1. Tryptic Soy Broth (TSB; Becton Dickinson and Co.).
2. 20% Tween-80 (Fisher Scientific; dissolved vol:vol in pre-warmed deionized H_2O, filter sterilized through an 0.2-μM filter, and stored at RT).
3. Tryptic Soy Agar (TSA; TSB containing 1.5% agar).
4. Sterile ink well bottles.
5. Sterile 250-mL glass flask
6. Centrifuge for 50-mL Falcon tubes at $3000 \times g$.
7. Hygromycin B (Roche Diagnostics). Comes as stock at 50 mg/mL in PBS. Store at 4°C and protect from light. Use at 100 μg/mL in mycobacteria.
8. MP buffer: 50 mM Tris/HCl containing 150 mM NaCl, 10 mM $MgCl_2$, and 2 mM $CaCl_2$; pH 7·6. The buffer is filter sterilized through an 0.2-μM filter and stored at RT.

22.2.4 CESTET

1. 20% (w/v) acetamide (Sigma) in deionized H_2O and filtered through an 0.2-μM filter. Use as a final concentration of 0.2% (w/v).
2. Tryptic Soy Broth (TSB; Becton Dickinson and Co.).
3. Tryptic Soy Agar (TSA; TSB containing 1.5% agar).
4. TSA plates containing 20 μg/mL kanamycin, 100 μg/mL hygromycin, and 0.2% acetamide.
5. Hygromycin B (Roche Diagnostics). Comes as stock at 50 mg/mL in PBS. Store at 4°C and protect from light. Use at 150 μg/mL in *E. coli* and 100 μg/mL in mycobacteria.
6. Kanamycin sulfate is dissolved in deionized water to make a 20 mg/mL stock, filter sterilized, and stored in aliquots at –20°C. A final concentration of 20 μg/mL is used in TSB or TSA for selection of *M. smegmatis*.
7. 2-mL cryovials.
8. Sterile 50% glycerol (v/v).

22.3 Methods

22.3.1 Generation of phΔpeg Mycobacteriophage Particles

22.3.1.1 Generation of phΔ*peg* Phasmid DNA

1. Grow *E. coli* HB101 overnight in LB supplemented with 10 mM $MgSO_4$ and 0.2% maltose at 37°C shaking (200 rpm).
2. Inoculate fresh 10 mL of supplemented LB with 0.5 mL of overnight culture. Shake culture at 200 rpm at 37°C until OD_{600} reaches 0.8 to 1.0.

3. Transfer cells to a 50-mL conical tube and pellet cells in a tabletop centrifuge at 3000 × g for 10 min at 4°C.
4. Decant supernatant and resuspend cell pellet in 5 mL of MP buffer.
5. Store cells at 4°C until ready for use.
6. Digest the *peg* AES plasmid (1 to 5 µg) with *Pac*I in a total volume of 20 µL.
7. Digest phAE159 (5 µg) with *Pac*I in a total volume of 20 µL.
8. Heat-inactivate *Pac*I by heating the DNA digestion reactions at 65°C for 10 min.
9. Set up 10 µL ligation reaction using 2 to 3 µL of *Pac*I-digested AES, 2 to 3 µL *Pac*I-digested phAE159, and T4 DNA ligase, and incubate at 30°C for 1 h (or overnight at 14°C).
10. Inactivate the T4 DNA ligase at 65°C for 10 min.
11. Thaw a vial of MaxPlax lambda *in vitro* packaging extract on ice and add 5 µL of ligation mixture to the packaging extract and mix gently by tapping lightly with finger.
12. Incubate at RT for up to 2 h.
13. Add 400 µL MP buffer to stop the packaging reaction. Then add 500 µL HB101 host cells (from step 5) and incubate at RT for 30 min.
14. Spin in microcentrifuge at 13,000 × g for 1 min.
15. Discard the supernatant and resuspend pellet in 500 µL LB broth. Incubate at 37°C for 1 h.
16. Plate different volume aliquots (ranging from 10 µL to 100 µL) of the resuspended cells on LB agar containing 150 µg/mL hygromycin and incubate at 37°C overnight.
17. Inoculate 4 to 6 hygromycin-resistant (Hyg^R) transformants obtained on above plates in 3 mL LB broth containing 150 µg/mL hygromycin and incubate at 37°C until OD_{600} is 0.8 to 1.
18. Use the culture for isolating plasmid DNA (referred to as phasmid) using the QIA-Prep Spin plasmid mini prep kit (QIAGEN) or a similar plasmid isolation kit (*see* **Note 5**).
19. Digest 5 µL of the phasmid DNA with *Pac*I restriction enzyme in a total reaction volume of 20 µL and analyze the digest by agarose gel electrophoresis.
20. Digests of the *bona fide* phΔ*peg* phasmids will show the presence of two bands: a >40-kb band of phage genome DNA and a smaller band of the linearized AES plasmid.

22.3.1.2 Generation of High Titers of phΔ*peg* Mycobacteriophage

1. Grow up *M. smegmatis* mc²155 in TSB containing 0.05% Tween-80 by inoculating the culture 24 h prior to the start of the next step.
2. Electroporate *M. smegmatis* mc²155 electrocompetent cells (procedure for electroporation of *M. smegmatis* is described in Chapter 13) with phΔ*peg* phasmid DNA (from step 20 in Section 22.3.1.1).

3. After electroporation, incubate *M. smegmatis* mc^{2}155 cells in 1 mL TSB for at least 4 h at 30°C.
4. Mix 100 µL of the above electroporated cells with 0.5 mL of *M. smegmatis* mc^{2}155 (pregrown in TSB to a OD$_{600}$ of 0.8 to 1) and then add to a sterile 14-mL tube containing 4 mL molten top agar (cooled to 50°C). Similarly, mix the remaining electroporated cells with 100 µL of pregrown *M. smegmatis* mc^{2}155 and add to another tube containing 4 mL molten top agar.
5. Overlay the top agar with cells on 7H9 basal agar plates.
6. Incubate the plates at 30°C for 2 to 3 days until plaques are visible on the matt growth of *M. smegmatis* mc^{2}155.
7. Using a sterile pipette tip, remove 1 to 2 plaques into 200 µL MP buffer.
8. Recover phage in the MP buffer by incubating at RT for at least 3 h or overnight at 4°C.
9. Add 5 µL of the above phage each to a 14-mL tube containing 4 mL molten top agar (cooled to 50°C) and 200 µL *M. smegmatis* mc^{2}155 culture. Mix well and overlay on 7H9 basal agar.
10. In parallel, do the same as in step 9 using a 1:5 dilution (in MP buffer) of the phage.
11. Incubate the overlay plates at 30°C for 3 days until plaques are visible.
12. Select the plate with approximately 1000 plaques ("lacy" pattern) for high-titer phage extraction: add 5 mL MP buffer to the plate and incubate for at least 4 h at RT or overnight at 4°C.
13. Filter the overlaying MP buffer through an 0.2-µM filter into a sterile 14-mL tube. Store the lysate at 4°C.
14. Make 10-fold serial dilutions of the phage lysate from above step in MP buffer.
15. Add 100 µL of each dilution to 200 µL of freshly grown *M. smegmatis* mc^{2}155 in 4 mL molten top agar and overlay on 7H9 basal agar plates.
16. Incubate the plates at 30°C for 2 days until plaques are visible. Count the plaques on each plate to calculate the plaque forming units (PFU) titer of the lysate obtained in step 13.

22.3.2 *Construction of a Merodiploid Strain*

22.3.2.1 Construction of a Single-Copy Integrative Vector Containing a Cloned Copy Acetamidase Promoter–Driven *peg*

1. Amplify gene of interest (*peg*).
2. Clone into an appropriate PCR-cloning vector.
3. Isolate plasmid DNA.
4. Sequence *peg* to check for errors.
5. Digest the *peg* ORF with one of the following enzyme combinations: *Bam*HI-*Cla*I, *Bam*HI-*Eco*RV, or *Bam*HI-*Pvu*II.
6. Clone into the corresponding enzyme sites of pSD26.

7. Digest the pSD26-*peg* construct with *Xba*I and one of the following: *Cla*I, *Eco*RV, *Pvu*II, *Hin*dIII, or *Eco*RI.
8. Clone into similarly digested pMV306 to give pMV306P$_{acetamidase}$*peg*.

22.3.2.2 Generation of a mc^{2}155 Strain Containing an Integrated Copy of pMV306P$_{acetamidase}$*peg*

1. Introduce pMV306P$_{acetamidase}$*peg* into *M. smegmatis* strain mc^{2}155 by electroporation [11] (*see* Chapter 13).
2. Select the merodiploid, kanamycin resistant (Kan^R) transformants on TSA containing 20 μg/mL kanamycin.
3. Inoculate between 5 and 10 individual Kan^R colonies into 5 mL TSB cultures containing 0.05% Tween-80 and 20 μg/mL kanamycin. Grow shaking at 37°C until OD$_{600}$ reaches between 0.8 to 1.
4. Mix the culture 1:1 (v/v) with sterile 50% glycerol and aliquot into 1.8-mL cryovials. Store frozen at –80°C. Use as inoculum for step 1 of Section 22.3.3.

22.3.3 Growth and Processing of M. smegmatis Cultures for Specialized Transduction

1. Inoculate one Kan^R *M. smegmatis* merodiploid strain from a frozen stock (from step 4 of Section 22.3.2.2) into 5 mL TSB containing 0.05% Tween-80 and 20 μg/mL kanamycin in an ink well bottle and grow shaking at 37°C until OD$_{600}$ reaches 0.6 to 0.8.
2. Use this culture as a 1/20 inoculum into a larger-sized culture for specialized transduction: you will eventually need 10 mL of culture per transduction (*see* **Note 6**). Incubate the cultures with shaking at 37°C until the OD$_{600}$ is 0.8 to 1.
3. Spin down 50 mL of culture in a 50-mL Falcon tube for 15 min at 4000 × g at RT. Resuspend the pellet in 5 mL MP buffer using a pipette until the consistency is even. Add 45 mL MP buffer to the suspension and mix well.
4. Spin for 15 min at 4000 × g at RT. Decant supernatant and resuspend the pellet again in MP buffer as above and spin for 15 min at 4000 × g at RT. Decant the supernatant and resuspend pellet in 5 mL MP buffer.
5. Mix 1 mL cell suspension with 1 mL high-titer *peg*-knockout-phage (phΔ*peg*; from step 13 of Section 22.3.1.2) lysate (~10^{10} PFU/mL) in a 15-mL tube and incubate at 37°C (nonpermissive temperature for phage replication; also *see* **Note 7**) for 1 to 2 h to allow for infection. As a negative control, mix 1 mL bacterial cell suspension with 1 mL MP buffer.
6. Spin tubes for 15 min at 4000 × g at RT. Decant supernatant and resuspend each pellet in 2 mL of TSB + 0.05% Tween-80. The pellets may have a "gloopy" consistency and will initially be difficult to resuspend.

7. Incubate the cultures at 37°C for 3 to 4 h.
8. Spin tubes for 15 min at 4000 × g at RT. Decant supernatant and resuspend each pellet in 400 µL TSB + 0.05% Tween-80.

22.3.4 CESTET

22.3.4.1 Essentiality Testing

1. Split the bacterial suspension (approximately 400 µL) from step 8 in Section 22.3.3 into four equal parts. Spread 100 µL each on two TSA plates containing kanamycin and hygromycin; and 100 µL each on two TSA plates containing kanamycin, hygromycin, and acetamide.
2. Incubate the plates at 37°C for 3 days before scoring for colonies. If hygromycin resistant (Hyg[R]) and Kan[R] colonies are obtained only in the presence of acetamide in the plates, then the targeted gene is essential for growth on TSA.
3. Streak out one of the Hyg[R] Kan[R] colonies obtained on a fresh TSA plate containing kanamycin, hygromycin, and acetamide and incubate at 37°C for 3 days to obtain single colonies. Pick up a well-isolated colony and inoculate into 5 mL TSB containing 0.5% Tween-80, kanamycin, hygromycin, and acetamide. Incubate until OD_{600} reaches 0.8 to 1. Aliquot the culture into 1.8-mL cryovials and mix with sterile 50% glycerol. Store at –80°C until further use.

22.3.4.2 Growth of Conditional Mutants of Essential Genes for Monitoring Loss of Function

1. Inoculate 100 µL from a frozen glycerol stock (step 3 in Section 22.3.4.1) into 5 mL TSB containing 0.05% Tween-80, kanamycin, hygromycin, and acetamide and grow at 37°C to an OD_{600} of 0.5.
2. Spin down 2 mL of culture at 4000 × g for 15 min and then wash cells twice in 10 mL TSB + 0.05% Tween-80 to remove traces of acetamide and resuspend in 2 mL TSB + 0.05% Tween-80.
3. Mix the washed cell suspension with 8 mL TSB containing 0.05% Tween-80, kanamycin, and hygromycin for overnight (12 h) growth at 37°C. This step is necessary to deplete intracellular levels of the protein encoded by *peg* in the bacterial cells due to acetamide-induced overexpression (*see* **Note 8**).
4. Use this overnight, *peg*-depleted culture to inoculate TSB (containing 0.05% Tween-80, kanamycin, and hygromycin) with or without acetamide. Incubate for up to 12 h at 37°C.
5. Start monitoring conditional mutant phenotype in the culture grown in the absence of acetamide from 3 h onwards. The culture grown in the presence of acetamide is used as a control for comparison. Biochemical profiles, transcriptional profiles (both specific and global), morphologic changes,

and/or loss of viability are among the different properties that can be monitored after conditional depletion of an essential gene function using this method.

22.4 Notes

1. Though not tested with specialized transduction, other vectors (both acetamidase promoter–based and those based on other promoters) may also function in the essentiality test [12, 13, 14]. The final aim of all strategies is to obtain a single-copy integrating vector containing a cloned copy of *peg* under the control of an inducible promoter that can be electroporated into a wild-type strain to generate a merodiploid strain. In all cases, it is important to choose a vector with a selection marker that is compatible with the hygromycin resistance cassette that is subsequently used for replacing the native copy of *peg*.

2. An alternate cloning strategy would be to obtain the acetamidase promoter–driven clone of *peg* in pMV306 by ligation of pMV306 with the fragments containing *peg* (from the sequencing vector clone) and the acetamidase promoter (from pSD26) in a triple ligation.

3. The AES plasmid is obtained by cloning PCR-amplified sequences (approximately 800 to 1000 bp) upstream and downstream of *peg* into multiple cloning sites flanking either side of a hygromycin resistance gene in the plasmid pYUB854 or its newer version pJSC347 [1].

4. pSD26 is used for inducible protein expression in *M. smegmatis* [10] (Fig. 22.1). It is a shuttle vector, capable of autonomous replication in *E. coli* and mycobacteria. The hygromycin resistance gene can be used for selection (100 μg/mL for *M. smegmatis* and 150 μg/mL for *E. coli*).

5. To maximize phasmid DNA yields, elute plasmid DNA in 50 μL of manufacturer-supplied EB-Buffer which has been preheated to 65°C.

6. To ensure proper aeration, do not use medium occupying more than 1/5 the volume of the culture flask (e.g., use a maximum of 50 mL TSB in a 250-mL flask).

7. A high-titer mycobacteriophage suspension is toxic *per se* to mycobacterial cells and thus it is important not to exceed the recommended times for phage-bacterium coincubation.

8. The acetamidase promoter is highly inducible in the presence of acetamide and, for this reason, is a promoter of choice for overexpression of proteins in *M. smegmatis*. For CESTET, the "starvation" step is thus crucial as it helps dilute-out the high levels of protein present in the cells, which may normally mask the effects of conditional loss of the target gene function. The length of time required for "acetamide-starvation" will also depend on the half-life of the protein encoded by the tested gene. The incubation time for "starvation" mentioned here is for genes encoding enzymes involved in mycolic acid biosynthesis [2]. When applied to other genes, the incubation times can be made shorter or longer depending on the stability of the protein encoded by the tested gene. This can be determined by monitoring levels of the protein by Western blot after acetamide "starvation."

References

1. Bardarov, S., Bardarov S. Jr., Pavelka M. S. Jr., Sambandamurthy, V., Larsen, M., Tufariello, J., Chan, J., Hatfull, G. & Jacobs W. R. Jr. (2002). Specialized transduction: an efficient method for generating marked and unmarked targeted gene disruptions in *Mycobacterium tuberculosis*, *M. bovis* BCG and *M. smegmatis*. *Microbiology* **148**, 3007–3017.

2. Bhatt, A., Kremer, L., Dai, A. Z., Sacchettini, J. C. & Jacobs, W. R. Jr. (2005). Conditional depletion of KasA, a key enzyme of mycolic acid biosynthesis, leads to mycobacterial cell lysis. *J Bacteriol* **187**, 7596–7606.
3. Mahenthiralingam, E., Draper, P., Davis, E. O. & Colston, M. J. (1993). Cloning and sequencing of the gene which encodes the highly inducible acetamidase of *Mycobacterium smegmatis*. *J Gen Microbiol* **139**, 575–583.
4. Parish, T., Mahenthiralingam, E., Draper, P., Davis, E. O. & Colston, M. J. (1997). Regulation of the inducible acetamidase gene of *Mycobacterium smegmatis*. *Microbiology* **143** (Pt 7), 2267–2276.
5. Dziadek, J., Rutherford, S. A., Madiraju, M. V., Atkinson, M. A. & Rajagopalan, M. (2003). Conditional expression of *Mycobacterium smegmatis ftsZ*, an essential cell division gene. *Microbiology* **149**, 1593–1603.
6. Gomez, J. E. & Bishai, W. R. (2000). *whmD* is an essential mycobacterial gene required for proper septation and cell division. *Proc Natl Acad Sci U S A* **97**, 8554–8559.
7. Lamichhane, G., Zignol, M., Blades, N. J., Geiman, D. E., Dougherty, A., Grosset, J., Broman, K. W. & Bishai, W. R. (2003). A postgenomic method for predicting essential genes at subsaturation levels of mutagenesis: application to *Mycobacterium tuberculosis*. *Proc Natl Acad Sci U S A* **100**, 7213–7218.
8. Sassetti, C. M., Boyd, D. H. & Rubin, E. J. (2001). Comprehensive identification of conditionally essential genes in mycobacteria. *Proc Natl Acad Sci U S A* **98**, 12712–12717.
9. Stover, C. K., de la Cruz, V. F., Bansal, G. P., Hanson, M. S., Fuerst, T. R., Jacobs, W. R. Jr. & Bloom, B. R. (1992). Use of recombinant BCG as a vaccine delivery vehicle. *Adv Exp Med Biol* **327**, 175–182.
10. Daugelat, S., Kowall, J., Mattow, J., Bumann, D., Winter, R., Hurwitz, R. & Kaufmann, S. H. (2003). The RD1 proteins of *Mycobacterium tuberculosis*: expression in *Mycobacterium smegmatis* and biochemical characterization. *Microbes Infect* **5**, 1082–1095.
11. Snapper, S. B., Melton, R. E., Mustafa, S., Kieser, T. & Jacobs, W. R. Jr. (1990). Isolation and characterization of efficient plasmid transformation mutants of *Mycobacterium smegmatis*. *Mol Microbiol* **4**, 1911–1919.
12. Blokpoel, M. C., Murphy, H. N., O'Toole, R., Wiles, S., Runn, E. S., Stewart, G. R., Young, D. B. & Robertson, B. D. (2005). Tetracycline-inducible gene regulation in mycobacteria. *Nucleic Acids Res* **33**, e22.
13. Carroll, P., Muttucumaru, D. G. & Parish, T. (2005). Use of a tetracycline-inducible system for conditional expression in *Mycobacterium tuberculosis* and *Mycobacterium smegmatis*. *Appl Environ Microbiol* **71**, 3077–3084.
14. Ehrt, S., Guo, X. V., Hickey, C. M., Ryou, M., Monteleone, M., Riley, L. W. & Schnappinger, D. (2005). Controlling gene expression in mycobacteria with anhydrotetracycline and Tet repressor. *Nucleic Acids Res* **33**, e21.

Chapter 23
Gene Switching and Essentiality Testing

Amanda Claire Brown

Abstract The identification of essential genes is of major importance to mycobacterial research, and a number of essential genes have been identified in mycobacteria, however confirming essentiality is not straightforward, as deletion of essential genes results in a lethal phenotype. In this chapter, protocols are described that can be used to confirm gene essentiality using gene switching, following the construction of a delinquent strain. Because deletion mutants cannot be created for essential genes, a second gene copy is introduced via an integrating vector, which allows the chromosomal gene copy to be deleted. The integrated vector can then be replaced using the gene switching method; where no transformants are obtained, essentiality is confirmed. This technique can also be used to confirm functionality of gene homologues and to easily identify essential operon members.

Keywords delinquent strain · essentiality testing · gene homologues · gene switching · homologous recombination · integrating vector · operon

23.1 Introduction

Essential genes are, by definition, those that are required for survival, and which thus cannot be deleted from the chromosome. Determining gene essentiality is a powerful tool in understanding the biology of any organism. Much of current mycobacterial research (especially in the case of the pathogenic mycobacteria) has focused on the identification of essential genes and establishing their function. This is of importance because it indicates the biological significance of the gene products/genes, and in the case of infectious agents, it can be used as the first step in identifying and validating drug targets.

Homologous recombination methods that allow the construction of marked or unmarked mutants in mycobacteria have been established (*see* Chapters 19

A.C. Brown

Institute of Cell and Molecular Science, Barts and the London, Queen Mary's School of Medicine and Dentistry, 4 Newark Street, Whitechapel, London, E1 2AA, UK

e-mail: a.c.brown@qmul.ac.uk

T. Parish, A.C. Brown (eds.), *Mycobacteria Protocols*,
doi: 10.1007/978-1-59745-207-6_23, © Humana Press, Totowa, NJ 2008

and 20), using both plasmid and phage systems. These have led to a wealth of knowledge regarding the roles of individual gene within the bacteria. Unmarked mycobacterial deletion mutants can be constructed by homologous recombination using various systems [1, 2, 3, 4]. Because recombination occurs at a low frequency in mycobacteria, two-step strategies are usually more successful in obtaining mutant strains [3]: the first step involves the creation of a single crossover strain (SCO) generated by a single recombination event. The SCO can then be grown to allow a second recombination event to occur, resulting in a double crossover (DCO) strain. For nonessential genes, this represents a successful strategy to obtain biological information about their role. However, for essential genes, more lengthy and sophisticated techniques are required in order to probe function.

We have developed and used a method called gene switching [5]. This method can be used both to prove or confirm gene essentiality and also to study the function of essential genes. The method first requires that a two-step homologous recombination procedure is followed to generate a strain with the chromosomal copy of the gene removed and the only functional copy of the gene provided on a integrated vector (a "delinquent" strain) derived from the L5 mycobacteriophage.

23.1.1 Generation of "Delinquent" Strain

The first step in the procedure is to generate a delinquent strain. This is a strain that carries a deleted copy of the gene in the normal chromosomal locus, generated by homologous recombination, and a functional copy in an integrated vector, generated by site-specific recombination. The procedure to generate this strain begins by constructing a SCO strain by transforming a suicide (nonreplicating) vector carrying the deletion allele (unmarked, in-frame deletion) into the wild-type strain. The SCO thus selected can then be used to generate DCOs (see Chapter 20) for nonessential genes; for essential genes, no deletion strains will be isolated, providing an indication of essentiality, although it is not formal proof. In order to demonstrate this, a second functional copy of the gene is introduced on an L5 integrating vector to make a merodiploid strain. If deletion DCOs can be obtained in this background, this is formal genetic proof of essentiality. In mycobacteria, it is sometimes problematic to obtain the deletion DCO from the merodiploid, possibly because of low recombination frequencies. In cases where only one or a few deletion DCOs are isolated, gene switching can be used to confirm essentiality.

23.1.2 Gene Switching to Test Essentiality

The switching method involves the removal of the resident integrated vector and its replacement by an alternative version carrying a different antibiotic selection marker (Fig. 23.1). This occurs at a high frequency [5].

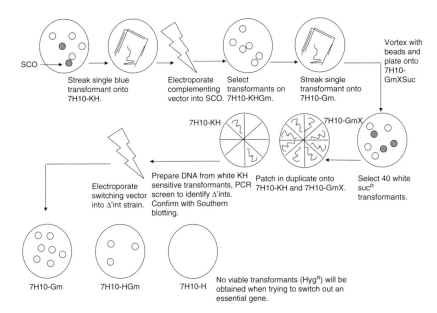

Fig. 23.1 Step–by-step schematic of the gene switching method for identifying essential genes, starting with an SCO strain made by the introduction of a suicide vector containing the flanking regions of the gene being deleted. The complementing vector (integrating vector containing a functional copy of the gene) is introduced via electroporation into the SCO; using sucrose selection, a DCO is obtained where the chromosomal copy is deleted (delinquent strain). The delinquent can be electroporated (switched) with an empty vector to confirm essentiality

Transformation of a delinquent strain with an empty vector (i.e., lacking the essential gene) will result in excision of the resident vector, but the resultant cells will not be viable and no transformants will be obtained. In contrast, transformation of a vector carrying a functional copy of the gene will result in a large number of transformants. The inability to switch out a vector carrying the only functional copy of the gene is further proof or confirmation of essentiality.

Gene switching can be also be used to great effect to rapidly screen all the genes in an operon for essentiality (Fig. 23.2). If the entire operon is deleted from the chromosome with a functional operon integrated in the delinquent strain, it is a simple matter to switch in different versions of the operon (e.g., lacking one gene only) and thus test which parts of the operon are required. The major advantage here is that it is much less time consuming and laborious than making individual deletion mutants for each gene.

23.1.3 Further Applications of Gene Switching

The great benefit of gene switching is that once a delinquent strain has been constructed, any number of different combinations of genes can be "switched"

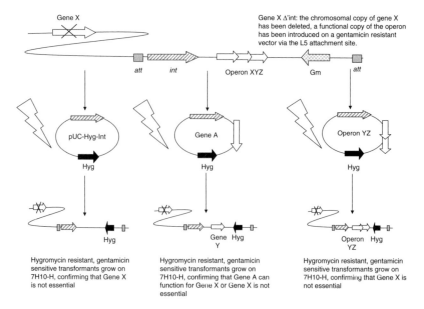

Fig. 23.2 Diagram of gene switching: in the case of a nonessential gene, the integrated gene copy (Gene X) can be removed and replaced with the hygromycin resistance gene, a gene homologue, or a whole operon

in, allowing for the study of essential genes. We have used this method for a number of different approaches including

Testing deletion variants of genes for functionality
Testing point mutations of genes for functionality
Testing alternative alleles
Testing essentiality on different media

These approaches can be valuable in determining the biology of the gene, as straightforward deletions cannot be made. For example, gene homologues from other mycobacteria or other bacteria, such as *Escherichia coli,* or even eukaryotes, can be "switched" in to identify if these are able to complement the deleted gene [5]. This can establish whether genes that appear to be homologues by sequence comparison truly are and enable greater understanding of gene function. This is particularly important as currently only ~40% of the protein encoding genes of *M. tuberculosis* (the most genetically studied member of the mycobacteria) have been ascribed a function [6], and despite a revisiting and reannotating of the genome [7], the majority of these are based solely on sequence identity to other proteins rather than on experimental evidence [8]. Because gene switching is efficient—we routinely get 10^6 switched transformants per μg vector DNA if viable—it can also be used to switch in a library of alleles generated by mutagenesis.

In this chapter, we outline the protocol for gene switching, beginning with a wild-type strain. Thus we detail the method for generating SCO, merodiploid, and delinquent strains and the switching methods itself.

23.2 Materials

23.2.1 Generating the Single Crossover Strain

1. Mycobacterial genomic DNA.
2. Primers designed to amplify the gene to be deleted.
3. PCR reagents: including proof reading polymerase.
4. PCR cleanup or gel extraction Kit , for example, QiaQuick (Qiagen, West Sussex, UK) or similar.
5. A-tailing reagents (dATP, *Taq* polymerase) (Promega, Southampton, UK) or similar.
6. pSC-A cloning vector (Stratagene, Amsterdam Zuidoost, The Netherlands) or similar PCR product cloning system.
7. Ampicillin (sodium salt). Make stock at 50 mg/mL in distilled water, filter sterilize and store in aliquots at –20°C. Use at 100 μg/mL.
8. 5-Bromo-4-chloro-3-indolyl-B-D-galatopyranoside (X-gal), make stock solution of 50 mg/mL in dimethyl sulfoxide (DMSO) and store at –20°C. Use at a working concentration of 50 μg/mL.
9. Kanamycin (sodium salt). Make stock at 50 mg/mL in distilled water, filter sterilize and store in aliquots at –20°C. Use at 50 μg/mL for *E. coli* and at 20 μg/mL for mycobacteria.
10. Luria-Bertani (LB) media (Difco, SLS, Nottingham, UK): To distilled water add 10 g tryptone (pancreatic digest of casein), 5 g yeast extract, 5 g NaCl and make up to 1 L. Autoclave to sterilize. Allow to cool and pour onto Petri dishes.
11. LA-K: To LB media add 15 g Bacto Agar before autoclaving. Make up to 1 L with distilled water. Autoclave to sterilize. Allow to cool, add 50 μg/mL kanamycin, mix well and pour onto Petri dishes.
12. LA-KX: To LB media add 15 g Bacto Agar before autoclaving. Make up to 1 L with distilled water. Autoclave to sterilize. Allow to cool, add 50 μg/mL kanamycin and 50 μg/mL X-gal, mix well and pour onto Petri dishes.
13. Competent *E. coli* strains suitable for recombinant genetic work (e.g., DH5α or XL1).
14. Plasmid purification kit (Qiagen) or similar.
15. p1NIL or p2NIL cloning vector (obtainable from Prof. T Parish, Barts and the London, 4 Newark St., London E1 2 AT, UK).
16. Restriction enzymes for cloning into p1NIL (*Pst*I, *Hind*III, *Xho*II, *Not*I) or p2NIL (*Pst*I, *Hind*III, *Pml*I, *Sal*I, *Bam*HI, *Kpn*I, *Not*I).
17. T4 DNA ligase or Rapid Ligation Kit (Roche, East Sussex, UK).

18. pGOAL19 (obtainable from Prof. T Parish, Barts and the London, 4 Newark St., London E1 2AT, UK).
19. *Pac*I.
20. Hygromycin B (Roche). Supplied as 50 mg/mL stock solution. Store at 4°C in light protected contained. Use at 50 μg/mL.
21. *M. tuberculosis* H37Rv culture.
22. 490-cm^3 roller bottles (Corning, Fisher Scientific, Leicestershire, UK).
23. 2 M glycine. Filter sterilize. Store at 4°C.
24. Oleic albumin dextrose catalase (OADC) supplement (Becton Dickinson, Oxford, UK). Store at 4°C and use at a final concentration of 10% v/v.
25. Glycerol. Make a 10% v/v stock in water and filter sterilize. Store at 4°C.
26. UV Stratalinker 1800 (Stratagene).
27. 7 H9-Tw medium: Dissolve 4.7 g Middlebrook 7H9 broth base (Difco) in 900 mL deionized water. Add 5 mL 20% (v/v) Tween 80 (final concentration of 0.05% v/v). Mix well and filter sterilize or autoclave. Add 100 mL OADC immediately before use.
28. 7H10-KHX plates: Dissolve 19 g Middlebrook 7H10 medium (Difco) in 900 mL deionized water. Autoclave. When media is cooled to ~50°C, add 100 mL OADC enrichment. Add kanamycin stock to give final concentration of 20 μg/mL, hygromycin to give final concentration of 50 μg/mL, and X-gal to give a final concentration of 50 μg/mL. Mix well and pour in standard plastic Petri dishes.
29. Plasmid DNA (up to 5 μg in a volume not exceeding 5 μL).
30. Electroporation cuvettes with a 2-mm gap.
31. Electroporation equipment capable of producing a pulse of 2.5 kV, 1000 Ω, 25 μF.
32. Recovery media: 10 mL 7H9-Tw in 50 mL sterile conical tube.

23.2.2 Construction of a Merodiploid Strain

23.2.2.1 Expression from the Ag85A Promoter

1. pAPA3 [9] (obtainable from Prof. T Parish, Barts and the London, 4 Newark St., London E1 2AT, UK).
2. *Pac*I.
3. Mycobacterial genomic DNA.
4. Primers designed to amplify the gene to be deleted.
5. PCR reagents.
6. PCR clean up or gel extraction Kit, for example, QiaQuick (Qiagen) or similar.
7. pSC-A cloning vector (Stratagene) or similar.
8. Competent *E. coli* strains suitable for recombinant genetic work (e.g., DH5α or XL1).
9. Plasmid purification kit (Qiagen) or similar.

10. Ampicillin. Make stock at 50 mg/mL in distilled water, filter sterilize and store in aliquots at –20°C. Use at 200 μg/mL.
11. 5-Bromo-4-chloro-3-indolyl-B-D-galatopyranoside (X-gal), make stock solution of 50 mg/mL in DMSO and store at –20°C. Use at a working concentration of 50 μg/mL.
12. Gentamicin: Make a stock solution of 50 mg/mL in distilled water. Filter sterilize and store at –20°C. Use at a working concentration of 10 μg/mL.

23.2.2.2 Expression from the Native Promoter

1. pUC-Gm-Int [10].
2. Mycobacterial genomic DNA.
3. Primers designed to amplify the gene being complemented.
4. PCR reagents.
5. PCR clean up or gel extraction Kit, for example, QiaQuick (Qiagen) or similar.
6. pSC-A cloning vector (Stratagene) or similar.
7. Competent *E. coli* strains suitable for recombinant genetic work (e.g., DH5α or XL1).
8. Plasmid purification kit (Qiagen) or similar.
9. Ampicillin. Make stock at 50 mg/mL in distilled water, filter sterilize and store in aliquots at –20°C. Use at 200 μg/mL.
10. 5-Bromo-4-chloro-3-indolyl-B-D-galatopyranoside (X-gal), make stock solution of 50 mg/mL in DMSO and store at –20°C. Use at a working concentration of 50 μg/mL.
11. Gentamicin: Make a stock solution of 50 mg/mL in distilled water. Filter sterilize and store at –20°C. Use at a working concentration of 10 μg/mL.
12. LA-Gm: To LB media add 15 g Bacto Agar before autoclaving. Make up to 1 L with distilled water. Autoclave to sterilize. Allow to cool, add 10 μg/mL gentamicin, mix well and pour onto Petri dishes.

23.2.2.3 Transformation of the Integrating Plasmid and Generation of the Delinquent

1. *M. tuberculosis* SCO strain (from Section 23.3.1).
2. 490-cm^3 roller bottles (Corning, Fisher Scientific).
3. 2 M glycine. Filter sterilize. Store at 4°C.
4. Oleic albumin dextrose catalase (OADC) supplement (Becton Dickinson). Store at 4°C and use at a final concentration of 10% v/v.
5. Glycerol. Make a 10% v/v stock in water and filter sterilize. Store at 4°C.
6. 7H9-Tw medium: Dissolve 4.7 g Middlebrook 7H9 broth base (Difco) in 900 mL deionized water. Add 5 mL 20% (v/v) Tween 80 (final concentration of 0.05% v/v). Mix well and filter sterilize or autoclave. Add 100 mL OADC immediately before use.

7. 7H9-Tw-KH medium: Dissolve 4.7 g Middlebrook 7H9 broth base (Difco) in 900 mL deionized water. Add 5 mL 20% (v/v) Tween 80 (final concentration of 0.05% v/v). Mix well and filter sterilize or autoclave. Add 100 mL OADC and kanamycin stock to give final concentration of 20 µg/mL, and hygromycin to give final concentration of 50 µg/mL, immediately before use.

8. 7H10-KHGm plates: Dissolve 19 g Middlebrook 7H10 medium (Difco) in 900 mL deionized water. Autoclave. When media is cooled to ~50°C, add 100 mL OADC enrichment. Add kanamycin stock to give final concentration of 20 µg/mL, hygromycin to give final concentration of 50 µg/mL, and gentamicin to give a final concentration of 10 µg/mL. Mix well and pour in standard plastic Petri dishes.

9. Plasmid DNA (up to 1 µg in a volume not exceeding 5 µL).

10. Electroporation cuvettes with a 2-mm gap.

11. Electroporation equipment capable of producing a pulse of 2.5 kV, 1000 Ω, 25 µF.

12. Recovery media: 10 mL 7H9-Tw in 50-mL sterile conical tube for each electroporation.

13. Sucrose: make a 50% (w/v) stock solution in distilled water. Filter sterilize. Use at a final concentration of 2% for *M. tuberculosis*.

14. 7H10-GmXSuc plates: Dissolve 19 g Middlebrook 7H10 medium (Difco) in 900 mL deionized water. Autoclave. When media is cooled to ~50°C, add 100 mL OADC enrichment. Add gentamicin to give a final concentration of 10 µg/mL, X-gal to give a final concentration of 50 µg/mL, and sucrose to give a final concentration of 2%. Mix well and pour in standard plastic Petri dishes.

15. 3 mL 7H9-Tw in 50-mL conical tubes with 1 mL 1-mm glass beads.

16. Primers designed to overlap the deletion region.

17. 2-mL screw-cap tubes.

18. 10 mM Tris (pH 8.0).

19. 0.2 µM, 15-mm filter units (Corning, Fisher Scientific).

20. 3-mL Luer Lok syringes (Becton Dickinson).

21. PCR reagents.

22. PCR cleanup or gel extraction Kit, for example, QiaQuick (Qiagen) or similar.

23.2.3 Confirming Essentiality by Gene Switching

1. Delinquent strain (from Section 23.3.2).

2. pUC-Hyg-Int [10].

3. 490-cm^3 roller bottles (Corning, Fisher Scientific).

4. Roller incubator (set at 37°C).

5. 2 M glycine. Filter sterilize. Store at 4°C.

6. Oleic albumin dextrose catalase (OADC) supplement (Becton Dickinson). Store at 4°C and use at a final concentration of 10% v/v.

7. Glycerol. Make a 10% v/v stock in water and filter sterilize. Store at 4°C.
8. 7H9-Tw-Gm medium: Dissolve 4.7 g Middlebrook 7H9 broth base (Difco) in 900 mL deionized water. Add 5 mL 20% (v/v) Tween 80 (final concentration of 0.05% v/v). Mix well and filter sterilize or autoclave. Add 100 mL OADC and gentamicin to give a final concentration of 10 μg/mL immediately before use.
9. 7H10-HGm plates: Dissolve 19 g Middlebrook 7H10 medium (Difco) in 900 mL deionized water. Autoclave. When media is cooled to ~50°C, add 100 mL OADC enrichment. Add hygromycin to give final concentration of 50 μg/mL, and gentamicin to give a final concentration of 10 μg/mL. Mix well and pour in standard plastic Petri dishes.
10. 7H10-Gm plates: Dissolve 19 g Middlebrook 7H10 medium (Difco) in 900 mL deionized water. Autoclave. When media is cooled to ~50°C, add 100 mL OADC enrichment. Add gentamicin to give a final concentration of 10 μg/mL. Mix well and pour in standard plastic Petri dishes.
11. 7H10-HGm plates: Dissolve 19 g Middlebrook 7H10 medium (Difco) in 900 mL deionized water. Autoclave. When media is cooled to ~50°C, add 100 mL OADC enrichment. Add hygromycin to give final concentration of 50 μg/mL. Mix well and pour in standard plastic Petri dishes.

23.3 Methods

23.3.1 Generating the Single Crossover Strain

1. Design PCR primers to amplify 1-kb regions flanking the gene of interest required with enzyme sites included at the 5′ end (see **Note 1**).
2. Amplify the flanking regions using a proofreading polymerase (see **Note 2**) and standard PCR protocol (see **Note 3**).
3. Check PCR products using agarose gel electrophoresis.
4. Isolate PCR products with PCR extraction kit/method or gel extract band, if required, and clean using gel extraction kit/method.
5. A-tail cleaned PCR products, incubate at 72°C for 20 min (see **Note 4**).
6. Ligate into pSC-A cloning vector (see **Note 5**).
7. Transform into competent *E. coli*.
8. Screen transformants for correct size of insert.
9. Isolate plasmid DNA from overnight cultures of successful transformants, check insert integrity by sequencing.
10. Recover inserts from cloning vector using enzymes appropriate for the sites added to the 5′ primer regions.
11. Ligate both flanking regions into digested p1NIL or p2NIL. Transform into competent *E. coli* and select on LA-K.
12. Purify plasmid DNA from individual transformants. Check for correct insert size and orientation using diagnostic restriction digests and sequencing to confirm that no mutations have been introduced.

13. Digest confirmed construct with *Pac*I and introduce the pGOAL19 cassette.
14. Transform into *E. coli* competent cells and select on LA-KX. The correct transformants will be blue.
15. Purify plasmid DNA from transformants and establish the orientation of the GOAL19 cassette by restriction digests (*see* **Note 6**).
16. Inoculate 1 to 3 mL *M. tuberculosis* H37Rv into 100 mL 7H9-Tw-KH in a roller bottle.
17. Incubate for 6 days at 37°C with rolling.
18. On day 6, add 10 mL 2 M glycine solution to the roller culture. Return to incubator overnight.
19. Pellet 50 mL of cells at 3000 x *g* at RT for 10 min.
20. Wash cells in 10 mL 10% glycerol.
21. Pellet 10 mL of cells at 3000 x *g* at RT for 10 min.
22. Wash cells in 5 mL 10% glycerol.
23. Pellet 5 mL of cells at 3000 x *g* at RT for 10 min.
24. Resuspend cells in 1 mL 10% glycerol.
25. Expose plasmid DNA to 100 mJ cm^{-2} of UV energy using the Stratalinker 1800 (*see* **Note 7**).
26. Electroporate plasmid DNA 1 µg into 200 µL cells, using a pulse of 2.5 kV, 1000 Ω, 25 µF (*see* **Note 8**).
27. Transfer cuvette contents to 10 mL recovery media. Incubate at 37°C overnight.
28. Centrifuge transformations for 10 min at 3000 x *g* to collect cells. Resuspend in 1 mL liquid broth and plate out onto 2 large 7H10-KHX plates.
29. Incubate at 37°C, colonies should be apparent after 3 to 5 weeks.

23.3.2 Construction of a Merodiploid Strain (see Notes 9 and 10)

23.3.2.1 Expression from the Ag85a Promoter (*see* Note 11)

1. Design PCR primers to amplify the region of DNA required with *Pac*I sites included to the 5′ end (*see* **Note 12**).
2. Amplify gene of interest using a proof reading polymerase and standard PCR protocol (*see* **Notes 2** and **3**).
3. Check PCR product using agarose gel electrophoresis and isolate.
4. A-tail cleaned PCR product (*see* **Note 4**).
5. Ligate into pSC-A cloning vector (*see* **Notes 5** and **3**) and transform into competent *E. coli*.
6. Screen transformants for correct size of insert. Isolate plasmid DNA and sequence.
7. Excise insert from cloning vector with *Pac*I, gel extract appropriate band, and ligate into *Pac*I-digested pAPA3.
8. Transform into competent *E. coli* and select on LA-ampicillin (*see* **Note 14**). Check for correct insert size and orientation using diagnostic restriction digests and sequencing to confirm that no mutations have been introduced.

23.3.2.2 Expression from the Native Promoter

1. Design PCR primers, ensuring that the region amplified contains the promoter sequence of the gene of interest
2. Amplify gene of interest using a proof reading polymerase (*see* **Note 2**) and standard PCR protocol (*see* **Note 3**).
3. A-tail cleaned PCR product (*see* **Note 4**) and ligate into T-A cloning vector (*see* **Note 5**). Transform into competent *E. coli*.
4. Screen transformants for correct insert. Isolate plasmid DNA and sequence.
5. Digest 1 µg of the recombinant plasmid with insert with *Hind*III.
6. Digest pUC-Gm-Int with *Hind*III and recover the Gm-Int cassette—approximately 3 kb.
7. Ligate plasmid with Gm-Int cassette. Transform into competent *E. coli* and select on LA plus ampicillin (*see* **Note 14**).
8. Patch transformants (~40) onto LA-Gm to select for the presence of the Gm-Int cassette.
9. Prepare plasmid DNA. Check for correct insert size using diagnostic restriction digests and sequencing to confirm that no mutations have been introduced (*see* **Note 15**).

23.3.2.3 Transformation of the Integrating Plasmid and Generation of the Delinquent

1. Inoculate 1 to 3 mL SCO mycobacterial strain (from Section 23.3.1) into 100 mL 7H9-Tw-KH in a roller bottle.
2. Incubate for 6 days at 37°C with rolling.
3. On day 6, add 10 mL 2 M glycine solution to the roller culture. Return to incubator overnight.
4. Pellet 50 mL of cells at 3000 x *g* at RT for 10 min.
5. Wash cells in 10 mL 10% glycerol.
6. Pellet 10 mL of cells at 3000 x *g* at RT for 10 min.
7. Wash cells in 5 mL 10% glycerol.
8. Pellet 5 mL of cells at 3000 x *g* at RT for 10 min.
9. Resuspend cells in 1 mL 10% glycerol.
10. Electroporate 1 µg plasmid DNA (from Section 23.3.2) into 200 µL cells, using a pulse of 2.5 kV, 1000 Ω, 25 µF.
11. Transfer cuvette contents to 10 mL recovery media. Incubate at 37°C overnight.
12. Centrifuge transformations for 10 min at 3000 x *g* to collect cells. Resuspend in 1 mL liquid broth and plate out as a dilution series (10^{-1} to 10^{-6}) onto 7H10-KHGm.
13. Incubate at 37°C, colonies should be apparent after 3 weeks.
14. Streak 40 transformants onto 7H10-Gm, incubate for 1 to 2 weeks at 37°C.

15. Scrape a loopful of cells from the plate into a tube containing 1 mL of 1-mm sterile glass beads and 3 mL 7H9-Tw. Vortex vigorously for 1 min, allow to stand for 10 min.
16. Plate serial dilutions onto 7H10-GmXSuc plates. Use several plates per dilution.
17. Incubate at 37°C for 4 to 6 weeks.
18. Pick white suc^R colonies, patch 40 individual transformants onto 7H10-GmX and 7H10-KHGm.
19. Incubate at 37°C for 2 to 3 weeks.
20. Score streaks for kanamycin, hygromycin sensitivity.
21. Carry out colony PCR on all Kan^S, Hyg^S colonies with primers that span the deletion (*see* **Note 16**) to identify delinquent strains. Reincubate plates to allow regrowth.
22. Confirm the correct genotype by Southern blotting (*see* **Note 17**).

23.3.3 Confirming Essentiality by Gene Switching

1. Inoculate 1 to 3 mL delinquent mycobacterial strain (from Section 23.3.2.3) into 100 mL 7H9-Tw-Gm in a roller bottle.
2. Incubate for 6 days at 37°C with rolling.
3. On day 6, add 10 mL 2 M glycine solution to the roller culture. Return to incubator overnight.
4. Pellet 50 mL of cells at 3000 x g at RT for 10 min.
5. Wash cells in 10 mL 10% glycerol.
6. Pellet 10 mL of cells at 3000 x g at RT for 10 min.
7. Wash cells in 5 mL 10% glycerol.
8. Pellet 5 mL of cells at 3000 x g at RT for 10 min.
9. Resuspend cells in 1 mL 10% glycerol.
10. Electroporate 1 µg plasmid DNA into 200 µL cells, using a pulse of 2.5 kV, 1000 Ω, 25 µF (*see* **Note 18**).
11. Transfer cuvette contents to 10 mL recovery media. Incubate at 37°C overnight.
12. Centrifuge transformations for 10 min at 3000 x g to collect cells. Resuspend in 1 mL liquid broth and plate out as a dilution series (10^{-1} to 10^{-6}) onto 7H10-H and 7H10-GmH.
13. Incubate at 37°C; colonies should be apparent after 3 weeks.
14. Count the transformants. If switching has occurred, there should be 10^4 to 10^6 HygR transformants, but no $Hyg^R Gm^R$ transformants (*see* **Note 19**).
15. Patch 24 Hyg^R transformants for Gm^R by patching onto 7H10-Gm (*see* **Note 20**).
16. Screen Gm^S, Hyg^R transformants by colony PCR and Southern blotting to confirm the expected genotype.

23.4 Notes

1. Suicide vectors lack a mycobacterial origin of replication and are pretreated, with either UV irradiation or alkali prior to transformation, to stimulate homologous recombination [11]. To create an unmarked mutant, a suicide vector, based on a p1NIL or p2NIL backbone, with the hygR, lacZ, sacB cassette from pGOAL19, is constructed to carry approximately 1 kb of flanking DNA from the gene of interest. The final suicide vector construct contains selectable antibiotic (hygromycin and kanamycin) markers; both are required as the frequency of spontaneous mutation in mycobacteria can be equal to the frequency of recombination. The lacZ gene is included to act as a visible indicator of transformants that carry the plasmid, which allows SCO and DCO transformants to be identified; and sacB, which confers sucrose sensitivity, DCOs that have lost the suicide vector will therefore be able to grow on media containing sucrose and will appear white in color [3].

2. For amplification of GC-rich regions of mycobacterial DNA by PCR, we have found that Pfu (Promega, Southampton, UK) gives good results. However, products must always be completely sequenced to confirm that no mutations have been introduced during amplification.

3. Typical PCR reaction mixes contain 5 μL 10X Pfu Buffer, 5 μL DMSO, 2.5 μL forward primer, 2.5 μL reverse primer (both at 10 pmol/mL), 1 μL dNTP mix (10 mM of each dNTP), 0.5 μL Pfu, 10 to 100 ng DNA, and SDW to 50 mL total volume. Typical PCR cycling reaction of 94°C for 2 min, followed by 35 cycles of 94°C for 30 s, annealing of typically 2°C to 4°C lower than lowest primer Tm for 1 min, and extension of 72°C for 1 min per kb of amplified product.

4. The products from proofreading polymerase PCR reactions typically are blunt ended and therefore require A-tailing prior to introduction into T-A cloning vectors, by adding 5 μL of 10x Taq buffer, 1 μL dATP, 5 U Taq to PCR product to give final volume of 50 μL. Incubate at 72°C for 20 min.

5. pSC-A is a very useful vector for this purpose. We have found that cloning mycobacterial DNA into this vector is more efficient and reliable than in other systems. The vector has an unique HindIII site that allows for the introduction of the Gm-int or Hyg-int cassettes.

6. Orientation of the GOAL19 cassette is not critical, but it can be useful to know when performing Southern analysis later on.

7. If using a different UV source, the pretreatment conditions will need to be worked out empirically.

8. If using different electroporation equipment, the pulse conditions will need to be worked out empirically.

9. The complementing vector must be an L5-derived integrating vector for gene switching to work. The L5 integrase catalyzes site-specific recombination between an attP site on the plasmid and the chromosomal attB site [12]. This type of vector can be removed from the chromosome either using the L5 excisionase (excision) or be replaced by an alternative version (switching). This method is described pictorially in Fig. 23.1 and has been used for many essential mycobacterial genes, (for examples, see Refs. 13 and 14).

10. Replicating plasmids, which express genes under the control of either a constitutive promoter, such as hsp60 [15], or an inducible promoter, such as the acetamidase promoter region from *Mycobacterium smegmatis* [16], and the tetracycline inducible P_{TET} system [17] can be used to express the additional functional gene copy. However, problems with stability have been reported, especially when introduced into *Mycobacterium tuberculosis*, which make these systems unreliable [15, 18, 19, 20]. Replicating

vectors can be used to determine essentiality in the simplest cases, but they cannot be where gene switching is required.

11. We generally use one of two different L5 plasmids for cloning. If the promoter of the gene is unknown, we use pAPA3 [9], which has the antigen 85A promoter for expression. If the native promoter is to be included, we would normally clone the gene plus promoter into *E. coli* cloning vector and then add the Gm-Int cassette from pUC-Gm-Int (gentamicin resistance marker, L5 integrase, *att*P) as a *Hind*III fragment. Both vectors have ampicillin and gentamicin resistance markers, L5 integrase and the attP site.

12. The addition of *Pac*I sites (5′-TTAATTAA-3′) to the 5′ end of primers facilitates cloning into pAPA3/pHAPY1; this restriction site is absent from the *M. tuberculosis* genome.

13. Cloning the insert into a vector such as pSCA allows us to see that the *Pac*I restriction digest has successfully excised the fragment and allows for ease of cloning into pAPA3.

14. We have found that direct selection of *E. coli* transformants on LA-Gm gives a low efficiency of transformation. Therefore, we plate transformants onto LA-Amp first and patch transformants for Gm^R. The same holds true for selection on hygromycin.

15. Orientation of the insert and the Gm-int cassette is not critical, but it can be useful to know when performing Southern analysis later on.

16. Crude extracts of DNA can be prepared for colony PCR as follows: take 1 loop of cells into a screw-cap tube containing 0.5 mL TE buffer. Cap the tube and heat at 95°C on a heating block for 20 min, allow to cool for 10 min at room temperature, and then filter through a 0.2-µM syringe filter with a 2-mL syringe, collecting sterile extract in a 1.5-mL Eppendorf tube; 4 µL of this extract in a 20 µL PCR normally yields good products. For the PCR screening, primers that span the deletion are very useful.

17. Proof of essentiality is obtained if no DCOs are obtained in the wild-type background but can be obtained in the merodiploid background. Further confirmation of essentiality is made when switching shows that the integrated copy is required for cell viability (e.g., it cannot be removed). This can be very useful when the delinquents are only isolated at a low frequency, as in these cases the difference between obtaining for example only 1 DCO in the merodiploid background is not statistically significant using Fisher's exact test.

18. For a simple test of essentiality, we use pUC-Hyg-Int as the switching vector. However, integrating vectors carrying other selection markers (e.g., kanamycin or streptomycin resistance) can also be used. The only restriction is that the incoming marker must differ from the resident plasmid. Some cross-resistance between gentamicin and kanamycin can occur, so that higher concentrations of antibiotics are needed for selection than normal. Alternative alleles can be tested for functionality by cloning them into the appropriate integrating vector and transforming the cells. In the case of essential genes, no switching should occur when transformed with an empty vector, but switching should occur at a high frequency with a control vector containing the gene. For nonessential genes, or where the gene being switched is fully functional, transformants should be obtained at a high frequency.

19. Hyg^{R}Gm^S transformants are the cells in which switching has occurred, where the original Gm plasmid has been replaced by the incoming Hyg plasmid. Cointegration of plasmids is sometimes seen, but it occurs at a much lower frequency, approximately 10^{-4} to 10^{-5} compared with switching. If the efficiency of transformation is high (above 10^6), cointegrants may be obtained. Cointegrant transformants will be Hyg^{R}Gm^R as they carry both plasmids.

20. If switching has occurred, the transformants will be Gm^S. Normally, 23 to 24 of the patches will be Hyg^{R}Gm^S confirming switching has occurred.

References

1. Stephan, J., Stemmer, V. & Niederweis, M. (2004). Consecutive gene deletions in *Mycobacterium smegmatis* using the yeast FLP recombinase. *Gene* **343**, 181–190.
2. Pavelka, M. S. Jr. & Jacobs, W. R. Jr. (1999). Comparison of the construction of unmarked deletion mutations in *Mycobacterium smegmatis*, *Mycobacterium bovis* Bacillus Calmette-Guerin, and *Mycobacterium tuberculosis* H37Rv by allelic exchange. *J Bacteriol* **181**, 4780–4789.
3. Parish, T. & Stoker, N. G. (2000). Use of a flexible cassette method to generate a double unmarked *Mycobacterium tuberculosis* tlyA plcABC mutant by gene replacement. *Microbiology-UK* **146**, 1969–1975.
4. Knipfer, N., Seth, A. & Shrader, T. E. (1997). Unmarked gene integration into the chromosome of *Mycobacterium smegmatis* via precise replacement of the *pyrF* gene. *Plasmid* **37**, 129–140.
5. Pashley, C. A. & Parish, T. (2003). Efficient switching of mycobacteriophage L5-based integrating plasmids in *Mycobacterium tuberculosis*. *FEMS Microbiol Lett* **229**, 211–215.
6. Cole, S. T. (1999). Learning from the genome sequence of *Mycobacterium tuberculosis* H37Rv. *FEBS Lett* **452**, 7–10.
7. Camus, J.-C., Pryor, M. J., Medigue, C. & Cole, S. T. (2002). Re-annotation of the genome sequence of *Mycobacterium tuberculosis* H37Rv. *Microbiology* **148**, 2967–2973.
8. Strong, M., Mallick, P., Pellegrini, M., Thompson, M. & Eisenberg, D. (2003). Inference of protein function and protein linkages in *Mycobacterium tuberculosis* based on prokaryotic genome organization: a combined computational approach. *Genome Biol* **4**, R59.
9. Parish, T., Roberts, G., Laval, F., Schaeffer, M., Daffe, M. & Duncan, K. (2007). Functional complementation of the essential gene fabG1 of *Mycobacterium tuberculosis* by *Mycobacterium smegmatis* fabG, but not *Escherichia coli* fabG. *J Bacteriol* **189**, 3721–3728.
10. Mahenthiralingam, E., Marklund, B.-I., Brooks, L. A., Smith, D. A., Bancroft, G. J. & Stokes, R. W. (1998). Site-directed mutagenesis of the 19-kilodalton lipoprotein antigen reveals no essential role for the protein in the growth and virulence of *Mycobacterium intracellulare*. *Infect Immun* **66**, 3626–3634.
11. Hinds, J., Mahenthiralingam, E., Kempsell, K. E., Duncan, K., Stokes, R. W., Parish, T. & Stoker, N. G. (1999). Enhanced gene replacement in mycobacteria. *Microbiology-UK* **145**, 519–527.
12. Parish, T., Lewis, J. & Stoker, N. G. (2001). Use of the mycobacteriophage L5 excisionase in *Mycobacterium tuberculosis* to demonstrate gene essentiality. *Tuberculosis* **81**, 359–364.
13. Parish, T. & Stoker, N. G. (2000). glnE is an essential gene in *Mycobacterium tuberculosis*. *J Bacteriol* **182**, 5715–5720.
14. Parish, T. & Stoker, N. G. (2002). The common aromatic amino acid biosynthesis pathway is essential in *Mycobacterium tuberculosis*. *Microbiology* **148**, 3069–3077.
15. Al-Zarouni, M. & Dale, J. W. (2002). Expression of foreign genes in *Mycobacterium bovis* BCG strains using different promoters reveals instability of the hsp60 promoter for expression of foreign genes in *Mycobacterium bovis* BCG strains. *Tuberculosis* **82**, 283–291.
16. Triccas, J. A., Parish, T., Britton, W. J. & Gicquel, B. (1998). An inducible expression system permitting the efficient purification of a recombinant antigen from *Mycobacterium smegmatis*. *FEMS Microbiol Lett* **167**, 151–156.
17. Carroll, P., Muttucumaru, D. G. N. & Parish, T. (2005). Use of a tetracycline-inducible system for conditional expression in *Mycobacterium tuberculosis* and *Mycobacterium smegmatis*. *Appl Environ Microbiol* **71**, 3077–3084.

18. De Smet, K. A. L., Kempsell, K. E., Gallagher, A., Duncan, K. & Young, D. B. (1999). Alteration of a single amino acid residue reverses fosfomycin resistance of recombinant *Mur*A from *Mycobacterium tuberculosis*. *Microbiology* **145**, 3177–3184.
19. Brown, A. C. & Parish, T. (2006). Instability of the acetamide-inducible expression vector pJAM2 in *Mycobacterium tuberculosis*. *Plasmid* **55**, 81–86.
20. Chawla, M. & Das Gupta, S. K. (1999). Transposition-induced structural instability of *Escherichia coli*-Mycobacteria shuttle vectors. *Plasmid* **41**, 135–140.

Chapter 24
Insertion Element IS*6110*-Based Restriction Fragment Length Polymorphism Genotyping of *Mycobacterium tuberculosis*

Robin M. Warren, Paul D. van Helden and Nicolaas C. Gey van Pittius

Abstract DNA fingerprinting techniques are based on genome variation and form the basis of molecular epidemiology studies of tuberculosis. A number of markers are in use for the molecular differentiation of *Mycobacterium tuberculosis* isolates by DNA fingerprinting. One of these markers is the IS*6110* insertion element, which may be present in up to 25 copies per *M. tuberculosis* genome. Variation in both the number and location of the IS*6110* elements makes it a very useful marker of strain genotype. IS*6110*-based DNA fingerprinting is globally considered as the reference genotyping technique for *M. tuberculosis* isolates. This method is based on visualization of restriction fragment length polymorphisms using a labeled probe derived from IS*6110*. In this chapter, the method of IS*6110* DNA fingerprinting is explained in such a way that it can be easily duplicated by molecular epidemiologists and will give reproducible results.

Keywords culture · DNA extraction · ECL labeling · IS*6110* insertion element · RFLP · Southern hybridization · transposition

24.1 Introduction

Various *Mycobacterium tuberculosis* genomic markers have been identified, including the direct repeat (DR) region [1,2], the polymorphic GC-rich sequences (PGRS) [3], and the variable number tandem repeats (VNTR) sequences (or mycobacterial interspersed repetitive units; MIRUs) [4]. However, the insertion sequence IS*6110* [5] has proved to be the most useful for the

R.M. Warren
DST/NRF Centre of Excellence in Biomedical Tuberculosis Research / MRC Centre for Molecular and Cellular Biology, Division of Molecular Biology and Human Genetics, Department of Biomedical Sciences, Faculty of Health Sciences, Stellenbosch University, PO Box 19063, Tygerberg, South Africa, 7505
e-mail: rw1@sun.ac.za

T. Parish, A.C. Brown (eds.), *Mycobacteria Protocols*,
doi: 10.1007/978-1-59745-207-6_24, © Humana Press, Totowa, NJ 2008

study of the disease dynamics of tuberculosis by identifying and tracking specific strains within a community [6]. IS*6110* DNA fingerprinting is the "gold standard" for molecular epidemiologic studies of *M. tuberculosis* [7]. The IS*6110* insertion element is 1355 bp in length with imperfect 28-bp terminal inverted repeats and contains two transposase open reading frames (ORF 1 and 2) that overlap by 1 bp. It is capable of transposition and thus of inserting copies of itself anywhere in the genome [8]. This marker can be used to determine whether *M. tuberculosis* isolates cultured from different patients are genotypically identical or different by determining the number (between 0 and 25 copies) and location of the element. *M. tuberculosis* isolates with identical IS*6110* DNA fingerprints are interpreted to infer transmission between individuals, whereas isolates with unique DNA fingerprints are thought to reflect reactivation of a prior infection in the patient [9, 10].

This method of typing has provided new insights into the epidemiology of tuberculosis and has challenged dogmas based on classic epidemiologic studies. Some of the major advances that have been achieved through IS*6110* DNA fingerprinting include (1) identifying the population structure of *M. tuberculosis* in different settings [11, 12]; (2) quantification of the contribution of casual contact to ongoing transmission [9, 10, 11]; (3) defining the mechanism leading to recurrent tuberculosis [13]; (4) identification of outbreaks [14]; (5) demonstration that drug-resistant strains can be transmitted [14]; (6) challenging the interpretation of drug surveillance studies [15]; (7) understanding human-animal and animal-human transmission of tuberculosis [16]; (8) understanding where transmission occurs [17]; (9) proving nosocomial transmission of *M. tuberculosis* [18]; (10) proving multiple infection in patients in a high-incidence community [19]; and (11) demonstrating laboratory cross-contamination [20]. It is envisaged that this methodology will continue to provide novel insights into the disease dynamics of the tuberculosis epidemic and will be essential for shaping future treatment protocols and determining the efficacy of novel vaccines and the efficacy of new antituberculosis drugs. Despite this, there are some negative aspects to the method, which continually drives the search for newer and better techniques. These include the large amount of isolate material required (when compared with for example PCR-based tests), the time-delay in waiting for cultures to grow, contamination of cultures due to lengthy procedures, the inability to perform this technique on nonviable cultures, and the large effort required for DNA isolation. Notwithstanding this, IS*6110* fingerprinting remains, for the time being, the only internationally standardized and accepted technique for robust *M. tuberculosis* strain typing.

IS*6110*-based DNA fingerprinting analyses produce a complex hybridization pattern that cannot be converted into a simple mathematical term. This is in contrast with other typing techniques like spoligotyping and MIRU analysis, both of which produce results that can be easily translated into specific numbering systems. The IS*6110* pattern is usually comparable

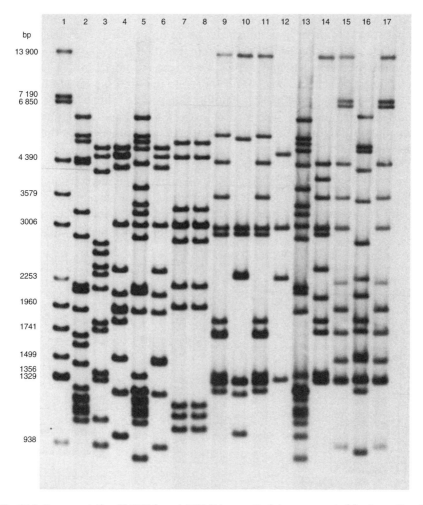

Fig. 24.1 Representative IS6110-based RFLP image. Isolates represented by **lanes 7** and **8**, **9** and **11**, and **15** and **17**, respectively, have the same pattern and can thus be epidemiologically linked. **Lane 1** shows the external marker Mt14323 used for intergel comparisons

between samples on the same membrane (indicating same or different strains; Fig. 24.1), but could differ between different membranes due to differences in conditions and running of the gels. This causes problems with regard to intergel comparison of strain patterns. To enable intergel comparisons, sophisticated gel-analysis computer programs are available for normalizing patterns within and between different membranes making use of the positions of internal and external markers. The internal marker is used to normalize the electrophoretic mobility of each band in each lane of the gel. The internal marker runs with the digested DNA, thereby standardizing the progress of the electrophoresis in each lane and allowing intersample comparisons. Southern blots thus need to be probed with both the IS*6110* and the internal marker probes. The external marker is used to normalize for differences between gels and allows for intergel comparisons. The DNA fingerprint images are scanned onto the computer, and the IS*6110* banding patterns and the internal marker banding patterns are aligned according to the orientation markers. The size of each IS*6110* band is automatically calculated and digitized and can be compared with other bands through a variety of different algorithms.

The methods described here outline the entire procedure for generating reproducible IS*6110* DNA fingerprints. A summary of this method is illustrated in Figure 24.2 and briefly entails (1) the extraction of genomic DNA from LJ slant cultures of *M. tuberculosis* strains, (2) restriction endonuclease digestion and gel electrophoresis of the DNA, (3) Southern transfer of the DNA,

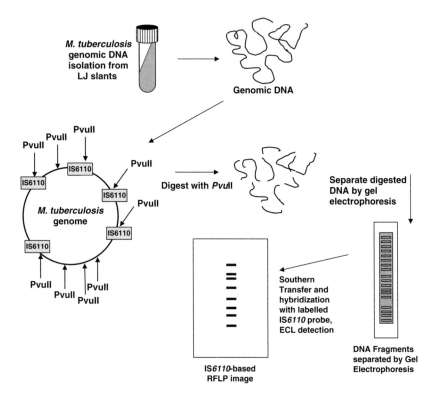

Fig. 24.2 Brief overview of the procedures for the IS*6110* DNA fingerprinting method. The genomic DNA is extracted from LJ slant cultures of *M. tuberculosis* strains; DNA is digested with *Pvu*II restriction endonuclease, followed by gel electrophoresis; the separated DNA is transferred to a nitrocellulose membrane by Southern transfer, after which the membrane is prehybridized, hybridized with the labeled IS*6110* and internal marker probes, and the signals detected and analyzed

(4) preparation and labeling of the IS*6110* and internal marker probes, (5) prehybridization, hybridization, detection, and analysis, (6) and lastly stripping and reprobing of the blots.

24.2 Materials

24.2.1 Extraction of M. tuberculosis Genomic DNA

1. Lowenstein-Jensen slant cultures of *M. tuberculosis* (*see* **Note 1**).
2. Fan oven (able to heat up to at least 80°C).
3. 50-mL polypropylene tubes (Falcon tubes; BD Biosciences, San Jose, CA, USA).
4. 5-mm glass beads.
5. Extraction buffer (pH 7.4): 5% monosodium glutamate (MSG), 50 mM Tris-HCl (pH 7.4), 25 mM EDTA. Adjust pH using HCl and store at room temperature.
6. Sterile disposable 10-μL plastic loops.
7. Lysozyme (Roche, Basel, Switzerland): dissolve to a final concentration of 50 mg/mL in sterile H_2O. Make up freshly before use.
8. RNaseA (Roche): dissolve to a final concentration of 10 mg/mL in 10 mM Tris-HCl (pH 7.5), 15 mM NaCl. Heat to 100°C for 15 min to remove DNAses, allow to cool slowly to room temperature, and store in aliquots at –20°C.
9. Proteinase K buffer (10X) (pH 7.8): 5% sodium dodecyl sulfate, 100 mM Tris-HCl (pH 7.8), 50 mM EDTA. Adjust pH using HCl and store at room temperature.
10. Proteinase K (Roche): dissolve to a final concentration of 10 mg/mL in sterile H_2O. Store in aliquots at –20°C.
11. Chloroform/isoamyl alcohol (24:1 v/v).
12. Phenol/chloroform/isoamyl alcohol (25:24:1 v/v).
13. 3 M sodium-acetate (pH 5.2). Adjust pH with glacial acetic acid and store at room temperature.
14. Isopropanol (100%). Store at –20°C, use ice-cold.
15. Thin glass rods (*see* **Note 2**).
16. 1.5-mL tubes.
17. Ethanol (70%). Store at –20°C, use ice-cold.
18. TE (pH 8.0): 10 mM Tris-HCl, 1 mM EDTA. Adjust to pH 8.0 using HCl, sterilize by autoclaving, and store at room temperature.

24.2.2 PvuII Restriction Endonuclease Digestion

1. *Pvu*II restriction endonuclease (10 U/μL) (New England Biolabs, Ipswich, MA, USA).
2. Restriction buffer NEBuffer 2 (10X) (New England Biolabs).
3. Sterile nuclease-free H_2O.

24.2.3 Gel Electrophoresis of Restricted DNA (Test Gel)

1. 0.5-mL tubes.
2. TE (pH 8.0): 10 mM Tris-HCl, 1 mM EDTA. Adjust to pH 8.0 using HCl, sterilize by autoclaving, and store at room temperature.
3. Loading Buffer (6X) (pH 8.0): 30 mL glycerol, 60 mg bromophenol blue, and 0.6 g sodium dodecyl sulfate (SDS). Make up to 100 mL using TE, and store at room temperature.
4. Agarose D-1 LE (Hispanagar, Burgos, Spain).
5. TBE (5X) (pH 8.3): 0.45 M Tris, 0.44 M boric acid, 10 mM EDTA. Store at room temperature.
6. Gel electrophoresis equipment: gel dimensions 20 cm × 25 cm, gel comb teeth 0.5-cm wide.
7. Ethidium bromide: dissolve to a final concentration of 10 mg/mL stock solution. Store in dark container (*see* **Note 3**).
8. UV light box and photography equipment.

24.2.4 Gel Electrophoresis of Restricted DNA (Final Gel)

1. 3 M sodium acetate (pH 5.2). Adjust pH with glacial acetic acid and store at room temperature.
2. Ethanol (100%). Store at –20°C.
3. Ethanol (70%). Store at –20°C.
4. TE (pH 8.0): 10 mM Tris-HCl, 1 mM EDTA. Adjust to pH 8.0 using HCl, sterilize by autoclaving, and store at room temperature.
5. Loading buffer (6X) (pH 8.0): 30 mL 100% glycerol, 60 mg bromophenol blue, and 0.6 g sodium dodecyl sulfate (SDS). Make up to 100 mL using TE, and store at room temperature.
6. Marker X (Roche) (250 ng/μL).
7. Loading buffer/Internal Molecular Weight Marker (IMWM) mix (1X): 6 mL TE (pH 8.0), 2 mL loading buffer, 6.6 μL Marker X (*see* **Note 4**). Store at –20°C.
8. Agarose SeaKem LE (Cambrex Bio Science, Rockland, Maine, USA).
9. *Pvu*II restriction endonuclease digested external marker MTB14323 DNA (*see* **Note 5**).
10. TBE (5X) (pH 8.3): 0.45 M Tris, 0.44 M boric acid, 10 mM EDTA. Store at room temperature.
11. Ethidium bromide: dissolve to a final concentration of 10 mg/mL stock solution. Store in dark container (*see* **Note 3**).
12. Gel electrophoresis equipment: gel dimensions 20 cm × 25 cm, gel comb teeth 0.5-cm wide.
13. UV light box and photography equipment.

24.2.5 Southern Transfer of the Fingerprinting Gel

1. Flat plastic container.
2. Denaturing buffer: 1.5 M NaCl, 0.5 M NaOH.
3. Venturi pump (HR&L3512; R&L Enterprises Limited, Bramley, Leeds, United Kingdom).
4. Neutralizing buffer: 1.5 M NaCl, 0.5 M Tris-HCl (pH 7.5).
5. Nylon membrane (Hybond N+) (Amersham Biosciences, Piscataway, NJ, USA).
6. Marker X (Roche) (250 ng/µL) (*see* **Note 4**).
7. *Pvu*II restriction endonuclease digested external marker MTB14323 DNA (*see* **Note 5**).
8. TE (pH 8.0): 10 mM Tris-HCl, 1 mM EDTA. Adjust pH to 8.0 using HCl, sterilize by autoclaving, and store at room temperature.
9. 4 M NaOH. Store at room temperature.
10. Orientation marker: 2 µL Marker X (250 ng/µL), 20 µL MTB14323 DNA (1.25 µg/µL) (*see* **Note 5**), 45 µL 0.8 M NaOH, 23 µL TE, pH 8.0.
11. SSPE solution (20X) (pH 7.4): 3 M NaCl, 0.2 M NaH$_2$PO$_4$, 20 mM EDTA. Adjust pH with NaOH pellets and store at room temperature.
12. Sterile H$_2$O.
13. Whatman 3 MM paper.
14. Parafilm.
15. 10-mL pipette.
16. Glass plate.
17. 1 kg weight.
18. Plastic sleeves.

24.2.6 Preparing the IS6110 Probe by PCR Amplification

1. *M. tuberculosis* DNA (0.2 mg per reaction).
2. PCR amplification reagents, for example, HotStarTaq DNA Polymerase system (Qiagen, Hilden, Germany) containing 5X Q-solution, 10X PCR Buffer, 25 mM MgCl$_2$, HotStarTaq DNA Polymerase (5 U/µL) (*see* **Note 6**).
3. 10 mM dNTPs.
4. IS*6110* primers. Forward 5'-TCG GTC TTG TAT AGG CCG TTG-3' and Reverse 5'-ATG GCG AAC TCA AGG AGC AC-3', 50 pmol/mL.
5. Sterile nuclease-free H$_2$O.
6. Agarose D-1 LE (Hispanagar, Spain).
7. Ethidium bromide: dissolve to a final concentration of 10 mg/mL stock solution. Store in dark container (*see* **Note 3**).
8. PCR amplification cleanup kit, for example, Wizard SV Gel and PCR Clean-Up System (Promega, Madison, WI, USA).
9. Gel electrophoresis equipment.
10. UV light box and photography equipment.

24.2.7 ECL Labeling of IS6110 and Internal Marker Probes

1. PCR amplified IS6110 DNA (200 ng).
2. Marker X (200 ng) (*see* **Note 4**).
3. 1.5-mL tubes (for example Eppendorf tubes, Eppendorf AG, Hamburg, Germany).
4. Nucleic acid labeling and detection system, for example, Amersham ECL Direct Nucleic Acid Labeling and Detection System (Amersham Bioscience), containing the following reagents (sufficient for labeling 5 to 10 μg nucleic acid and detecting 2000 to 4000 cm^2 of membrane): labeling reagent (horse radish peroxidase), cross-linker (glutaraldehyde solution), control DNA, blocking agent, ECL Detection Reagents, and ECL Gold Hybridization Buffer.
5. Sterile nuclease-free H$_2$O.

24.2.8 Prehybridization and Hybridization

1. Plastic bag (25 cm × 35 cm).
2. Distilled H$_2$O.
3. ECL Gold Hybridization Buffer (Amersham Bioscience). Prepare according to the manufacturer's instructions and store in aliquots at –20°C. Thaw before use.
4. Heat sealer.
5. 10-mL pipette.
6. Flat plastic container.
7. Shaking water bath.
8. Standard saline citrate (SSC) (20X) (pH 7.0): 3 M NaCl, 0.3 M tri-sodium citrate (dihydrate). Store at room temperature and dilute to 2X as required.
9. Primary wash buffer: 720 g urea, 8 g sodium dodecyl sulfate (SDS), 25 mL 20X SSC. Make up to 2 L and store at room temperature.

24.2.9 Detection of Hybridization

1. Plastic bag (25 cm × 35 cm).
2. 10-mL pipette.
3. McCartney bottle.
4. X-ray film and X-ray cassettes.
5. ECL Detection Reagents (Amersham Bioscience). Store at 4°C.
6. Darkroom, developer, stop solution, and fixing solution for developing x-rays.
7. Gel analysis software (e.g., Gelcompar; Applied Maths, Kortrijk, Belgium).

24.2.10 Stripping the Membrane

1. Flat plastic container.
2. Sodium dodecyl sulfate (SDS) (10% w/v). Weigh out required amount in a fume hood wearing mask and gloves and dissolve in H_2O. Store at room temperature and dilute to 0.1% before use.

24.3 Methods

24.3.1 Extraction of M. tuberculosis Genomic DNA

1. A confluent Lowenstein-Jensen slant culture of *M. tuberculosis* (*see* **Note 7**) is required for a good DNA preparation.
2. Heat inactivate cultures by incubating Lowenstein-Jensen slants in a preheated fan oven at 80°C for 1 h (*see* **Note 8**).
3. Label 50-mL polypropylene tubes containing approximately 20×5-mm glass beads (enough to fill the conical bottom part of the tube) with the corresponding isolate number (labeling should be both on the lid and the tube wall).
4. Allow the heat-inactivated *M. tuberculosis* culture to cool to room temperature (usually 5 to 10 min). Add 3 mL extraction buffer and wet entire slant by gently inverting. Carefully scrape off the bacteria without loosening the medium using a sterile disposable 10-µL plastic loop (*see* **Note 9**).
5. Pour the bacterial suspension into the corresponding labeled sterile 50-mL polypropylene tube containing the 5-mm glass balls (*see* **Note 10**).
6. Add an additional 3 mL extraction buffer to the slant. Scrape any remaining bacteria gently off the Lowenstein-Jensen slant using a sterile 10-µL loop and transfer the bacterial suspension to the corresponding 50-mL propylene tube (*see* **Note 10**).
7. Vortex vigorously for approximately 2 min (*see* **Note 11**).
8. Add 500 µL lysozyme and 2.5 µL RNaseA to each bacterial suspension (*see* **Note 12**).
9. Gently mix by inversion and incubate at 37°C for 2 h in a preheated oven.
10. Add 600 µL 10X Proteinase K buffer and 150 µL Proteinase K (10 mg/mL) (*see* **Note 12**).
11. Gently mix by inversion and incubate at 45°C for 16 h in a preheated oven.
12. Add 5 mL phenol/chloroform/isoamylalcohol (25:24:1) and gently mix by inversion (*see* **Notes 13** and **14**). Repeat gentle mixing every 30 min for 2 h at room temperature.
13. To ensure complete phase separation, centrifuge each sample at $1800 \times g$ for 20 min at room temperature.
14. Aspirate the top phase carefully (without collecting any interface) and transfer to a new 50-mL polypropylene tube containing 5 mL chloroform/ isoamylalcohol (24:1) (*see* **Note 15**).

15. Mix gently by inversion and centrifuge at $1800 \times g$ for 20 min at room temperature.
16. Aspirate the top phase carefully (without collecting any interface) and transfer to a new 50-mL polypropylene tube containing 600 μL 3 M sodium-acetate pH 5.2 (*see* **Note 15**).
17. Precipitate the DNA by adding 7 mL ice-cold isopropanol (*see* **Note 16**).
18. Collect the precipitated DNA immediately on a thin glass rod (*see* **Note 2**). Incubate the rod in a 1.5-mL tube containing 1 mL 70% ethanol for 10 min (*see* **Note 17**).
19. Transfer the glass rod with the precipitated DNA to a new labeled 1.5-mL tube and incubate at room temperature until the DNA is dry (approximately 2 to 3 h).
20. Rehydrate the DNA by adding 300 to 600 μL TE (pH 8.0) and release from the glass rod by gentle mixing.
21. Allow the DNA to redissolve by incubating at 4°C overnight or at 65°C for 2 h. Store at –20°C.

24.3.2 Pvu*II Restriction Endonuclease Digestion*

1. Determine the DNA concentration spectrophotometrically (*see* **Note 18**).
2. Perform restriction enzyme digestion of the DNA in a total volume of 100 μL in a 1.5-mL tube as follows:

> 10 μL 10X restriction buffer M
> 6 μg DNA
> 30 units of *Pvu*II restriction endonuclease
> sterile ddH$_2$O to a final volume of 100 μL

3. Mix the contents of the tube and incubate at 37°C for 3 to 16 h (*see* **Note 19**).
4. Heat-inactivate the *Pvu*II enzyme by incubating at 65°C for 10 min.

24.3.3 Gel Electrophoresis of Restricted DNA (Test Gel)

1. To test that the *Pvu*II digestion is complete, remove an 8-μL aliquot of the digestion reaction (from Section 24.3.2) and transfer to a 0.5-mL tube. Add 4 μL 6x loading buffer for a 2x final concentration and mix by repeated pipetting.
2. Prepare a 1% agarose gel by dissolving 3 g of agarose in 300 mL of 1x TBE (pH 8.3). Gently heat in a microwave oven until all of the agarose is dissolved and then cast into a 20 cm × 25 cm gel electrophoresis tray containing a gel comb. Allow to cool to room temperature before removing the comb.

3. Separate the *Pvu*II-digested DNA by electrophoresis in the 1% agarose gel in 1X TBE at 1.5 V/cm for 16 h or 4 V/cm for 4 h. It is not necessary to load a DNA molecular weight marker, as only the amount of digested DNA will be judged, not the size of the bands.

4. To enable detection of the digested DNA, stain the gel in 500 mL 1X TBE containing 50 μL of a 10 mg/mL ethidium bromide stock solution (*see* **Note 3**) with shaking for 30 min at room temperature.

5. Visualize the DNA by trans-illumination at 245 nm on a UV light box (*see* **Note 20**) and capture the image by photography.

6. Determine whether the restriction digestion is complete; digested DNA will show a smear with an evident banding pattern. The intensity of the digested DNA in each lane can be used to compare the samples and to estimate whether equal amounts of DNA are present in each digestion reaction.

24.3.4 Gel Electrophoresis of Restricted DNA (Final Gel)

1. Precipitate the remaining *Pvu*II-digested DNA (92 μL) by adding 9 μL 3 M sodium acetate, pH 5.2, and 300 μL ice-cold 100% ethanol. Mix and incubate at −20°C for 16 h.

2. Pellet the *Pvu*II-digested DNA by centrifugation at 10,000 × g for 30 min at 4°C.

3. After centrifugation, aspirate the supernatant to 50 μL without disturbing the DNA pellet.

4. Wash the DNA pellet with 500 μL ice-cold 70% ethanol and centrifuge at 10,000 × g for 30 min at 4°C. Aspirate the supernatant to 50 μL without disturbing the DNA pellet.

5. Dry the DNA pellet at room temperature for 16 h.

6. Redissolve the DNA in 1X loading buffer/Internal Molecular Weight Marker (IMWM) mix at 4°C for 16 h or at 65°C for 4 h with mixing every hour (*see* **Note 21**). The volume of 1X loading buffer/IMWM mix is determined according to the intensity of the DNA detected in step 6 of Section 24.3.3. Samples with low-intensity bands according to the test gel should be redissolved with a proportionally lower volume of 1X loading buffer/IMWM mix to allow equal amounts of DNA to be loaded in each lane of the final gel.

7. Prepare an 0.8% agarose gel by dissolving 2.4 g of agarose in 300 mL 1x TBE (pH 8.3).

8. Separate a 10-μL aliquot of *Pvu*II-digested DNA, as well as a 10-μL aliquot of external marker MTB14323 in separate lanes by electrophoresis at 2 V/cm for 16 h.

9. Visualize the bands by staining the agarose gel with ethidium bromide and photography .

24.3.5 Southern Transfer of the Fingerprinting Gel

1. Invert the gel and place into a flat plastic container. Denature the DNA in the gel by incubating in 500 mL denaturing buffer at 25°C for 30 min with gentle shaking.
2. Aspirate the denaturing solution using a Venturi pump. Thereafter, neutralize the gel with 500 mL neutralizing buffer and incubate at 25°C for 30 min with gentle shaking.
3. Label the nylon membrane (Hybond N+) (*see* **Note 22**) with a black ballpoint pen to allow future recognition. Spot 0.2-μL aliquots of orientation marker onto the membrane at six different positions on the upper and lower edges of the membrane to allow for future alignment of resultant autoradiographs.
4. Hydrate the membrane briefly in H_2O and transfer and equilibrate in 20X SSPE solution. (*see* **Note 23**).
5. Set up the Southern transfer by placing the inverted agarose gel onto Whatman 3 MM paper presoaked in 20X SSPE (*see* **Note 24**).
6. Remove air bubbles from under the gel by gently rolling a 10-mL pipette over the gel. Place strips of Parafilm around the gel to ensure that the fluid flows through the gel during Southern transfer.
7. Place the membrane onto the agarose gel with orientation markers facing the agarose gel and remove air bubbles by gently rolling with a 10-mL pipette over the membrane (*see* **Note 25**).
8. Place two layers of Whatman 3 MM blotting paper (*see* **Note 26**) prewetted in 20X SSPE (pH 7.4) onto the nylon membrane and again remove air bubbles.
9. To ensure fluid flow through the agarose gel, stack folded paper towels onto the Whatman 3 MM papers and ensure contact with the addition of a glass plate and a 1-kg weight.
10. Fill the blotting tray with 20X SSPE (pH 7.4). Don't overfill!
11. Allow the Southern transfer to proceed for 16 h.
12. After transfer, remove the nylon membrane and wash in 2X SSPE for 10 min. Bake the nylon membrane at 80°C for 2 h (between two sheets of Whatman 3 MM blotting paper) to fix the DNA to the membrane. Seal the membrane in a plastic sleeve and store at 4°C until further use.

24.3.6 Preparing the IS6110 Probe by PCR Amplification

1. Set up a PCR amplification reaction mixture containing:

    0.2 mg *M. tuberculosis* DNA template
    5 μL Q-Solution
    2.5 μL 10X PCR Buffer
    2 μL 25 mM $MgCl_2$

4 µL 10 mM dNTPs
1 µL of each IS*6110* primer (50 pmol/µL)
0.125 µL HotStarTaq DNA polymerase.

Make up to 25 µL with sterile nuclease-free H_2O (*see* **Note 27**).

2. Initiate amplification by incubation at 95°C for 15 min, followed by 35 to 45 cycles of 94°C for 1 min, 62°C for 1 min, and 72°C for 1 min. After the last cycle, incubate the samples at 72°C for 10 min.
3. Check the efficiency of amplification by electrophoretically fractionating a 5-µL aliquot in a 2% agarose gel (1 x TBE, pH 8.3) followed by staining with ethidium bromide (*see* **Note 3**).
4. Purify the IS*6110* PCR amplification product using a PCR amplification clean-up kit (e.g., the Wizard SV Gel and PCR Clean-Up System) according to the manufacturer's instructions, and elute the DNA in nuclease-free H_2O. Determine the concentration of the IS*6110* probe spectrophotometrically (*see* **Note 18**) and store at –20°C.

24.3.7 ECL Labeling of IS6110 and Internal Marker Probes

1. Add 200 ng of probe DNA (PCR amplified IS6110 DNA or Marker X; *see* **Note 4**) to an 0.5-mL tube and make up to 15 µL with nuclease-free H_2O (*see* **Note 28**).
2. Denature the probe DNA by incubating at 100°C for 5 min, followed by snap cooling on ice (to 4°C) for 5 min (*see* **Note 29**).
3. Label the probe using a nucleic acid labeling system (e.g., the Amersham ECL Direct Nucleic Acid Labeling and Detection System) according to the manufacturer's instructions. For the Amersham ECL Direct Nucleic Acid Labeling and Detection System, add 15 µL of the labeling mix to the probe and mix well (*see* **Note 30**). Add 15 µL glutaraldehyde solution, mix well, and incubate at 37°C for 10 min (*see* **Note 31**). The labeled probe should be used immediately to ensure optimal detection.

24.3.8 Prehybridization and Hybridization

1. Rehydrate the nylon membrane by incubating in 500 mL distilled H_2O and seal in a plastic bag.
2. Add 48 mL of ECL Gold Hybridization Buffer to the membrane in the plastic bag. Remove all air bubbles and seal the bag with a heat sealer (*see* **Note 32**).
3. Spread the ECL Gold Hybridization Buffer by rolling with a 10-mL pipette and then prehybridize in a flat plastic container in a shaking water bath at 90 rpm and 42°C for at least 60 min (*see* **Note 33**).

4. To initiate hybridization, remove the plastic bag containing the membrane from the water bath and cut off one corner. Add the labeled probe directly to the ECL Gold Hybridization Buffer (*see* **Note 34**). Remove all air bubbles and reseal the bag (*see* **Note 35**).
5. Hybridize at 42°C for 16 h with shaking at 90 rpm.
6. After hybridization, remove the nylon membrane from the hybridization bag, place in a clean flat plastic container, and wash twice in 400 mL prewarmed Primary Wash at 42°C for 20 min at 90 rpm. (*see* **Note 36**).
7. Wash the membrane twice in 400 mL 2X SSC for 5 min at room temperature on the shaker.

24.3.9 Detection of Hybridization

1. Place the membrane in a new plastic bag and remove all excess liquid by rolling with a 10-mL pipette.
2. Mix 4 mL of each of the two Amersham ECL Detection Reagents in a McCartney bottle and add the mixture to the membrane in the bag. Spread the detection reagent over the membrane continuously for 90 s by rolling with a 10-mL pipette (*see* **Note 37**).
3. Remove all excess detection reagent by rolling with a 10-mL pipette and seal the plastic bag.
4. In a darkroom, expose the membrane to x-ray film for an optimum time period (from 1 min to 2 h) (*see* **Note 38**).
5. Develop the x-ray film according to the standardized method.
6. Analyze the IS*6110* RFLP patterns using a gel-analysis computer program.

24.3.10 Stripping the Membrane

1. After hybridization and detection, remove the probe by placing the membrane in a flat plastic container containing 400 mL of boiling 0.1% SDS (*see* **Note 39**).
2. Incubate with shaking until the solution reaches room temperature (approximately 60 min).
3. Place the nylon membrane into a plastic bag and roll out the excess fluid by using a 10-mL pipette.
4. Seal the plastic bag. This can be stored for years at 4°C in a fridge *or* can be used immediately for a second prehybridization and hybridization (*see* Section 24.3.8).

24.4 Notes

1. Lowenstein-Jensen medium is used primarily for the isolation and propagation of *Mycobacterium* spp. and is a relatively simple solid medium supplemented with glycerol and egg mixture, which provide fatty acids and protein required for the metabolism of

mycobacteria. The coagulation of the egg albumin during sterilization provides a solid medium for inoculation purposes. This medium also contains malachite green as an indicator and inhibitor to microorganisms other than acid-fast bacilli. In order to isolate enough DNA, each clinical isolate of *M. tuberculosis* is inoculated under biosafety level 3 conditions onto two Lowenstein-Jensen slants (10-mL Lowenstein-Jensen medium in 25-mL Bijou bottles) and incubated at 37°C with weekly aeration until confluent growth is observed (up to 8 weeks).

2. Thin glass rods can be easily constructed from sterile glass Pasteur pipettes by flaming the open tip of the pipette in a Bunsen burner in order to melt and close the tip.

3. Ethidium bromide is carcinogenic and toxic. Always wear gloves and work in designated areas. Discard ethidium bromide waste in specially provided containers.

4. *Pvu*II digested supercoiled ladder (Gibco BRL, Invitrogen, Carlsbad, California, USA) could be used as an alternative internal molecular weight marker [7].

5. Reference strain *M. tuberculosis* Mt14323 (available on request from the Unit of Molecular Microbiology, National Institute of Public Health and Environmental Protection, P.O. Box 1, 3720 BA, Bilthoven, The Netherlands): When digested with *Pvu*II and probed with IS6110, this strain gives 13 approximately evenly spaced bands of around 13.9, 7.2, 6.9, 4.4, 3.6, 3.0, 2.3, 2.0, 1.7, 1.5, 1.4, 1.3, and 1.0 kb (*see* Fig. 24.1, lane 1) [7].

6. We recommend the use of the Qiagen HotStarTaq DNA Polymerase system, due to the fact that it provides high specificity in hot-start PCR, as well as the fact that it makes use of the Q-solution, which enhances PCR over the high G+C rich regions of the *M. tuberculosis* genome.

7. The quality of the IS6110 DNA fingerprint depends on the quality of the *M. tuberculosis* culture. DNA should be extracted from fresh cultures, as DNA degradation occurs during prolonged storage. Isolation of genomic DNA from liquid cultures of *M. tuberculosis* is extremely poor and is discouraged.

8. The outer surface of each Lowenstein-Jensen slant bottle should be sterilized by swabbing with a suitable disinfectant, after which it is placed in a stainless steel rack within an unsealed biosafety autoclave bag and transferred to the 80°C prewarmed circulating fan oven (580 × 540 × 510 mm) for 1 h to ensure heat killing. This method allows for the efficient killing of *M. tuberculosis* without compromising DNA integrity for subsequent molecular investigation [21].

9. To ensure safety, all the following procedures should be done in a safety cabinet (see local regulations). All waste should be sterilized by autoclaving. Gloves should be worn at all times to prevent contamination of the samples.

10. If more than one Lowenstein-Jensen culture per isolate is being used, first transfer the scraped bacterial suspension to the second Lowenstein-Jensen culture. Carefully scrape off the bacteria and then transfer the combined bacterial suspension to the labeled 50-mL polypropylene tube.

11. Vortexing the 50-mL tubes with the glass balls is performed to break the cells apart as they tend to clump together. This ensures a homogeneous cell suspension.

12. *M. tuberculosis* is a Gram-positive organism, with a thick cell wall containing peptidoglycan. Proteinase K digests the proteins in the sample, RNAse A digests the RNA, and lysozyme digests the cell wall by targeting the peptidoglycan component.

13. Phenol is a strong denaturing agent and can cause severe burns. Wear gloves and protective eyewear.

14. These and the following steps should be done in a standard fume hood.

15. Transfer the marked tops from the old 50-mL tubes to the new 50-mL tube. It is very important to transfer the labeled top to the correct tube.

16. The sodium acetate will change the pH and the charge of the DNA so that it is slightly hydrophobic and will precipitate once the isopropanol is added. It is important to gently invert the tube to allow for gentle mixing and gradual precipitation of the DNA.

This also allows for a cleaner banding pattern when run on an agarose gel. Do not shake vigorously as this may shear DNA.

17. If there is insufficient DNA to be fished out, place tube in the freezer at $-20°C$ overnight, then centrifuge at $10,000 \times g$ for 20 min at $4°C$. Carefully pour off the supernatant and wash precipitate with 10 mL of cold 70% ethanol. Repeat centrifugation as above and thereafter pour off the ethanol, invert the tube on a blotting paper or paper towel. Allow DNA to dry at room temperature and then rehydrate and dissolve the DNA in 200 µL TE pH 8.0.

18. An optical density (OD) of 1 at 260 nm is equivalent to 50 µg/uL of double-stranded DNA. The purity of the DNA can be assessed by looking at the relationship of the values of OD_{260}/OD_{280}. This relationship value should be between 1.8 and 2.0.

19. The digestion should be incubated in an oven to reduce evaporation and condensation.

20. UV light is dangerous, and protective eyewear should be worn at all times.

21. The internal marker is used to normalize the electrophoretic mobility of each band in each lane of the gel. The internal marker runs with the digested DNA, thereby standardizing the progress of the electrophoresis in each lane and allowing intersample comparisons. Southern blots thus need to be probed with both the IS6110 and the internal marker probes. The external marker is used to normalize for differences between gels and allows for intergel comparisons.

22. Gloves should be worn when handling the nylon membrane to prevent skin oils from preventing transfer of the DNA onto the membrane.

23. It is very important to hydrate the membrane in sterile water. If you hydrate the membrane in a salt solution like SSPE, the membrane will not rehydrate completely.

24. The edges of the gel, as well as the wells, should be removed with a surgical blade.

25. If the air bubbles are not removed, successful transfer of the DNA cannot occur.

26. The Whatman 3 MM paper should overlap the agarose gel by 5 to 10 mm on each side.

27. Guidelines to prevent cross-contamination during PCR amplification: (1) use a specific, dedicated area, containing a laminar flow hood, where the PCR reactions are made up, (2) clean the working area (laminar flow hood) first with 10% bleach followed by 70% ethanol before and after setting up a PCR reaction, (3) use dedicated, cleaned pipettes (cleaned with 10% bleach and 70% ethanol), (4) wear gloves at all times and change regularly, (5) open and close tubes carefully to prevent the formation of aerosols, (6) use a separate, dedicated area for adding DNA to the mix, do not add DNA in the same place where the reaction mix is made up, (7) use a clean lab coat for every area entered, (8) do not take apparatus from one lab to another, (9) use positive and negative controls (no DNA) to all reaction sets.

28. The probe DNA should be free of any primary amines (i.e., Tris), which will quench the glutaraldehyde cross-linking activity, thereby inhibiting the cross-linking of the horse radish peroxidase (HRP) to the probe DNA.

29. Heating to $100°C$ and rapid cooling causes DNA to denature and remain in single strands. Single-stranded DNA is required for labeling and hybridization.

30. Horse radish peroxidase (HRP) binds to DNA through electrostatic interactions.

31. Glutaraldehyde cross-links horse radish peroxidase (HRP) to DNA via a Schiff base intermediate.

32. ECL Gold Hybridization Buffer contains a blocking agent, gelatin, which binds to all the sites not occupied by the transferred DNA, preventing nonspecific binding of the probe to the membrane; and NaCl to increase the stringency of the hybridization.

33. To ensure even distribution of the prehybridization solution across the nylon membrane, a second plastic bag containing 500 mL H_2O can be placed on top of the bag containing the nylon membrane. When shaken, the weight of the water in the top bag gently massages the prehybridization solution covering the membrane in the bottom bag.

34. The undiluted probe should not be allowed to come in contact with the nylon membrane as this may result in nonspecific binding and poor DNA fingerprint resolution.
35. The membrane must not fold or bend at any stage of the hybridization. Avoid air bubbles as these will limit contact between the probe and the substrate at the position of the bubbles.
36. The second primary wash buffer can be reused for a subsequent first primary wash.
37. The detection reagents contain the substrate, Luminol, which is oxidized by the horse radish peroxidase (HRP) in the presence of H^+ to produce light.
38. Ensure that the membrane is orientated such that the surface containing the DNA faces the X-ray film.
39. The solution must be boiling in order to denature the DNA and separate the probe from the DNA fixed to the membrane. It is important to remove all hybridized probe from the membrane before rehybridization. Failure to remove the first probe may result in background on subsequent detection.

Acknowledgments We would like to thank Ms. Talita Lotz for providing the IS6110-based RFLP image.

References

1. Hermans, P. W., van Soolingen, D., Bik, E. M., de Haas, P. E., Dale, J. W., and van Embden, J. D. (1991). Insertion element IS987 from *Mycobacterium bovis* BCG is located in a hot-spot integration region for insertion elements in *Mycobacterium tuberculosis* complex strains. *Infect. Immun.* **59**, 2695–2705.
2. Groenen, P. M., Bunschoten, A. E., van Soolingen, D., and van Embden, J. D. (1993). Nature of DNA polymorphism in the direct repeat cluster of *Mycobacterium tuberculosis*; application for strain differentiation by a novel typing method. *Mol. Microbiol.* **10**, 1057–1065.
3. Ross, B. C., Raios, K., Jackson, K., and Dwyer, B. (1992). Molecular cloning of a highly repeated DNA element from *Mycobacterium tuberculosis* and its use as an epidemiological tool. *J. Clin. Microbiol.* **30**, 942–946.
4. Supply, P., Mazars, E., Lesjean, S., Vincent, V., Gicquel, B., and Locht, C. (2000). Variable human minisatellite-like regions in the *Mycobacterium tuberculosis* genome. *Mol. Microbiol.* **36**, 762–771.
5. Thierry, D., Brisson-Noel, A., Vincent-Levy-Frebault, V., Nguyen, S., Guesdon, J. L., and Gicquel, B. (1990). Characterization of a *Mycobacterium tuberculosis* insertion sequence, IS6110, and its application in diagnosis. *J. Clin. Microbiol.* **28**, 2668–2673.
6. Kremer, K., van Soolingen, D., Putova, I., and Kubin, M. (1996). Use of IS6110 DNA fingerprinting in tracing man-to-man transmission of *Mycobacterium tuberculosis* in the Czech Republic. *Cent. Eur. J. Public Health* **4**, 3–6.
7. van Embden, J. D., Cave, M. D., Crawford, J. T., Dale, J. W., Eisenach, K. D., Gicquel, B. et al. (1993). Strain identification of *Mycobacterium tuberculosis* by DNA fingerprinting: recommendations for a standardized methodology. *J. Clin. Microbiol.* **31**, 406–409.
8. Dale, J. W. (1995). Mobile genetic elements in mycobacteria. *Eur. Respir. J. Suppl* **20**, 633s–648s.
9. Alland, D., Kalkut, G. E., Moss, A. R., McAdam, R. A., Hahn, J. A., Bosworth, W. et al. (1994). Transmission of tuberculosis in New York City. An analysis by DNA fingerprinting and conventional epidemiologic methods. *N. Engl. J. Med.* **330**, 1710–1716.

10. Small, P. M., Hopewell, P. C., Singh, S. P., Paz, A., Parsonnet, J., Ruston, D. C. et al. (1994). The epidemiology of tuberculosis in San Francisco. A population-based study using conventional and molecular methods. *N. Engl. J. Med.* **330**, 1703–1709.

11. Warren, R., Hauman, J., Beyers, N., Richardson, M., Schaaf, H. S., Donald, P. et al. (1996). Unexpectedly high strain diversity of *Mycobacterium tuberculosis* in a high-incidence community. *S. Afr. Med. J.* **86**, 45–49.

12. Hermans, P. W., Messadi, F., Guebrexabher, H., van Soolingen, D., de Haas, P. E., Heersma, H. et al. (1995). Analysis of the population structure of *Mycobacterium tuberculosis* in Ethiopia, Tunisia, and The Netherlands: usefulness of DNA typing for global tuberculosis epidemiology. *J. Infect. Dis.* **171**, 1504–1513.

13. van Rie, A., Warren, R., Richardson, M., Victor, T. C., Gie, R. P., Enarson, D. A. et al. (1999). Exogenous reinfection as a cause of recurrent tuberculosis after curative treatment. *N. Engl. J. Med.* **341**, 1174–1179.

14. van Rie, A., Warren, R. M., Beyers, N., Gie, R. P., Classen, C. N., Richardson, M. et al. (1999). Transmission of a multidrug-resistant *Mycobacterium tuberculosis* strain resembling "strain W" among noninstitutionalized, human immunodeficiency virus-seronegative patients. *J. Infect. Dis.* **180**, 1608–1615.

15. van Rie, A., Warren, R., Richardson, M., Gie, R. P., Enarson, D. A., Beyers, N. et al. (2000). Classification of drug-resistant tuberculosis in an epidemic area. *Lancet* **356**, 22–25.

16. Michalak, K., Austin, C., Diesel, S., Bacon, M. J., Zimmerman, P., and Maslow, J. N. (1998). *Mycobacterium tuberculosis* infection as a zoonotic disease: transmission between humans and elephants. *Emerg. Infect. Dis.* **4**, 283–287.

17. Verver, S., Warren, R. M., Munch, Z., Richardson, M., van der Spuy, G. D., Borgdorff, M. W. et al. (2004). Proportion of tuberculosis transmission that takes place in households in a high-incidence area. *Lancet* **363**, 212–214.

18. Heyns, L., Gie, R. P., Goussard, P., Beyers, N., Warren, R. M., and Marais, B. J. (2006). Nosocomial transmission of *Mycobacterium tuberculosis* in kangaroo mother care units: a risk in tuberculosis-endemic areas. *Acta Paediatr.* **95**, 535–539.

19. Richardson, M., Carroll, N. M., Engelke, E., van der Spuy, G. D., Salker, F., Munch, Z. et al. (2002). Multiple *Mycobacterium tuberculosis* strains in early cultures from patients in a high-incidence community setting. *J. Clin. Microbiol.* **40**, 2750–2754.

20. Small, P. M., McClenny, N. B., Singh, S. P., Schoolnik, G. K., Tompkins, L. S., and Mickelsen, P. A. (1993). Molecular strain typing of *Mycobacterium tuberculosis* to confirm cross-contamination in the mycobacteriology laboratory and modification of procedures to minimize occurrence of false-positive cultures. *J. Clin. Microbiol.* **31**, 1677–1682.

21. Warren, R., de Kock, M., Engelke, E., Myburgh, R., Gey, v. P., Victor, T. et al. (2006). Safe *Mycobacterium tuberculosis* DNA extraction method that does not compromise integrity. *J. Clin. Microbiol.* **44**, 254–256.

Chapter 25
Typing *Mycobacterium tuberculosis* Using Variable Number Tandem Repeat Analysis

T.J. Brown, V.N. Nikolayevskyy and F.A. Drobniewski

Abstract DNA-based typing has contributed to the understanding of *M. tuberculosis* epidemiology and evolution. IS6110 RFLP was the first method described and has been used in many epidemiologic investigations. Technological difficulties have hampered the widespread establishment of this method, and it has been found to be of little use in evolutionary studies. PCR-based methods such as spoligotyping and variable number tandem repeat (VNTR) analysis largely overcome these difficulties. Spoligotyping alone is of limited value in epidemiologic investigations due to low discrimination but can be useful in evolutionary studies. Panels of VNTR loci selected from the 59 polymorphic VNTRs described to date have been shown to be useful in both epidemiologic and evolutionary studies. A VNTR type is identified by, first, amplifying a series of PCR fragments each encompassing a different VNTR locus and, second, determining the PCR fragment sizes from which the number of repeats present is calculated. The repeat number present at a series of loci is used as numerical code to describe a type. This chapter describes a high-throughput automated method for VNTR analysis at 15 loci using a capillary fragment analyzer and a manual method using agarose gel analysis.

Keywords epidemiology · ETR · fragment analysis · MIRU · *Mycobacterium tuberculosis* · typing · VNTR

25.1 Introduction

25.1.1 Purpose of Typing

DNA typing has made a major contribution to the understanding of both *Mycobacterium tuberculosis* epidemiology and evolution. In general, high levels

T.J. Brown
HPA MRU, Queen Mary's School of Medicine and Dentistry, 2 Newark Street,
London, E1 2AT, UK
e-mail: t.brown@qmul.ac.uk

T. Parish, A.C. Brown (eds.), *Mycobacteria Protocols*,
doi: 10.1007/978-1-59745-207-6_25, © Humana Press, Totowa, NJ 2008

of strain discrimination are required in clinical settings where questions concerning transmission of tuberculosis (TB) between patients, reactivation versus reinfection in an individual, or laboratory cross-contamination are asked. Lower levels of strain discrimination are required when studying *M. tuberculosis* evolution where robust markers identifying significant genetic events are required for construction of phylogenetic lineages.

25.1.2 History of Typing Tuberculosis

In 1991, the typing of TB strains using IS6110 restriction fragment length polymorphism (RFLP) analysis was described and quickly became the standard for epidemiologic investigations but in the format used had limited value for investigating the phylogeny of *M. tuberculosis* [1]. Although IS6110 RFLP typing has been shown to be a powerful epidemiologic tool, it has four major problems: first, the technique is relatively technically demanding; second, large quantities (microgram amounts) of highly purified DNA from this slow-growing organism are required; third, the data is in the form of banding patterns meaning that comparison of data between laboratories may be unreliable or at least difficult; and fourth, discrimination is poor when the technique is applied to strains containing the IS6110 element at low copy number (<5).

The use of *spacer oligonucleotide* or *spoligotyping* as described in 1997 [2] for the typing of *M. tuberculosis* is a PCR-based method that circumvents the practical difficulties associated with IS6110 RFLP typing, requiring small amounts of template DNA and producing digital data allowing simple data analysis and data sharing between laboratories. Spoligotype data has been found to be useful in distinguishing between the members of the *M. tuberculosis* complex and defining *M. tuberculosis* clades and families in phylogenetic studies [3]. Spoligotyping has been shown to be of limited value for epidemiological transmission studies due to its low discriminative power when used alone [4]. IS6110 RFLP analysis and spoligotyping can be combined to provide a useful epidemiologic tool, where spoligotyping is used in addition to IS6110 RFLP for strains containing IS6110 in low copy number.

25.1.3 Variable Number Tandem Repeat Analysis

The analysis of repeated DNA motifs often referred to as minisatellites or variable number tandem repeats (VNTRs) for human fingerprinting was first described in 1985 [5]. Similar features were identified in *M. tuberculosis* [6], and initially six VNTR loci, designated ETR A to E, were shown to be simple to amplify by PCR and polymorphic, but their level of discrimination was not sufficient to be useful for *M. tuberculosis* epidemiologic investigations [4]. Subsequently with the publication of the *M. tuberculosis* H37Rv genome, a total of 41 VNTR loci were identified [7] and on this occasion designated mycobacterial interspersed repetitive units (MIRU) 1 to 41; 12 of which were shown to be polymorphic in *M. tuberculosis* strains. This panel of 12 MIRU/ VNTR loci have been repeatedly shown to be useful in epidemiologic investigations [8, 9]. The discriminative power of a series of VNTR locus panels consisting of various numbers and combinations of the 59 polymorphic VNTR loci that have been described to date (shown in Table 25.1) [6, 7, 10, 11, 12, 13, 14, 15] have been evaluated against both test collections of *M. tuberculosis* strains [10, 12, 13, 16] and in population studies [17]. These studies have shown that levels of discrimination can be achieved that are in excess of those seen with the original 12 MIRU/VNTR panel. In studies where VNTR data has been compared directly with IS6110 RFLP data, the discrimination of VNTR approaches or equals that seen with RFLP, but it is still not clear whether the highest resolution VNTR analysis gives the same level of discrimination as that seen with IS6110 RFLP when performed on strains containing the insertion sequence at high copy number (>5). In addition to its value as an epidemiologic tool, VNTR data has been shown to be a useful tool in phylogeny studies [18]. It appears therefore that the high levels of strain discrimination and the presence of phylogenetically useful markers coupled with ease of data generation and handling now make VNTR analysis an attractive approach for most epidemiologic studies.

25.1.4 VNTR Methodology

The enumeration of the repeats at a given locus is performed in two stages. The first stage is a PCR for which primers are designed that anneal to regions flanking the tandem repeats. At the second stage, the PCR product is sized and the repeat number calculated by subtracting the length of the flanking regions and dividing the resultant fragment length by the size of the repeat. This is shown graphically in Figure 25.1 for MIRU40. The repeat numbers from a series of loci can be put together in a string giving a numerical type designation. These numerical codes are simple to store and manipulate in widely available software such as Excel but perhaps most importantly are portable so that data can easily be shared between laboratories. This property of VNTR data has already resulted in the construction of Web-based databases, for example that found at http://www.hpa-bioinformatics.org.uk/bionumerics/prototype/ home.php. Sizing can be performed in simple agarose or polyacrylamide gels where PCR products are compared with size standards, and as the product sizes vary in increments of the tandem repeat size, where the tandem repeat is of at least 50 bp, size estimation of ±15 bp is adequate to identify the number

Table 25.1 VNTR Reported in the Literature

Locus name	H37Rv genome location	VNTR panels*	Repeat size (bp)	Alternative names given in other publications						
				MIRU [7]	ETR [6]	Mtb-v [15]	Mtub [10]	QUB [12, 13]	[14]	[11]
24	24648		18			Mtb-v20	Mtub01			
55	55481		3			Mtb-v1				
79	79503		9				Mtub02			
154	154073	1,3	53	MIRU2						
241	241423		11			Mtb-v3				
424	424010		51				Mtub4			
472	472658		9			Mtb-v4				
531	531430		15		MPTR-A					
566	566196		18			Mtb-v5				
569	569819		56						VNTR0569	
577	577172	1,2,3,4	58		ETR-C					
580	580578	1,2,3	77	MIRU4	EDR-D					
595	595334		58					QUB0595C		Msx.4
802	802429	1,2,3,4	54	MIRU40						
917	917609		58						VNTR0917	
960	960173	1,2,3,4	53	MIRU10				QUB-0960C		
1121	1121658		15				Mtub12			
1122	1122852		6			Mtb-v6				
1281	1281895		64					QUB-1281C		
1305	1305493		62						VNTR1305	
1443	1443417		56				Mtub16			
1451	1451778		57					QUB-1451		
1612	1612529		21					QUB-23		
1644	1644261	1,2,3	53	MIRU16						
1895	1895344		57					QUB-1895		

Table 25.1 (continued)

| Locus name | H37Rv genome location | VNTR panels* | Repeat size (bp) | Alternative names given in other publications | | | | | | |
				MIRU [7]	ETR [6]	Mtb-v [15]	Mtub [10]	QUB [12, 13]	[14]	[11]
1907	1907458		56							VNTR1907
1955	1955580	4	57				Mtub21			
1982	1982887	4,5	78					QUB-18		
2059	2059441	1,3	77	MIRU20						
2074	2074431		56				Mtub24	QUB-2074		
2163a	2163323	4,5	69					QUB-11a		
2163b	2163741	2,3	69					QUB-11b		
2165	2165223	1,2,3,4	75		ETR-A					
2347	2347393	3	57				Mtub29			
2372	2372435		57						VNTR2372	
2687	2687128	1,3	54	MIRU24						
2401	2401815	2,3	58		ETR-B		Mtub30	QUB-2401		
2461	2461279	1,3,4	57							
2531	2531898	1,3	53	MIRU23				QUB-2531C		
2604	2604134		9			Mtb-v10				
2703	2703890		57						VNTR2703	
2990	2990582		55				Mtub31	QUB-2990		
2996	2996003	1,2,3,4	51	MIRU26						
3007	3007063	1,3,4	53	MIRU27						
3155	3155880		54					QUB-5		
3171	3171468	3	51				Mtub34	QUB-15		
3192	3192202	1,2,3,4	53	MIRU31	ETR-E					
3232	3232649	4,5	56					QUB-3232		
3239	3239469		79+55		ETR-F	Mtb-v15				

Table 25.1 (continued)

Locus name	H37Rv genome location	VNTR panels*	Repeat size (bp)	Alternative names given in other publications						
				MIRU [7]	ETR [6]	Mtb-v [15]	Mtub [10]	QUB [12, 13]	[14]	[11]
3336	3336499	4,5	59					QUB-3336		
3594	3594260		56						VNTR3594	
3690	3690947	4	58				Mtub39			
3820	3820389		57						VNTR3820	
4052	4052971	2,3	111					QUB-26		
4120	4120914		57						VNTR4120	
4155	4155457		57						VNTR4155	Mpp.8
4156	4156797	2,3	59					QUB-4156c		
4226	4226924		3			Mtb-v18				
4348	4348721	1,3	53	MIRU39						

The table shows the key features and aliases of the *Mycobacterium tuberculosis* VNTR loci described in the literature (*see* **Note 12**).
*Panel 1 comprises 15 VNTR loci and is routinely used for *M. tuberculosis* genotyping at the HPA MRU and is described in the current chapter. Panel 2 comprises 15 VNTR loci and has been recently described as an optimized panel for routine genotyping [22]. Panel 3 comprises 24 VNTR loci and has been recently described as a panel for phylogenetic studies [22]. Panel 4 comprises 14 VNTR loci and has been reported as highly discriminative for prospective epidemiologic studies [17]. Panel 5 comprises 4 VNTR loci that have been reported as highly discriminative for Beijing strains [23].

A

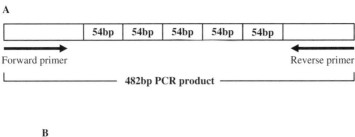

B

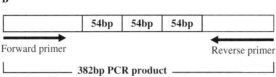

C

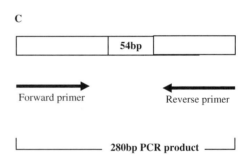

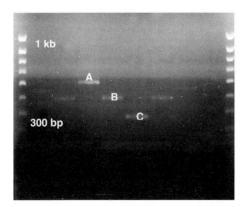

Fig. 25.1 Outline of VNTR analysis. The PCR product in **(A)** corresponds with the length of the flanking regions of MIRU40 plus five tandem repeats of 54 bp. The PCR product in **(B)** corresponds with the length of the flanking regions plus three tandem repeats of 54 bp, and the PCR product in **(C)** corresponds with the length of the flanking regions plus one tandem repeat of 54 bp. This image shows fragments A–C stained with ethidium bromide in an agarose gel

of repeats present. High throughput and automation can be achieved by using the fragment analysis features of DNA sequencing systems where data output is in a digital format that can interface easily with analytical software. A further high-throughput approach has been described where PCR products are sized using denaturing high-performance liquid chromatography (HPLC) [19].

In this chapter, we describe two methods for the analysis of a panel of 15 VNTR loci, this being a composite of the five ETR/VNTR panel described by Frothingham and MacConell [6] and the 12 MIRU/VNTR panel described by Supply [20]. The first high-throughput method analyzes appropriately labeled PCR products using automated sequencing equipment to determine product size. The CEQ8000 Genetic Analyser (Beckman Coulter, High Wycombe, UK) used in the method described below will accurately size fragments between 60 and 640 bp, therefore primers were designed to give expected fragment sizes in this range. This instrument will generate fragment lists from samples loaded into a 96-well PCR tray. Samples are loaded automatically into an array of eight capillaries containing an acrylamide gel matrix through which DNA fragments labeled with fluorescent dyes are separated by electrophoresis. When labeled fragments pass a laser, they are excited and emissions are detected. (Size of fragments and intensity of emission are recorded as a trace as shown later in Fig. 25.3). When this is compared against a trace generated by size standards, fragment sizes can be calculated.

Suitable primers for the 12 MIRU/VNTR loci were described by Kwara [21] and for the ETR/VNTR A-C loci have been described by Frothingham and MacConell [6]. Forward primers in each set are labeled with Beckman dyes 2, 3, or 4 and the reverse primers are used unlabeled. The Beckman dye 1 is used for labeling the size markers. A series of nine PCR reactions are carried out six of which are duplex reactions and three of which are simple, as shown in Table 25.2. Upon completion, they are combined into three pools: pool A and B contain six PCR products and pool C contains three products. The PCR products in each pool and the dyes with which they are labeled are shown in Table 25.3. Each pool is analyzed in one capillary on the eight-capillary instrument, allowing 32 isolates to be analyzed in a 96-well sample plate that can be processed in an overnight run. The resulting fragment peaks are analyzed, sized, and repeat number allotted using the CEQ8000 instrument software. The output file contains allele names, fragment sizes, and repeat numbers in CSV format.

The second method described is a manual method that uses agarose gel electrophoresis to estimate PCR product size. Fifteen PCR reactions are carried out, each containing a single unlabeled primer set. Upon completion, the PCR products are separated on an agarose gel alongside a 100 bp ladder. Fragment sizes are estimated and repeat numbers allotted according to the expected sizes given in Table 25.3.

Table 25.2 PCR Primer Designs for VNTR/MIRU Analysis

Primer sets	Markers			Primer sequences		5' Label automated system
	H37Rv locus name	MIRU system	ETR system	Forward	Reverse	
1	0580	MIRU4	EDR-D	5'-GTC AAA CAG GTC ACA ACG AGA GGA A-3'	5'-CCT CCA CAA TCA ACA CAC TGG TCA T-3'	WellRED Dye 2
	1644	MIRU16		5'-CGG GTC CAG TCC AAC TAC CTC AAT-3'	5'-GAT CCT CCT GAT TGC CCT GAC CTA-3'	WellRED Dye 2
2	4348	MIRU39		5'- CGG TCA AGT TCA GCA CCT TCT ACA TC-3'	5'- GCG TCC GTA CTT CCG GTT CAG-3'	WellRED Dye 2
	2165		ETR-A	5'-AAA TCG GTC CCA TCA CCT TCT TAT-3'	5'-CGA AGC CTG GGG TGC CCG CGA TTT-3'	WellRED Dye 2
3	2059	MIRU20		5'-CCC CTT CGA GTT AGT ATC GTC GGT T-3'	5'-CAA TCA CCG TTA CAT CGA CGT CAT C-3'	WellRED Dye 2
4	0154	MIRU2		5'- CAG GTG CCC TAT CTG CTG ACG-3'	5'-GTT GCG TCC GGC ATA CCA AC-3'	WellRED Dye 3
	2387	MIRU24		5'-GAA GGC TAT CCG TCG ATC GGT T-3'	5'-GGG CGA GTT GAG CTC ACA GAA C-3'	WellRED Dye 3
5	3192	MIRU31	ETR-E	5'-CGT CGA AGA GAG CCT CAT CAA TCA T-3'	5'- AAC CTG CTG ACC GAT GGC AAT ATC-3'	WellRED Dye 3
	0802	MIRU40		5'-GAT TCC AAC AAG ACG CAG ATC AAG A-3'	5'-TCA GGT CTT TCT CTC ACG CTC TCG-3'	WellRED Dye 3
6	3006	MIRU27		5'- TCT GCT TGC CAG TAA GAG CCA-3'	5'-GTG ATG GTG ACT TCG GTG CCT T-3'	WellRED Dye 3

Table 25.2 (continued)

| Primer sets | Markers | | | Primer sequences | | 5′ Label automated system |
	H37Rv locus name	MIRU system	ETR system	Forward	Reverse	
7	0959	MIRU10		5′-ACC GTC TTA TCG GAC TGC ACT ATC AA-3′	5′- CAC CTT GGT GAT CAG CTA CCT CGA T-3′	WellRED Dye 4
	2531	MIRU23		5′-CGA ATT CTT CGG TGG TCT CGA GT-3′	5′-ACC GTC TGA CTC ATG GTG TCC AA-3′	WellRED Dye 4
8	2461		ETR-B	5′-GCG AAC ACC AGG ACA GCA TCA TG-3′	5′-GGC ATG CCG GTG ATC GAG TGG-3′	WellRED Dye 4
	0577		ETR-C	5′-GTG AGT CGC TGC AGA ACC TGC AG-3′	5′-GGC GTC TTG ACC TCC ACG AGT G-3′	WellRED Dye 4
9	2996	MIRU26		5′-GCG GAT AGG TCT ACC GTC GAA ATC-3′	5′- TCC GGG TCA TAC AGC ATG ATC A-3′	WellRED Dye 4

Table 25.3 Table of Apparent Fragment Sizes

Copy no.	Pool A						Pool B						Pool C		
	D2		D3		D4		D2		D3		D4		D2	D3	D4
	4	16	2	24	10	23	39	A	31	40	C	B	20	27	26
0	103	367	189	325	219	78	194	195	106	229	—	121	215	272	244
1	189	420	238	375	272	131	243	270	159	280	133	174	292	325	295
2	264	473	287	425	325	183	292	346	212	331	187	227	369	378	344
3	339	526	336	475	378	235	341	422	265	382	242	280	446	431	393
4	414	579	385	525	431	287	390	499	318	433	297	333	523	484	442
5	489	632	434	575	484	339	439	570	371	484	350	386	600	537	491
6	564		483	625	537	391	488	645	424	535	405	439	677	590	540
7	638		532		590	443	537	720	477	586	460	492		643	589
8	713		581		643	495	586	795	530	637	515	545			638
9			630			547	635	870	583		570	598			677
10						599			636		625	651			
11						651									

The table shows the apparent size of the PCR products and repeat numbers designated by the CEQ8000 system (*see* **Note 13**).

25.2 Materials

25.2.1 Sample Preparation

1. Plastic 1.5-mL screw-cap tubes with caps containing "O" rings.
2. TE buffer: 1 mM Tris-HCl, pH 8.0, 1 mM EDTA in purified water.
3. Chloroform.
4. Class 1 or class 2 Biology safety cabinet.
5. Covered water-bath capable of 80°C.

25.2.2 Automated Fragment Analysis

1. 2X reaction buffer: 2 mL 10x Ammonium PCR buffer, (Bioline, London, UK), 600 μL 50 mM $MgCl_2$, 7.24 mL purified water, 40 μL of dATP, dTTP, dGTP, and dCTP each at 100 mM (Bioline). Store in 1520-μL aliquots at –20°C.
2. Purified water (molecular grade).
3. TE buffer.
4. A stock set of forward primers with 5'-WellRED dye labeling as shown in Table 25.2 (Sigma-Proligo, France). Dilute to 200 μM in TE buffer and store at –20°C. Labeled primers should be stored in amber (lightproof) tubes and the number of freezing-thawing cycles should be minimized.
5. A stock set of unlabeled reverse primers as shown in Table 25.2. Dilute to 200 μM in TE buffer and store at –20°C.
6. Sets of 9 primer mixes at working concentration. These are prepared by labeling nine 1.5-mL plastic tubes 1 to 9. Add 796 μL of purified water to tubes 3, 6, and 9 and 792 μL to the remainder. Add 2.0 μL of labeled or unlabeled stock primers according to the scheme shown in Table 25.2. Dispense the nine primer mixes in 160-μL aliquots resulting in 5 complete sets of the nine primer mixes. These are stored at –20°C.
7. BIOTAQ DNA Polymerase (Bioline; Cat. No. 21040).
8. Dimethyl sulfoxide (DMSO), molecular grade (Sigma, Dorset, UK).
9. 50 mM $MgCl_2$ solution (Bioline).
10. 0.2-mL 96-well PCR plates.
11. Plate centrifuge capable of 1000 × g.
12. Rubber 96-well PCR plate sealing mats (Applied Biosystems).
13. Thermal cycler (96-well format). Beckman Coulter CEQ8000 Automated genetic analyzer.
14. Separation Capillary Array 33-75B (Beckman Coulter, Cat. No. 608087).
15. CEQ Sample Loading Solution (Beckman Coulter, Cat. No. 608082).
16. CEQ Separation Gel (Beckman Coulter, Cat. No. 608010).
17. CEQ DNA Size Standard Kit 600 bp (Beckman Coulter, Cat. No. 608095).
18. Sample plates (Beckman Coulter, Cat. No. 609801).

19. 96-well plates (Beckman Coulter, Cat. No. 609844).
20. Mineral oil, molecular biology grade.
21. Multichannel pipette 0.5 to 10 μL.
22. Purified water (molecular grade).

25.2.3 Manual Fragment Analysis

1. 2X reaction buffer: 2 mL 10x Ammonium PCR buffer, (Bioline, London, UK), 600 μL 50 mM MgCl$_2$, 7.24 mL purified water, 40 μL of dATP, dTTP, dGTP, and dCTP each at 100 mM (Bioline). Store in 1520-μL aliquots at –20°C.
2. Purified water (molecular grade).
3. Unlabeled primers : 15 pairs (Table 25.2). Make stock solutions of these primers at 200 μM in TE buffer and store at –20°C.
4. Prepare primer mixes for each locus at working concentration by combining 0.5 μL of the forward and reverse primer stock solution in each set (Table 25.2) with 199 μL water.
5. BIOTAQ DNA Polymerase (Bioline).
6. Dimethyl sulfoxide (DMSO), molecular grade .
7. 50 mM MgCl$_2$ solution (Bioline,).
8. 0.2-mL 96-well PCR plates.
9. Plate centrifuge capable of 1000 × g.
10. Rubber 96-well PCR plate sealing mats (Applied Biosystems).
11. Thermal cycler (96-well format).
12. Beckman Coulter CEQ8000 automated genetic analyzer.
13. Agarose gel electrophoresis equipment, gel tanks and gel casting trays of at least 20 cm.
14. High-resolution low-melting-point agarose.
15. Ethidium bromide 1 mg/mL solution.
16. 100-bp molecular weight markers (Bioline HyperLadder II, Hyper-Ladder IV).
17. Gel loading buffer (Sigma).
18. 10X TAE buffer (Invitrogen).
19. Transilluminator; 306-nm wavelength.

25.3 Methods

25.3.1 Sample Preparation

1. Suspend a loopful of visible mycobacterial growth (from solid media) in 100 μL of TE buffer or in the case of liquid cultures transfer 100 μL into a 1.5-mL screw-cap plastic tube (*see* **Note 1**).

2. Add an equal volume of chloroform, vortex for 30 s, and heat tubes in a waterbath or dry heating block to 80°C for 20 min.
3. Place the tubes in the freezer at –20°C for 5 min.
4. Remove the tubes from the freezer, thaw and centrifuge for 3 min at 12,000 × g prior to adding to the PCR mixture (see **Note 2**).

25.3.2 Automated Fragment Analysis

25.3.2.1 PCR for Automated Fragment Analysis

1. On the eight-capillary CEQ 8000 instrument, it is most efficient to analyze 32 samples in each run. If smaller runs are required, samples may be run in multiples of eight and volumes must be scaled down accordingly (see **Note 3**).
2. Thaw a 1520-μL aliquot of 2X reaction buffer and a set of primer aliquots (9 × 160 μL) (see **Note 4**). Add 28 μL DMSO and 30.4 μL Taq to the 2X reaction buffer and mix by gently inverting the tube (see **Note 5**).
3. Aliquot 160 μL of this mix into each of 9 labeled tubes. Add 9.6 μL 50 mM MgCl$_2$ to tube 1. Add 160 μL primer mixes 1 to 9 to the tubes 1 to 9. Mix by gently inverting.
4. Label four 96-well PCR plates 1 to 4. Add 9 μL of PCR mix 1 to column 1 of each tray, then 9 μL of PCR mix 2 to column 2 and so on until 9 μL of PCR mix 9 has been added to column 9 in each tray.
5. Centrifuge the DNA preparations (from Section 25.3.1) at 12,000 × g for 1 min. Add 1 μL of the aqueous phase of the DNA preparation to the first 9 wells in a row on the PCR plate. Repeat until 32 DNA samples have been added, one to each row.
6. Centrifuge the four plates at 1500 × g for 1 min. Cover the plates with a rubber sealing lid. Load plates onto the thermal cycler and amplify using the program, 95°C for 3 min, then 35 cycles of 95°C for 30 s, 60°C for 30 s, 72°C for 60 s with a final hold of 5 min at 72°C.
7. Load PCR products onto the sequencer immediately (Section 25.3.3) or store frozen at –20°C for up to 1 week. For storage, plates should be sealed with the aluminum or acetate sealing tape.

25.3.2.2 Analysis of Automated Fragments

1. Add 88 μL of molecular grade water to each well in columns 1, 2, and 3 of the PCR plates from Section 25.3.2.
2. Combine and dilute the labeled PCR products using a multichannel pipette as follows. Add 1 μL from column 4 to C1, 1 μL from C5 to C2, and 1 μL from C6 to C3. Then add 0.5 μL from C7 to C1, 0.5 μL from C8 to C2, and 0.5 μL from C9 to C3. This is shown graphically in Figure 25.2. After each transfer, pump the pipette several times to mix the contents of the wells before discarding the tips.

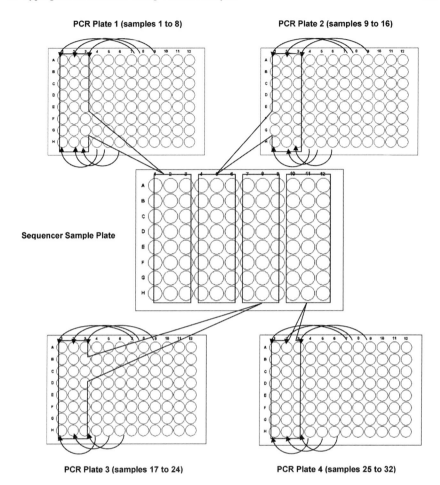

Fig. 25.2 Scheme for diluting and combining PCR products. The illustration shows a scheme for combining the PCR products generated in PCR plates 1 to 4 into a single plate for multiplex analysis on the fragment analyzer. Products are transferred using a multichannel pipette from columns 4–6 to columns 1–3, then from columns 7–9 to columns 1–3. These columns are transferred to the analyzer plate at the center

3. Centrifuge the trays for 1 min at 1500 × g. This tray now contains 3 pools of PCR products for each isolate. The contents of pools A to C are shown in Table 25.3.

4. In a 1.5-mL tube, mix 625 μL of Sample Loading Solution with 2.5 μL of the 600-bp size standard (*see* **Note 6**), invert the tube gently several times, and centrifuge at 8000 × g for 1 min.

5. Dispense 25 μL of the Sample Loading Solution and size standard mixture into each well of a CEQ8000 sample plate.

6. Using the multichannel pipette, add 1 μL of each PCR product combination to a well of the sample plate as shown in Figure 25.3.

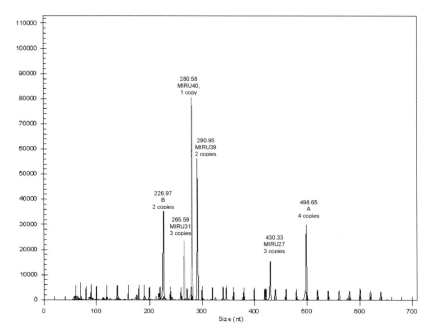

Fig. 25.3 Analyzed CEQ8000 trace. The illustration shows the graphical output of the fragment analyzer. Labels indicate the molecular weight of the fragment and automated allele assignation and calculated number of copies within the locus. These data are downloaded in tabular format

7. Overlay each well of the sample plate with a drop of mineral oil and centrifuge at $1500 \times g$ for 1 min.
8. Run the CEQ8000 programme and open the Database Manager Module.
9. Create a database and specify it as the working database and close the Database manager.
10. Open the Setup Module and create a new sample plate.
11. Enter the sample names followed by suffix A in columns 1, 4, 7, and 10, the suffix B in columns 2, 5, 8, and 11, and suffix C in columns 3, 6, 9, and 12.
12. Create a separation method with the following parameters:

- capillary temperature to 60°C
- denature 90°C for 120 s
- inject 2.0 kV for 30 s
- separation voltage to 6.0 kV for 60 m

13. Save the method as "MIRU15."
14. Choose the method MIRU15 for all the columns on the plate then save the sample plate and close the Setup Module.
15. Open the Run Module.
16. Load the CEQ analyser using recommendations from the Operators' Manual.

25.3.2.3 Setting the System for Automated Data Analysis

1. Open the Fragment Analysis module in the CEQ8000 software and select Analysis parameters. Select "Default fragment analysis parameter" and "Edit."
2. The parameters for peak recognition are set in the General parameters window. Set the slope threshold parameter to 2, the relative peak height threshold parameter to 2%, check "Identify STR alleles in allele identification," and set the confidence level to 95%.
3. Open "Analysis Method" inset and select Size Standard 600 from the drop-down menu and specify the "Quartic" model ensuring all 33 peaks of 600-bp standard are selected. Save the parameter set.
4. Open the STR locus tags inset. In order to create 15 VNTR/MIRU locus tags for automated assignation of allele and copy, a total of 9 locus tag sets must be created, 3 tag sets for dyes 2, 3, and 4 in each of capillaries A, B, and C.
5. Select "New locus" to open the locus tag editor. Assign a name to the tag, in the form D2-15A indicating Dye 2 in capillary A. Specify Dye 2, set repeat unit length and number of repeats in the shortest allele to 1, and then select "Nominal Size."
6. Refer to the calling list (Table 25.3). Tube A contains 2 PCR fragments labeled with Dye 2 (loci MIRU4 and MIRU16). Specify the shortest allele size (100 bp) and click on the "Generate Allele list" button. The list will appear on the bottom of the active window.
7. Using the calling table, enter locus designation and number of repeats in the ID column of allele list in the following format: X","N, where is X is a locus name (e.g., 4,10,A etc.) and N is a number of copies corresponding with the size of the fragment. Allow tolerances ±2 bp for fragments <200 bp, ±3 bp for fragments 200 to 400 bp, and ±4 bp for fragments exceeding 400 bp. Leave remaining cells in the ID column blank.
8. Open the "Advanced" inset and select "PA ver.1" in the Dye mobility calibration section. Make sure that option "use calculated dye spectra" is selected. Save the parameter set.
9. Open the inset "Allele ID criteria" in the locus tag editor and select all three available options in the stutter definition section. Enter stutter detection window width as 3 repeats and maximum relative stutter peak height as 99%. Select "Detect spurious peaks" and specify the maximum height = 20%. Select "Overwrite system confidence interval" and set it to 0.490. Selecting OK will save all changes you have made for locus tag D2-15A.
10. Repeat steps 5 to 7 for remaining 2 D2 tags (tubes B and C), three D3 tags (tubes A, B, and C), and three D4 tags (tubes A, B, and C). Save the tags with names in the form given in step 8. Before closing the window, check all 9 tags are shown in the list of available locus tags.
11. In the Analysis Parameter window, select "Default fragment analysis parameter" and "Edit." Open the STR Locus Tags inset again and select the 3 appropriate locus tags (e.g., D2-15A, D3-15A, D4-15A) and save as 15 VNTR/MIRU-A. Repeat for pool B and pool C.

25.3.2.4 Running the Automated Data Analysis

1. Open the fragment analysis module within the CEQ8000 software and create a new study using raw data by selecting the appropriate options from the menu.
2. Using the suffix A added in step 11 of Section 25.3.2, identify and select the pool A PCR product data set for each of 32 isolates.
3. In the next window, select the MIRU15A method and start analysis. This analysis will take up to 5 min.
4. When the analysis is complete, open the Fragment view window and use the Exclusion filter set tool to remove the size standards from the list of called fragments. Close the Fragment view window.
5. By selecting Add raw data to the study in the "File" menu, repeat steps 2 and 3 for pool B, using the MIRU15B method and pool C, using the MIRU15C method.
6. From within the results view window, open each result set and check all peaks have been called. Remove any erroneous calls and manually enter calls for missed peaks using Table 25.3. We strongly recommend that peaks entered manually are identified as such in the comments field.
7. Open the "Fragment List View" window and group the fragments according to their sample numbers by clicking on the header row in the window. Using the select column window, select the columns Sample No, Dye, Estimated Fragment Size, Comment, and Allele ID. Make sure that the Allele ID column is last in the list. The table can now be exported using the "Export Grid" tool.
8. Examine the exported data and identify the allele assignations that have been entered manually before checking the given fragment sizes against the repeat numbers given in Table 25.3 (*see* **Notes 7, 8** and **9**).
9. Files exported from the Beckman Coulter software in this manner are in comma delimited format (CSV) and are compatible with other software such as word MS Excel and Access (*see* **Note 10**).

25.3.3 Manual Analysis

25.3.3.1 PCR for Manual Analysis

1. For each of 15 VNTR loci, calculate the total volume of each PCR mastermix by multiplying the constituent volumes by the number of isolates to be analyzed and combine them in a labeled tube. For each reaction combine 5 μL of 2X reaction buffer, 0.5 μL DMSO, 2.1 μL of the appropriate Primer Mix (prepared as described in Section 25.2.3), 0.1 μL BIOTAQ polymerase, and 2.3 μL purified water. Mix by gently inverting the tubes.
2. Add 9 μL of PCR mastermix to wells of 96-well PCR tray. For smaller numbers of samples, individual 0.2 mL PCR tubes or strips of PCR tubes can be used.

3. Add 1 μL of the aqueous phase of the DNA preparation made in Section 25.3.1 to each well or PCR tube.
4. Centrifuge tubes (or plates) for 1 min, load into the thermal cycler, and amplify using program 95°C for 3 min, then 35 cycles of 95°C for 30 s, 60°C for 30 s, 72°C for 60 s, then finally hold for 5 min at 72°C.
5. Analyze PCR products immediately or stored frozen at –20°C. Where the products are stored, the plates must be sealed with aluminum or acetate sealing tape.

25.3.3.2 Analysis of Manual Fragments

1. Prepare a 1.5% agarose gel by adding 1.5 g of agarose to 100 mL of 1X TAE buffer. Melt the agarose by heating the open vessel in a microwave oven.
2. Assemble the gel former and comb before adding the cooled but still liquid agarose.
3. Once the gel has solidified, remove comb and place the gel into the tank with 1x TAE buffer.
4. Load the appropriate amount of 100-bp molecular weight ladder into the outermost wells and also every 5 wells across the gel (this allows the accurate sizing of PCR fragments).
5. Mix 5 μL of the PCR product with 1 μL of the loading buffer before loading one sample per well in the gel.
6. Separate at 120 V until the loading dye has run al least 10 cm. The time is dependent upon the equipment used.
7. Stain the gel by immersing in an ethidium bromide solution made by diluting 1 μL of 10 mg/mL ethidium bromide solution in 100 mL distilled water for 15 min. Wash twice with distilled water.
8. Transfer the gel to the UV trans-illuminator. Record an image of the gel for analysis.

25.3.3.3 Manual Data Analysis

1. Identify each of the size standard fragments in the standard lane and the sample and VNTR locus being analyzed in each test lane on the gel.
2. Allele assignation is performed using the calling table shown in Table 25.3, using the biases shown for agarose electrophoresis (*see* **Notes 8** and **11**).

25.4 Notes

1. As with many PCR-based methods, VNTR analysis can be performed on crude DNA extracts. These may be obtained either from cultures grown on solid or in liquid media. It is often useful to perform the DNA amplification and analysis outside the confines of a containment level III laboratory. In the presence of chloroform, the heat treatment described here fully inactivates *M. tuberculosis* in our laboratory. Care must be taken to ensure the entire vessel including the lid reaches the target temperature. As materials and

equipment may vary, it is advisable to demonstrate deactivation of *M. tuberculosis* in any laboratory before undertaking this work.

2. DNA extracts with chloroform can be stored at –20°C for several months. For longer storage, centrifuge for 5 min at 12,000 × g to separate phases and transfer the aqueous phase (containing DNA) into a new tube.

3. Before loading and running the CEQ8000 instrument, operators should familiarize themselves with the instrument and its operation. Full general operating instructions and troubleshooting details can be found in the operator's manual or at http://www.beckman-coulter.com/customersupport/default.asp.

4. The fluorescently labeled PCR primers are stable for up to 6 months when stored protected from light at –20°C. After this time, a reduction of signal is seen. Repeated freeze-thaw cycles should also be avoided by making the working concentration primer mixes described in step 5 of Section 25.2.2 in the largest batches possible.

5. When performing a high-throughput PCR procedure such as that described here, amplicon contamination can lead to erroneous results. The risk of this occurring can be minimized by the use of a three-area workspace model for PCR. All PCR reagents are stored and manipulated in a clean area, all DNA is extracted, manipulated, and stored in a specimen preparation area, and all PCR products are analyzed and stored in a dirty area. It is advisable to use separate equipment in each area and maintain a workflow from the clean area through the specimen preparation area to the PCR dirty area.

6. For 15 MIRU automated analysis, usage of proprietary Beckman Coulter size standards labeled with D1 dye (600 bp Size Standard Kit, Cat. No. 608095) is recommended. This standard covers the range 60 to 640 bp and allows unambiguous calling of the majority of the commonly seen allelic variants of 15 MIRU-ETR loci. In <5% strains, the ETR-A locus contains >6 repeats where fragment sizes are greater than the 640-bp maximum standard size. In this case where the repeat number cannot be estimated using stutter peaks, repeat number must be determined using manual analysis.

7. Approximately 90% of PCR products are automatically sized and the allele designated by the CEQ8000 software. A proportion of peaks and therefore fragments are absent due to PCR failure. A further proportion is present but falls outside the parameters defining peak height and shape. Where peaks have been excluded due to low height, the fragment size is automatically calculated but allele designation must be entered manually. Where peak shape has excluded a fragment, the raw data may be reanalyzed using a threshold parameter of 40 or 80. This is particularly useful if peaks within the size standards have been excluded as this leads to the analysis of a pool failing. In some cases where peaks are excluded due to shape, their size can be estimated by comparison with the size standard in the raw data window. Here allele designation is entered into the exported data grid described in Section 25.3.2.4. Where multiple alleles of the one VNTR are seen in the same pool, this indicates the presence multiple strains. This is interpreted as evidence of bacteriological cross-contamination of the original isolate or multiple infection within the patient from which the isolate was derived.

8. Stutter is a PCR artifact of amplified VNTR in which a series of minor additional PCR products are seen corresponding with the major PCR product size minus increments of the repeat size. If the stutter is of great enough magnitude, each peak is automatically and erroneously called. These calls must be removed. Stutter however can be beneficial to the analysis where fragments larger than the size standard ladder can be designated with an allele by counting the stutter peaks beyond the largest within the size standard range.

9. The main sources of error in the 15 VNTR/MIRU data sets are manual data input and manually assigning alleles. This can be minimized by rechecking data against the source list before finishing the data input operation. If manually assigned alleles are marked as suggested in Section 25.3.2.4, these too can be rechecked. In population studies where clustering of strains is important, inclusion of errors in the final data set will usually lead to the exclusion of a strain from a cluster; therefore the error rate will be reflected in an

underestimate of cluster rate. In small studies, replicate analyses can be performed on every strain allowing errors to be identified and corrected. In large retrospective or prospective studies comprising hundreds or thousands of samples, replicate analysis may not be practical and if no epidemiologic data is available at the time of analysis, identification of errors is difficult. Under these circumstances, the use of a quality assurance system is required. The authors recommend that a reference strain with a known VNTR profile such as H37Rv is analyzed at the start of a study and then periodically throughout the study. In addition, 10% of strains in the study should be reanalyzed in order to estimate the error rate.

10. Data in the CSV format can be directly imported into Excel or Access for storage and simple data analysis. For more complex analyses such as the construction of dendrograms for the inference of relationships and comparison of multiple data types, Bionumerics (http://www.applied-maths.com) software is useful. When using this software, it is convenient to use scripts in the Bionumerics macro language to both import data and for checking and reformatting data in the CSV format.

11. For manual VNTR analysis, the determination of repeat numbers can be simplified using custom size standards for each VNTR locus. These can be generated by assembling a collection of strains containing the maximum number of different repeat numbers at each locus. The manual simplex PCR protocol above can be used to generate PCR products for each allele. These can be combined as shown in Figure 25.4. This is particularly useful

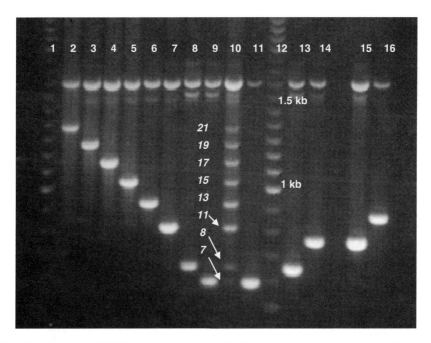

Fig. 25.4 Analysis of VNTR in an agarose gel. The illustration shows the analysis of higher-weight fragments in an agarose gel and the construction of an actual fragment size ladder for locus 3232. **Lanes 1** and **12** show 100-bp molecular weight ladder, **lane 10** a custom molecular weight ladder for locus 3232 [23] consisting of 21 repeats (**lane 2**), 19 repeats (**lane 3**), 17 repeats (**lane 4**), 15 repeats (**lane 5**), 13 repeats (**lane 6**), 11 repeats (**lane 7**), 8 repeats (**lane 8**), 7 repeats (**lane 9**). **Lanes 11** and **13** to **16** show unknown PCR products containing 7, 8, 10, 10, and 12 repeats, respectively

for the analysis of larger PCR products and fragments containing partial repeats. For the automated fragment analysis of loci giving fragments 640 to 1000 bp, custom size standards can be obtained from BioVentures (Murfreesboro, TN). The analysis method for the CEQ8000 data must be amended to include the additional size standards. Automated analysis of fragments of >1000 bp using the CEQ8000 is not recommended.

12. In addition to the 15 VNTR set described here, a further 44 loci have been described in the literature as shown in Table 25.1. Panels of these VNTR can be analyzed using the platforms described here by substituting the amplification primers and loading the relevant STR tags to the CEQ8000 software. When comparing new data with existing data, it is important to ensure that any partial repeats have been considered in the same manner in both sets of calculated repeat numbers. Panels including products up to 1000 bp can be analyzed using the 50- to 1000-bp size standard set mentioned in **Note 10**. It is advisable when designing an analysis for a VNTR set to start with simplex PCR. Care should be taken when pooling PCR products with the same dye that the potential fragment sizes for each VNTR are at least 10 bp apart.

13. The migration characteristics of DNA fragments during electrophoresis may vary depending on the analysis system used. Parameters such as gel matrix type and extent of denaturization are important. This results in discrepancy between apparent size of repeats and actual repeat size established by DNA sequencing. This is apparent in Table 25.3, which shows apparent fragment sizes for the automated system described here.

Acknowledgments We would like to thank Philip Supply for his advice and stimulating discussions.

References

1. van Soolingen, D., Hermans, P. W., de Haas, P. E., Soll, D. R. & van Embden, J. D. (1991). Occurrence and stability of insertion sequences in Mycobacterium tuberculosis complex strains: evaluation of an insertion sequence-dependent DNA polymorphism as a tool in the epidemiology of tuberculosis. *J Clin Microbiol* **29**, 2578–2586.
2. Kamerbeek, J., Schouls, L., Kolk, A., van Agterveld, M., van Soolingen, D., Kuijper, S., Bunschoten, A., Molhuizen, H., Shaw, R., Goyal, M. & van Embden, J. (1997). Simultaneous detection and strain differentiation of Mycobacterium tuberculosis for diagnosis and epidemiology. *J Clin Microbiol* **35**, 907–914.
3. Brudey, K., Driscoll, J., Rigouts, L., Prodinger, W., Gori, A., Al-Hajoj, S., Allix, C., Aristimuno, L., Arora, J., Baumanis, V., Binder, L., Cafrune, P., Cataldi, A., Cheong, S., Diel, R., Ellermeier, C., Evans, J., Fauville-Dufaux, M., Ferdinand, S., de Viedma, D., Garzelli, C., Gazzola, L., Gomes, H., Guttierez, M. C., Hawkey, P., van Helden, P., Kadival, G., Kreiswirth, B., Kremer, K., Kubin, M., Kulkarni, S., Liens, B., Lillebaek, T., Ly, H., Martin, C., Martin, C., Mokrousov, I., Narvskaia, O., Ngeow, Y., Naumann, L., Niemann, S., Parwati, I., Rahim, Z., Rasolofo-Razanamparany, V., Rasolonavalona, T., Rossetti, M. L., Rusch-Gerdes, S., Sajduda, A., Samper, S., Shemyakin, I., Singh, U., Somoskovi, A., Skuce, R., van Soolingen, D., Streicher, E., Suffys, P., Tortoli, E., Tracevska, T., Vincent, V., Victor, T., Warren, R., Yap, S., Zaman, K., Portaels, F., Rastogi, N. & Sola, C. (2006). Mycobacterium tuberculosis complex genetic diversity: mining the fourth international spoligotyping database (SpolDB4) for classification, population genetics and epidemiology. *BMC Microbiol* **6**, 23.
4. Kremer, K., van Soolingen, D., Frothingham, R., Haas, W. H., Hermans, P. W., Martin, C., Palittapongarnpim, P., Plikaytis, B. B., Riley, L. W., Yakrus, M. A., Musser, J. M. & van Embden, J. D. (1999). Comparison of methods based on different molecular

epidemiological markers for typing of Mycobacterium tuberculosis complex strains: inter-laboratory study of discriminatory power and reproducibility. *J Clin Microbiol* **37**, 2607–2618.

5. Jeffreys, A. J., Wilson, V. & Thein, S. L. (1985). Individual-specific 'fingerprints' of human DNA. *Nature* **316**, 76–79.

6. Frothingham, R. & Meeker-O'Connell, W. (1998). Genetic diversity in the Mycobacterium tuberculosis complex based on variable numbers of tandem DNA repeats. *Microbiology* **144**, 1189–1196.

7. Supply, P., Mazars, E., Lesjean, S., Vincent, V., Gicquel, B. & Locht, C. (2000). Variable human minisatellite-like regions in the Mycobacterium tuberculosis genome. *Mol Microbiol* **36**, 762–771.

8. Mazars, E., Lesjean, S., Banuls, A. L., Gilbert, M., Vincent, V., Gicquel, B., Tibayrenc, M., Locht, C. & Supply, P. (2001). High-resolution minisatellite-based typing as a portable approach to global analysis of Mycobacterium tuberculosis molecular epidemiology. *Proc Natl Acad Sci U S A* **98**, 1901–1906.

9. Hawkey, P. M., Smith, E. G., Evans, J. T., Monk, P., Bryan, G., Mohamed, H. H., Bardhan, M. & Pugh, R. N. (2003). Mycobacterial interspersed repetitive unit typing of Mycobacterium tuberculosis compared to IS6110-based restriction fragment length polymorphism analysis for investigation of apparently clustered cases of tuberculosis. *J Clin Microbiol* **41**, 3514–3520.

10. Le Fleche, P., Fabre, M., Denoeud, F., Koeck, J.-L. & Vergnaud, G. (2002). High resolution, on-line identification of strains from the Mycobacterium tuberculosis complex based on tandem repeat typing. *BMC Microbiol* **2**, 37.

11. Namwat, W., Luangsuk, P. & Palittapongarnpim, P. (1998). The genetic diversity of *Mycobacterium tuberculosis* strains in Thailand studied by amplification of DNA segments containing direct repetitive sequences. *Int J Tuberculosis Lung Dis* **2**, 153–159.

12. Roring, S., Scott, A., Brittain, D., Walker, I., Hewinson, G., Neill, S. & Skuce, R. (2002). Development of variable-number tandem repeat typing of Mycobacterium bovis: comparison of results with those obtained by using existing exact tandem repeats and spoligotyping. *J Clin Microbiol* **40**, 2126–2133.

13. Skuce, R. A., McCorry, T. P., McCarroll, J. F., Roring, S. M. M., Scott, A. N., Brittain, D., Hughes, S. L., Hewinson, R. G. & Neill, S. D. (2002). Discrimination of Mycobacterium tuberculosis complex bacteria using novel VNTR-PCR targets. *Microbiology* **148**, 519–528.

14. Smittipat, N., Billamas, P., Palittapongarnpim, M., Thong-On, A., Temu, M. M., Thanakijcharoen, P., Karnkawinpong, O. & Palittapongarnpim, P. (2005). Polymorphism of variable-number tandem repeats at multiple loci in Mycobacterium tuberculosis. *J Clin Microbiol* **43**, 5034–5043.

15. Spurgiesz, R. S., Quitugua, T. N., Smith, K. L., Schupp, J., Palmer, E. G., Cox, R. A. & Keim, P. (2003). Molecular typing of mycobacterium tuberculosis by using nine novel variable-number tandem repeats across the Beijing family and low-copy-number IS6110 isolates. *J Clin Microbiol* **41**, 4224–4230.

16. van Deutekom, H., Supply, P., de Haas, P. E. W., Willery, E., Hoijng, S. P., Locht, C., Coutinho, R. A. & van Soolingen, D. (2005). Molecular typing of Mycobacterium tuberculosis by mycobacterial interspersed repetitive unit-variable-number tandem repeat analysis, a more accurate method for identifying epidemiological links between patients with tuberculosis. *J Clin Microbiol* **43**, 4473–4479.

17. Gopaul, K. K., Brown, T. J., Gibson, A. L., Yates, M. D. & Drobniewski, F. A. (2006). Progression toward an improved DNA amplification-based typing technique in the study of Mycobacterium tuberculosis epidemiology. *J Clin Microbiol* **44**, 2492–2498.

18. Gibson, A., Brown, T., Baker, L. & Drobniewski, F. (2005). Can 15-locus mycobacterial interspersed repetitive unit-variable-number tandem repeat analysis provide insight into the evolution of Mycobacterium tuberculosis? *Appl Environ Microbiol* **71**, 8207–8213.

19. Evans, J. T., Hawkey, P. M., Smith, E. G., Boese, K. A., Warren, R. E. & Hong, G. (2004). Automated high-throughput mycobacterial interspersed repetitive unit typing of Mycobacterium tuberculosis strains by a combination of PCR and nondenaturing high-performance liquid chromatography. *J Clin Microbiol* **42**, 4175–4180.

20. Supply, P., Lesjean, S., Savine, E., Kremer, K., van Soolingen, D. & Locht, C. (2001). Automated high-throughput genotyping for study of global epidemiology of Mycobacterium tuberculosis based on mycobacterial interspersed repetitive units. *J Clin Microbiol* **39**, 3563–3571.

21. Kwara, A., Schiro, R., Cowan, L. S., Hyslop, N. E., Wiser, M. F., Roahen Harrison, S., Kissinger, P., Diem, L. & Crawford, J. T. (2003). Evaluation of the epidemiologic utility of secondary typing methods for differentiation of Mycobacterium tuberculosis isolates. *J Clin Microbiol* **41**, 2683–2685.

22. Supply, P., Allix, C., Lesjean, S., Cardoso-Oelemann, M., Rusch-Gerdes, S., Willery, E., Savine, E., de Haas, P., van Deutekom, H., Roring, S., Bifani, P., Kurepina, N., Kreiswirth, B., Sola, C., Rastogi, N., Vatin, V., Gutierrez, M. C., Fauville, M., Niemann, S., Skuce, R., Kremer, K., Locht, C. & van Soolingen, D. (2006). Proposal for standardization of optimized mycobacterial interspersed repetitive unit-variable number tandem repeat typing of Mycobacterium tuberculosis. *J Clin Microbiol* **44**, 4498–4510.

23. Nikolayevskyy, V., Gopaul, K., Balabanova, Y., Brown, T., Fedorin, I. & Drobniewski, F. (2006). Differentiation of tuberculosis strains in a population with mainly Beijing-family strains. *Emerg Infect Dis* **12**, 1406–1413.

Chapter 26
Molecular Detection of Drug-Resistant *Mycobacterium tuberculosis* with a Scanning-Frame Oligonucleotide Microarray

Dmitriy V. Volokhov, Vladimir E. Chizhikov, Steven Denkin and Ying Zhang

Abstract The increasing emergence of drug-resistant *Mycobacterium tuberculosis* poses significant threat to the treatment of tuberculosis (TB). Conventional drug susceptibility testing is time-consuming and takes several weeks because of the slow growth rate of *M. tuberculosis* and the requirement for the drugs to show antimycobacterial activity. Resistance to TB drugs in *M. tuberculosis* is caused by mutations in the corresponding drug resistance genes (e.g., *katG, inhA, rpoB, pncA, embB, rrs, gyrA, gyrB*), and detection of these mutations can be a molecular indicator of drug resistance. In this chapter, we describe the utility of a microarray-based approach exploiting short overlapping oligonucleotides (sliding-frame array) to rapidly detect drug resistance–associated mutations (substitutions, deletions, and insertions) in the *pncA* gene responsible for resistance of *M. tuberculosis* to pyrazinamide (PZA) as an example for this approach. Hybridization of *pncA*-derived RNA or DNA with the microarray enables easy and simple screening of nucleotide changes in the *pncA* gene. Sliding-frame microarrays can be used to identify other drug-resistant TB strains that have mutations in relevant drug resistance genes.

Keywords drug resistance · microarray · mutations · tuberculosis

26.1 Introduction

Tuberculosis (TB), which is caused by *Mycobacterium tuberculosis*, is a leading infectious disease worldwide with 9 million new cases and about 1.7 million to 2 million deaths every year [1, 2, 3]. The increasing emergence of drug-resistant TB (multidrug resistant, MDR-TB; and most recently extensively drug resistant, XDR-TB) and HIV coinfection, which compromises the host defense and

Y. Zhang
Department of Molecular Microbiology and Immunology, Bloomberg School of Public Health, Johns Hopkins University, 615 N. Wolfe Street, Baltimore, Maryland 21205, USA
e-mail: yzhang@jhsph.edu

allows latent infection to reactivate or render individuals more susceptible to TB, pose an enormous challenge for effective disease control. Drug resistance in *M. tuberculosis* is caused by mutations in chromosomal genes rather than by plasmids or transposons [4]. The MDR phenotype results from sequential accumulation of spontaneous mutations in different genes involved in individual drug resistance due to inappropriate treatment or poor compliance of patients to treatment. The increasing drug-resistant TB problem has highlighted the importance of understanding the mechanisms of resistance in *M. tuberculosis*. A great deal of progress has been made in this area in the past decade [5, 6, 7, 8, 9]. Mechanisms of resistance to all first-line TB drugs such as isoniazid (INH), rifampin (RMP), pyrazinamide (PZA), ethambutol (EMB), and streptomycin (SM) and also several second-line drugs have been identified (Table 26.1).

Rapid molecular detection of drug-resistance mutations is critical for effective monitoring of drug-resistant *M. tuberculosis* and can provide useful clinical

Table 26.1 Mechanisms of Drug Resistance in *M. tuberculosis*

Antibiotics	MIC (μg/mL)	Mechanism of action	Gene(s) involved in resistance	Role	Frequency (%) of mutation among resistant strains
Isoniazid	0.06–0.2	Inhibition of mycolic acid biosynthesis and other multiple effects on DNA, lipids, carbohydrates, and NAD metabolism	*katG* *inhA* *ahpC*	Prodrug conversion Drug target Marker of resistance	42–58 21–34 10–15
Rifampin	0.5–2	Inhibition of transcription	*rpoB*	Drug target	96
Pyrazinamide	16–50 (pH5.5)	Depletion of membrane energy and	*pncA*	Prodrug conversion	72–97
Ethambutol	1–5	Inhibition of arabinogalactan synthesis	*embCAB*	Drug target	47–65
Streptomycin	2–8	Inhibition of protein synthesis	*rpsL* *rrs (16S rRNA)*	Drug target Drug target	52–59 8–21
Amikacin/ kanamycin	2–4	Inhibition of protein synthesis	*rrs (16S rRNA)*	Drug target	76
Quinolones	0.5–2.5	Inhibition of DNA gyrase	*gyrA, gyrB*	Drug target	75–94
Ethionamide	2.5–10	Inhibition of mycolic acid biosynthesis	*etaA* *inhA*	Prodrug conversion Drug target	37 56

guidance for the treatment of the disease [10, 11, 12]. Molecular detection of mutations in the genes associated with drug resistance has the main advantage of being rapid and eliminating the need for the time-consuming phenotype-based susceptibility testing [13]. Although various molecular methods, such as polymerase chain reaction (PCR) single-stranded conformation polymorphism (SSCP) [14], restriction fragment length polymorphism (PCR-RFLP) [15], dideoxy fingerprinting [16], heteroduplex formation (denaturing high performance liquid chromatography; DHPLC) [17], amplification refractory mutation system (ARMS-PCR) [18], and PCR–peptide nucleic acid (PNA) based enzyme-linked immunosorbent assay (ELISA) [19], confocal single-molecule fluorescence spectroscopy [20], and dot blot hybridization [21] have been previously developed for rapid screening of potential drug-resistant strains, these techniques are still tedious and do not demonstrate the required sensitivity and high-throughput sample screening capability. Another significant limitation of these techniques is that they only detect known mutations associated with drug resistance and cannot be used to detect unknown mutations that can occur in drug-resistance genes. To date, the most accurate and reliable method for detection of mutations in the target genes is DNA sequencing. However, this method is still relatively expensive particularly when multiple genes are involved in resistance or resistance mutations are not clustered and randomly distributed along the target gene.

Hybridization of genetic samples with miniature glass microchips containing oligonucleotide probes was previously demonstrated to be a valuable tool for detecting minor genetic changes in different microorganisms [6, 22, 23, 24, 25, 26]. The feasibility of microarrays for analysis of isoniazid or rifampin resistance determinants in *M. tuberculosis* strains has been evaluated [27, 28, 29, 30, 31, 32, 33, 34, 35]. The results of these studies demonstrated the potential of microarray for rapid and accurate detection of selected mutations known to be responsible for *M. tuberculosis* drug resistance.

In this chapter, we provide a detailed protocol of the sliding-frame oligonucleotide microarray for rapid screening of spontaneous mutations in the *pncA* gene that cause resistance to pyrazinamide in *M. tuberculosis* [4, 7, 9]. We designed overlapping microarray oligonucleotide probes (oligoprobes) on the basis of the nucleotide sequence of the complete *pncA* gene of *M. tuberculosis* H37Rv (GenBank accession number U59967). The whole set of oligoprobes for analysis of the *pncA* gene includes 79 short oligoprobes with basic melting temperature about 47°C to 51°C (Table 26.2). An additional oligoprobe QCprb with the sequence of the forward primer was included in the microarray to monitor the synthesis of full-length hybridization target (ssRNA or ssDNA). A schematic diagram showing how this technology works is shown in Figure 26.1.

The advantage of the sliding-frame oligonucleotide microarray is that it can detect unknown mutations in given drug-resistance genes. The cost of this technology is still high, and it is likely that the technology will find wider application when the costs of microarrays decrease. The same protocol and principle can be applied to detect mutations involved in other drug resistances

Table 26.2 *pncA*-Specific Oligoprobes for the Scanning-Frame Oligonucleotide Microarray [6, 22]

Name	Sequence	Length (in nucleotides)	G+C (%)	T_m* (°C)
T7-Myc-Rev	CGTTAATACGACTCACTATAGGGC CAACAGTTCATCCCGGTTCGGC	46	52	72
Myc-Forw	CTGCCGCGTCGGTAGGCAAACTGC	24	67	64
A1	CCGGGCAGTCGCCC	14	86	52
A2	GTCGCCCGAACGTATGG	17	65	52
A3	AACGTATGGTGGACGTATG	19	47	49
A4	TGGACGTATGCGGGCG	16	69	51
A5	TGCGGGCGTTGATCATC	17	59	49
A6	TTGATCATCGTCGACGTGC	19	53	51
A7	GTCGACGTGCAGAACGA	17	59	49
A8	GCAGAACGACTTCTGCG	17	59	49
A9	ACTTCTGCGAGGGTGGC	17	65	52
A10	GAGGGTGGCTCGCTG	15	73	50
A11	GCTCGCTGGCGGTAAC	16	69	51
A12	GCGGTAACCGGTGGC	15	73	50
A13	CCGGTGGCGCCGCG	14	93	55
A14	CGCCGCGCTGGCCC	14	93	55
A15	CTGGCCCGCGCCAT	14	79	49
A16	GCGCCATCAGCGACTAC	17	65	52
A17	AGCGACTACCTGGCCG	16	69	51
A18	CCTGGCCGAAGCGG	14	79	49
A19	GAAGCGGCGGACTACC	16	69	51
A20	GGACTACCATCACGTCG	17	59	49
A21	ATCACGTCGTGGCAACC	17	59	49
A22	GTGGCAACCAAGGACTTC	18	56	50
A23	AAGGACTTCCACATCGACC	19	53	51
A24	CACATCGACCCGGGTG	16	69	51
A25	CCCGGGTGACCACTTC	16	69	51
A26	ACCACTTCTCCGGCACA	17	59	49
A27	CTTCTCCGGCACACCG	16	69	51
A28	TCCGGCACACCGGAC	15	73	50
A29	GGCACACCGGACTATTC	17	59	49
A30	CACCGGACTATTCCTCG	17	59	49
A31	TATTCCTCGTCGTGGCC	17	59	49
A32	GTCGTGGCCACCGC	14	79	49
A33	CCACCGCATTGCGTCAG	17	65	52
A34	TTGCGTCAGCGGTACTC	17	59	49
A35	GCGGTACTCCCGGC	14	79	49
A36	TCCCGGCGCGGACT	14	79	49
A37	GCGGACTTCCATCCCAG	17	65	52
A38	CCATCCCAGTCTGGACA	17	59	49
A39	GTCTGGACACGTCGGC	16	69	51
A40	ACGTCGGCAATCGAGGC	17	65	52

Table 26.2 (continued)

Name	Sequence	Length (in nucleotides)	G+C (%)	T_m* (°C)
A41	AATCGAGGCGGTGTTCTAC	19	53	51
A42	GGTGTTCTACAAGGGTGC	18	56	50
A43	CAAGGGTGCCTACACCG	17	65	52
A44	CCTACACCGGAGCGTAC	17	65	52
A45	GGAGCGTACAGCGGC	15	73	50
A46	ACAGCGGCTTCGAAGGA	17	59	49
A47	TTCGAAGGAGTCGACGAG	18	56	50
A48	GTCGACGAGAACGGCAC	17	65	52
A49	GAACGGCACGCCACTG	16	69	51
A50	CGCCACTGCTGAATTGG	17	59	49
A51	CTGAATTGGCTGCGGCA	17	59	49
A52	TGGCTGCGGCAACGCG	16	75	54
A53	CAACGCGGCGTCGATG	16	69	51
A54	CGTCGATGAGGTCGATG	17	59	49
A55	AGGTCGATGTGGTCGGT	17	59	49
A56	GTGGTCGGTATTGCCAC	17	59	49
A57	TATTGCCACCGATCATTGTG	20	45	50
A58	GATCATTGTGTGCGCCAG	18	56	50
A59	TGTGCGCCAGACGGC	14	73	50
A60	AGACGGCCGAGGACG	15	73	50
A61	CGAGGACGCGGTACG	15	73	50
A62	GCGGTACGCAATGGCTT	17	59	49
A63	CAATGGCTTGGCCACCA	17	59	49
A64	TGGCCACCAGGGTGC	15	73	50
A65	CAGGGTGCTGGTGGAC	16	69	51
A66	TGGTGGACCTGACAGCG	17	65	52
A67	CTGACAGCGGGTGTGTC	17	65	52
A68	GGGTGTGTCGGCCGA	15	73	50
A69	TCGGCCGATACCACCG	16	69	51
A70	TACCACCGTCGCCGC	15	73	50
A71	GTCGCCGCGCTGGA	14	79	49
A72	CGCTGGAGGAGATGCG	16	69	51
A73	GAGATGCGCACCGCC	15	73	50
A74	GCACCGCCAGCGTC	14	79	49
A75	CAGCGTCGAGTTGGTTTG	18	56	50
A76	GTTGGTTTGCAGCTCCTG	18	56	50
A77	CAGCTCCTGATGGCACC	17	65	52
A78	CGTCGGTAGGCAAACTGC	18	61	53
A79-3	AGTCGCCCGAACGTATGGTGGA	22	59	59
A80-8	AGAACGACTTCTGCGAGGGT	20	55	54
QCprb	TGGCAGAAGCTATGAAACGATATGGG	27	44	58
Cy3-QC	CCCATATCGTTTCATAGCTTCTGCCA	26	46	58

*Basic melting temperature (T_m) was calculated with the oligonucleotide properties calculator (http://www.basic.northwestern.edu/biotools/OligoCalc.html).

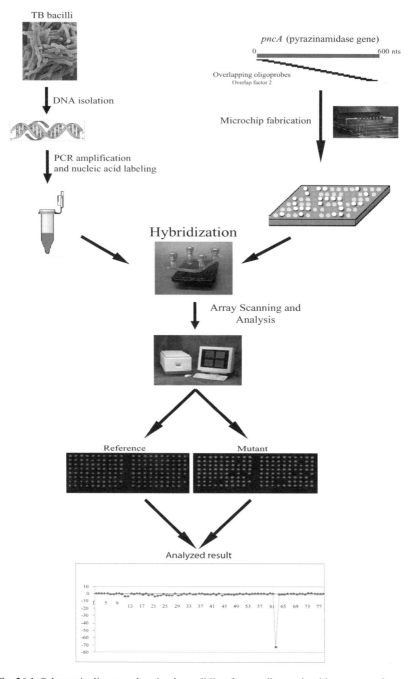

Fig. 26.1 Schematic diagram showing how sliding frame oligonucleotide array works

so that a single chip can contain oligoprobes for all the drug resistance genes for detection of all TB drug resistance by microarray.

Oligonucleotide microarray-based hybridization analysis is a novel technology that potentially allows for rapid and cost-efficient screening for all possible mutations and sequence variations in genomic DNA [36, 37]. Because microarray-based analysis for mutation detection has a potential for clinical use, performance of such tests should be thoroughly evaluated [38]. As a rule, the extensive evaluation of the efficiency of microarray-based tests is hampered by a limited number of reference strains with known mutations in the target gene. To overcome the limitation and enable an explicit evaluation of the capability of microarray oligoprobes to readily detect mutations in the respective sequence regions, we propose using DNA samples of the target gene with mutations artificially generated using an error-prone PCR [39]. Error-prone PCR protocol represents a modification of standard PCR where magnesium is replaced with manganese, which significantly enhances the error rate of Taq polymerase [40] and facilitates generation of DNA samples with numerous randomly distributed mutations within a target gene. The use of the set of genetically mutated variants of the target gene will therefore allow accurate evaluation of the efficiency of any novel microarray probes without the requirement for spontaneous natural mutants of the target gene. We therefore present a method for achieving this.

26.2 Materials

26.2.1 Microarray Fabrication

1. Oligonucleotide probes (*see* Table 26.2 for details of the oligonucleotide probes used for detection of *pncA* mutations by microarray). The 5' end of each microarray oligoprobe should be modified during the synthesis using the TFA Aminolink CE reagent (PE Applied Biosystems, Foster City, CA) to allow immobilization to the surface of CodeLink Activated Slides.
2. 6× CodeLink Print buffer: 300 mM sodium phosphate buffer pH 8.5. Dissolve the following in 90 mL of nuclease-free distilled water: 0.41 g NaH_2PO_4, 3.785 g Na_2HPO_4. Adjust the pH to 8.5 using 1 M NaOH or 1 M HCl. Bring the final volume to 100 mL with nuclease-free distilled water. Store aliquots at –20°C.
3. 384-well PCR plate (Marsh Bio Products, Rochester, NY).
4. CodeLink Activated Slides (GE Healthcare BioSciences Corp., Piscataway, NJ). Store desiccated at room temperature (*see* Note 1).
5. Contact microspotting robotic system PIXSYS 5500 (Cartesian Technologies, Inc., Ann Arbor, MI).
6. Microspotting pin CMP7 (ArrayIt, Sunnyvale, CA).
7. Dry-seal desiccator (PGC Scientific, Frederick, MD).
8. CodeLink Blocking solution: 0.1 M Tris, 50 mM ethanolamine (pH 9.0). Dissolve the following in 900 mL of nuclease-free distilled water: 6.055 g Trizma

Base, 7.88 g Trizma HCl, 3.05 g (3.0 mL) of ethanolamine. Mix thoroughly. Adjust the pH to 9.0 using 6 M HCl. Bring the final volume to 1 L with nuclease-free distilled water. Store at room temperature for up to 1 month.

9. 20X SSC (Dissolve 175.3 g NaCl and 88.2 g sodium citrate in 800 mL distilled H$_2$O. Adjust the pH to 7.0 with a few drops of 1 M HCl. Bring the final volume to 1 L with additional ddH$_2$O. Store at room temperature for up to 1 month.

10. 4X SSC (sodium chloride-sodium citrate), 0.1% (w/v) SDS (sodium dodecyl sulphate).

11. Eppendorf centrifuge 5810R (Eppendorf AG, Hamburg, Germany) or equivalent.

26.2.2 Isolation of Genomic DNA from M. tuberculosis

1. *M. tuberculosis* strains (Table 26.3).
2. Middlebrook 7H9 liquid medium with 10% (v/v) albumin-dextrose-catalase enrichment (Difco, Detroit, MI).
3. Refrigerated microcentrifuge, for example, Eppendorf 5415R (Eppendorf AG, Hamburg, Germany).
4. TE buffer (Make from 1 M stock of Tris-Cl (pH 7.5) and 500 mM stock of EDTA (pH 8.0). 1X TE buffer contains 10 mM Tris-Cl and 1 mM EDTA. Adjust the pH to 7.5. Bring the final volume to 1 L with additional ddH$_2$O. Store at room temperature for up to 1 month.
5. Lysozyme: 50 mg/mL.
6. RNaseA (Ambion, Austin, TX).
7. SDS (Sodium dodecyl sulfate, Sigma-Aldrich Co., St. Louis, MO).
8. Proteinase K: 10 mg/mL
9. Phenol/chloroform/isoamyl alcohol (25:24:1).
10. Chloroform/isoamyl alcohol (49:1).
11. 3 M sodium acetate pH 5.2.
12. Isopropyl alcohol.
13. 70% (v/v) ethanol.
14. UV/visible spectrophotometer, for example, Ultraspec 2100 *pro* (GE Healthcare Bio-Sciences Corp., Piscataway, NJ).

26.2.3 PCR Amplification of the pncA Gene

1. *HotStarTaq* DNA Polymerase 10X buffer and 2.5. mM MgCl$_2$ (QIAgene, Chatsworth, CA).
2. *HotStarTaq* DNA Polymerase (QIAgene, Chatsworth, CA). Store at –20°C.
3. 10 mM deoxynucleotide triphosphates (dNTPs) (Invitrogen, Carlsbad, CA). Store at –20°C.
4. Forward and reverse primers (T7-Myc-Rev and Myc-Forw; *see* Table 26.2).

Table 26.3 *M. tuberculosis* Strains Analyzed by Scanning-Frame Oligonucleotide Microarray [6]

Strain	*pncA* gene changes
H37Rv	No changes (reference strain)
BCG (*M. bovis*)	C to G at 169; His57Asp
M43548	C to A at 153; His51Gln
BD195	G to A at 394; Gly132Ser
VA205	G to A at 394; Gly132Ser
CDCBP98	G to C at 415; Val139Leu
M4812	G to C at 415; Val139Leu
29	T to G at 488; Val163Arg
21	T to G at 269; Iso90Ser
W76757	A to C at 422; Gln141Pro
W57575	A to C at 422; Gln141Pro
31	T to C at 355; Tyr118Arg
8989	C to T at 137; Ala46Val
10426	C to T at 401; Ala134Val
2721-1	C to G at 206; Pro69Arg
F43948	C to G at 206; Pro69Arg
H3652	T to C at 254; Leu85Pro
27795	T to C at 254; Leu85Pro
T5721	T to C at 254; Leu85Pro
H3628	A to C at 35; Asp12Ala
M52997	G to A at 289; Gly97Ser
T7527	T to C at 214; Cys72Arg
9953	T to C at 214; Cys72Arg
A7153	A to G at 245; His82Arg
17	C to A at 285; Tyr95Stop
25	G to T at 233; Gly78Val
9579	C to A at 418; Arg140Ser; 8 bp deletion at 446
10347	C to A at 418; Arg140Ser; 8 bp deletion at 446
10800	C to A at 418; Arg140Ser; 8 bp deletion at 446
11041	C to A at 418; Arg140Ser; 8 bp deletion at 446
11135	C to A at 418; Arg140Ser; 8 bp deletion at 446
9131	C to A at 418; Arg140Ser; 8 bp deletion at 446
9132	C to A at 418; Arg140Ser; 8 bp deletion at 446
9769	C to A at 418; Arg140Ser; 8 bp deletion at 446
9811	C to A at 418; Arg140Ser; 8 bp deletion at 446
10348	C to A at 418; Arg140Ser; 8 bp deletion at 446
10350	C to A at 418; Arg140Ser; 8 bp deletion at 446
11243	C to A at 418; Arg140Ser; 8 bp deletion at 446
11823	C to A at 418; Arg140Ser; 8 bp deletion at 446
W296	G deletion at 443
H4171	11 bp deletion at start codon
10467	C deletion at 514
955293	G deletion at 71
T61823	GG insertion at 392
11627	A insertion at 193

Table 26.3 (continued)

Strain	*pncA* gene changes
9869	-11 promoter mutation A to G
APZAR6	T to C at 40; Cys14Arg
RK56	T to C at 40; Cys14Arg
3	No changes found
H2374	No changes found
37	No changes found
11552	No changes found
11830	No changes found
941392	No changes found
APZAR1	No changes found
APZAR3	No changes found

5. UltraPure DEPC-Treated Water (Invitrogen). Store at room temperature.
6. 1% 8 well TAE agarose precast gels with ethidium bromide (Cambrex Corporation, East Rutherford, NJ). Store at room temperature.
7. UltraPure 10× TAE Buffer (Invitrogen). Store at room temperature.
8. UltraPure 10 mg/mL Ethidium Bromide (Invitrogen). Store at room temperature.

26.2.4 Preparation of Fluorescently Labeled Hybridization Targets

26.2.4.1 Single-Stranded RNA Synthesis

1. MEGAscript T7 High Yield Transcription Kit (Ambion, Austin, TX). Store at – 20°C.
2. UltraPure DEPC-Treated Water (Invitrogen). Store at room temperature.
3. Centrisep Spin Columns (Princeton Separations, Adelphia, NJ).

26.2.4.2 Single-Stranded DNA Preparation

1. *HotStarTaq* DNA Polymerase 10X buffer and 2.5. mM MgCl$_2$ (QIAgene, Chatsworth, CA).
2. *HotStarTaq* DNA Polymerase (QIAgene, Chatsworth, CA). Store at –20°C.
3. 10 mM deoxynucleotide triphosphates (dNTPs) (Invitrogen, Carlsbad, CA). Store at –20°C.
4. Reverse primer T7-Myc-Rev. (*See* Table 26.2. Adjust the concentration of primer stock to 200 µM and store the stock at –20°C; Prepare the primer working stock solution with concentration 20 µM and keep the stock at –20°C; use this primer in your PCR reaction at concentration of 2.0 to 2.5 µM.)
5. UltraPure DEPC-Treated Water (Invitrogen). Store at room temperature.
6. Centrisep Spin Columns (Princeton Separations, Adelphia, NJ).

26.2.4.3 Fluorescent Labeling of Microarray Hybridization Target

1. MICROMAX ASAP RNA Labeling Kit (PerkinElmer, Boston, MA).
2. Centrisep Spin Columns (Princeton Separations, Adelphia, NJ).
3. UV/visible spectrophotometer, for example, Ultraspec 2100 *pro* (GE Healthcare Bio-Sciences Corp., Piscataway, NJ).

26.2.5 Microarray Hybridization

1. Glass coverslip (LifterSlip coverslip (Eire Scientific Company, Portsmouth, NH).
2. ArrayIt Hybridization Cassette chamber (ArrayIt, Sunnyvale, CA).
3. 4X SSC, 0.1% (w/v) SDS solution.
4. 2X SSC solution.
5. 1X SSC solution.
6. Eppendorf centrifuge 5810R (Eppendorf AG, Hamburg, Germany) or equivalent.

26.2.6 Signal Detection and Data Processing

1. ScanArray 5000 (PerkinElmer).
2. ScanArray Express software (PerkinElmer).

26.2.7 Generation of Random Mutations in the Target DNA Using Error-Prone PCR

1. *HotStarTaq* DNA Polymerase 10X buffer (QIAgene, Chatsworth, CA).
2. *HotStarTaq* DNA Polymerase (QIAgene, Chatsworth, CA). Store at –20°C.
3. 2.5 mM $MnCl_2$
4. 10 mM deoxyribonucleotide triphosphates (dNTPs) (Invitrogen, Carlsbad, CA). Store at –20°C.
5. Forward and reverse primers (T7-Myc-Rev and Myc-Forw; *see* Table 26.2).
6. UltraPure DEPC-Treated Water (Invitrogen). Store at room temperature.
7. 1% 8-well TAE agarose precast gels with ethidium bromide (Cambrex Corporation, East Rutherford, NJ). Store at room temperature.
8. UltraPure 10× TAE Buffer (Invitrogen). Store at room temperature.
9. UltraPure 10 mg/mL Ethidium Bromide (Invitrogen). Store at room temperature.
10. QIAquick PCR Purification Kit (QIAgene) or similar. Store at room temperature.

11. TOPO XL PCR Cloning Kit with One Shot TOP10 Chemically Competent *E. coli* (Invitrogen) or similar. Store at –80°C.
12. imMedia Kan Agar and imMedia Kan liquid (Invitrogen). Store at room temperature.
13. QIAprep Spin Miniprep Kit (QIAgene) or similar. Store at room temperature.

26.3 Methods

The microarray protocol presented here was established and optimized during investigations aimed at the development of rapid methods for detection of spontaneous mutations in the genes responsible for the drug resistance of *M. tuberculosis* [6, 22].

26.3.1 Microarray Fabrication (see Notes 2 and 3)

1. Adjust the concentration of each oligoprobe to 150 µM in 1× CodeLink Print buffer. Add the quality control oligonucleotide (QCprb) to a final concentration of 10 µM.
2. Transfer 10 µL of each oligoprobe solution into a 384-well PCR plate.
3. Deposit the *pncA*-specific oligoprobes onto the surface of CodeLink Activated Slides using a contact microspotting robotic system equipped with a microspotting pin (see **Note 4**).
4. After spotting, place the slides into a dry-seal desiccator with the constant humidity of 80% (see **Note 5**).
5. Block residual reactive groups on the slide surface by treating the slides with CodeLink Blocking solution (prewarmed to 50°C) for 30 min.
6. Remove unbound oligonucleotides by washing the slides once with 4× SSC, 0.1% SDS (50 mL of prewarmed to 50°C per slide) for 30 to 60 min followed by five washes with distilled water for 1 min each.
7. Remove traces of water from the slide surface by centrifugation of slides at 128 × g for 3 min.

26.3.2 Isolation of Genomic DNA from M. tuberculosis

1. Grow *M. tuberculosis* strains in Middlebrook 7H9 liquid medium with albumin-dextrose-catalase enrichment at 37°C for 2 to 3 weeks (see **Note 6**).
2. Inactivate the bacteria by heating 1 mL of culture at 80°C for 20 min.
3. Spin down cells at 3000 × g and remove supernatant. Resuspend the pellet in 200 µL of 1× TE buffer. Add 50 µL of lysozyme and 5 µL of RNase A and incubate at 37°C for 2 h. Mix occasionally during the incubation.

4. Add 50 µL SDS and 20 µL Proteinase K solution. Mix gently and incubate at 45°C overnight.
5. Add an equal volume of phenol/chloroform/isoamyl alcohol, 25:24:1 (v/v), pH 7.9, and vortex tubes intermittently for 3 to 5 min (make sure the tubes are tightly closed and do not leak).
6. Spin at 3000 × *g* at room temperature for 10 min.
7. Transfer upper aqueous phase to a fresh tube.
8. Add equal volume of chloroform/isoamyl alcohol, 49:1 (v/v), and mix gently. Spin at 3000 × *g* for 20 min. Transfer the upper aqueous phase to a fresh tube and repeat this step one more time to make sure the interface is clean.
9. Transfer the upper aqueous phase to a fresh tube, and add 0.1 volume of 3 M sodium acetate, pH 5.2.
10. Add 0.6 volume of isopropyl alcohol. Invert the tube gently. Leave at –20°C for 30 min and spin at 3000 × *g* for 10 min.
11. Discard supernatant, and wash the DNA pellet with 70% ethanol. Spin at 3000 × *g* in Eppendorf centrifuge 5415R for 10 min. Carefully remove supernatant.
12. Spin again for a few seconds, and remove the supernatant with a pipette. Air dry the DNA pellet for 15 min.
13. Add 100 µL of 0.1× TE buffer to the DNA pellet, mix gently by tapping the tube. Leave at 37°C for 30 min to dissolve large molecular weight DNA.
14. Quantitate the DNA concentration by diluting the DNA as 1/50 dilution in water and measure A_{260}.

26.3.3 PCR Amplification of the pncA Gene

1. Set up a 50 µL PCR mixture containing:

   5 µL *HotStarTaq* DNA Polymerase buffer
   2 units of *HotStarTaq* DNA Polymerase
   2.5 mM $MgCl_2$ (supplied with *HotStarTaq* DNA Polymerase)
   200 µM of each dNTP (dATP, dGTP, dCTP, and dTTP)
   500 nM of each forward and reverse primers
   1 µL DNA template (approximately 0.1 µg of DNA template)
   UltraPure DEPC-Treated Water to a final volume of 50 µL (add first)

2. Perform the PCR amplification using a thermocycler with the following thermal cycle conditions:

   Initial denaturing at 95°C for 15 min
   Followed by 40 cycles at 94°C for 40 s, 55°C for 40 s, 72°C extension for 40 s
   Final extension at 72°C for 10 min

3. Store the PCR reaction at 4°C until use.
4. Check the presence of amplified PCR products by electrophoresis in 1% 8-well TAE agarose precast gels containing ethidium bromide followed by ultraviolet visualization (see **Note 7**). The expected *pncA* amplicon size should be approximately 650 bp.

26.3.4 Preparation of Fluorescently Labeled Hybridization Targets

26.3.4.1 Single-Stranded RNA Synthesis

1. Set up a 30-µL single-stranded RNA reaction mixture as following:

   $1\times$ MEGAscript T7 reaction buffer
   2 µL of MEGAscript T7 Enzyme Mix.
   5 mM of ATP, UTP, CTP, and GTP each
   0.1 to 0.5 µg of the *pncA* PCR product as template
   UltraPure DEPC-Treated Water to a final volume of 30 µL (add first)

2. Incubate the reaction mixture at 37°C for 2 to 4 h. Hold the reaction at 4°C until used.
3. Remove the unincorporated rNTPs using a Centrisep Spin Columns according to the manufacturer's protocol.

26.3.4.2 Single-Stranded DNA Preparation

1. Set up a 50-µL single-stranded DNA reaction mixture as following:

   $1\times$ *HotStarTaq* DNA Polymerase buffer
   2 units of *HotStarTaq* DNA Polymerase
   2.5 mM $MgCl_2$ (supplied with *HotStarTaq* DNA Polymerase)
   200 µM of each dNTP (dATP, dGTP, dCTP, and dTTP)
   600 nM of reverse primer T7-Myc-Rev
   0.1 to 0.5 µg of the *pncA* PCR product as template
   UltraPure DEPC-Treated Water to a final volume of 50 µL (add first)

2. Run the PCR amplification using a thermocycler with the following cycle conditions:

   Initial denaturing at 95°C for 15 min
   40 cycles at 94°C for 40 s, 55°C for 40 s, 72°C extension for 40 s
   Final extension at 72°C for 10 min

3. Hold the reaction at 4°C.
4. Remove the unincorporated dNTPs by purification through the Centrisep Spin Columns according to the manufacturer's protocol.

26.3.4.3 Fluorescent Labeling of Microarray Hybridization Target

Use the MICROMAX ASAP RNA Labeling Kit for equally efficient incorporation of Cy5 fluorophore into either ssRNA or ssDNA molecules.

1. Set up a 50-μL PCR mixture as follows:

   10 μg of ssRNA or ssDNA sample (from Sections 26.3.4.1 or 26.3.4.2).
   1 μL of ASAP Cyanine-5 Labeling Reagent
   ASAP Labeling Buffer to the final volume of 50 μL (add first)

2. Run the labeling reaction using a thermocycler at 85°C for 40 min.
3. Hold the samples at 4°C until used.
4. Remove the unincorporated ASAP labeling reagent by purification through the Centrisep Spin Columns according to the manufacturer's protocol.
5. Quantitate the DNA concentration by diluting the nucleic acid as 1/50 dilution in water and measure A_{260}.
6. Adjust the nucleic acid concentration to 0.5 to 1.0 μM in the MICROMAX Hybridization Buffer III.

26.3.5 Microarray Hybridization

Hybridization between microarray oligoprobes and fluorescently labeled ssRNA (ssDNA) samples is conducted in the MICROMAX Hybridization Buffer III at 60°C for 1 h (see **Notes 8, 9** and **10**).

1. Mix 10 μM of Cy5-labeled ssRNA (or ssDNA) sample (from Section 26.3.4) with a Cy3-QC probe (see Table 26.2) at molar ratio 10:1. Denature at 95°C for 1 min.
2. Place the hybridization mixture on microarray slide and cover it with a glass coverslip to prevent evaporation of the solution during incubation.
3. After hybridization, wash the slides for 15 min in 4× SSC with 0.1% SDS (prewarmed to 60°C), then in 2× SSC buffer for 5 min, and in 1× SSC buffer for 5 min.
4. Remove traces of buffer from the slide surface by centrifugation at $128 \times g$ for 3 min.

26.3.6 Signal Detection and Data Processing

1. Generate the fluorescent images of processed microarray slides using the array scanner.
2. Measure and compare the fluorescent signals from each spot using ScanArray Express software (*see* **Note 11**).
3. Identify the position(s) of the mutation(s) in *pncA* by dividing the numeric intensities of fluorescent signals from each array element measured for the reference wild-type *pncA* gene to those of the analyzed isolate. An alteration

of more than twofold in the ratio is considered as indication of a mutation in the oligoprobe binding region within the *pncA* gene. An example is shown in Figure 26.2, where the mutations in *pncA* in several PZA-resistant *M. tuberculosis* strains (isolates W296, M43548, 21, 8989, 10467, and T7527; *see* Table 26.3 for mutation profile of these strains) can be detected by the sliding frame microarray.

26.3.7 Generation of Random Mutations in the Target DNA Using Error-Prone PCR

In theory, the use of oligonucleotide probes and microarrays can be used for any gene. The criteria used to design microarray oligoprobes were described in details elsewhere [6]. Once oligonucleotides have been synthesized, error-prone PCR can be used to evaluate the discriminatory efficiency of the microarray.

1. Set up a 50 μL error-prone PCR mixture as follows:

   1× *HotStarTaq* DNA Polymerase buffer
   2 units of *HotStarTaq* DNA Polymerase
   2.5 mM $MnCl_2$
   200 μM of each dNTP (dATP, dGTP, dCTP, and dTTP)
   200 nM of each forward and reverse primers
   1 μL DNA template (approximately 0.1 μg of DNA template)
   UltraPure DEPC-Treated Water added to the final volume of 50 μL (add first)

2. Perform the PCR amplification using a thermocycler with the following cycle conditions:

   Initial denaturing at 95°C for 15 min
   40 cycles at 94°C for 40 s, 55°C for 40 s, 72°C extension for 40 s
   The final extension at 72°C for 10 min

3. Hold at 4°C. Check the presence of amplified PCR products by electrophoresis in 1% TAE agarose precast gels with ethidium bromide followed by ultraviolet amplicon visualization.
4. Purify the PCR from the agarose gel using the Qiagen gel-purification kit according to manufacturer's instructions.
5. Clone PCR products into pCR-XL-TOPO plasmid using the TOPO XL PCR Cloning Kit with One Shot TOP10 Chemically Competent cells according to the recommended protocol.
6. Randomly select white transformant colonies and grow overnight at 37°C using imMedia Kan Agar plates.
7. Isolate plasmid DNA using QIAprep Spin MiniPrep kit and confirm the presence of the insertion in the plasmid by 0.8% agarose gel electrophoresis, restriction enzyme digestion, or DNA sequencing.

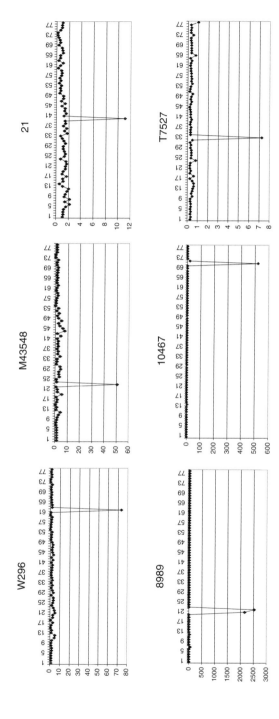

Fig. 26. 2 Evaluation of microarray discrimination efficiency using several *M. tuberculosis* strains containing known single-point mutations in the *pncA* gene. Discriminatory capability of sliding-frame microchip was evaluated using hybridization with several *M. tuberculosis* isolates containing known mutations in the *pncA* gene (isolates W296, M43548, 21, 8989, 10467, and T7527) characterized in our previous studies (*see* Table 26.3 for mutation profile of these strains). Ratios of normalized fluorescence signals of the reference strain H37Rv and strains containing mutations in the *pncA* gene are shown. (Adapted from Wade et al., *Diagn. Microbiol. Infect. Dis.*, 2004;49:89–97. Permission is granted by Elsevier.)

8. Identify sequences of plasmid insertions for multiple clones.
9. Select plasmids with insertions containing different random mutations. Amplify the insertions using PCR primers shown in Table 26.2 or custom designed PCR primers.
10. Prepare hybridization targets as described above.

26.4 Notes

1. CodeLink Activated Slides are compatible with most available systems for microarray manufacturing and scanning. CodeLink Activated Slides are coated with a three-dimensional surface chemistry composed of a long-chain, hydrophilic polymer containing amine-reactive groups. The cross-linked polymer is covalently attached to the surface of the slide. The use of the slides with the polymer minimizes the effect of the glass surface on hybridization process. It also eliminates the need in the use of poly(dT) or PEG (polyethylene glycol) spacers. All this results in lower fluorescent background of processed slides.

2. Quadruplication of each oligoprobe on microchip increases the reliability of mutation detection, reduces the probability of misinterpretation of microarray data that could occur due to potential failure during microarray fabrication, and allows for application of statistical approaches for analysis of microarray data.

3. To increase the confidence of mutation detection in the *pncA* gene and to avoid the misinterpretation of microarray results, which could be caused by irregularities during the slide preparation, each array is recommended to be represented by four identical subarrays containing the complete set (79 oligoprobes) of oligoprobes required for the analysis of the *pncA* gene sequence. The size of the CodeLink microscopic slide enables one to place five separate microarray arrays, which allows for simultaneous analysis of five different samples.

4. During the spotting procedure, the humidity inside the microspotter environmental chamber should be maintained at 75% to 80%. The settings of the spotting program should be chosen to provide 450-μm intervals between spots (center to center). Training with technical experts in the field of microarray printing, slide scanning, and image analysis may be necessary. Alternatively, microarray slides may be customary printed using commercially available microarray service.

5. The required percent of humidity can be provided by the presence of saturated NaCl aqueous solution on the bottom of a desiccator.

6. Whereas the method described here for DNA isolation is appropriate for a relatively small number of strains, this method may not be adequate when a large number of strains are used for mutation analysis. In this case, the DNA can be isolated by heating the bacterial cell suspension (100 to 200 μL) from liquid cultures grown in BACTEC or MGIT media or from several loopfuls of solid growth on Lowenstein-Jensen medium resuspended in distilled water (100 to 200 μL) at 80°C for 20 min in the presence of equal volume of chloroform followed by vortex for 5 min and then centrifugation. Alternatively, 0.1-mm glass beads can be added to a small volume (100 to 200 μL) of cell suspension (the amount of glass beads should not exceed half volume of cell suspension) followed by vortex for 5 min and boiling for 15 min. After centrifugation, the supernatant (5 to 10 μL) containing genomic DNA can be used directly for PCR amplification.

7. To determine the sequence, PCR-amplified DNA fragments may be separated by using the 1% agarose gel electrophoresis followed by extraction of amplicons from the gel using a gel-purification kit (Qiagen, Chatsworth, CA) according to the manufacturer's instructions. The PCR products may be directly sequenced using commercially available DNA sequencing service.

A   **BCG (RNA, silylated slide)**

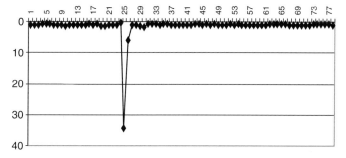

B   **BCG (RNA, CodeLink slide)**

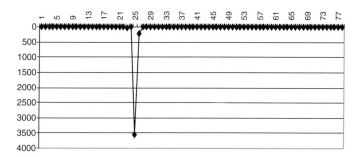

C   **BCG (ssDNA, silylated slide)**

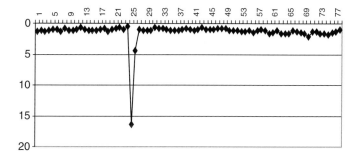

Fig. 26.3 Dependence of microarray discrimination efficiency on the slide surface coating and the nature of fluorescently labeled sample. **(A)** Hybridization profile of RNA sample hybridized with the microarray fabricated using silylated slide. **(B)** Profile of the sample but hybridized with microarray immobilized on the CodeLink slide. **(C)** Hybridization profile of ssDNA sample hybridized with the microarray fabricated using silylated surface. (From Wade et al., *Diagn. Microbiol. Infect. Dis.*, 2004;49:89–97. Permission is granted by Elsevier.)

8. The set of sliding oligoprobes capable of detecting any potential mutation in the mycobacterial *pncA* gene was designed using empirical criteria described previously [36]. In total, 79 oligoprobes covering the entire coding and promoter region of the *pncA* gene, plus one additional oligonucleotide identical to the sequence of the forward primer, are included in the final microarray to monitor the emergence of spontaneous alterations in the sequence of the *pncA* gene.

9. We have screened different commercially available brands of microarray slides for immobilization of short oligoprobes to identify the slide coating that is the most suitable for accurate detection of minor genetic changes with high sensitivity and specificity. CodeLink Activated (CLA) slides are coated with a novel three-dimensional surface chemistry comprised of a long-chain, hydrophilic polymer containing amine-reactive groups. The cross-linked polymer, combined with end-point attachment, orients the immobilized DNA and holds it away from the surface of the slide. This type of immobilization excludes the need to use additional spacers for oligoprobes. Simultaneous comparison of silylated (CEL Associates, Inc., Pearland, TX) and CLA slides showed that the discriminatory power (ratio of signals from the same oligoprobe hybridized with the fluorescent samples complementary to the probe or containing one nucleotide mismatch) of CLA slides is higher than that of silylated slides (Fig. 26.3).

10. Hybridization protocols for CodeLink activated slides were optimized to guarantee high confidence of mutation detection. The use of a fluorescently labeled ssRNA sample instead of ssDNA can substantially improve the discriminatory power of the assay (Fig. 26.3). Partial degradation of ssRNA by treatment with 0.1 M NaOH did not exhibit a noticeable effect on discrimination. The concentration of ssRNA was found to be crucial for the consistency of microarray data. To make a linear regression model applicable for the normalization of hybridization data, the concentrations of the reference strain H37Rv and the mutant BCG fluorescently labeled samples should not differ significantly. We found that the concentration of fluorescently labeled ssRNA samples in the range 0.5 to 1.0 μM is most reliable for the confident detection of any minor genetic changes in the target sequences.

11. Training with the technical experts of PerkinElmer may be necessary before performing microarray data analysis using the ScanArray 5000 and ScanArray Express software systems.

Acknowledgments This work was supported by NIH grants AI44063, AI/HL49485, and AI61908.

References

1. Brudey, K., Driscoll, J. R., Rigouts, L., Prodinger, W. M., Gori, A., Al-Hajoj, S. A., Allix, C., Aristimuno, L., Arora, J., Baumanis, V., Binder, L., Cafrune, P., Cataldi, A., Cheong, S., Diel, R., Ellermeier, C., Evans, J. T., Fauville-Dufaux, M., Ferdinand, S., Garcia de Viedma, D., Garzelli, C., Gazzola, L., Gomes, H. M., Guttierez, M. C., Hawkey, P. M., van Helden, P. D., Kadival, G. V., Kreiswirth, B. N., Kremer, K., Kubin, M., Kulkarni, S. P., Liens, B., Lillebaek, T., Ho, M. L., Martin, C., Martin, C., Mokrousov, I., Narvskaia, O., Ngeow, Y. F., Naumann, L., Niemann, S., Parwati, I., Rahim, Z., Rasolofo-Razanamparany, V., Rasolonavalona, T., Rossetti, M. L., Rusch-Gerdes, S., Sajduda, A., Samper, S., Shemyakin, I. G., Singh, U. B., Somoskovi, A., Skuce, R. A., van Soolingen, D., Streicher, E. M., Suffys, P. N., Tortoli, E., Tracevska, T., Vincent, V., Victor, T. C., Warren, R. M., Yap, S. F., Zaman, K., Portaels, F., Rastogi, N., Sola, C. (2006) *Mycobacterium tuberculosis* complex genetic diversity: mining the fourth

international spoligotyping database (SpolDB4) for classification, population genetics and epidemiology. *BMC Microbiol* **6**, 23.

2. Rusen, I. D., Enarson, D. A. (2006) FIDELIS–innovative approaches to increasing global case detection of tuberculosis. *Am J Public Health* **96**, 14–16.

3. Zignol, M., Hosseini, M. S., Wright, A., Weezenbeek, C. L., Nunn, P., Watt, C. J., Williams, B. G., Dye, C. (2006) Global incidence of multidrug-resistant tuberculosis. *J Infect Dis* **194**, 479–485.

4. Zhang, Y., C. Vilcheze, W. R. Jacobs Jr. (2005) Mechanisms of drug resistance in *Mycobacterium tuberculosis*. In *Tuberculosis and the Tubercle Bacillus*, ed. Stewart T. Cole, K. D. E., David N. McMurray, William R. Jacobs, Jr., pp. 115–140. Washington, D.C: ASM Press.

5. Telenti, A., Iseman, M. (2000) Drug-resistant tuberculosis: what do we do now? *Drugs* **59**, 171–179.

6. Wade, M. M., Volokhov, D., Peredelchuk, M., Chizhikov, V., Zhang, Y. (2004) Accurate mapping of mutations of pyrazinamide-resistant *Mycobacterium tuberculosis* strains with a scanning-frame oligonucleotide microarray. *Diagn Microbiol Infect Dis* **49**, 89–97.

7. Wade, M. M., Zhang, Y. (2004) Mechanisms of drug resistance in *Mycobacterium tuberculosis*. *Front Biosci* **9**, 975–994.

8. Zhang, Y. (2005) The magic bullets and tuberculosis drug targets. *Annu Rev Pharmacol Toxicol* **45**, 529–564.

9. Zhang, Y., Amzel, L. M. (2002) Tuberculosis drug targets. *Curr Drug Targets* **3**, 131–154.

10. Zhang, Y., Post-Martens, K., Denkin, S. (2006) New drug candidates and therapeutic targets for tuberculosis therapy. *Drug Discov Today* **11**, 21–27.

11. Garcia de Viedma, D. (2003) Rapid detection of resistance in *Mycobacterium tuberculosis*: a review discussing molecular approaches. *Clin Microbiol Infect* **9**, 349–359.

12. Musser, J. M. (1995) Antimicrobial agent resistance in *mycobacteria*: molecular genetic insights. *Clin Microbiol Rev* **8**, 496–514.

13. Telenti, A., Persing, D. H. (1996) Novel strategies for the detection of drug resistance in *Mycobacterium tuberculosis*. *Res Microbiol* **147**, 73–79.

14. Cardoso, R. F., Cooksey, R. C., Morlock, G. P., Barco, P., Cecon, L., Forestiero, F., Leite, C. Q., Sato, D. N., Shikama Mde, L., Mamizuka, E. M., Hirata, R. D., Hirata, M. H. (2004) Screening and characterization of mutations in isoniazid-resistant *Mycobacterium tuberculosis* isolates obtained in Brazil. *Antimicrob Agents Chemother* **48**, 3373–3381.

15. Kiepiela, P., Bishop, K. S., Smith, A. N., Roux, L., York, D. F. (2000) Genomic mutations in the *katG, inhA* and *aphC* genes are useful for the prediction of isoniazid resistance in *Mycobacterium tuberculosis* isolates from Kwazulu Natal, South Africa. *Tuber Lung Dis* **80**, 47–56.

16. Felmlee, T. A., Liu, Q., Whelen, A. C., Williams, D., Sommer, S. S., Persing, D. H. (1995) Genotypic detection of *Mycobacterium tuberculosis* rifampin resistance: comparison of single-strand conformation polymorphism and dideoxy fingerprinting. *J Clin Microbiol* **33**, 1617–1623.

17. Shi, R., Otomo, K., Yamada, H., Tatsumi, T., Sugawara, I. (2006) Temperature-mediated heteroduplex analysis for the detection of drug-resistant gene mutations in clinical isolates of *Mycobacterium tuberculosis* by denaturing HPLC, SURVEYOR nuclease. *Microbes Infect* **8**, 128–135.

18. Johnson, R., Jordaan, A. M., Pretorius, L., Engelke, E., van der Spuy, G., Kewley, C., Bosman, M., van Helden, P. D., Warren, R., Victor, T. C. (2006) Ethambutol resistance testing by mutation detection. *Int J Tuberc Lung Dis* **10**, 68–73.

19. Bockstahler, L. E., Li, Z., Nguyen, N. Y., Van Houten, K. A., Brennan, M. J., Langone, J. J., Morris, S. L. (2002) Peptide nucleic acid probe detection of mutations in *Mycobacterium tuberculosis* genes associated with drug resistance. *Biotechniques* **32**, 508–510, 512, 514.

20. Marme, N., Friedrich, A., Muller, M., Nolte, O., Wolfrum, J., Hoheisel, J. D., Sauer, M., Knemeyer, J. P. (2006) Identification of single-point mutations in mycobacterial 16S

rRNA sequences by confocal single-molecule fluorescence spectroscopy. *Nucleic Acids Res* **34**, e90.

21. Van Rie, A., Warren, R., Mshanga, I., Jordaan, A. M., van der Spuy, G. D., Richardson, M., Simpson, J., Gie, R. P., Enarson, D. A., Beyers, N., van Helden, P. D., Victor, T. C. (2001) Analysis for a limited number of gene codons can predict drug resistance of *Mycobacterium tuberculosis* in a high-incidence community. *J Clin Microbiol* **39**, 636–641.

22. Denkin, S., Volokhov, D., Chizhikov, V., Zhang, Y. (2005) Microarray-based pncA genotyping of pyrazinamide-resistant strains of *Mycobacterium tuberculosis*. *J Med Microbiol* **54**, 1127–1131.

23. Ivshina, A. V., Vodeiko, G. M., Kuznetsov, V. A., Volokhov, D., Taffs, R., Chizhikov, V. I., Levandowski, R. A., Chumakov, K. M. (2004) Mapping of genomic segments of influenza B virus strains by an oligonucleotide microarray method. *J Clin Microbiol* **42**, 5793–5801.

24. Sergeev, N., Volokhov, D., Chizhikov, V., Rasooly, A. (2004) Simultaneous analysis of multiple staphylococcal enterotoxin genes by an oligonucleotide microarray assay. *J Clin Microbiol* **42**, 2134–2143.

25. Volokhov, D. V., George, J., Liu, S. X., Ikonomi, P., Anderson, C., Chizhikov, V. (2006) Sequencing of the intergenic 16S-23S rRNA spacer (ITS) region of *Mollicutes* species and their identification using microarray-based assay and DNA sequencing. *Appl Microbiol Biotechnol* **71**, 680–698.

26. Neverov, A. A., Riddell, M. A., Moss, W. J., Volokhov, D. V., Rota, P. A., Lowe, L. E., Chibo, D., Smit, S. B., Griffin, D. E., Chumakov, K. M., Chizhikov, V. E. (2006) Genotyping of measles virus in clinical specimens on the basis of oligonucleotide microarray hybridization patterns. *J Clin Microbiol* **44**, 3752–3759.

27. Caoili, J. C., Mayorova, A., Sikes, D., Hickman, L., Plikaytis, B. B., Shinnick, T. M. (2006) Evaluation of the TB-Biochip oligonucleotide microarray system for rapid detection of rifampin resistance in *Mycobacterium tuberculosis*. *J Clin Microbiol* **44**, 2378–2381.

28. Aragon, L. M., Navarro, F., Heiser, V., Garrigo, M., Espanol, M., Coll, P. (2006) Rapid detection of specific gene mutations associated with isoniazid or rifampicin resistance in *Mycobacterium tuberculosis* clinical isolates using non-fluorescent low-density DNA microarrays. *J Antimicrob Chemother* **57**, 825–831.

29. Gryadunov, D., Mikhailovich, V., Lapa, S., Roudinskii, N., Donnikov, M., Pan'kov, S., Markova, O., Kuz'min, A., Chernousova, L., Skotnikova, O., Moroz, A., Zasedatelev, A., Mirzabekov, A. (2005) Evaluation of hybridisation on oligonucleotide microarrays for analysis of drug-resistant *Mycobacterium tuberculosis*. *Clin Microbiol Infect* **11**, 531–539.

30. Tang, X., Morris, S. L., Langone, J. J., Bockstahler, L. E. (2005) Microarray and allele specific PCR detection of point mutations in *Mycobacterium tuberculosis* genes associated with drug resistance. *J Microbiol Methods* **63**, 318–330.

31. Yoshikawa, Y., Ichihara, T., Suzuki, Y. (2003) Detection of drug-resistant *Mycobacterium tuberculosis* isolates using DNA microarray. *Rinsho Biseibutshu Jinsoku Shindan Kenkyukai Shi* **14**, 45–50.

32. Yue, J., Shi, W., Xie, J., Li, Y., Zeng, E., Liang, L., Wang, H. (2004) Detection of rifampin-resistant *Mycobacterium tuberculosis* strains by using a specialized oligonucleotide microarray. *Diagn Microbiol Infect Dis* **48**, 47–54.

33. Strizhkov, B. N., Drobyshev, A. L., Mikhailovich, V. M., Mirzabekov, A. D. (2000) PCR amplification on a microarray of gel-immobilized oligonucleotides: detection of bacterial toxin- and drug-resistant genes and their mutations. *Biotechniques* **29**, 844–848, 850–842, 854 passim.

34. Tillib, S. V., Strizhkov, B. N., Mirzabekov, A. D. (2001) Integration of multiple PCR amplifications and DNA mutation analyses by using oligonucleotide microchip. *Anal Biochem* **292**, 155–160.

35. Sivkov, A., Boldyrev, A. N., Azaev, M., Bodnev, S. A., Medvedeva, E. V., Baranova, O. I., Ivlev-Dantau, A. P., Blinova, L. N., Pasechnikov, A. D., Tat'kov, S. I. (2006) Evaluation of

reasons of the MDR *M. tuberculosis* strains dissemination by analysis of the rifampicin and/or isoniazid resistant isolates. *Mol Gen Mikrobiol Virusol* **2**, 20–25.

36. Cherkasova, E., Laassri, M., Chizhikov, V., Korotkova, E., Dragunsky, E., Agol, V. I., Chumakov, K. (2003) Microarray analysis of evolution of RNA viruses: evidence of circulation of virulent highly divergent vaccine-derived polioviruses. *Proc Natl Acad Sci U S A* **100,** 9398–9403.

37. Heller, M. J. (2002) DNA microarray technology: devices, systems, and applications. *Annu Rev Biomed Eng* **4,** 129–153.

38. Rudert, F. (2000) Genomics and proteomics tools for the clinic. *Curr Opin Mol Ther* **2,** 633–642.

39. Cirino, P. C., Mayer, K. M., Umeno, D. (2003) Generating mutant libraries using error-prone PCR. *Methods Mol Biol* **231,** 3–9.

40. Pritchard, L., Corne, D., Kell, D., Rowland, J., Winson, M. (2005) A general model of error-prone PCR. *J Theor Biol* **234,** 497–509.

Chapter 27
*Myco*DB: An Online Database for Comparative Genomics of the Mycobacteria and Related Organisms

Roy R. Chaudhuri

Abstract *Myco*DB (http://myco.bham.ac.uk) is an online resource designed to facilitate genomic analyses of *Mycobacterium* spp. and related genera. Regions of interest can be found by searching the annotation, BLAST searching against the sequence data, or specifying genomic coordinates. Tools are provided to access the primary sequence data and annotation, design primers, and view the equivalent region in other genomes, as determined using MUMmer. The whole chromosome can also be displayed and genes "painted" using data such as their GC content. The site also acts as a portal, allowing the user to access other relevant online resources.

Keywords bioinformatics · BLAST · comparative genomics · genome alignments · genome browser · MUMmer · *Myco*DB · *x*BASE

27.1 Introduction

*Myco*DB (http://myco.bham.ac.uk) is an online resource designed to facilitate genomic analyses of *Mycobacterium* spp. and the related genera *Streptomyces*, *Tropheryma*, *Corynebacterium*, and *Clavibacter*. It is based on the *x*BASE template originally developed for the *E. coli* database *coli*BASE [1, 2]. The focus of the database is on comparative genomics, and the user interface has been designed with ease of use as a priority. The database provides tools for browsing genome annotation, viewing genome alignments, displaying information such as GC content by "gene painting," and retrieving primary sequence data. It also acts as a portal, providing centralized access to other relevant internet resources.

R.R. Chaudhuri
Department of Veterinary Medicine, University of Cambridge, Madingley Road,
Cambridge, CB3 0ES, UK
e-mail: rrc22@cam.ac.uk

T. Parish, A.C. Brown (eds.), *Mycobacteria Protocols*,
doi: 10.1007/978-1-59745-207-6_27, © Humana Press, Totowa, NJ 2008

27.2 Materials

27.2.1 Software Requirements

To obtain the most out of *Myco*DB, it should be accessed using a modern Internet browser such as Mozilla Firefox (http://www.getfirefox.com), Microsoft Internet Explorer (http://www.microsoft.com/windows/ie), Opera (http://www.opera.com), or Safari (http://www.apple.com/safari). For full functionality, JavaScript and Java must be enabled (this is the default for most popular browsers).

27.2.2 Data in Myco*DB*

There are currently 12 complete genomes within *Myco*DB: *Corynebacterium diphtheriae* NCTC13129 (biotype gravis), *Corynebacterium efficiens* YS-314, *Corynebacterium glutamicum* ATCC13032, *Mycobacterium avium* subsp. paratuberculosis K-10, *Mycobacterium bovis* AF2122/97 (spoligotype 9), *Mycobacterium leprae* TN, *Mycobacterium tuberculosis* CDC1551, *M. tuberculosis* H37Rv, *Streptomyces avermitilis* MA-4680, *Streptomyces coelicolor* A3(2), *Tropheryma whipplei* TW08/27, and *Tropheryma whipplei* Twist. The plasmids SAP1, from *Streptomyces avermitilis* MA-4680, and SCP1 and SCP2, from *S. coelicolor* A3(2), are also included. Also available are two unfinished genome sequences, from *Mycobacterium marinum* and *Clavibacter michiganensis* subsp. sepedonicus ATCC33113. These preliminary sequence data were made available by the Wellcome Trust Sanger Institute.

27.3 Methods

27.3.1 Finding Genes

The center of *Myco*DB is the gene page. There are a number of ways of navigating to a gene of interest. The home page of *Myco*DB (Fig. 27.1) has a single search box, intended to be familiar to users of Internet search engines. This performs a search over all the genes from all the genomes in the database. By default, the search is run on the full annotation text, derived from both GenBank and Uniprot (formerly know as Swissprot/TrEMBL). This means that genes can be identified by name or by other keywords such as "recombination" or "virulence". Using the radio buttons below the search box allows the search to be restricted to just gene names or just accession numbers (these can be GenPept IDs, GI numbers, Uniprot accession numbers, or the internal *Myco*DB identifier numbers such as MY001234).

Fig. 27.1 The *Myco*DB home page and primary search interface

Searches are not case sensitive, and partial word matches are not returned (so searching for "putA" will match the gene *putA* but not any gene that happens to include the word "putative" in its annotation). As with most Internet search engines, an asterisk (*) can be used as a wild card, so "transpo*" can match genes containing any of the words "transport," "transporter," "transposon," or "transposase." If multiple search terms are entered, they are combined using a logical OR, so genes that match any of the terms are returned, but those that match all terms are ranked higher. Any phrase enclosed in double quotes is treated as a single search term. Full details of the search interface can be found at: http://myco.bham.ac.uk/searchhelp.

The search interface (Fig. 27.1) returns a list of genes that match the query and a brief description of their function (derived from the "product" annotation from GenBank). They are preceded by their *Myco*DB identifier, which is linked to the relevant gene page. Below the list is a second list, consisting of genes from the database that did not match the query but are putative orthologues of one of the genes that did match the query. Putative orthologues are identified by means of a precomputed all-against-all reciprocal blastp search between each pair of genomes. Genes that are mutual best hits, which show above 80% identity and for which the aligned region covers more than 90% of the length of the shorter sequence are considered to be putative orthologues.

There are a number of alternative methods of locating genes of interest in the database. The advanced search page (http://myco.bham.ac.uk/advsearch)

operates in a similar manner to the home page but allows searches to be restricted to a subset of the genomes in the database. The *Myco*BLAST interface (http://myco.bham.ac.uk/blast) allows genes to be identified by homology, using a protein or nucleotide query sequence. If "MycoDB genes" or "MycoDB proteins" are selected as the BLAST database, each BLAST hit will include a link to the appropriate gene page. Genes can also be identified using the Genome Browser interface (http://myco.bham.ac.uk/genome). This displays a map of the whole chromosome or plasmid that is clickable, allowing the user to focus in on any gene of interest. The Fragment Viewer (http://myco.bham. ac.uk/fragment) allows a region of interest to be selected by genomic coordinate and displayed using the "*Myco*Browser" interface. As with the Genome Browser, genes of interest can be selected by clicking on the image.

27.3.2 The MycoDB Gene Page

The gene page is the central hub of *Myco*DB, providing information about an individual gene and links to a range of tools for sequence analysis. At the top of the page is a diagrammatic representation of the Genomic Context; that is, the genes in the region surrounding the gene of interest (Fig. 27.2). By default, the region is 20 kb, but this can be adjusted. The image is interactive, allowing

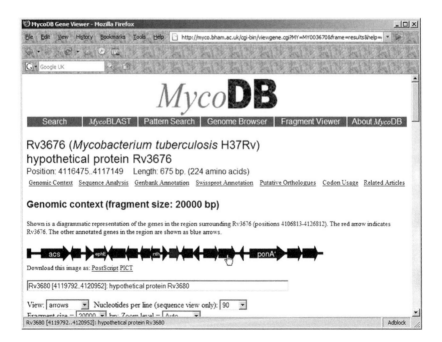

Fig. 27.2 The top of the gene page, showing the genomic context viewer

thc annotation of neighboring genes to be viewed by moving the mouse pointer over them (JavaScript must be enabled for this feature to be used; this is the default for most modern Web browsers). Clicking on a gene opens the appropriate gene page. Below the image are two links that allow the image to be downloaded in PostScript or PICT format. These are structured image formats that can be imported into, and modified in, external programs such as Corel Draw or Microsoft PowerPoint, allowing their incorporation into publications or presentations. In most browsers, the images can be downloaded by right-clicking (or ctrl-clicking on a Macintosh) and selecting the option "Save link target as ..." (the exact text of the option varies in different browsers).

Below the Genome Context image are a number of Sequence Analysis Tools. To run any of the analyses, click on the appropriate "Go" button. By default, the analysis launches in the same browser window, but there is an option that when activated allows analyses to be launched in new windows. This can be useful if the user wants to launch numerous analyses on the same gene. The analysis tools are divided into three sections. The first concerns options for viewing genome alignments and is discussed later.

The second section has a number of tools for analysis of the gene sequence and that of the encoded protein. The DNA or protein sequence can be retrieved in the standard Fasta format, along with (optionally) the sequence of predicted orthologues from other genomes within *Myco*DB. The DNA sequence of the gene can also be displayed with some of the flanking regions and optionally the complementary strand and/or the encoded protein sequence. This is useful if you are interested in elements such as promoters external to the coding sequence. Other tools allow the protein sequence to be used as the query in a BLAST search, either using the internal *Myco*BLAST, or the BLASTp interface at the NCBI, allowing searches against the nonredundant protein database (nr). Conserved domains in the encoded protein can also be identified by searching the CDD (Conserved Domain Database), also provided by the NCBI [3]. A further option allows *Myco*DB to act as a portal to other databases. These include organism-specific databases such as Tuberculist and Leproma and generic databases including KEGG [4], Entrez Proteins [5], and Uniprot [6].

The third section consists of a number of fragment tools for analysis of the entire region pictured in the genomic context viewer. The sequence can be displayed in Fasta format or together with its annotation in EMBL format. The annotated sequence can also be displayed in an applet version of Artemis [7] (Fig. 27.3). This tool, produced by the Wellcome Trust Sanger Institute, allows the annotation to be directly related to the raw sequence data. Because the tool is embedded in an applet, it can be run directly from any Java-enabled Web browser (most modern browsers make the process of running applets relatively straightforward). The software allows the user to dynamically zoom out to view the whole sequence or zoom in to show the raw sequence data and its translation in all six reading frames. Other useful tools include a "sliding window" plot of GC content.

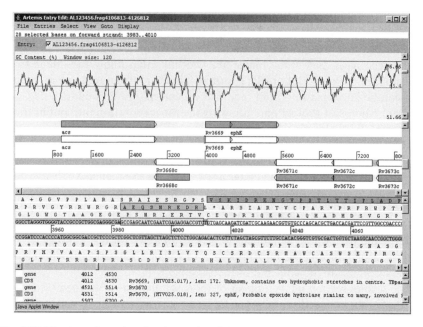

Fig. 27.3 The Artemis applet, showing a plot of GC variation across the region

Another piece of external software integrated into *Myco*DB is the PCR primer design software Primer3 (Fig. 27.4), originally produced by the Whitehead Institute [8]. The software is flexible and easy to use and is enhanced in *Myco*DB by the addition of an integrated primer BLAST interface. Primers are by default selected anywhere within the fragment sequence, although this can be refined by specifying a particular target that must be amplified by the designed primers. The system allows parameters such as the primer melting temperature, length, and self-complementarity to be specified by setting optimum, minimum, and maximum values. Once primers have been selected, clicking on each one launches a *Myco*BLAST form with options preset to be appropriate for use with PCR primers. Running the BLAST against the "MycoDB genomes" database allows the designed primers to be assessed, both to check that they target a unique sequence within the genome and (if necessary) that they target a well-conserved region and hence are suitable for use with a range of target genomes.

The final two analysis tools allow BLASTp to be run using every protein encoded by the fragment as a query sequence. The search can be run against either the local "MycoDB proteins" database or nr. Selecting the option starts the BLAST searches, and the page automatically reloads until the searches are complete. Once complete, the region is shown in the genomic context viewer. Moving the mouse over each gene displays the top BLAST results obtained using that gene as the query (Fig. 27.5). This is particularly useful for obtaining

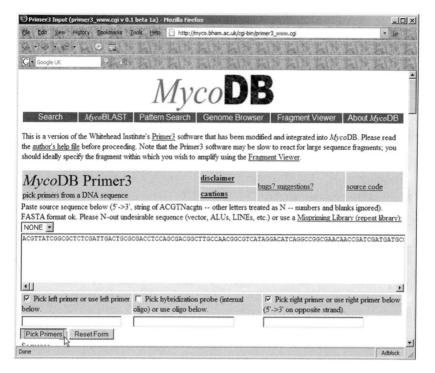

Fig. 27.4 The version of the primer design software Primer3 embedded in $Myco$DB

a "quick and dirty" annotation of a region of an incomplete genome or for verifying the annotation of a published genome.

Below the analysis tools is the Annotation section. This includes the annotation derived from both GenBank and Uniprot. Also shown is a list of the putative orthologues of the gene in $Myco$DB, with links to their respective gene pages, and base composition and codon usage data derived from CodonW [9]. These include the proportion of each individual base, and of G+C, calculated both throughout the gene and only at synonymous positions (sites that when mutated do not change the encoded amino acid and hence reflect any underlying bias without any influence of purifying selection to maintain the protein sequence). Also included are measures of GC and AT skew, which can be indicative of whether a gene is on the leading or lagging strand during DNA replication, if the organism has an asymmetric mutational bias. The effective number of codons (Nc) is included as a measure of the bias in codon usage [10], and the aromaticity and hydropathicity index scores of the encoded protein are also shown.

The final section of the gene page provides links to relevant articles from the literature, taken from Uniprot. These are accompanied by links to PubMed, both to the abstract of the article and to related articles. To further assist the user in locating published information about the gene, a search box is provided,

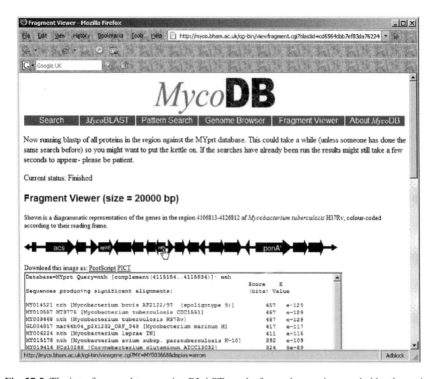

Fig. 27.5 The interface used to examine BLAST results for each protein encoded by the region

prefilled with all the alternative gene names (combined with a logical OR) and the species name. This can be used to search PubMed or the full-text journal search tools Google Scholar, HighWire, Scirius, and Crossref Pilot. These can be useful for identifying papers that mention a particular gene outside of the abstract, for example in the results table of a whole genome experiment such as a microarray.

27.3.3 Genome Alignments in MycoDB

Perhaps the most useful aspect of *Myco*DB is the ability to view comparisons between homologous regions from different genomes. Whole genome alignments between each pair of genomes are precalculated using the MUMmer package produced by TIGR [11]. Two different algorithms from this package are used. The first is the raw MUMmer algorithm. This identifies maximal regions of exact identity between the two genomes. The "maxmatch" option is applied, meaning that these matches need not be unique. As the matches are disrupted by single-base insertions, deletions, or substitutions, this algorithm is most suited for comparisons between closely related genomes, such as two strains of *Mycobacterium tuberculosis*. For more distant comparisons, for example between *M. tuberculosis* and *Streptomyces coelicolor*, the PROmer algorithm is more suitable. This translates the nucleotide sequences in all six reading frames and identifies maximal matches at the protein level (equivalent to the tblastx program from BLAST). Adjacent matches are then clustered to demark large blocks of homology.

The alignment viewer can be accessed from the gene pages or fragment viewer using a drop-down list of genomes for each algorithm. The list includes only those genomes for which a region homologous to that displayed was found by the alignment algorithm. This means that more comparison genomes are usually available for PROmer than for MUMmer. The comparison is displayed in the Alignment Viewer (Fig. 27.6). This shows the region of interest from the first genome and below any homologous regions from the comparison genome. The homologous "blocks" identified by the alignment algorithm are shown in red, with a black border indicating the end of the match due to a gap or substitution. For the MUMmer algorithm, it is common to get large black blocks; these simply indicate the presence of large numbers of small matches

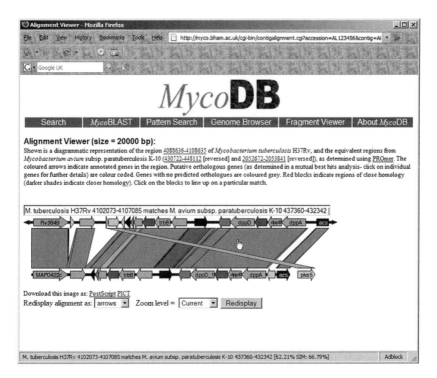

Fig. 27.6 A *Myco*DB genome alignment showing a region from *M. tuberculosis* H37Rv, and the equivalent regions from *M. avium* subsp. paratuberculosis K-10, as determined using PROmer

and indicate a homologous region that has accumulated many SNPs and/or indels. The genes are colored to indicate pairs of orthologous genes in the two genomes, as identified by the automated analysis detailed above. As with the gene page, moving the mouse over the image allows the annotation of genes to be viewed. If the pointer is positioned over a match, the coordinates and percentage identity of the match are displayed.

27.3.4 Whole Genome Displays and "Gene Painting"

The Genome Viewer allows a circular representation of a complete chromosome or plasmid to be displayed (Fig. 27.7). As with all the displays in *Myco*DB, moving the mouse over the genes displays their annotation, and clicking on them opens the appropriate gene page. *Myco*DB allows the genes to be "painted" according to various data. By default, the image is colored according the GC content, across a range ±2 standard deviations from the genomic mean. This allows regions of atypically high or low GC content to be easily identified. Alternative color schemes include GC or AT skew, useful for

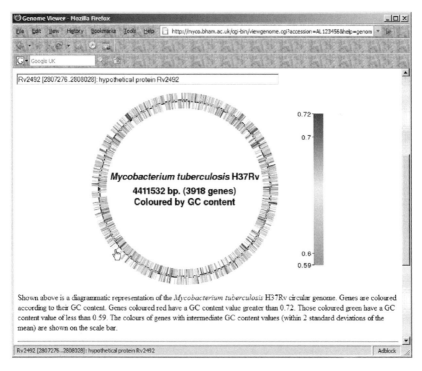

Fig. 27.7 The complete *M. tuberculosis* H37Rv chromosome, as displayed using the Genome Viewer

identifying the position of the start and end of replication, and coloring by orthologues. This allows the user to select a range of genomes for comparison and colors genes according to the occurrence of its orthologues in the selected range. Thus a gene that is conserved throughout all the genomes will be colored red, whereas a gene that is unique to the viewed genome will be colored green (note that the colors used for the display can be altered). It can be difficult to identify individual genes on a whole chromosome display, so the colors of the genes are maintained through to the individual gene pages when the image is clicked.

27.3.5 Pattern Matching

The final tool in *Myco*DB is the Pattern Search (Fig. 27.8), which acts as a wrapper around the program fuzznuc, part of the EMBOSS suite [12]. This is useful for searching for conserved motifs within a genome. The search is "fuzzy," allowing IUPAC degenerate nucleotide codes such as N, R, and Y, and a specified number of mismatches can be allowed. The search can be

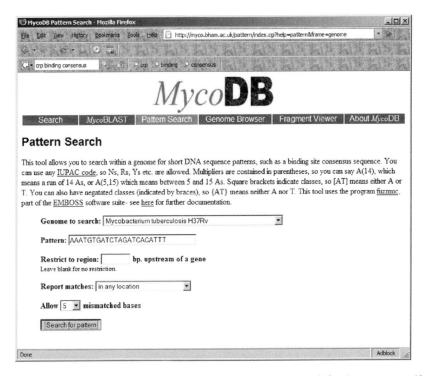

Fig. 27.8 The pattern search interface, which allows the user to search for degenerate motifs such as a transcription factor consensus binding site

restricted to intragenic or intergenic regions or to within a specified number of bases upstream of a gene (useful for identifying potential promoter binding sites from a consensus sequence). The search returns the coordinates of matches, their position relative to the nearest downstream gene (with links through to the appropriate gene page). Matches to the complementary strand are listed below the top strand matches.

27.4 The Future of *Myco*DB

Development of *Myco*DB in its current form has ceased. The current interface will continue to be available from http://myco.bham.ac.uk for the foreseeable future. However, the functionality is in the process of being transferred to the second generation of *x*BASE. The current model of prebuilt databases is becoming increasingly unsuitable as we move from the "multigenomic" era of a handful of genomes from each taxonomic group into the "polygenomic" era, where genome sequencing of hundreds of strains of any bacterial species will be commonplace. The new *x*BASE is designed to be self-updating and will perform alignments on demand rather than requiring lengthy calculations during each database build. This will mean that the end user will have access to a *Myco*DB-like interface for all bacterial genome sequences as soon as they are made publicly available, rather than relying on sequences being manually added to the database. The distinction between the component databases (*Myco*DB, *coli*BASE, etc.) will be removed, allowing comparisons between distantly related organisms, and there will be an overhaul of the user interface (although the design principles that have led to the success of *x*BASE version 1 will be maintained).

The initial release of *x*BASE 2 is scheduled for August 2007 and will be available from http://xbase.bham.ac.uk. Funding has recently been obtained from the BBSRC to allow continued active development of *x*BASE for a further 5 years.

Acknowledgments The author would like to thank Mark Pallen for management of the *x*BASE project and contribution to the interface design, Arshad Khan and Chris Bailey for assistance with server and database administration, and the BBSRC for funding the project (grant number EGA16107).

References

1. Chaudhuri, R. R., Khan, A. M. & Pallen, M. J. (2004). *coli*BASE: an online database for *Escherichia coli*, *Shigella* and *Salmonella* comparative genomics. *Nucleic Acids Res* **32** (Database issue), D296–9.
2. Chaudhuri, R. R. & Pallen, M. J. (2006). *x*BASE, a collection of online databases for bacterial comparative genomics. *Nucleic Acids Res* **34**, D335–7.

3. Marchler-Bauer, A., Anderson, J. B., Derbyshire, M. K., DeWeese-Scott, C., Gonzales, N. R., Gwadz, M., Hao, L., He, S., Hurwitz, D. I., Jackson, J. D., Ke, Z., Krylov, D., Lanczycki, C. J., Liebert, C. A., Liu, C., Lu, F., Lu, S., Marchler, G. H., Mullokandov, M., Song, J. S., Thanki, N., Yamashita, R. A., Yin, J. J., Zhang, D. & Bryant, S. H. (2007). CDD: a conserved domain database for interactive domain family analysis. *Nucleic Acids Res* **35**, D237–40.

4. Kanehisa, M., Goto, S., Hattori, M., Aoki-Kinoshita, K. F., Itoh, M., Kawashima, S., Katayama, T., Araki, M. & Hirakawa, M. (2006). From genomics to chemical genomics: new developments in KEGG. *Nucleic Acids Res* **34**, D354–7.

5. Wheeler, D. L., Barrett, T., Benson, D. A., Bryant, S. H., Canese, K., Chetvernin, V., Church, D. M., DiCuccio, M., Edgar, R., Federhen, S., Geer, L. Y., Kapustin, Y., Khovayko, O., Landsman, D., Lipman, D. J., Madden, T. L., Maglott, D. R., Ostell, J., Miller, V., Pruitt, K. D., Schuler, G. D., Sequeira, E., Sherry, S. T., Sirotkin, K., Souvorov, A., Starchenko, G., Tatusov, R. L., Tatusova, T. A., Wagner, L. & Yaschenko, E. (2007). Database resources of the National Center for Biotechnology Information. *Nucleic Acids Res* **35**, D5–12.

6. The UniProt Consortium. (2007). The Universal Protein Resource (UniProt). *Nucleic Acids Res* **35**, D193–7.

7. Rutherford, K., Parkhill, J., Crook, J., Horsnell, T., Rice, P., Rajandream, M. A. & Barrell, B. (2000). Artemis: sequence visualization and annotation. *Bioinformatics* **16**, 944–5.

8. Rozen, S. & Skaletsky, H. (2000). Primer3 on the WWW for general users and for biologist programmers. *Methods Mol Biol* **132**, 365–86.

9. Peden, J. (2005). CodonW Condon usage analysis package [online]. Available from http://condonw.sourceforge.net [cited 15 April 2005].

10. Wright, F. (1990). The 'effective number of codons' used in a gene. *Gene* **87**, 23–9.

11. Delcher, A. L., Phillippy, A., Carlton, J. & Salzberg, S. L. (2002). Fast algorithms for large-scale genome alignment and comparison. *Nucleic Acids Res* **30**, 2478–83.

12. Rice, P., Longden, I. & Bleasby, A. (2000). EMBOSS: the European Molecular Biology Open Software Suite. *Trends Genet* **16**, 276–7.

Chapter 28
Environmental Amoebae and Mycobacterial Pathogenesis

Melanie Harriff and Luiz E. Bermudez

Abstract Environmental amoebae have been shown to be a host to pathogenic mycobacteria. *Mycobacterium avium, Mycobacterium marinum,* and *Mycobacterium peregrinum* can all grow inside *Acanthamoeba* and other environmental amoebae. Once ingested by *Acanthamoeba, M. avium* upregulates a number of genes, many of them similar to genes upregulated upon phagocytosis of *M. avium* by macrophages. Mycobacteria ingested by amoebae grow intracellularly, acquiring an invasive phenotype, evident when the bacterium escapes the infected amoeba. Once inside of amoeba, it has been shown that mycobacteria are protected from antibiotics and disinfectants, such as chlorine. This chapter describes methods employed for the study of the interaction of *M. avium* and *Acanthamoeba.*

Keywords *Acanthamoeba* · increase of virulence · intracellular · macrophage · mucosal epithelial cell · *Mycobacterium avium* · *Mycobacterium marinum* · *Mycobacterium peregrinum*

28.1 Introduction

Mycobacterium avium is an environmental bacterium that infects immunosuppressed and immunocompetent individuals [1, 2]. In humans, *M. avium* can be associated with disseminated disease, as well as lung infection [3]. The majority of *M. avium* infections are acquired through the gastrointestinal tract, and once ingested, *M. avium* can survive the harsh environment of the stomach [4]. The bacterium has been shown to interact with the intestinal mucosa and invade it, as a step prior to dissemination. Bacterial genes associated with the ability to cross the intestinal mucosa have been identified [5, 6], and *M. avium* has been shown to survive at least for 6 months in mesenteric lymph node

L.E. Bermudez
Department of Biomedical Sciences, College of Veterinary Medicine, Oregon State University, 105 Dryden Hall, Corvallis, Oregon 97331, USA
e-mail: luiz.bermudez@oregonstate.edu

T. Parish, A.C. Brown (eds.), *Mycobacteria Protocols,*
doi: 10.1007/978-1-59745-207-6_28, © Humana Press, Totowa, NJ 2008

macrophages [7]. The bacteria also invade the bronchial mucosa, a process that seems to require the ability to form a biofilm prior to translocation [8]. Bacterial genes involved in biofilm formation have been described, and strategies to prevent the biofilm phase of infection are in the experimental phase of development [9, 10].

As an environmental organism, *M. avium* can potentially infect protozoa, such as amoebae. Studies carried out in mice, as well as in zebrafish, have demonstrated that passage of environmental mycobacteria such as *M. avium*, *Mycobacterium marinum,* and *Mycobacterium peregrinum,* in *Acanthamoeba castellanii* results in an increase in virulence [11, 12], as well as enhanced resistance to antibiotics [13]. Recently, we have shown that *Acanthamoeba* infection by *M. avium* is associated with the upregulation of a number of known and unknown virulence genes [14]. The majority of the genes regulated inside an amoeba are also the genes required for survival in macrophages [14, 15, 16].

To survive within *Acanthamoeba*, *M. avium* needs to upregulate virulence genes. Starting with the premise that *M. avium* genes required to replicate and survive within environmental amoebae should be partially common to the genes important for the survival within macrophages, we screened a green fluorescent protein (GFP)-promoter library of *M. avium* in *Mycobacterium smegmatis* upon uptake by *Acanthamoeba castellanii*. The *M. avium* genomic library in *M. smegmatis* offers a number of advantages over a GFP–*M. avium* library in *M. avium*. *M. smegmatis* transformation is significantly more efficient than is transformation of *M. avium*, which makes the effort of creating 10,000 clones much easier.

The methods outlined below describe (1) the infection of *Acanthamoeba* by mycobacteria, and (2) the construction and screening of a promoter GFP library to determine the effect of passage through amoebae on virulence. The screening of *M. avium* genes can be done in *M. avium* or *M. smegmatis* using a vector with a promoterless GFP gene downstream of the *M. avium* promoter library. Gene that are induced upon infection of macrophages and amoebae trigger GFP expression in the bacterium. *M. smegmatis* has a limited life in macrophages and could only be used up to 48 h after infection. In contrast, in an amoeba, *M. smegmatis* lives for longer periods of time; therefore, the assay can be used up to 72 h after amoeba infection.

28.2 Materials

28.2.1 Culture Preparation

1. *M. avium* MAC 104 (transformable strain) or *M. smegmatis* mc^{2}155 (transformable strain).
2. 20% (v/v) Tween 80: mix 20 mL Tween 80 with 80 mL distilled water, filter sterilize.

3. 7H10-OADC plates: Dissolve 19 g Middlebrook 7H10 medium (Difco, Detroit MI) in 900 mL deionized water. Add 8.4 mL 60% glycerol (v/v). Autoclave. When media is cooled to ~50°C, add 100 mL oleic albumin dextrose catalase (OADC) enrichment. Mix well and pour in standard plastic Petri dishes.
4. 7H9-Tw medium: Dissolve 4.7 g Middlebrook 7H9 broth base (Difco) in 900 mL deionized water and add 2 mL glycerol, mix well, and filter sterilize. Add 5 mL 20% (v/v) Tween 80 (final concentration of 0.05% v/v). Add 100 mL OADC immediately before use.
5. Phosphate-buffered saline (PBS). Make up 10X PBS as follows: dissolve 80 g NaCl, 2 g KCl, 1.4 g Na_2HPO_4, and 2.4 g KH_2PO_4 in 800 mL distilled water, pH to 7.4, adjust volume to 1 L. Autoclave to sterilize.
6. McFarland standards (PML Microbiologicals, Wilsonville, OR).
7. Incubator set at 37°C.
8. Centrifuge.

28.2.2 Acanthamoeba *Infection by Mycobacteria*

1. *Acanthamoeba castellanii* ATCC 30234.
2. 712PYG medium (*see* Section 28.2.1, step 9).
3. Tissue culture flasks: 25 to 75 cm^2.
4. Neubauer counting chamber or hemocytometer.
5. 6-well or 24-well, and 96-well tissue culture plates.
6. *M. avium* culture.
7. Triton-X 100 (Sigma, St Louis Mo): 10% (v/v) solution in distilled water, filter sterilize.
8. 23-gauge needles and syringes.
9. Falcon 2054 tubes (Marathon, San Jose, CA).
10. Sterile distilled water.
11. 7H10-OADC plates (*see* Section 28.2.1 step 3).
12. Dark incubator set at 24°C.
13. Microscope.

28.2.3 *Mycobacterial Gene Regulation Within Amoebae*

1. GTE buffer: 50 mM glucose, 10 mM 0.5 M EDTA, pH 8.0, 25 mM 1 M Tris-HCl, pH 8.0.
2. Lysozyme stock solution (50 mg/mL): dissolve 1 g in 20 mL distilled water. Filter sterilize. Store in 500-μL aliquots at –20°C, use in GTE buffer to give a final concentration of at 20 μg/mL.
3. Proteinase K stock solution (50 mg/mL): dissolve 1 g in 20 mL distilled water, filter sterilize. Store in 500-μL aliquots at –20°C, use in GTE buffer to give a final concentration of at 20 μg/mL.
4. 20% (w/v) SDS.

5. Cetyl trimethyl ammonium bromide (CTAB) extraction solution: 2% (w/v) CTAB (Genco Tech, St. Louis, MO), 100 mM Tris-Cl, pH 8.0, 20 mM EDTA, pH 8.0, 1.4 M NaCl.
6. 5 M NaCl.
7. Phenol:chloroform:isoamyl alcohol (25:24:1).
8. GFP plasmid, pEMC1. The pEMC1 plasmid contains a kanamycin-resistance gene, and the promoterless GFP gene, downstream of a cloning site. The plasmid also contains a stop site upstream of the cloning site (Fig. 28.1) [5]. (Plasmid available from Bermudez Lab, Department of Biomedical Sciences, Oregon State University, Corvallis, Oregon 97331).
9. *Sau*3A1 and *Bam*HI restriction enzymes.
10. Rapid Ligation Kit (Roche, San Diego, CA) or similar.
11. Agarose gels and electrophoresis equipment.
12. Gel extraction kit (Qiagen, Valencia, CA or similar).
13. *Escherichia coli* competent cells.
14. LB media: To 800 mL distilled water, add 10 g tryptone (pancreatic digest of casein), 5 g yeast extract, 5 g NaCl, and make up to 1 L. Autoclave to sterilize. For solid medium (LB-agar), add 15 g agar before autoclaving. Allow to cool and pour onto Petri dishes.
15. HiSpeed Plasmid Midi Kit (Qiagen) or similar.

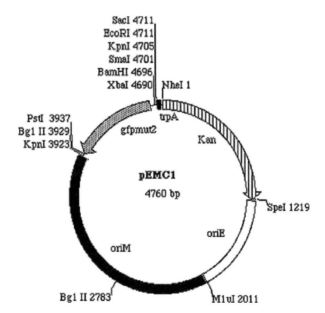

Fig. 28.1 Plasmid map of pEMC1. This plasmid is a cloning site upstream of a promoterless GFP gene for transcriptional fusions. *oriE*, *E. coli* origin of replication; *oriM*, mycobacterial origin of replication; *GFPmut2*, gene encoding green fluorescent protein; *trpA*, transcriptional terminator

16. Kanamycin stock solution (50 mg/mL): dissolve 1 g in 20 mL distilled water, filter sterilize. Store in 500-μL aliquots at –20°C, use in media to give a final concentration of 20 μg/mL for mycobacteria, 50 μg/mL for *E. coli*.
17. Shaking incubator (~200 rpm) set at 37°C.
18. *M. smegmatis* competent cells (*see* Chapter 13).
19. 7H10-OADC plates (*see* Section 28.2.1, step 3).
20. 712PYG medium (*see* Section 28.2.1, step 9).
21. Tissue culture flasks ranging in size from 25 to 75 cm^2.
22. Neubauer counting chamber or hemocytometer.
23. Hank's balanced salt solution (HBSS).
24. 96-well tissue culture plate.
25. Cytofluorimeter.
26. RT-PCR reagents and equipment.
27. Heating block or water bath set to 65°C.
28. Incubators set at 24°C and 37°C.

28.2.4 Infection of Macrophages

1. THP-1 cells or other human monocyte-derived macrophage.
2. RPMI-1640 cell culture media (Invitrogen, Carlsbad, CA).
3. Hank's balanced salt solution (HBSS).
4. Tissue culture flasks ranging in size from 25 to 75 cm^2.
5. Triton-X 100 (Sigma): 10% (v/v) solution in distilled water, filter sterilize.
6. Sterile distilled water.
7. 24-well tissue culture plates.
8. 7H10-OADC plates (*see* Section 28.2.1, step 3).
9. Incubator set at 37°C with 5% CO_2.

28.3 Methods

28.3.1 Culture Preparation

1. Streak a loop of *M. avium* or *M. smegmatis* onto a 7H10-OADC plate.
2. Seal plate in plastic bag or with Parafilm.
3. Incubate at 37°C (*see* **Note 1**).
4. Inoculate a 10 mL 7H9 culture with a loopful of cells from the plate.
5. Incubate at 37°C (*see* **Note 1**).
6. Centrifuge liquid cultures at 3000 × *g* for 10 min at room temperature to pellet bacteria.
7. Wash pellet with the original culture volume of PBS.
8. Pellet bacteria at 3000 × *g* for 10 min at room temperature.
9. Resuspend cell pellet in 0.5 mL PBS (*see* **Note 2**).
10. Adjust cell suspension using the McFarland Standard 1.0, to give 1 × 10^7 bacteria/mL.

28.3.2 Acanthamoeba *Infection by Mycobacteria*

1. Inoculate a 75 cm^2 tissue culture flask containing 20 mL of 712PYG medium with *Acanthamoeba castellanii* (*see* **Note 3**).
2. Incubate flask in the dark at room temperature (*see* **Note 4**).
3. Detach adherent amoebae by tapping the side of the tissue culture flask a few times.
4. Count the number of amoebae in suspension with a counting chamber or hemocytometer.
5. Use 10^5 amoebae per well to seed a 24-well tissue culture plate in 1 mL of 712PYG medium.
6. Allow to sit overnight for adherence.
7. Infect amoebae with mycobacteria at a multiplicity of infection (MOI) of 0.1, 1, and 10.
8. After 1 to 2 h, wash the extracellular bacteria with HBSS and pipette (*see* **Note 5**).
9. Add 1 mL fresh 712PYG medium per well.
10. Place the amoebac monolayers at room temperature or 37°C for 2 to 4 days (*see* **Note 6**).
11. Add 100 μL 10% Triton X-100 per well to lyse amoebae.
12. Incubate at room temperature for up to 30 min.
13. Check wells under a microscope in order to ensure complete lysis of amoebae.
14. Pass well contents through a 23-gauge needle to ensure complete lysis (*see* **Note 7**).
15. Transfer well contents to separate Falcon 2054 tubes.
16. Plate the lysate and serial dilutions (made using sterile distilled water) onto 7H10 plates for quantification of the viable intracellular bacteria.
17. Seal plates in plastic bags or with Parafilm.
18. Incubate at 37°C (*see* **Notes 1** and **8**).

28.3.3 *Mycobacterial Gene Regulation Within Amoebae*

28.3.3.1 Construction of a *M. avium* GFP-Promoter Library in *M. smegmatis*

1. Extract genomic DNA from 10^8 to 10^9 *M. avium* as follows: resuspended the bacterial pellet in GTE containing 20 μg/mL lysozyme and incubate overnight at 37°C. Add 20 μg/mL proteinase K and 0.1% SDS and incubate for an additional 3 h at 37°C. Add 300 μL CTAB and 350 μL 5 M NaCl/CTAB; the lysate can now be used for DNA extraction using multiple rounds of phenol:chloroform:isoamyl alcohol (25:24:1) extraction.
2. Digest 1 μg of DNA with 2 μL of *Sau*3A1 overnight at 37°C.
3. Prepare a 1% agarose gel and separate digest fragments by electrophoresis.

4. Excise the segment of the gel containing the fragments from 300 to 1000 bp. Recover DNA with a DNA gel extraction kit (Qiagen or similar).
5. Digest 1 μg of pEMC1 with *Bam*HI overnight at 37°C.
6. Heat pEMC1 digest reaction mixture at 65°C for 20 min to denature the enzyme.
7. Ligate the gel extracted *Sau*3A1 digested DNA and pEMC1 using the Rapid Ligation Kit, as per the manufacturer's instructions.
8. Transform ligation mix into *E. coli* competent cells. Select on LB agar supplemented with 50 μg/mL of kanamycin.
9. Scrape transformants from plates and inoculate a 100 mL LB culture supplemented with 50 μg/mL of kanamycin. Incubate overnight at 37°C with shaking at ~200 rpm.
10. Extract plasmid DNA using HiSpeed Plasmid Midi Kit (Qiagen or similar).
11. Electroporate 1 μg plasmid DNA into competent *M. smegmatis* cells (*see* Chapter 13).
12. Plate the *M. smegmatis* transformants onto approximately 50 7H10 agar plates supplemented with 20 μg/mL kanamycin, such that >10,000 clones are obtained.
13. Store clones in pools of 5 in 10 96-well plates containing Middlebrook 7H9 broth with 50% glycerol.
14. Place 5 to 10 clones per well in a 96-well tissue culture plate with 7H9 broth.
15. Transfer clones to replicate 96-well tissue culture plates with 7H9 broth for screening.

28.3.3.2 Screening of an *M. avium* GFP-Promoter Library for Genes Expressed in Amoebae

1. Grow *Acanthamoeba* in a 25-cm^2 tissue culture flask containing 10 mL 712 PYG medium. Incubate for 3 to 4 days at 24°C in the dark.
2. Detach amoebae from the plastic by tapping the flask vigorously.
3. Count amoebae/mL, either using a Neubauer counting chamber or hemocytometer.
4. Seed 10^5 amoebae in each well of a 96-well tissue culture plate. Let them adhere overnight in the dark at room temperature.
5. Infect each well with pools of 5 to 10 *M. smegmatis* clones of the *M. avium* GFP-promoter library (from Section 28.3.3.1). You need to have approximately 10^6 to 10^7 bacteria per monolayer.
6. Incubate for 30 min to 1 h at 37°C.
7. Wash the monolayer twice with 0.2 mL HBSS.
8. Add 200 μL fresh 712 PYG medium and read the baseline emission in GFP band of a cytofluorimeter, 450 of excitation, 500 emission.
9. Place the plate in the dark for 15 min and repeat the reading for GFP expression. Repeat the reading at 30 min, 1 h, 2 h, 24 h, 48 h, or as long as it is desired.

10. Plate clone pools that have an increase in GFP expression at least 2 times above baseline, at desired time points, onto 7H10 plates containing 50 μg/mL of kanamycin to obtain single colonies.
11. From 7H10 plates, select 20 individual colonies arising from each well and seed them in a 96-well tissue culture plate with 7H9 broth.
12. Repeat the infection of amoebae using isolated colonies. Those clones that are associated with 2.5-fold increase or more, compared with the baseline, should be tested again to confirm the increase in GFP expression.
13. Sequence the isolated clones using primers specific for the plasmid regions upstream of the mycobacterial promoter and downstream of the GFP sequence [15].
14. Confirm the expression of the genes in *M. avium* (or another environmental bacteria) by using real-time PCR.

28.3.4 Infection of Macrophages

To determine if infection of amoebae modifies the interaction of mycobacteria with the second (following) host, bacteria obtained from the first host should not be plated or seeded in culture medium.

1. Obtain bacteria directly from the amoebae lysate or whole amoebae (as in Section 28.3.2, steps 11 to 14), without passage in medium, in order to maintain the phenotype.
2. Use bacteria grown *in vitro* (medium) as control. The MOI (multiplicity of infection) of the bacteria grown in medium and the bacteria from amoebae should be as close as possible.
3. Seed 5×10^5 cells macrophages (primary human, mouse, or THP-1 cells) in a 24-well tissue culture plate (*see* **Note 9**).
4. Use 3 wells per bacterial strain per time point (days 0, 3, 4, and 5)
5. Infect macrophage monolayers for 1 or 2 h, using an MOI of 10:1.
6. Remove extracellular bacteria by washing with 1 mL HBSS.
7. Add 1 mL sterile RPMI media to wells.
8. Incubate at 37°C with 5% CO_2.
9. Remove tissue culture media from day 0 wells.
10. Add 0.5 mL sterile water to wells to lyse day 0 cells.
11. Dilute the lysate (10^{-3}, 10^{-4}, 10^{-5}).
12. Plate 100 mL of the lysate dilutions onto 7H10 agar plates.
13. Incubate at 37°C (*see* **Note 1**).
14. The other macrophage monolayer wells should be lysed on days 3, 4, or 5 after infection.
15. The analysis of the data should be done comparing the growth of mycobacteria obtained from amoebae to mycobacteria grown *in vitro*. For each experimental group, growth at days 3, 4, or 5 should be compared with the number of bacteria inside macrophages at 1 h or 2 h (after infection).

28.4 Notes

1. *M. avium* will grow in approximately 10 days and *M. smegmatis* in 4.5 days, both in agar as well as in broth. The *M. avium* colonies on the plate should have a smooth, dome, transparent morphotype, or smooth opaque, with a small percent of the strains having a rough morphotype. The *M. smegmatis* phenotype is rough.
2. Hank's balanced salt solution (HBSS) can be used in place of PBS.
3. Amoebae can also be maintained in high-salt medium, which does not allow the protozoa to replicate, although they remain viable [11].
4. *Acanthamoeba* in culture will adhere to the bottom of the plastic flask. Amoebae should be observed daily and medium changed every 5 days. If left for too long without changing media, amoebae will encyst and additional time will be required for optimal infectious conditions.
5. Amoebae will not detach easily from plastic.
6. At room temperature, the bactericidal ability of amoebae will be maintained, but at 37°C, the ability to control bacterial growth will be impaired.
7. Triton X-100 has no harmful effect on *M. avium*, *M. marinum*, or *M. peregrinum*. There are other detergents that can be used for this purpose (e.g., SDS). If using another bacterial strain, the effect of the detergent on the bacteria should be tested first.
8. Depending on the bacteria, the outcome of the infection will vary. *Mycobacterium smegmatis* will be killed whereas *M. avium* will grow (Fig. 28.2).
9. THP-1 monocytes are to be cultured in RPMI-1640, and medium must be replenished every 3 days. Remove half of the content of the flask and replenish it with fresh RPMI-1640 supplemented with 10% FBS. When using THP-1 cells, phorbol-myristate acetate (PMA) should be used prior to infection to induce maturation of the monocytes. Add from 10 to 20 μg/mL per well for 4 to 24 h. Once the cells are mature (they adhere to the bottom of the well), they can be infected.

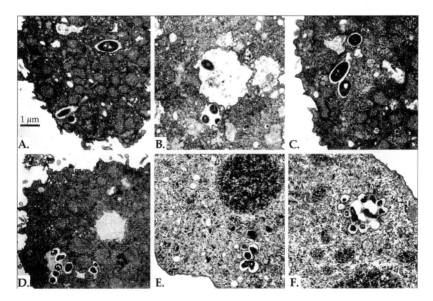

Fig. 28.2 Transmission electron micrographs of *M. avium* within *A. castellanii* at (**A**) 5 min, (**B**) 15 min, (**C**) 1 h, (**D**) 2 days, (**E**) 3 days, and (**F**) 5 days after infection. The 1-mm bar shown in (A) applies to all panels. (Reprinted with permission, *Infection and Immunity*.)

References

1. Inderlied, C. B., Kemper, C. A., and Bermudez, L. E. (1993) The *Mycobacterium avium* complex. *Clin Microbiol Rev* **6**, 266–310.
2. Wolinsky, E. (1979) Nontuberculous mycobacteria and associated diseases. *Am Rev Respir Dis* **119**, 107–159.
3. Aksamit, T. R. (2002) *Mycobacterium avium* complex pulmonary disease in patients with pre-existing lung disease. *Clin Chest Med* **23**, 643–653.
4. Bodmer, T., Miltner, E., and Bermudez, L. E. (2000) *Mycobacterium avium* resists exposure to the acidic conditions of the stomach. *FEMS Microbiol Lett* **182**, 45–49.
5. Dam, T., Danelishvili, L., Wu, M., and Bermudez, L. E. (2006) The fadD2 gene is required for efficient *Mycobacterium avium* invasion of mucosal epithelial cells. *J Infect Dis* **193**, 1135–1142.
6. Miltner, E., Daroogheh, K., Mehta, P. K., Cirillo, S. L., Cirillo, J. D., and Bermudez, L. E. (2005) Identification of *Mycobacterium avium* genes that affect invasion of the intestinal epithelium. *Infect Immun* **73**, 4214–4221.
7. Petrofsky, M., and Bermudez, L. E. (2005) CD4+ T cells but not CD8+ or gammadelta+ lymphocytes are required for host protection against *Mycobacterium avium* infection and dissemination through the intestinal route. *Infect Immun* **73**, 2621–2627.
8. Yamazaki, Y., Danelishvili, L., Wu, M., Hidaka, E., Katsuyama, T., Stang, B., Petrofsky, M., Bildfell, R., and Bermudez, L. E. (2006) The ability to form biofilm influences *Mycobacterium avium* invasion and translocation of bronchial epithelial cells. *Cell Microbiol* **8**, 806–814.
9. Carter, G., Young, L. S., and Bermudez, L. E. (2004) A subinhibitory concentration of clarithromycin inhibits *Mycobacterium avium* biofilm formation. *Antimicrob Agents Chemother* **48**, 4907–4910.
10. Yamazaki, Y., Danelishvili, L., Wu, M., Macnab, M., and Bermudez, L. E. (2006) *Mycobacterium avium* genes associated with the ability to form a biofilm. *Appl Environ Microbiol* **72**, 819–825.
11. Cirillo, J. D., Falkow, S., Tompkins, L. S., and Bermudez, L. E. (1997) Interaction of *Mycobacterium avium* with environmental amoebae enhances virulence. *Infect Immun* **65**, 3759–3767.
12. Harriff, M., Bermudez, L., and Kent, M. L. (2007) Experimental exposure of zebrafish (*Danio rerio* Hamilton) to *Mycobacterium marinum* and *Mycobacterium peregrinum* reveals the gastrointestinal tract as the primary route of infection: A potential model for environmental mycobacterial infection. *J Fish Dis* **29**, 1–13.
13. Miltner, E. C., and Bermudez, L. E. (2000) *Mycobacterium avium* grown in *Acanthamoeba castellanii* is protected from the effects of antimicrobials. *Antimicrob Agents Chemother* **44**, 1990–1994.
14. Tenant, R., and Bermudez, L. E. (2006) *Mycobacterium avium* genes upregulated upon infection of *Acanthamoeba castellanii* demonstrate a common response to the intracellular environment. *Curr Microbiol* **52**, 128–133.
15. Danelishvili, L., Poort, M. J., and Bermudez, L. E. (2004) Identification of *Mycobacterium avium* genes up-regulated in cultured macrophages and in mice. *FEMS Microbiol Lett* **239**, 41–49.
16. Hou, J. Y., Graham, J. E., and Clark-Curtiss, J. E. (2002) *Mycobacterium avium* genes expressed during growth in human macrophages detected by selective capture of transcribed sequences (SCOTS). *Infect Immun* **70**, 3714–3726.

Index